BIOPOLYMER MIXTURES

Proceedings of Previous Easter Schools in Agricultural Science, published by Butterworths, London

SOIL ZOOLOGY Edited by D.K. McL, Kevan (1955)
THE GROWTH OF LEAVES Edited by F.L. Milthorpe (1956)
CONTROL OF PLANT ENVIRONMENT Edited by J.P. Hudson (1957)
NUTRITION OF THE LEGUMES Edited by E.G. Hallsworth (1958)
THE MEASUREMENT OF GRASSLAND AND PRODUCTIVITY Edited by J.J. Ivins (1959)
DIGESTIVE PHYSIOLOGY AND NUTRITION OF THE RUMINANT Edited by D. Lewis (1960)
NUTRITION OF PIGS AND POULTRY Edited by J.T. Morgan and D. Lewis (1961)
ANTIBIOTICS IN AGRICULTURE Edited by M. Woodbine (1962)
THE GROWTH OF THE POTATO Edited by J.D. Ivins and F.L. Milthorpe (1963)
EXPERIMENTAL PEDOLOGY Edited by E.G. Hallsworth and D.V. Crawford (1964)
THE GROWTH OF CEREALS AND GRASSES Edited by F.L. Milthorpe and J.D. Ivins (1965)
REPRODUCTION IN THE FEMALE ANIMAL Edited by G.E. Lamming and E.C. Amoroso (1967)
GROWTH AND DEVELOPMENT OF MAMMALS Edited by G.A. Lodge and G.E. Lamming (1968)
ROOT GROWTH Edited by W.J. Whittington (1968)
PROTEINS AS HUMAN FOOD Edited by R.A. Lawrie (1970)
LACTATION Edited by I.R. Falconer (1971)
PIG PRODUCTION Edited by D.J.A. Cole (1972)
SEED ECOLOGY Edited by W Heydecker (1973)
HEAT LOSS FROM ANIMALS AND MAN: ASSESSMENT AND CONTROL Edited by J.L. Montieth and L.E. Mount (1974)
MEAT Edited by D.J.A. Cole and R.A. Lawrie (1975)
PRINCIPLES OF CATTLE PRODUCTION Edited by Henry Swan and W.H. Broster (1976)
LIGHT AND PLANT DEVELOPMENT Edited by H. Smith (1976)
PLANT PROTEINS Edited by G Norton (1977)
ANTIBIOTICS AND ANTIBIOSIS AGRICULTURE Edited by M Woodbine (1977)
CONTROL OF OVULATION Edited by D.B. Crighton, N.B. Haynes, G.R. Foxcroft and G.E. Lamming (1978)
POLYSACCHARIDES IN FOOD Edited by J.M.V. Blanshard and J.R. Mitchell (1979)
SEED PRODUCTION Edited by P.D. Hebblethwaite (1980)
PROTEIN DEPOSITION IN ANIMALS Edited by P.J. Buttery and D.B. Lindsay (1981)
PHYSIOLOGICAL PROCESSES LIMITING PLANT PRODUCTIVITY Edited by C. Johnson (1981)
ENVIRONMENTAL ASPECTS OF HOUSING FOR ANIMAL PRODUCTION Edited by J.A. Clark (1981)
EFFECTS OF GASEOUS AIR POLLUTION IN AGRICULTURE AND HORTICULTURE Edited by M.H. Unsworth and D.P. Ormrod (1982)
CHEMICAL MANIPULATION OF CROP GROWTH AND DEVELOPMENT Edited by J.S. McLaren (1982)
CONTROL OF PIG REPRODUCTION Edited by D.J.A. Cole and G.R. Foxcroft (1982)
SHEEP PRODUCTION Edited by W. Haresign (1983)
UPGRADING WASTE FOR FEEDS AND FOOD Edited by D.A. Ledward, A.J. Taylor and R.A. Lawrie (1983)
FATS IN ANIMAL NUTRITION Edited by J. Wiseman (1984)
IMMUNOLOGICAL ASPECTS OF REPRODUCTION IN MAMMALS Edited by D.B. Crighton (1984)
ETHYLENE AND PLANT DEVELOPMENT Edited by J.A. Roberts and G.A. Tucker (1985)
THE PEA CROP Edited by P.D. Hebblethwaite, M.C. Heat and T.C.K. Dawkins (1985)
PLANT TISSUE CULTURE AND ITS AGRICULTURAL APPLICATIONS Edited by Lindsay A. Withers and P.G. Alderson (1986)
CONTROL AND MANIPULATION OF ANIMAL GROWTH Edited by P.J. Buttery, N.B. Haynes and D.B. Lindsay (1986)
COMPUTER APPLICATIONS IN AGRICULTURAL ENVIRONMENTS Edited by J.A. Clark, K. Gregso and R.A. Saffell (1986)
MANIPULATION OF FLOWERING Edited by J.G Atherton (1987)
NUTRITION AND LACTATION IN THE DAIRY COW Edited by P.C. Garnsworthy (1988)
MANIPULATION OF FRUITING Edited by C.J. Wright (1989)
APPLICATIONS OF REMOTE SENSING IN AGRICULTURE Edited by M.D. Steven and J.A. Clark (1990)
GENETIC ENGINEERING OF CROP PLANTS Edited by G.W. Lycett and D. Grierson (1990)
FEEDSTUFF EVALUATION Edited by J.A. Wiseman and D.J.A. Cole (1990)
THE CONTROL OF FAT AND LEAN DEPOSITION Edited by K.N. Boorman, P.J. Buttery and D.B. Lindsay (1992)

Proceedings of Previous Easter Schools in Agricultural Science, published by Nottingham University Press

THE GLASSY STATE IN FOODS Edited by J.M.V. Blanshard and P.J. Lillford (1993)
RESOURCE CAPTURE BY CROPS Edited by J.L. Montieth, R.K. Scott and M.H. Unsworth (1994)
PRINCIPLES OF PIG SCIENCE Edited by D.J.A. Cole, J. Wiseman and M.A. Varley (1994)
ISSUES IN AGRICULTURAL BIOETHICS Edited by T.B. Mepham, G.A. Tucker andJ. Wiseman (1995)

Biopolymer Mixtures

S.E. HARDING, S.E. HILL and J.R. MITCHELL
University of Nottingham

NOTTINGHAM
University Press

Nottingham University Press
Manor Farm, Main Street
Thrumpton, Nottingham, NG11 0AX, UK

NOTTINGHAM

First published 1995

British Library Cataloguing in Publication Data
A catalogue record for this book is available from the British Library

ISBN 1-897676-247

Typeset by Nottingham University Press, Nottingham
Printed and bound by Redwood Books, Trowbridge, Wiltshire

PREFACE

The study of how macromolecules behave in mixtures - whether or not they interact chemically- underpins the whole of biological science since most biological structures and processes involve a combination of more than one macromolecular type. Applied Science in particular is replete of examples where a mixture of two or more macromolecules or a mixture of a macromolecule with a smaller molecule can be used to produce a favourable effect. For example in Pharmaceutical Science the success and efficiency with which a drug can be delivered can be greatly enhanced by the design of a suitable macromolecular carrier system which interacts with other macromolecules (mucus glycoproteins, plasma proteins etc.) in a favourable way. Or in Food Technology the ability of protein or polysaccharide to act as a gelling or thickening agent can be greatly enhanced by the addition of a different macromolecule or even a small sugar .

When two macromolecules are mixed together in an aqueous environment (either dilute or colloidal) one of three things can happen:

- nothing
- phase separation, due to thermodynamic incompatibility: i.e. molecules "pushing apart"
- the macromolecules interact non-covalently or covalently in either a reversible or non-reversible manner: i.e. molecules "sticking together"

Although there are cases where it is important to establish no effect on mixing two macromolecules together, (cf. Chapter 19) the greatest interest is in those cases where there is an effect, and the first part of this book is mainly concerned with the examination of phase separation phenomena, and the second part concentrates more on the analysis of interacting systems (simple associative or irreversible aggregative phenomena). This dividing line is not however rigorous, since several chapters consider both effects.

Part I, "*Phase separation phenomena*" aptly starts with a contribution from the pioneer in phase separation research, *Per-Åke Albertsson*, who reviews the principles of bioseparation phenomena. This theme is continued by fellow Swedes from Lund, *L. Picullel* and *K. Bergfeldt*, with *S. Nilsson* of Rogalind Research, Norway who examine the factors determining phase behaviour in mixed polymer systems. *A. Clark* describes the kinetics of demixing and this is followed by two chapters on rheological behaviour, the first by *C. Brown, T. Foster, I. Norton and J. Underdown* who investigate whether shear can affect the microstructure of mixed biopolymer systems and whether phase separated systems behave like oil-in water emulsions, and the second by *S. Ross-Murphy* who discusses the small deformation rheological properties of these mixed systems. One of the goals of these fundamental studies is the enhancement of our understanding of the design of fat replacers in low-fat food products. There then follows a series of chapters describing the application of a range of other techniques to the investigation of phase separation phenomena in mixed systems. *A. Donald, C. Durrani, Robin Jones, A. Rennie and R. Tromp* describe the use of Fourier Transform Infra Red (FTIR) spectroscopy, time-resolved small angle light scattering and phase contrast microscopy to examine phase separation phenomena and this is followed by chapters from *M. Williams, R. Keenan and T. Halstead* focusing on NMR methods and *R. Abeysekera and A. Robards* on the use of optical and electron microscopic methods, the latter for the examination of phase separating starch-gelatin systems. *P. Williams and N. Smith* then investigate, using a combination of osmotic pressure, dynamic light scattering and turbidity, how the addition of biopolymers can sterically stabilise dispersions by the method of depletion flocculation, and this is followed by chapters by *A. Clewlow, A. Rowe and M. Tombs* describing the use of ultracentrifuge based methods to look at the effects of polydispersity on phase separation phenomena (with specific reference to the pectin-gelatin system) and *S. Kasapis*, who describes the use of mechanical spectroscopy, differential scanning calorimetry and microscopic methods to investigate phase separations and considers the results in terms of the Takayanagi model. Part I finishes with a contribution by *A.-M. Hermansson, S. Kidman and K. Svegmark* who investigate specifically one macromolecular system - starch - and phase separation phenomena involving its components, amylose and amylopectin.

Part II, *"Interacting Systems"* concentrates more on molecules "sticking together" and the effects of these types of interaction phenomena on a range of processes from gel formation, rheology and the enhancement of drug delivery. In Chapter 13, *E. Morris* re-examines the definition of the term "synergism" in terms of these sticking-together phenomena, looking specifically at the area of polysaccharide-polysaccharide interactions and over twenty different mixed polysaccharide systems are considered. This theme of synergistic polysaccharide-polysaccharide interactions is continued in Chapter 14 by *V. Morris*, focusing on the interactions involving two particular classes of polysaccharide, the galactomannans and glucomannans. Before corresponding protein systems are considered, the bridge is provided by a chapter on protein-polysaccharide systems by *J.-L Doublier, C. Castelain, G. Lllamas and J. Lefebvre* (Chapter 15): synergistic interactions between these "unlike" classes of macromolecule are rare, whereas phase separation appears the norm. Certain mixed protein systems can also give rise to phase separation phenomena, many give rise to synergistic interactions(with compatible gel formation and foams) and sometimes aggregation or precipitation, and this is all considered in detail by *N. Howell* in Chapter 16. Mixed biopolymers and their organisation at interfaces are then considered by *E. Dickinson* (Chapter 17) with specific attention directed towards the role of protein-protein and protein-polysaccharide interactions in relation to the stability of emulsions and foams in food systems. The application of protein-polysaccharide (or mucoprotein-polysaccharide) interactions in another area of commercial importance - Pharmaceutical Science is then considered in Chapter 18 with an examination by I. Fiebrig, S. ("Bob") Davis and S. Harding on the use of polysaccharides (particularly the polycationic chitosans) as carrier mucoadhesives for increasing the transit time of orally administered drugs. Mucoadhesion is a medical phenomena where favourable interactions are sought. There are however medical phenomena where interactions are undesirable and *D. Fisher* and co-workers (Chapter 19) describe the procedure of "PEGylation" of proteins and liposomes to render them non-antigenic and non-immunogenic. The volume finishes with two chapters considering favourable interactions occurring between biopolymers and small sugars. *S. Hill and J. Mitchell* (Chapter 20) consider and describe the use of the Maillard reaction between proteins and reducing sugars (or proteins and polysaccharides) and the control of oxidative degradation of polysaccharides and finally the book closes with a consideration by *J. Blanshard* (Chapter 21) of low water content mixed biopolymer systems, and how developments in techniques such as differential scanning calorimetry, mechanical methods, nuclear magnetic resonance, infra-red spectroscopy and x-ray diffraction are giving an insight into the molecular interactions and behaviour that occur.

Throughout this book there are approximately 150 different combinations of biopolymer considered (a separate index to help the reader is provided) and over 30 different techniques described for their investigation. Gels, foams, emulsions, glassy states and plasticizers, self-association, complex formation phenomena, aggregation, precipitation and phase separation phenomena are all considered. A whole range of applications with particular relevance to the Food and Drug industries are covered and workers in these areas should find this book a useful reference source.

Finally, this volume has been produced based on the 56th "Easter School" held at the University of Nottingham between September 19-21st 1994 and attended by over 110 delegates. Special thanks to Mrs. Kath Brasnett without whose extensive help and patience with the eccentricities of the Editors neither this volume nor the Meeting would have been a possibility. Thanks also to Dr. John Horton and Karen ("Tranmere Rovers") Gray for type-editing the Chapters and also Sarah Keeling from Nottingham University Press. We hope readers of this book will find it as enjoyable as the Meeting has been.

Stephen Harding
Sandra Hill
John Mitchell

CONTENTS

PART I: PHASE SEPARATION PHENOMENA

AQUEOUS POLYMER PHASE SYSTEMS: PROPERTIES AND APPLICATIONS IN BIOSEPARATION

PER-ÅKE ALBERTSSON
Department of Biochemistry, Chemical Centre, University of Lund, P.O. Box 124, S-221 00 Lund, Sweden

Introduction

Phase separation in polymer mixtures is a very common phenomenon. In fact miscibility among polymers is an exception rather than the rule. This is due to the high molecular weight of the polymers which results in a relatively low entropy of mixing per unit weight. This in turn means that the interaction between different polymer segments is the dominating factor determining the free energy of mixing, and hence whether phase separation occurs or not. If the interaction between segments of two polymers is repulsive, phase separation occurs above certain polymer concentrations and the two polymers will collect separately in two opposite phases. If, however, the interaction between the two polymers is attractive, as is the case between two oppositely charged polyelectrolytes, the two polymers will collect in the same phase and the result is a phase system with one polymer-rich phase in equilibrium with a polymer-poor phase. The greater the interaction between two polymers the lower the concentration of polymers needed to obtain phase separation. To be fully miscible, however, the two polymers must be very similar in their properties. The theoretical background to phase separation is treated by Piculell in Chapter 2 in this book. For a general account of aqueous polymer phase systems and various applications in bioscience the reader is referred to the books by Walter, Brooks and Fisher (1985) and Albertsson (1986), and volume 228 of *Methods in Enzymology* (1994).

Phase Systems

In Table 1.1 are listed some pairs of water soluble polymers which have been used in constructing phase systems for use in bioseparation. Figure 1.1 shows a phase diagram of a dextran–polyethylene glycol (PEG)–H_2O system which has been widely used for separation of various biomolecules such as proteins and nucleic acids and cell organelles such as chloroplasts and mitochondria, membrane vesicles and even whole cells.

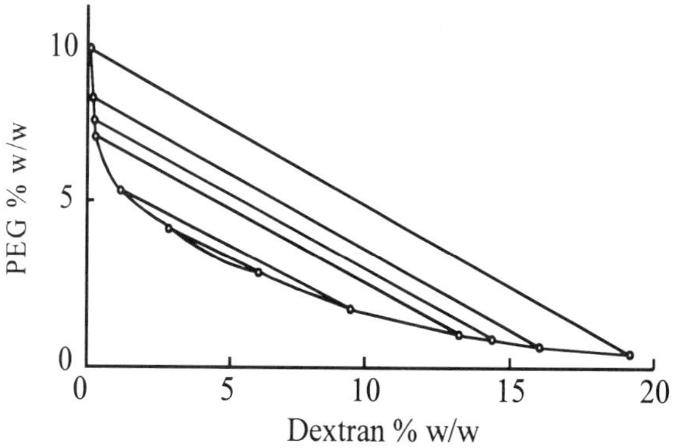

Figure 1.1 Phase diagram of a dextran–polyethylene glycol–water phase system at 20°C. Dextran 500 (Pharmacia), molecular weight 500 000 (weight average); polyethylene glycol (Carbowax PEG 6000), molecular weight 8000. Dextran is enriched in the lower phase and polyethylene glycol in the upper phase (Albertsson, 1986)

Table 1.1 SOME PAIRS OF WATER-SOLUBLE POLYMERS WHICH CAN FORM AQUEOUS POLYMER TWO-PHASE SYSTEMS

Dextran–polyethylene glycol
Dextran–methyl cellulose
Dextran–Ficoll
Dextran–hydroxypropyl dextran
Ficoll–polyethylene glycol
Hydroxypropylstarch–polyethylene glycol

The phase systems listed in Table 1.1 are all the result of "incompatibility" between the polymers, i.e. the two polymers collect in separate phases. A particularly interesting combination of two polymers is the pair dextran sulphate–dielthylaminoethyldextran (DEAE–dextran) (Albertsson, 1986). At low ionic strength, and at neutral pH, the interaction between these two polymers is attractive because dextran sulphate is negatively charged and DEAE–dextran positively charged. The two polymers will therefore collect in the same phase. At high ionic strengths this attraction is weakened and the hydrophobic diethyl groups of DEAE–

dextran will dominate the interaction between the two polymers and make it repulsive so that the two polymers will be incompatible and collect in opposite phases. At certain intermediate ionic strengths the attractive and repulsive forces will cancel and the two polymers are miscible. This phase system, therefore, displays all three phenomena which are possible when mixing two polymers and should be an interesting system for theoretical studies of phase separation in polymer mixtures.

A two-phase system can be described by a phase diagram where the compositions of the two conjugated phases in equilibrium are given for various mixtures of the two polymers in water (Figure 1.1). Phase diagrams of several other systems can be found in the literature (Albertsson, 1986). Aqueous polymer–polymer two-phase systems are characterized by the following properties (Albertsson, 1986). The difference in density as well as in refractive index between the phases is small. The interfacial tension between the phases is small, 0.1–100 N/m. The settling time (phase separation at normal gravity) is slow (15 min to several hours) due to the small density difference and the relatively high viscosity of the phases. The time taken for separation to occur can be reduced, however, by low speed centrifugation.

One can construct isopycnic phase systems (e.g. dextran–Ficoll–H_2O) with no density difference (Albertsson, 1986). In such phase systems one phase will eventually form a globular phase, immersed in the other, like an egg yolk in an egg. The behaviour is similar to that of other phase systems at zero gravity, which has been studied in one of the space shuttles (Van Alstine, Boyce, Harris, Bamberger, Curreri, Snyder and Brooks, 1986).

Three-phase systems are obtained if solutions of three different polymers, all mutually incompatible, are mixed. With four polymers a four-phase system is obtained, and so on and so forth (Albertsson, 1986). Many polymers form two-phase systems when combined with suitable salts (e.g. phosphates, sulphates, citrates). These systems have found use in the technical extraction of enzymes (Kula, Kroner and Hustedt, 1982).

An aqueous polymer two-phase system can also be obtained with only one polymer and water. For example, water solutions of random co-polymers of polyethylene glycol and polypropylene glycol form a two-phase system above certain temperatures and such systems have also been used in bioseparation (Alred, Kozlowski, Harris and Tjerneld, 1994).

Factors determining Partition

Partition depends on many factors originating from both of the phase polymers, the ionic composition and the partitioned substance. Among the phase polymer factors are type of polymers and their molecular weights as well as the presence of certain chemical groups on the polymer (e.g. ionized, hydrophobic, or biospecific groups).

The ionic composition is of paramount importance because the ions present in the system determine the sign and magnitude of the interfacial electric potential. Among the properties of the partitioned substance that determine its partition are size, charge and hydrophobic properties of the surface, and presence of receptors for biospecific ligands (attached to one of the phase polymers). The chirality of molecules can also be reflected in the partition (Ekberg, Sellergren and Albertsson, 1985). Generally, small molecules partition evenly between the phases unless there is a binding molecule in one of the phases.

Aqueous two-phase separations

The different factors that determine partition can be explored separately or in combination to achieve an effective separation. Some of the factors can be expanded so that they will dominate the partition behaviour. With respect to the partitioned substances, the following types of partition can be distinguished.

1. Size-dependent partition: molecular size or the surface area of the particles (or molecules) is the dominating factor.
2. Electrochemical partition: electrical potential between the phases is used to separate molecules or particles according to their charge.
3. Hydrophobic affinity partition: hydrophobic properties of a phase system are used for separation according to the hydrophobicity of molecules or particles.
4. Biospecific affinity partition: the affinity between sites on the molecules or particles and affinity ligands attached to one of the phase polymers is used for separation.
5. Conformation-dependent partition: conformation of the molecules and particles is the determining factor.
6. Chiral partition: enantiomeric forms are separated.

The partition of a solute (e.g. a protein) between the phases is described by a partition coefficient, K, defined as the ratio between the concentrations of solute in the upper and lower phases. Formally, we can resolve the partition coefficient into a number of factors:

$$K = K^{o}K_{el}K_{hfob}K_{biosp}K_{size}K_{conf} \tag{1}$$

where the indices el, hfob, biosp, size and conf stand for electrochemical, hydrophobic, size-dependent, and conformational contributions to the partition coefficient respectively. The K° part includes all other factors, such as the general relative solvation of the solute molecule in the phases. The logarithmic form of the equation above is especially useful when the various effects are studied:

$$\log K = \log K^{o} + \log K_{el} + \log K_{hfob} + \log K_{biosp} + \log K_{size} + \log K_{conf} \quad (2)$$

The effects of a number of parameters on the partition of proteins and nucleic acids are outlined below.

POLYMER CONCENTRATION

Increasing the polymer concentration moves the phase system further away from the critical point and the two phases become more different. As a result, soluble substances such as proteins partition more towards one of the phases, and often the separation factor (i.e. the ratio between the partition coefficients of two components) increases. The viscosity of the phases increases with increase in polymer concentration, which one has to consider in finding an optimal phase system.

Particles such as cell organelles or membrane particles usually adsorb more strongly to the interface when the polymer concentration is increased. This adsorption to the interface is selective and can be utilized for separation purposes. For example, plasma membranes can be purified from other cellular membranes by three repeated batch extractions in a dextran–polyethylene glycol phase system (Kjellbom and Larsson, 1986).

MOLECULAR WEIGHT OF POLYMERS

A general rule holds for the effect of the molecular weight of polymers on partition: partitioned molecules or particles will favour a phase more if the molecular weight of its polymer is reduced. For example, the partition coefficient of a protein in the dextran–polyethylene glycol system will increase if the molecular weight of polyethylene glycol is decreased or the molecular weight of dextran is increased. Conversely, the partition of the protein will decrease if the molecular weight of polyethylene glycol is increased or the molecular weight of the dextran is decreased.

This effect of the molecular weight of polymers depends in turn on the molecular weight of the partitioned protein. Proteins with higher molecular weights are more influenced by changes in the molecular weight of polymers than proteins with small molecular weights. Polymers with different molecular weights can therefore be used to optimize separation of protein molecules differing in size (Albertsson, Cajarville, Brooks and Tjerneld, 1987).

SALTS AND pH

The salt composition is of paramount importance in partitioning of all kinds of molecules and cell particles. Although salts partition fairly evenly between the phases, there are small but significant differences in the partition coefficients of different salts (Johansson, 1970, 1974a), meaning that different ions have different affinities for the two phases. An electrical potential between the phases is therefore created (Albertsson, 1986). For a salt, the ions of which have the charges Z^+ and Z^-, the interfacial potential is given by

$$\psi = \left[\frac{RT}{(Z^+ + Z^-)F}\right] \ln \frac{K_-}{K_+} \tag{3}$$

where R is the gas constant, F the Faraday constant, T the absolute temperature, and K_- and K_+ hypothetical partition coefficients of the ions in the absence of a potential. The interfacial potential will be larger the larger the K_-/K_+ ratio. A salt with two ions which have very different affinities for the two phases will generate a larger potential difference than a salt with ions that have similar affinities for the phases.

It can be shown that in the presence of excess salt a protein will partition as follows:

$$\ln K_p = K_p^{\circ} + \frac{FZ}{RT} \psi \tag{4}$$

where K_p is the partition coefficient of the protein and $K°_p$ is the value of this coefficient when the interfacial potential (generated by the excess salt) is zero or when the protein net charge Z is zero. Thus the difference in partition of the ions of the salt generates an electric potential difference according to eq. (1), which in turn affects the partition of the protein according to eq. (2). Even if ψ is small, it will strongly influence K_p because Z is a large number for most proteins. K_p changes exponentially with Z. Experimental results agree very well with eq. (4) (Johansson, 1974a, 1974b). The strong effect of the ionic composition on the partition holds for all charged macromolecules and particularly so for cell particles which each carry a vast number of charges.

The influence of different salts on the partition of proteins in the dextran–polyethylene glycol system has been studied in detail (Johansson, 1974a, 1974b; Albertsson, 1986). It is the ratio between different ions which is the most important factor, while the ionic strength *per se* is of minor influence. The cations decrease the partition coefficient of negatively charged proteins in the order $Li^+ < NH_4^+ < Na^+ < Cs^+ < K^+$ and the monovalent anions in the order $F^- < Cl^- < Br^- < I^-$. The divalent anions HPO_4^{2-}, SO_4^{2-}, and citrate^{2-} increase the partition coefficient relative to the monovalent anions. For positively charged proteins the orders given above are reversed. In summary, to achieve a high partition coefficient for negatively charged proteins, one should use lithium or sodium monohydrogen phosphate, sulphate, or citrate. On the other hand, to achieve a lower partition coefficient, one should use potassium chloride or bromide. For a positively charged protein, the opposite will hold.

Cross partition

The isoelectric point of proteins can be determined by a so-called cross partition (Albertsson, 1986). Using this method the partition coefficient is determined as a function of pH with two different salts (in excess over buffer). Two curves are obtained, one for each salt, when the partition coefficient is plotted versus pH. The two curves cross at the isoelectric point. This method can also be used for the determination of the isoelectric point of cell organelles, membrane vesicles and cells.

Charged polymers

A charged polymer concentrated in one of the phases shows a stronger effect on the partition of charged biomacromolecules than salts do. However, the concentration of low molecular electrolytes in the system must be kept low compared with the concentration of polymer-bound charged groups. Charged polyethylene glycol (PEG) has been used for directing the partition of proteins. Positively charged trimethylamino–PEG (TMA–PEG) extracts negatively charged proteins into the PEG-rich upper phase while positively charged proteins are excluded. Negatively charged PEG (e.g. PEG–sulphonate (S–PEG) or carboxymethyl–PEG (CM–PEG)) has the opposite effect. By changing the pH value in the system, the net charge of proteins present can be successively varied and their partition coefficients adjusted within extreme values (Johansson, Hartman and Albertsson, 1973; Johansson, 1994).

Hydrophobic groups

By introducing a low concentration (1 mM or less) of PEG-bound hydrophobic groups (e.g. palmitate), proteins with hydrophobic binding sites show increased affinity for the upper phase (Shanbhag and Axelsson, 1975; Johansson, 1994). Membrane vesicles and cells can also be extracted into the PEG-phase even at very low concentrations of the PEG–palmitate (Eriksson, Albertsson and Johansson, 1976).

Affinity ligands

Affinity ligands attached to one of the polymers have been used to extract ligand-binding proteins and nucleic acids into the corresponding phase. Triazine dye is particularly useful as an affinity ligand (Kopperschläger, 1994) The log K_{biosp} is obtained as the difference in the log K values observed with (excess) of ligand-PEG and with no ligand present. The efficiency in extraction measured in this way increases with the concentration of phase-forming polymers.

Temperature

The composition of the phases changes with temperature. Proteins usually partition more equally between the phases of a two-polymer systems when the temperature is increased. This effect can be counteracted by using higher concentrations of the polymers.

Applications

Phase partition using aqueous polymer systems has been applied to the separation of a large number of different molecules and particles of biological origin. The partition method is extremely versatile. It can be applied on a small scale (down to microlitres) as well as on a large industrial scale (volumes of cubic metres).

Batch procedures are applied for preparative purposes. These involve a few extraction steps either with a constant phase composition or by a "gradient" procedure, where the composition of the phase system is changed after each partition step in order to extract the different components of a mixture sequentially (Albertsson, 1986; Hammond, Majewski, Onorato, Brazy and Morré, 1993).

Multistage procedures such as countercurrent distribution or partition chromatography are used for analytical purposes when one is interested in the study of the heterogeneity or complexity of a mixture. Using these procedures one may also discover new molecules or particles. Here follow some examples of bioseparation which illustrate the versatility of the partition method:

1. Proteins: several enzymes have been purified by batch procedures using affinity partition where a specific ligand was bound to one of the polymers (Kopperschläger, 1994). The salt-PEG phase system has been used widely for large scale purification of enzymes from a crude lysate of bacteria and cell homogenates (Kula *et al.*, 1982). Of particular interest is the application of partition chromatography. One phase, the dextran-rich phase of the dextran–PEG system, is bound to a solid support, while the other, PEG-rich phase is mobile (Müller, 1994). This has recently been used in the analysis of immunoglobins and antigen–antibody complexes (Andersson, Wingren and Hansson, 1993).
2. Nucleic acids: these have been analysed by partition chromatography and can be separated according to both size and base composition (Müller, 1994).
3. Cell organelles and membrane vesicles: chloroplasts and mitochondria have been purified by either batch or countercurrent distribution procedures (López-Pérez, 1994; Larsson, 1994). Plasma membranes, from both animal and plant cells, have been purified by batch procedures involving three partition steps (Larsson, Sommarin and Widell, 1994; Morré, Reust and Morré, 1994). Specific lectins bound to the dextran in the bottom phase of the dextran–PEG system have been used to withdraw plasma membranes (which expose sialic acid groups on their outer surface) from other cellular membranes which, with a selected ionic composition, partition into the upper PEG phase. By such biospecific affinity partition, very selective and efficient purification can be obtained rapidly with only one or two partition steps (Persson and Jergil, 1994).

   Of particular interest is the combination of fragmentation with separation in order to analyse the structure of biological membranes (Albertsson, 1994; Albertsson, Andreasson, Stefánsson and Wollenberger, 1994). The photosynthetic membrane of chloroplasts has been fragmented by sonication followed by countercurrent distribution to separate the different membrane vesicles so obtained. Biochemical analysis of the different fragments and comparison with theoretical models has allowed the construction of realistic models for the structure and function of the photosynthetic membrane.
4. Virus: different virus particles have been enriched and concentrated from a solution by adding polymers, so that a phase system, with for example a very small bottom phase into which the virus partitions, is formed (Hammar, 1994). This is a very mild and efficient way of concentrating a virus from large volumes of culture media. This technique should be applicable to other particles too, since it is a general phenomenon that particles often partition one-sidedly into one of the phases.
5. Cells: various cells such as erythrocytes, leucocytes, cultured tissue cells, algae and bacteria have been partitioned, usually with the dextran–PEG phase

system. The partition depends on the surface properties of the cells and these can be separated according to their stage in the cell's life cycle (Walter, 1994) Some mutants of bacteria show different partition behaviour due to changes in the composition of their surface layer. The isoelectric points and the hydrophobicity of the bacterial surface have also been determined and compared with the virulence of pathogenic bacteria (Magnusson, 1994).

Conclusion

A large number of data have now been collected concerning the behaviour of biomacromolecules and cell particles. These show that reproducible partition can be achieved if conditions such as temperature and phase composition are carefully controlled. We also know most of the factors which determine partition. Future research should be directed towards finding new phase systems with increased selectivity and polymers which allow an economical application on a technical scale.

References

Albertsson, P.-Å. (1986). *Partition of Cell Particles and Macromolecules*, 3rd edition, New York: Wiley

Albertsson, P.-Å. (1988) Analysis of the domain of membranes by fragmentation and separation in aqueous polymer two-phase systems. *Quarterly Review of Biophysics*, **21**, 61–98

Albertsson, P.-Å. (1994). In *Methods in Enzymology*, Vol. 228, edited by Walter, H. and Johansson, G., New York: Academic Press, pp 640–658

Albertsson, P.-Å., Andreasson, E., Stefánsson, H. and Wollenberger, L. (1994). In *Methods in Enzymology*, Vol. 228, edited by Walter, H. and Johansson, G., New York: Academic Press, pp 469–482

Albertsson. P.-Å., Cajarville, A., Brooks, D.E. and Tjerneld, F. (1987) Partition of proteins in aqueous polymer two-phase systems and the effect of molecular weight of the polymer. *Biochemica et Biophysica Acta*, **926**, 87–93

Alred, P.A., Kozlowski, A., Harris, J.M. and Tjerneld, F. (1994) Application of temperature-induced phase partitioning at ambient temperature for enzyme purification. *Journal of Chromatography*, **659**, 289–298

Andersson, K., Wingren, C. and Hansson, U.-B. (1993) Liquid–liquid partition chromatography as a method to examine surface properties of antibodies and antigen–antibody complexes. *Scandinavian Journal of Immunology*, **38**, 95–101

Ekberg, B., Sellergren, B. and Albertsson, P.-Å. (1985) Direct chiral resolution in an aqueous two-phase system using the counter-current distribution principle. *Journal of Chromatography*, **333**, 211–214

Eriksson, E., Albertsson, P.-Å. and Johansson, G. (1976) Hydrofobic surface properties of erythrocytes studied by affinity partition in aqueous two-phase systems. *Molecular and Cellular Biochemistry*, **10**, 123–128

Hammar, L. (1994). In *Methods in Enzymology*, Vol. 228, edited by Walter. H. and Johansson, G., New York: Academic Press, pp 640–658

Hammond, T.G., Majewski, R.R., Onorato, J.J., Brazy, P.C. and Morré, D.J. (1993) Isolation and characterization of renal cortical membranes using an aqueous two-phase partition technique. *Biochemical Journal*, **292**, 743–748

Johansson, G. (1970) Partition of salts and their effects on partition of proteins in a dextran–poly (ethylene glycol) water two-phase system. *Biochimica et Biophysica Acta*, **221**, 387–390

Johansson, G., Hartman, A. and Albertsson, P.-Å. (1973) Partition of proteins in two-phase systems containing charged poly(ethylene glycol). *European Journal of Biochemistry*, **33**, 379–386

Johansson, G. (1974a) Effects of on the partition of proteins in aqueous polymeric two-phase systems. *Acta Chemica Scandinavia*, **B28**, 873–882

Johansson, G. (1974b) Partition of proteins and micro-organisms in aqueous biphasic systems. *Molecular and Cellular Biochemistry*, **4**, 169–180

Johansson, G. (1994). In *Methods in Enzymology*, Vol. 228, Edited by Walter, H. and Johansson, G., New York: Academic Press, pp 64–74

Kopperschläger, G. (1994). In *Methods in Enzymology*, Vol. 228, edited by Walter, H. and Johansson, G., New York: Academic Press, pp 121–143

Kula, M.-R., Kroner, K.H. and Hustedt, H. (1982) Purification of enzymes by liquid–liquid extraction. In *Advances in Biochemical Engineering*, Vol. 24, edited by Fiechter, A., Berlin: Springer Verlag, pp 73–118

Larsson, C. (1994). In *Methods in Enzymology*, Vol. 228, edited by Walter, H. and Johansson, G., New York: Academic Press, pp 419–424

Larsson, C., Sommarin, M. and Widell, S. (1994) Isolation of highly purified plant plasma membranes and separation of inside-out and right-side-out vesicles. In *Methods in Enzymology*, Vol. 228, edited by Walter, H. and Johansson, G., New York: Academic Press, pp 451–469

López-Pérez, M.J. (1994). In *Methods in Enzymology*, Vol. 228, edited by Walter, H. and Johansson, G., New York: Academic Press, pp 403–411

Magnusson, K.-E. (1994) Testing for charge and hydrophobicity correlates in cell–cell adhesion. In *Methods in Enzymology*, Vol. 228, edited by Walter, H. and Johansson, G., New York: Academic Press, pp 326–334

Morré, D.J., Reust, T. and Morré, D.M. (1994). In *Methods in Enzymology*, Vol. 228, edited by Walter, H. and Johansson, G., New York: Academic Press, pp

448–495

Müller, W. (1994) Separation of proteins and nucleic acids. In *Methods in Enzymology*, Vol. 228, edited by Walter, H. and Johansson, G., New York: Academic Press, pp 193–206

Persson, A. and Jergil, B. (1994). In *Methods in Enzymology*, Vol. 228, edited by Walter, H. and Johansson, G., New York: Academic Press, pp 489–496

Shanbhag, V.P. and Axelsson, V.P. (1975) Hydrofobic interaction determined by partition in aqueous two-phase systems. Partition of proteins in systems containing fatty acid esters of poly(ethylene glycol). *European Journal of Biochemistry*, **60**, 17–22

Van Alstine, J.M., Boyce, J., Harris, M., Bamberger, S., Curreri, P.A., Snyder, R.S. and Brooks, D.E. (1986). *Proceedings of an NSF Workshop on Interfacial Phenomena*, Space Science Laboratory preprint series No. 86-139 (NASA)

Walter, H., Brooks, D.E. and Fisher, D. (Editors) (1985). *Partitioning in Aqueous Two-Phase Systems: Theory, Methods, Uses, and Applications to Biotechnology*, Orlando (Florida): Academic Press

Walter, H. (1994) In *Methods in Enzymology*, Vol. 228, edited by Walter, H. and Johansson, G., New York: Academic Press, pp 299–326

2

FACTORS DETERMINING PHASE BEHAVIOUR OF MULTI COMPONENT POLYMER SYSTEMS

L. PICULELL and K. BERGFELDT
Physical Chemistry 1, University of Lund, Chemical Centre, P.O. Box 124, S-22100 Lund, Sweden

S. NILSSON
Rogaland Research, P.O. Box 2503, Ullandhaug, N-4004 Stavanger, Norway

Introduction

Why should one be interested in the phase behaviour of mixed polymer solutions? Possible answers to this question include both fundamental and practical aspects. For fundamental reasons, a limited miscibility is more common in mixtures involving polymers than in mixtures of small molecules. Therefore, phase separation is a phenomenon which is quite likely to be encountered when one deals with mixed polymer solutions. In practical applications, such mixtures are quite common, either because the system naturally contains mixed polymers (such as in food applications), or because mixtures are used in order to optimise certain (often rheological) properties. Sometimes, indeed, the applications exploit the phase separation phenomena, such as in aqueous two-phase partitioning (Albertsson, 1986 and Chapter 1) or in certain mixed polymer gels (Tolstoguzov, 1986; Morris, 1990). At other times, single-phase solutions are desired, and a phase separation would be an unwanted side-effect. However, regardless of whether one wishes to accomplish phase separation or to avoid it, it is essential to know something about the underlying mechanisms, i.e. the molecular interactions that give rise to the phase behaviour. In fact, the desire to extract information regarding molecular interactions (to be used in the interpretation of more complex mixtures, such as mixed polymer gels) is often the motive behind studies of the phase behaviour of mixed polymer solutions.

The purpose of this contribution, then, is to broadly review our current understanding of the molecular mechanisms underlying the *equilibrium* phase behaviour of mixed solutions of two polymers, one or both of which may be a polyelectrolyte. (Note that the structure of non-equilibrium systems, such as certain mixed polymer gels, may or may not be simply related to the equilibrium phase behaviour of the mixtures.) Specifically, we will deal with flexible polymers dissolved in water, although some of the conclusions presented are quite general.

All our considerations are restricted to phase equilibria where isotropic liquid phases are formed; cases involving crystalline or liquid crystalline phases will not be discussed. The interpretations of the phase behaviour will be in terms of two types of mechanisms, i.e. *short-range interactions* and *polyelectrolyte effects*. We will not be concerned with the nature of the short-range interactions (e.g. hydrogen bonds, van der Waals forces), but only with their relative magnitudes. To illustrate our points, we will use both theoretical results, based on the Flory–Huggins model, and experimental results on simple model systems.

Various aspects of our topic have been discussed before, by others as well as by ourselves (Tompa, 1956; Patterson, 1982; Khokhlov and Nyrkova, 1992; Piculell and Lindman, 1992; Piculell, Iliopoulos, Linse, Nilsson, Turquois, Viebke and Zhang, 1994). Nevertheless, our ambitions here are to present a complete – if sketchy – picture, rather than to restrict ourselves to recent progress in the area. This approach seems justified in the present context. Hopefully it will also increase the usefulness of the text.

Classification of the phase behaviour of polymer–polymer–solvent mixtures

The top phase diagrams in Figure 2.1 illustrate the two types of phase separation most commonly encountered in mixtures of two polymers (P_2 and P_3) in a common solvent (S_1). Figure 2.1(a) shows a case where the two equilibrium phases have similar overall concentrations of polymers, but where each phase is enriched in one of the polymeric components. The best studied example of this type of phase behaviour in aqueous solution is probably that of poly(ethylene oxide), PEO, and dextran (Albertsson, 1986). In contrast, Figure 2.1(b) shows a type of phase separation where one of the two phases is enriched in both of the polymeric components. This behaviour is most commonly encountered for oppositely charged polyelectrolytes in salt-free solution. A classical example is that of oppositely charged gelatin and gum arabic (Bungenberg de Jong, 1949), and a very recent example is that of oppositely charged copolymers of acrylamide (Frugier and Audebert, 1994). On the basis of the partitioning of the polymers between the separating phases, it is convenient to refer to these two types of phase separation as *segregative* and *associative* respectively (Piculell and Lindman, 1992). In an associative phase separation both polymers are enriched in one of the separating phases, whereas, in a segregative phase separation, the two polymers are enriched in separate phases.

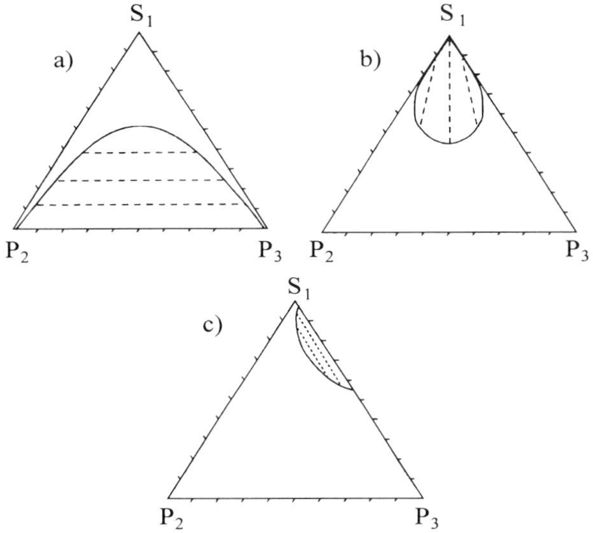

Figure 2.1 Schematic illustrations of segregative (a), associative (b) and borderline (c) phase separation in polymer–polymer–solvent mixtures

Segregative and associative phase diagrams need not necessarily be symmetrical with respect to the bisector of the solvent corner, as is the case in Figures 2.1(a) and 2.1(b); the criteria given above also suffice to distinguish between segregation and association in the general (asymmetrical) case. Only in the very special case where one of the polymers is distributed exactly equally between two phases will the phase separation be neither associative nor segregative. This case will be called a *borderline* phase separation, and is exemplified in Figure 2.1(c). The characteristic feature of the borderline phase diagram is that the tie-lines are exactly parallel to one of the binary polymer–solvent axes. The slightest deviation from exact parallelism would, in terms of our definition, yield either an associative or a segregative phase separation. The borderline case is interesting, since under certain conditions it represents a state of maximum miscibility of the system.

A segregative phase separation is often referred to as "polymer incompatibility", or, in the older literature, as a "simple coacervation" (Bungenberg de Jong, 1949). An associative phase separation involving oppositely charged polymers, on the other hand, is an example of what Bungenberg de Jong (1949) called "complex coacervation", i.e. a liquid–liquid phase separation that may be attributed to the presence of polymer charges. We prefer to avoid these terms in the classification of polymer–polymer–solvent phase diagrams, since they are based on (assumptions regarding) the interactions underlying the phase behaviour, rather than on the compositions of the separating phases. Thus the term polymer incompatibility implies that the segregation in a common solvent is caused by the same mechanism as in a binary polymer blend, i.e. an effective repulsion between unlike polymers. However, as has been pointed out previously (Zeman and Patterson, 1972; Hsu and Prausnitz, 1974; Patterson, 1982), and as we shall see in more detail below, a segregative phase separation may also be driven by a difference in the solubility of the two polymers. An associative phase separation, on the other hand, does not always involve oppositely charged polyelectrolytes. Indeed, it may even occur for non-ionic polymers in non-aqueous solvents (Djadoun, Goldberg and Morawetz, 1977). Moreover, as will be exemplified below, it is possible to obtain an associative phase separation even without an effective attraction between different polymers.

Transitions between association and segregation

In a number of instances, more or less subtle chemical changes in one or both of the polymers, or in the composition of the (mixed) solvent, have been shown to lead to transitions from one type of phase behaviour to the other. These possibilities of manipulating the phase behaviour, by relatively simple means, may have important practical consequences. Most of the cases studied have involved associating pairs of oppositely charged polyelectrolytes, where a reversal of the charge of one of the polymers (such as for gelatin in mixtures with gum arabic (Bungenberg de Jong, 1949)), or the addition of high concentrations of salt (such as for mixtures of oppositely charged dextran derivatives (Albertsson, 1986)), result in a transition from association to segregation. Thalberg, Lindman and Karlström (1991) made a detailed study of a transition of the latter type, which occurs when increasing concentrations of NaBr are added to mixtures containing sodium hyaluronate and the cationic surfactant tetradecyltrimethylammonium bromide.

Transitions from segregation to association can also occur, however, in cases where the association is of a different nature. Thus, Perrau, Iliopoulos and Audebert (1989) demonstrated that mixtures of poly(vinyl pyrrolidone) (PVP) with poly(acrylic acid) (PA) phase separate segregatively at high degrees of neutralisation, α, of PA, whereas a strongly associative phase separation occurs at low α. At intermediate degrees of neutralisation, complete miscibility was found. Recently, two of us (Bergfeldt, Piculell and Tjerneld, 1995) found a similar transition in mixtures of PA with poly(styrene sulphonate) (PSS) in the presence of 1 M NaCl. Experimental phase diagrams of this system are shown in Figure 2.2. Here, the phase behaviour changes continuously from segregation to association, as α of PA is changed from unity (where all monomers are ionized) to zero. Furthermore, the solvent conditions become increasingly poorer as is decreased, so that, below α = 0.30, there is some demixing in the mixture of (quasi)binary PA and 1 M NaCl. (The latter effect is not apparent in Figure 2.2(c) since, owing to an unequal distribution of salt between the separating phases, the salt concentration was always appreciably lower than 1 M in the PA-rich phases.) Figure 2.2 demonstrates that remarkably large effects may result from such a small molecular event as the protonation of a carboxylic acid group.

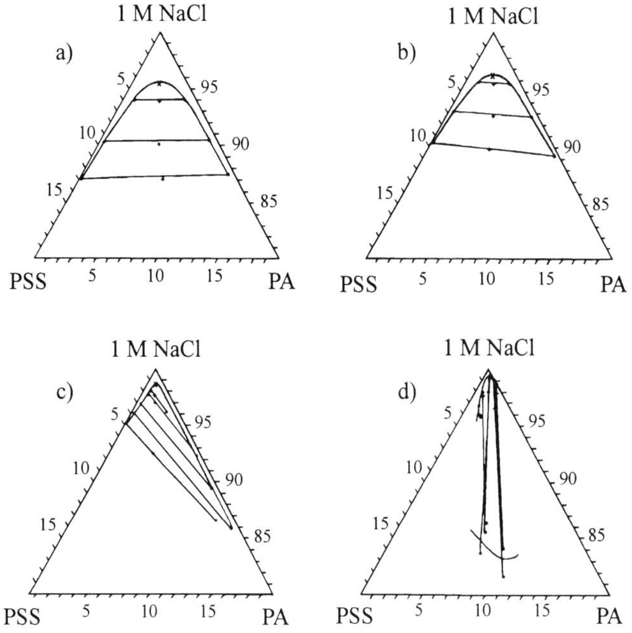

Figure 2.2 Experimental phase diagrams for mixtures of PSS (M_w = 800,000) and PA (M_w = 150,000) in 1 M NaCl for α(PA) = 1 (a), 0.43 (b), 0.29 (c) and 0 (d). Total compositions (+) of phase-separated samples, and cloud-points (×) of mixtures with equal (by weight) concentrations of PA and PSS, are indicated. For α(PA) = 0.29 and 0, a limited miscibility was found in the quasi-binary solutions of PA in 1 M NaCl

In the absence of added salt, a segregative phase separation was still found in PA–PSS mixtures at high α, but the associative phase separation at low α disappeared. Instead, a highly enhanced viscosity was found at low α in concentrated mixtures. We will frequently return to the complex behaviour of PA–PSS mixtures for purposes of illustrating how the interplay between the various interactions may influence the phase behaviour of mixed polymer solutions. A detailed account of our findings on this system is given elsewhere (Bergfeldt, Piculell and Tjerneld, 1995).

Modelling phase behaviour

In order to gain insight into the phase behaviour of mixed polymer solutions, the Flory–Huggins lattice model (Flory, 1953; Tompa, 1956) is most useful. We will now briefly review the model and recall some key results. To appreciate the origin of polymer effects, it is instructive to first consider the cases of simple *binary* mixtures of two small molecules, a polymer and a small molecule, or two polymers. Most of the results quoted in this section will concern such binary mixtures.

In the Flory–Huggins treatment, the polymer solution is modelled as a lattice, where each lattice site is occupied by either a solvent molecule or a polymer segment. (Note that, in the modelling of a real system, a segment is not necessarily equal to the repeating unit of a polymer; rather it is a portion of the polymer molecule that occupies a reference volume, corresponding to the volume of a lattice site, which is commonly taken as the partial volume of a solvent molecule.) A polymer is taken to be flexible, so that segment number k + 1 of a given polymer may, with equal probability, occupy any one of the sites around segment number k, except for the site already occupied by segment number k - 1.

The configurational contribution to the entropy of mixing is given by the number of distinguishable ways in which the polymer molecules can be laid out onto the lattice. This means that all configurations are considered as equally probable, irrespective of molecular interactions. For a process where a random binary mixture of components 1 and 2 (which may or may not be polymeric) is obtained from the pure liquid systems, the resulting entropy (expressed per lattice site in the mixture) is given by (Flory, 1953)

$$\Delta S_{mix,site} = -k\left[\frac{\phi_1}{r_1}\ln\phi_1 + \frac{\phi_2}{r_2}\ln\phi_2\right] \tag{1}$$

Here ϕ_1 is the volume fraction and r_1 the relative length (i.e. the number of segments per molecule) of species 1. Eq. (1) shows that the configurational entropy of mixing decreases strongly as the lengths of the molecules increase; tying segments together severely restricts the number of possible arrangements of the molecules on the lattice. This is, in fact, the *only* effect of the degree of polymerisation in the model. We will now look specifically at three cases, i.e. the "monomeric" mixture ($r_1 = r_2 = 1$), the polymer solution ($r_1 = 1$, $r_2 >> 1$), and the polymeric mixture (r_1, $r_2 >> 1$). For the monomeric mixture, Eq. (1) yields the familiar expression for an ideal mixture (note that, for $r_1 = r_2 = 1$, volume fractions are equal to molar fractions). For the polymer solution, the entropy of dissolving a liquid flexible polymer into a solvent is much smaller, as the second term within brackets vanishes when r_2 becomes large. Finally, Eq. (1) shows that the entropy of mixing becomes very small for a polymeric mixture when both r_1 and r_2 are large.

Short-range interactions are, in the model, restricted to pair interactions between nearest neighbours on the lattice. The effective interaction between components i and j in the mixture is described by the dimensionless pair interaction parameter

$$\chi_{ij} = \frac{z}{kT}\left(w_{ij} - \frac{w_{ij} + w_{ij}}{2}\right) \tag{2}$$

where z is the number of nearest neighbours adjacent to any lattice site, and w_{ij} (etc.) are the (free) energies of interaction between segments of species i and j when these occupy neighbouring positions in the lattice. It is important to understand the significance of the pair interaction parameter. The expression within brackets is the difference between, on the one hand, the pair interaction between two unlike segments and, on the other hand, the average of the pair interactions between like segments. This difference is multiplied by the number of nearest neighbours to a segment, and normalised with respect to the "thermal energy", kT. Since attractive pair interactions are negative, χ_{ij} will be positive when like molecules prefer like. In this case, the "effective" pair interaction between i and j is said to be repulsive. When $\chi_{ij} < 0$, the effective interaction is attractive, and it is athermal when $\chi_{ij} = 0$. At normal temperatures kT is of the same order of magnitude as the strength of typical intermolecular "bonds", such as a hydrogen bond or a van der Waals bond between atoms. The number of nearest neighbours (or number of bonds) is of the order of 4–12 in a liquid. This means that small relative differences in the pair interaction strengths are sufficient to yield interaction parameters of a magnitude of unity or less; values of this magnitude should therefore be common.

For simple systems, where only van der Waals forces operate, the W_{ij} are essentially pure interaction *energies*. Moreover, it is possible to show that, for simple systems, the attractions between like molecules are generally stronger than between unlike molecules (Hildebrand and Scott, 1962), which means that χ_{ij} should then be positive. For application of the theory to a structured solvent such as water, however, it is necessary to let the W_{ij} describe *free* energies of interaction, including both energetic and entropic effects. This is because the pair interactions in this case also affect the orientations of the molecules (e.g. in the formation of hydrogen bonds), and thereby the entropy. (These entropic effects provide a contribution to the entropy of mixing in addition to the configurational contribution discussed above.) Predictions of the signs and magnitudes of the interaction parameters are therefore considerably more difficult in aqueous solutions. Nevertheless, the qualitative predictions of the Flory–Huggins model should still be valid (Gustafsson, Wennerström and Tjerneld, 1986).

In a binary mixture of components 1 and 2, a limited miscibility (i.e. a two-phase region) appears when the interaction parameter exceeds a critical value, given by (Scott, 1949)

$$\chi_{c12} = \frac{(\sqrt{r_1} + \sqrt{r_2})^2}{2r_1r_2} \tag{3}$$

The critical interaction parameter is a measure of the amount of effective segment–segment repulsion that a mixture can tolerate before phase separation occurs. Note that the value of χ_{c12} only depends on the relative lengths of the components. This dependence is entirely due to the configurational entropy of mixing, given by Eq. (1) above. As the relative length increases, the entropy of mixing becomes smaller, and less effective repulsion is required to generate phase separation. For a monomeric mixture, $\chi_{c12} = 2$, whereas, for a polymer solution, χ_{c12} rapidly approaches 0.5 for long polymers. Finally, for a polymeric mixture, χ_{c12} approaches zero when both r_1 and r_2 approach infinity. The latter result provides the physical explanation of polymer incompatibility in a binary polymer blend; even a very slight effective repulsion is sufficient to generate a demixing of very long polymers.

From the above it is clear that the *number of particles* in the system has very important consequences for the entropy of mixing. The greatly enhanced solubility and, in certain cases, miscibility of polyelectrolytes is best understood from this point of view. A polyelectrolyte, when dissolved in a solvent with a high dielectric constant, dissociates into one polyion and n counterions, where, typically, $n \gg 1$. The entropy of mixing strongly favours that these counterions should distribute evenly throughout the whole system. (This is, in fact, the reason that they dissociate from the polyion in the first place.) However, owing to the condition of electroneutrality, which requires that each equilibrium phase must contain an equivalent concentration of polyions and counterions, a polyelectrolyte must, in a mixture with otherwise non-ionic components, still partition as a single component. In terms of the phase behaviour then, a polyelectrolyte solution is in many ways analogous to a solution of a simple salt, an analogy worth keeping in mind.

In the presence of excess salt, however, it becomes possible to form a concentrated, polyelectrolyte-rich phase without also creating a grossly uneven distribution of the small ions. This is why certain polyions may be salted out at comparatively low concentrations of salt. In this context, it is useful to distinguish between intrinsically water-soluble and intrinsically water-insoluble polyelectrolytes: an intrinsically water-insoluble polyelectrolyte (e.g. a hydrophobic, rod-like or compact globular polymer) derives its solubility solely from the entropy of mixing of the counterions, and is salted out on addition of salt. In contrast, an intrinsically water-soluble polyelectrolyte is soluble also in the absence of the dissociating counterions, owing to sufficiently favourable short-range interactions, and is not salted out (except, possibly, at very high concentrations of added salt, because of a different mechanism). Similarly, we may talk of a polyelectrolyte pair as intrinsically segregating or associating, referring to the properties in the absence of polyelectrolyte effects.

The above reasoning suggests that a simple way of incorporating polyelectrolyte effects in the Flory–Huggins treatment is to let the counterion be an additional component of the system, which mixes randomly in each single phase and contributes to the entropy of mixing, but which is subject to the additional requirement that each macroscopic phase must be electroneutral (Nilsson). (This model is similar in spirit, but differs in important details from the treatment of polyelectrolyte–polymer mixtures recently presented by Khokhlov and Nyrkova (1992).) The electroneutrality condition, allowing for an electrostatic potential difference between the phases, is the only way in which electrostatic interactions are taken into account in this treatment. The model captures, as we shall see, many important features of mixed polymer–polyelectrolyte solutions, and, also, mixed solutions of similarly charged polyelectrolytes. It does not, however, predict an associative phase separation in solutions of oppositely charged polyelectrolytes, nor the effects of added salt in such systems, although these effects may, indeed, be understood in terms of the entropy of mixing of the counterions, as we shall see later. A crude way of describing the association between oppositely charged polyelectrolytes within the Flory–Huggins model is to let the polymer–polymer interaction parameter be strongly attractive (Thalberg *et al.*, 1990, 1992).

Phase behaviour governed by short range interactions

In this section we will review the types of phase behaviour expected for mixtures of two polymers (P_2 and P_3) in a common solvent (S_1), in cases where only short-range interactions are operative. These cases are relevant for non-ionic polymers, or for polyelectrolytes in the presence of sufficient amounts of salt. The predictions are those given by the Flory–Huggins model, with various choices of the interaction parameters. We will also use this model in an attempt to interpret the phase behaviour of a real system, i.e. the mixtures of PSS, PA and 1 M NaCl shown in Figure 2.2 above. In all calculated examples the polymers are of equal length ($r_1 = r_2 = r$), and we have used r = 1000 throughout. The polymer–solvent interaction parameters, χ_{ps}, are chosen to represent solvent conditions that are "good" (i.e. χ_{ps} appreciably lower than the critical value of the binary polymer–solvent mixture, χ_{cps}), "marginal" (χ_{ps} just below χ_{cps}) or "poor" (χ_{ps} above χ_{cps}). For r = 1000, $\chi_{cps} = 0.532$.

THE SYMMETRICAL CASE

The symmetrical case (Figure 2.3) is the special and simple case where the polymer–solvent interaction parameters are equal for both polymers ($\chi_{12} = \chi_{13}$). A segregative phase separation in this case can only be obtained if χ_{23} exceeds the critical value for the binary polymer mixture, χ_{cpp}. (For r = 1000, $\chi_{cpp} = 0.002$.) As χ_{23} approaches χ_{cpp} from above, the two-phase area shrinks towards the P_2–P_3 axis (Figures 2.3(a) and 2.3(b)). The phase behaviour, in fact, only depends on the product $r\chi_{23}$ (which has to exceed 2 for phase separation to occur), and is independent of the value of the polymer–solvent interaction parameter, as long as this is smaller than χ_{cps}. The effect of the solvent in this special case is merely to dilute the system, hence increasing the miscibility. The symmetrical segregative case is the classical case of polymer incompatibility in a mixed solvent, first treated by Scott (1949).

In contrast to the segregation, the associative phase separation (Figures 2.3(c) and 2.3(d)) in symmetrical mixed polymer solutions is strongly dependent on the polymer–solvent interaction parameter. This is because the formation of a concentrated phase is unfavourable under good solvent conditions, but is facilitated as the solvent conditions become worse. It is thus clear that an association will occur even for an athermal polymer mixture (χ_{23} = 0), if the polymer–solvent

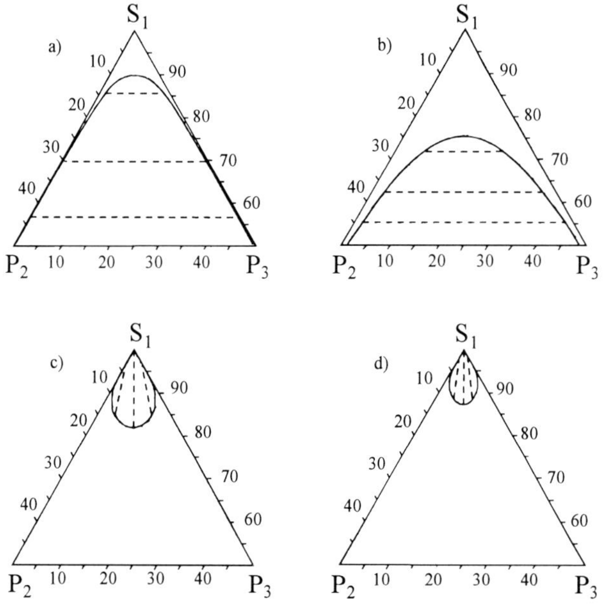

Figure 2.3 Calculated phase diagrams for symmetrical ($\chi_{12} = \chi_{13}$) mixtures of non-ionic polymers (r = 1000) in a common, good or marginal, solvent (χ_{12}, $\chi_{13} < \chi_{cps}$). Interaction parameter values are χ_{23} = 0.020 (a); χ_{23} = 0.008 (b); $\chi_{12} = \chi_{13}$ = 0.525, χ_{23} = - 0.202 (c); and $\chi_{12} = \chi_{13}$ = 0.504, χ_{23} = -0.202 (d)

interaction parameter exceeds χ_{cps}. For better solvent conditions, however, when there is complete miscibility in the binary polymer–solvent mixtures, an associative phase separation always requires an effective polymer–polymer attraction (χ_{23} < 0). The cases shown in Figures 2.3(c) and 2.3(d) are for marginal solvent conditions, where only a moderately strong effective attraction is required for phase separation to occur.

THE ASYMMETRICAL CASE

In the (more general) asymmetrical case (Figure 2.4), $\chi_{12} \neq \chi_{13}$. Segregation in this case has been studied by Zeman and Patterson (1972), Hsu and Prausnitz (1974) and Patterson (1982). They showed that a small difference in the polymer–solvent interactions may give rise to a considerable increase in the segregative two-phase area. Compare Figures 2.3(a) and 2.4(a), which show the segregation for the same value of χ_{23} in the symmetrical and strongly asymmetrical cases respectively. In fact, it is possible to generate a segregative phase separation from this effect alone, for an athermal, or, indeed, even for a slightly attractive effective polymer–polymer interaction (Figure 2.4(b)). This purely "solvency-driven" segregation – which is different from the "incompatibility-driven" segregation is characterised by a two-phase area that does not extend all the way to the P_2–P_3 axis, but forms a closed loop inside the phase triangle. A closed loop appears as soon as $\chi_{23} < \chi_{cpp}$, when there is complete miscibility in the binary P_2–P_3 system. Experimentally, however, this criterion for recognising a solvency-driven segregation is of limited value, since the polymer-rich parts of the phase diagram are often very difficult to explore. We conclude that *it is not generally safe to interpret a segregative phase separation in terms of an effective polymer–polymer repulsion.*

For the associative phase separation, the effects of asymmetry are not so pronounced. As in the symmetrical case, our calculations indicate that an effective

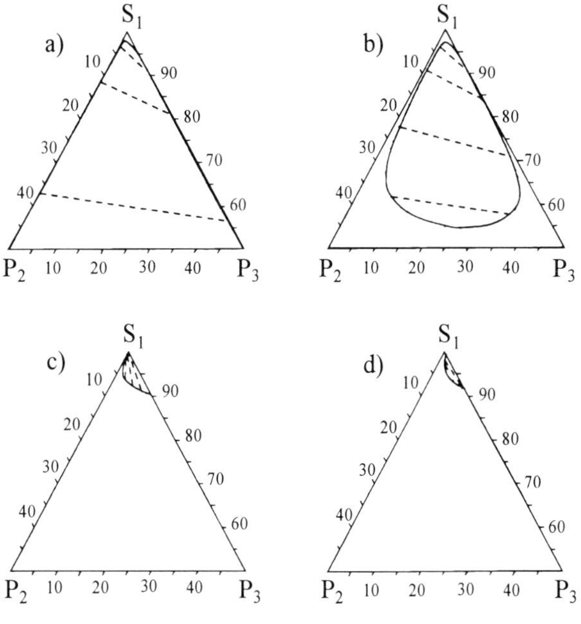

Figure 2.4 Calculated phase diagrams for asymmetrical mixtures of non-ionic polymers (r = 1000) in a common solvent. Interaction parameter values are $\chi_{12} = 0.404$, $\chi_{13} = 0.525$, $\chi_{23} = 0.020$ (a); $\chi_{12} = 0.404$, $\chi_{13} = 0.525$, $\chi_{23} = -0.004$ (b); $\chi_{12} = 0.529$, $\chi_{13} = 0.545$, $\chi_{23} = 0$ (c); and $\chi_{12} = 0.404$, $\chi_{13} = 0.541$, $\chi_{23} = -0.081$ (d)

attraction is necessary to generate an associative phase separation, as long as both of the polymer-solvent interaction parameters are below χ_{cps}. The magnitude of the effective attraction required is roughly the same as for a symmetrical mixture where the polymer-solvent interaction parameter is equal to the average of χ_{12} and χ_{13}. An interesting special case, however, is the "nearly symmetrical" case where one of the polymer–solvent interaction parameters is slightly above, and the other slightly below χ_{cps}. Here we may generate an associative phase separation also when there is no net attraction between the different polymers (Figure 2.4(c)). (In fact, calculations showed that an associative phase separation may result even with a slight effective repulsion between P_2 and P_3.) This means that *it is not always safe to interpret an associative phase separation obtained, say, as a result of adding a second polymer to a polymer solution, in terms of an effective polymer–polymer attraction*.

In systems where only short-range interactions are operative, a borderline phase separation seems only to be possible when one of the polymer–solvent interaction parameters is above, and the other below χ_{cps}. One example of phase separation close to the borderline case is shown in Figure 2.4(d). For this type of system, it is possible to obtain a continuous change from segregation to association, via a borderline phase separation, simply by making the polymer–polymer interaction continuously more attractive. Calculations indicate (Bergfeldt, Piculell and Linse) that for such a transition, the borderline phase separation represents the case with the smallest two-phase area. Thus, the approach towards borderline phase separation, rather than any particular value of the polymer–polymer interaction parameter, represents the condition for maximum miscibility. In the borderline case, there is a balance between the segregative and associative tendencies. (In a symmetrical system, this balance always results in complete miscibility.) For instance, in Figure 2.4(d), a solvency-driven segregation (a difference between χ_{12} and χ_{13}) is balanced by a combination of an effective polymer–polymer attraction (a negative χ_{23}) and a poor solvency of P_3 ($\chi_{13} > \chi_{cps}$).

INTERPRETATION OF THE PA–PSS–1 M NaCl SYSTEM

The complex behaviour of PA–PSS–1 M NaCl mixtures provides a challenge for model interpretations. Since large concentrations of salt are present, it is expected that polyelectrolyte effects (cf below) may be neglected. Therefore, we found it interesting to test if the Flory–Huggins theory, assuming only short–range interactions, could reproduce this behaviour at all, and, if so, whether it would give a unique qualitative interpretation of the underlying interactions. Figure 2.5 shows model calculations (Bergfeldt, Piculell and Tjerneld, 1995) where PSS is modelled as component 2, PA as component 3 and 1 M NaCl as (pseudo)component 1.

Gratifyingly, the model succeeds in capturing all the qualitative features of the real system, and only with quite restricted choices of the interaction parameters (shown in Figure 2.6). This demonstrates the utility of the Flory–Huggins approach. We will now discuss in some detail the analysis that led to the α-dependence of the interaction parameters shown in Figure 2.6.

In the modelling, it is required that the PSS–solvent interaction parameter χ_{12} be independent of α and lower than χ_{cps}, since the PSS and solvent components

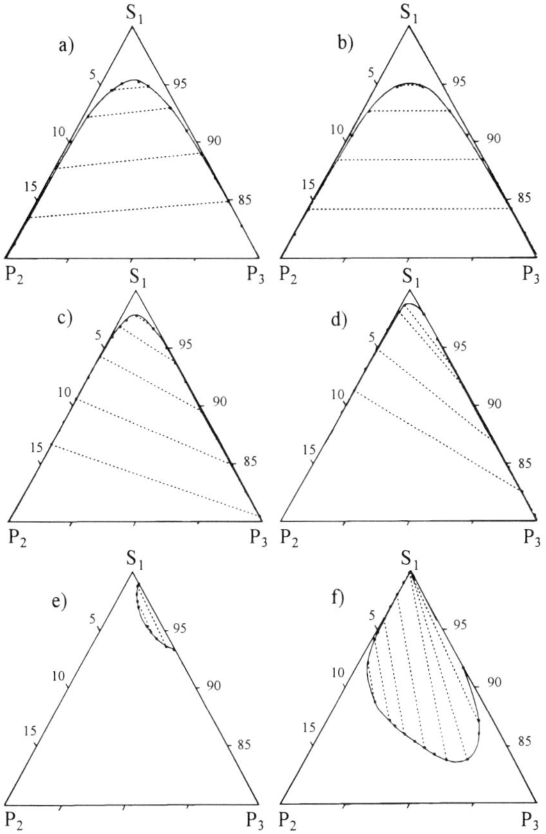

Figure 2.5 Calculated phase diagrams for mixtures of non-ionic polymers (r = 1000) in a common solvent, mimicking the behaviour of PA–PSS–1M NaCl mixtures. The degree of neutralisation of PA (=P_2) was $\alpha = 1$ (a), 0.8 (b), 0.20 (c), 0.04 (d), 0.02 (e), and 0 (f). Interaction parameter values are given in Figure 2.6

are constant and completely miscible. All observed effects must therefore be due to changes in the interactions involving PA, i.e. χ_{13} and/or χ_{23}. Moreover, these interaction parameters should change in a *monotonic* way as α is varied, since the structure of polymer 3 is changed in a monotonic fashion. Furthermore, to account for the observed limited solubility of PA in 1 M NaCl at low degrees of neutralisation, the PA–solvent interaction must change from below χ_{cps} at high α to above χ_{cps} at low α. A subtlety is that χ_{13} is taken to have a value below χ_{12} at $\alpha = 1$. This

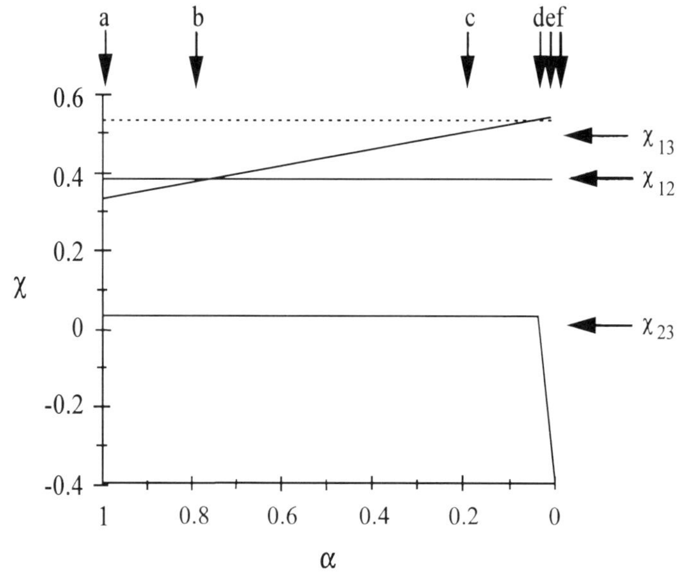

Figure 2.6 Interaction parameter values chosen to represent mixtures of PSS (=P_2) and PA (=P_3) in 1 M NaCl (=S_1), at various degrees of neutralisation of PA. Dashed line represent χ_{cps}. Top arrows indicate parameter sets corresponding to the various phase diagrams of Figure 2.5.

results in a minimum in the extension of the segregative two-phase area around $\alpha = 0.8$, where χ_{13} crosses χ_{12} and a symmetrical phase diagram results. On either side of this point, the two-phase area increases, as a result of increasing differences in the polymer–solvent interactions, but with different slopes of the tie-lines. These features were clearly seen experimentally (Bergfeldt, Piculell and Tjerneld, 1995), although the minimum extension of the two-phase area is not apparent in the phase diagrams of Figure 2.2. Finally, it was found to be impossible to reproduce the phase behaviour without having χ_{23} change from effective repulsion at high α to effective attraction at low α. At high α, where $\chi_{12} \approx \chi_{13}$, the segregation must be due to an effective repulsion. At low α, a strong attraction was required in order to generate the observed association. The only questionable feature of the modelling is that χ_{23} changes sharply from repulsion to attraction over a narrow interval of α, rather than linearly over the entire range. This feature was found necessary in order to reproduce the slow change of the phase behaviour at high degrees of neutralisation, followed by a rapid changeover to attraction at low α. It seems that the model, which always assumes random mixing in each single phase, is unable to capture the cooperative nature of the polymer association in the real system, if the interaction parameter is varied linearly over the entire range of α. We note that a sharp disappearance of the association, when only a small proportion of non-associating monomers is introduced on one of the polymers, has also been found for other associating polymer pairs (Perrau *et al.*, 1989; Iliopoulos and Audebert, 1991).

The model interpretation given above provides important information regarding the interactions in the system. Thus, the analysis clearly indicates that both a poor solvency of PA and a PA–PSS attraction are essential in order to explain the association at low α. The requirement of a PA–PSS attraction is important, since its existence is rather unexpected. The PA–PSS attraction does not readily fall into any of the categories (i.e. polyelectrolyte complex, hydrogen bonding complex, stereocomplex or charge transfer complex) that are commonly recognised among associating polymer pairs (Tsuchida and Abe, 1982). It could therefore have been suspected that the association was solvency-driven, rather than driven by a polymer—polymer attraction. We were, however, unable to reproduce the strong association obtained experimentally with such an assumption. The attraction is also supported by the enhanced viscosity of salt-free mixtures of PSS with PA seen at low α (cf. below).

Effects of dissociating counterions

As was emphasised above, the dissociated counterions of a polyelectrolyte enhance the entropy of mixing of that polyelectrolyte considerably (Khokhlov and Nyrkova, 1992; Piculell and Lindman, 1992; Piculell *et al.*, 1994). This effect is general and independent of the type of the polymer. Thus, the presence of ionic groups affects the solubility and miscibility of globular, rod-like or flexible macromolecules qualitatively in the same way. Indeed, the counterions will dominate the entropy of mixing in all of these cases.

MIXTURES OF A POLYMER AND A POLYELECTROLYTE

In mixed polymer solutions, the introduction of ionic groups on *one* of the polymers will generally dramatically reduce the tendency towards phase separation, both in the segregative and the associative cases, owing to the entropy of mixing of the counterions. For the segregative case, this effect has been demonstrated in illustrative experiments by Iliopoulos and co-workers (Perrau *et al.*, 1989; Iliopoulos, Frugier and Audebert, 1989). Some of their results are reproduced in Figure 2.7. Figure 2.7(a) shows that the miscibility of the intrinsically segregating pair of poly(acrylamide), PAm, and PEO increases dramatically when only a small fraction, f_c, of charged residues is introduced on the PAm polymer. In mixtures of PEO with completely neutralised PA (Figure 2.7(b)), the phase separation in the absence of added salt is quite insensitive to the degree of polymerisation of PA. This demonstrates that, as expected, it is the numerous sodium counterions, rather than the much fewer PA anions, that determine the entropy of mixing of the polyelectrolyte. On adding salt, however, the counterion effects disappear, and the normal molecular-weight dependence of the phase diagram reappears. Figure 2.7(c), finally, shows predictions from the simple Flory–Huggins approach, outlined above. Clearly, the model succeeds in predicting both the increased miscibility and the shift of the two-phase area towards the polyelectrolyte–solvent axis, as charged groups are introduced.

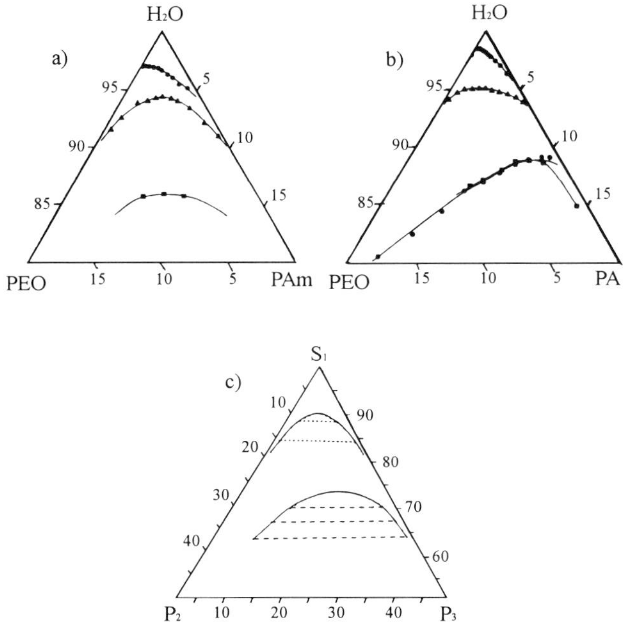

Figure 2.7 Phase diagrams for polymer—polyelectrolyte—solvent mixtures. (a) Experimental diagrams for mixtures of PEO with PAm where f_c(PAm) = 0 (●), 0.018 (▲), or 0.034 (■). (b) Experimental diagrams for mixtures of PEO with fully neutralised PA of M_w = 10,000 (⬣,▲) or 185,000 (■,●) in water (lower curves) and in 1 M NaCl (upper curves). (c) Calculated diagrams for r = 200, $\chi_{12} = \chi_{13} = 0$, $\chi_{23} = 0.1$, $f_c(P_2) = 0$, and $f_c(P_3) = 0$ (dotted tie-lines) or 0.02 (dashed tie-lines)

The associative case of a polymer–polyelectrolyte mixture is well illustrated by mixtures of non-neutralised PA with PSS. In salt-free solutions, the strongly associative phase separation which was found in the presence of salt (Figure 2.2(d)) more or less disappears; cf. Figure 2.8(a). This is due primarily to the large entropic cost of creating a concentrated phase containing most of the PSS counterions. (To a lesser extent, the greater solvency of PA in water, compared to 1 M NaCl, also contributes to the difference.) The association between unlike polymers still remains in the salt-free system, however, as is demonstrated by a strong synergistic viscosity increase on mixing PA and PSS (Figure 2.8(b)). This example, in fact, illustrates a good (and common) recipe for making a viscous polymer mixture or gel, that is, mix in water an associative pair of macromolecules, one of which is a polyelectrolyte. The association provides a mechanism for network formation, while the dissociated counterions of the polyelectrolyte component prevent the network from collapsing into a concentrated phase. Other examples of such networks include mixtures of ionic acrylic acid copolymers with PEO or PVP (Iliopoulos and Audebert, 1991), mixtures of certain neutral polymers with ionic surfactants (Carlsson, Karlström and Lindman, 1986) and mixtures of ionic polymers with neutral surfactants (Iliopoulos and Olsson, 1994). As expected, such systems generally have a tendency to phase separate associatively when salt is added (cf the PA–PSS case) or if the ionic groups are removed.

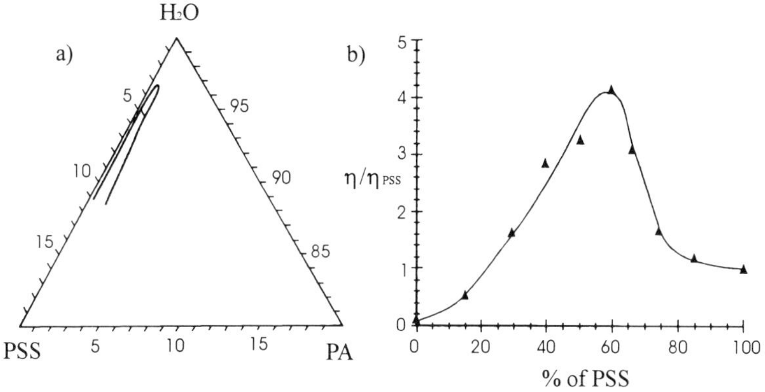

Figure 2.8 Experimental data for mixtures of PA (α=0) and PSS in water. (a) Phase diagram. (b) Normalised viscosities for mixtures with a total polymer concentration of 2%, at different relative amounts (in weight % of total polymer) of PSS.

MIXTURES OF SIMILARLY CHARGED POLYELECTROLYTES

For an intrinsically associating pair of similarly charged polyelectrolytes, the phase behaviour is expected to be qualitatively the same as for the polymer–polyelectrolyte mixture: the dissociating counterions provide a strong thermodynamic force that opposes the formation of a concentrated polymer phase. In the segregative case, however, the situation is quite different. The presence of charges on both polyions now makes it possible for these to segregate into different phases, without creating an unfavourably enhanced concentration of the counterions in any phase. Similarly charged segregating polyelectrolytes are, therefore, expected to behave quite similarly to mixtures of segregating neutral polymers. This prediction is confirmed by the behaviour of mixtures of gum arabic with gelatin at a pH above the isoelectric point of the latter component (Bungenberg de Jong, 1949), PSS with dextran sulphate (Piculell *et al.*, 1991), and, also, PSS with fully neutralised PA (Bergfeldt, Piculell and Tjerneld, 1995). The phase behaviour of the latter mixture in pure water is shown in Figure 2.9(a), and it may be seen that it is quite similar to the behaviour of the same mixture in 1 M NaCl (Figure 2.2(a)).

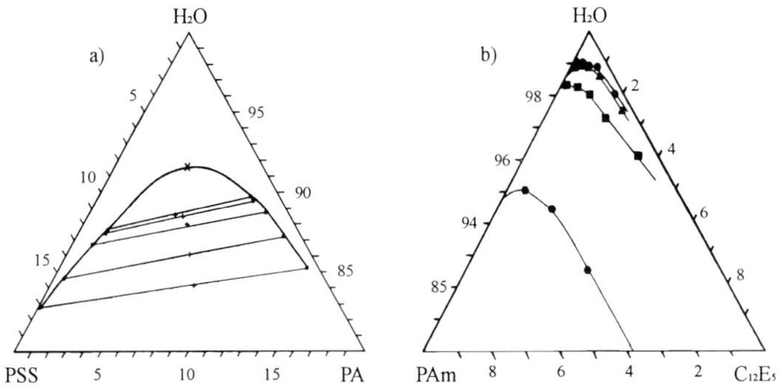

Figure 2.9 Experimental phase diagrams of aqueous mixtures of similarly charged polyelectrolytes. (a) PSS and PA(α = 1). (b) Partially charged PAm with partially charged $C_{12}E_5$ surfactant micelles. The uncharged mixture (●) is compared with mixtures where f_c(PAm) = 0.034 and $f_c(C_{12}E_5)$ = 0 (⬢), 0.018 (▲), or 0.034 (■)

It is expected that the extent of the segregation in a mixture of similarly charged polyelectrolytes will depend also on how well the charge densities, or fractions of charged groups, balance each other in the two polyelectrolytes. This point is illustrated by experiments on a segregating mixture of PAm with micellar aggregates of a non-ionic surfactant (Piculell *et al.*, 1994), shown in Figure 2.9(b). The fraction of charged residues on the latter "macromolecule" may be varied conveniently by the addition of small amounts of an ionic surfactant, in this case sodium dodecyl sulphate (SDS). Compared to the neutral system, the two-phase area decreases when charges are added to the PAm molecule, as expected. On adding a small fraction of SDS molecules to the system containing the charged PAm, the segregation increases, as is also expected. In the case where 1.8 mole % of SDS is added, a phase diagram is obtained which happens to almost exactly match that of the neutral system. However, on further addition of SDS, the two-phase area decreases again. These experiments indicate that the charge densities of the polyelectrolytes may be balanced to a point where a maximum segregation occurs, and that this maximal segregation roughly corresponds to the phase separation in the neutral reference system.

What is, then, the condition for balancing charge densities of a segregating polymer pair? In the symmetrical case, the condition must obviously be that f_c is the same for both polyelectrolytes. In the asymmetrical case, the answer is less obvious, but Flory–Huggins calculations provide interesting insights. Figure 2.10 shows calculations (Piculell *et al.*, 1994) for a strongly asymmetrical intrinsically segregating polymer pair. The dashed reference phase diagram shows the segregation for the uncharged polymers, whereas the solid diagrams are for the corresponding polyelectrolyte mixtures where f_c of P_2 is varied, but kept constant for P_3. When the fractions of charged residues are the same on both polyelectrolytes, Figure 2.10(a), the tie-lines of the two-phase region become more horizontal – reflecting the importance of the entropy of mixing of the counterions – but the two-phase area is smaller, compared to the non-ionic mixture. On increasing f_c of PE_1, the slope of the tie-lines changes continuously, as a consequence of the continuously increasing concentration of counterions in the P_2-rich phase, but the extension of the two-phase area changes in a non-monotonic fashion. A maximum two-phase area occurs when the slope of the tie-lines in the polyelectrolyte system is the same as the slope in the reference system (Figure 2.10(c)). At this point the phase diagrams of the non-ionic and polyelectrolytic mixtures are almost identical. This provides us with a clue to the condition of maximum segregation in an intrinsically segregating mixture of similarly charged polyelectrolytes: the segregation becomes maximal when the fractions of charged residues on the polyelectrolytes are chosen such that *the dissociating counterions do not perturb the distribution of the polymers from that which would have been obtained in the absence of charges*. Thus, in the balanced case, a phase separation of the polyelectrolytes according to the neutral polymer phase diagram also gives the same activities of the counterions in both phases. A consequence of this is that the polymer that has the highest affinity for the solvent in the neutral mixture must also have the highest charge density in the balancedpolyelectrolyte mixture.

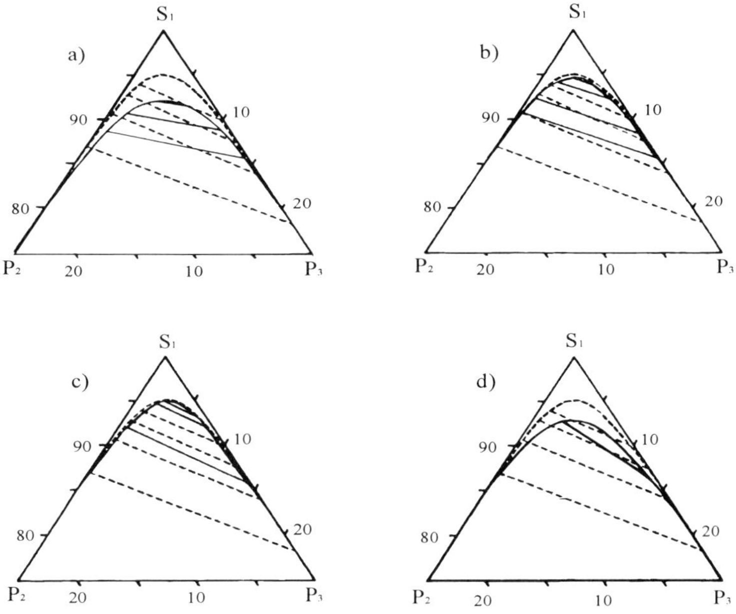

Figure 2.10 Calculated phase diagrams for aqueous mixtures of intrinsically segregating polyelectrolytes. r = 200, $\chi_{12} = 0$, $\chi_{13} = 0.4$, $\chi_{23} = 0.08$. Dashed diagram for $f_c(P_2) = f_c(P_3) = 0$, solid diagrams for $f_c(P_3) = 0.1$ and $f_c(P_2) = 0.1$ (a), 0.15 (b), 0.2 (c) or 0.3 (d)

Needless to say, the above predictions are only valid in cases where the introduction of charged groups does not change the short-range interaction parameters significantly. The latter condition is clearly not satisfied for the mixtures of PA and PSS shown above.

MIXTURES OF OPPOSITELY CHARGED POLYELECTROLYTES

For mixtures of oppositely charged polyelectrolytes (Smid and Fish, 1988), the phase behaviour may again be understood in terms of the entropy of mixing of the counterions. In this case, however, we need to look more closely on the distributions of these ions around the polyions in order to understand the effect. Owing to the attractive Coulomb interaction, the concentration of counterions is generally enhanced near the polyion. From an entropic point of view, this departure from a uniform distribution is unfavourable. In a mixed solution of oppositely charged polyelectrolytes, however, a more uniform distribution of both kinds of counterions may be attained simply by letting the polyions associate. The association of the polyions is certainly entropically unfavourable, but this effect is outweighed by the favourable entropy contribution from the more numerous small ions.

On addition of salt to such an associative system, the effects of the counterions will, as usual, disappear. The counterion distribution will become more uniform, regardless of the state of association of the polyions. The phase diagram will, therefore, again depend on the intrinsic phase behaviour of the mixture, determined by the short range interactions. If the polyions are intrinsically soluble, the associative phase separation will disappear. If, in addition, the polyions are an intrinsically segregating pair, a segregative phase separation will eventually appear. However, if the polyions are intrinsically water-insoluble, a (solubility-driven) association will persist even with added salt.

Conclusions

On the basis of the distribution of the polymers between the separating phases, the phase behaviour of polymer–polymer–water solutions may be divided into two main categories, i.e. associative or segregative phase separation. The mechanisms behind the phase behaviour may be understood in terms of short-range interactions and polyelectrolyte effects.

As regards the short-range interactions, Flory–Huggins calculations show that both association and segregation are influenced by the polymer–solvent interactions, in addition to the polymer–polymer interactions. In the associative case, the polymer–solvent interactions are always important, increasing the tendency towards association as the solvent conditions become poorer. In the segregative case, one may speak of two independent mechanisms; one caused by polymer incompatibility (effective polymer–polymer repulsion), and one caused by differences in the polymer–solvent interactions. The interplay between all pair interactions can give rise to quite complex phase behaviour, and it is generally not safe to draw conclusions regarding polymer–polymer interactions (effective attraction or repulsion) from isolated measurements of phase compositions (association or segregation) in phase separation experiments.

The main effects arising from polymer charges (polyelectrolyte effects) may be understood in terms of the entropy of mixing of the dissociated counterions. In the absence of added salt, this effect generally perturbs the phase behaviour from the "intrinsic" phase behaviour given by the short-range interactions alone, except in the very special case of similarly charged intrinsically segregating polyelectrolytes with balancing charge densities. For mixed similarly charged polyelectrolytes in general, and for polymer–polyelectrolyte mixtures, the effect of the dissociating counterions is always to increase the miscibility (decrease the two-phase area) of the system. In these cases, the qualitative effect may be introduced quite simply in the Flory–Huggins model. For mixtures of oppositely charged polyelectrolytes, the phase separation is a result of the more uniform distributions of the small counteranions and countercations that results from the association of the polyions.

Acknowledgements

We are grateful to Ilias Iliopoulos for numerous stimulating discussions on the subject of this contribution. This work was funded by grants from the Swedish National Board for Industrial and Technical Development (N.U.T.E.K.), and from the Swedish Natural Science Research Council (N.F.R.).

References

Albertsson, P.-Å. (1986). *Partition of Cell Particles and Macromolecules*, 3rd edition, New York: Wiley

Bergfeldt, K., Piculell, L and Linse, P. Unpublished

Bergfeldt, K., Piculell, L and Tjerneld, F. (1995) Phase separation phenomena and viscosity enchancements in aqueous mixtures of poly(styrenesulfonate) with poly(acrylic acid) at different degrees of neutralization. *Macromolecules,* **28**, 336–3370

Bungenberg de Jong (1949). In *Colloid Science*, Vol. II, edited by Kruyt, H.R., New York: Elsevier, pp 232–258 and pp 335–384

Carlsson, A., Karlström, G. and Lindman, B. (1986) Synergistic surfactant electrolyte effect in polymer solutions. *Langmuir*, **2**, 536–537

Djadoun, S., Goldberg, R.N. and Morawetz, H. (1977). Ternary systems containing an acid copolymer, a basic copolymer and a solvent. 1. Phase equilibria. *Macromolecules*, **10**, 1015–1020

Flory, P.J. (1953). *Principles of Polymer Chemistry*, Ithaca: Cornell University Press

Frugier, D. and Audebert, R. (1994) Interaction between oppositely charged low ionic density polyelectrolytes: complex formation or simple mixture? In *Macromolecular Complexes in Chemistry and Biology*, edited by Dubin, P., Bock, J., Davis, R., Schulz, D.N. and Thies, C., Heidelberg: Springer-Verlag, pp 135–149

Gustafsson, Å., Wennerström, H. and Tjerneld, F. (1986) The nature of phase separation in aqueous two-polymer solutions. *Polymer*, **27**, 1768–1770

Hildebrand, J.H. and Scott, R.L. (1962). *Regular Solutions*, Englewood Cliffs, N.J.: Prentice-Hall

Hsu, C.C. and Prausnitz, J.M. (1974). Thermodynamics of polymer compatibility in ternary systems. *Macromolecules*, **7**, 320–324

Iliopoulos, I., Frugier, D. and Audebert, R. (1989). Phase separation in mixtures of two polymers in water. *Polymer Preprints*, **30**, 371–372

Iliopoulos, I. and Audebert, R. (1991) Complexation of acrylic acid copolymers with polybases – Importance of co-operative effects. *Macromolecules*, **24**, 2566–2575

Iliopoulos, I. and Olsson, U. (1994) Polyelectrolyte association to micelles and bilayers. *Journal of Physical Chemistry*, **98**, 1500–1505

Khokhlov, A.R. and Nyrkova, I.A. (1992) Compatability enhancement and microdomain structuring in weakly charged polyelectrolyte mixtures. *Macromolecules*, **25**, 1493–1502

Morris, E.R. (1990). In *Food Gels*, edited by Harris, P., London: Elsevier Applied Science, pp 291–359

Nilsson, S. Unpublished

Patterson, D. (1982) Polymer compatibility with and without a solvent. *Polymer Engineering and Science*, **22**, 64–73

Perrau, M.B., Iliopoulos, I. and Audebert, R. (1989) Phase separation of polyelectrolyte nonionic polymer systems in aqueous solution – Effects of salt and charge density. *Polymer*, **30**, 2112–2117

Piculell, L., Nilsson, S., Falck, L. and Tjerneld, F. (1991) Phase separation in aqueous mixtures of similarly charged polyelectrolytes. *Polymer Communications*, **32**, 158–160

Piculell, L. and Lindman, B. (1992) Association and segregation in aqueous polymer/polymer, polymer/surfactant, and surfactant/surfactant mixtures – Similarities and differences. *Advances in Colloid and Interface Science*, **41**, 149–178

Piculell, L., Iliopoulos, I., Linse, P., Nilsson, S., Turquois, T., Viebke, C and Zhang, W. (1994) Association and segregation in ternary polymer solutions and gels. In *Gums and Stabilisers for the Food Industry 7*, edited by Phillips, G.O., Williams, P.A. and Wedlock, D.J., Oxford: Oxford University Press, pp 309–322

Scott, R.L. (1949) The thermodynamics of high polymer solutions. V. Phase equilibria in the ternary system: polymer 1–polymer 2–solvent. *Journal of Chemical Physics*, **17**, 279–284

Smid, J. and Fish, D. (1988). In *Encyclopaedia of Polymer Science and Engineering*, 2nd edition, Vol. 11, edited by Mark, H.F., Bikales, N.M., Overberger C.G. and Menges, G., New York: Wiley, pp 720–739

Thalberg, K., Lindman, B. and Karlström, G. (1990) Phase diagram of a system of cationic surfactant and anionic polyelectrolyte – Tetradecyl trimethylammonium bromide–hyaluronan–water. *Journal of Physical Chemistry*, **94**, 4289–4295

Thalberg, K., Lindman, B. and Karlström, G. (1991) Phase behaviour of a system of cationic surfactant and anionic polyelectrolyte – The effect of salt. *Journal of Physical Chemistry*, **95**, 6004–6011

Tolstoguzov, V.B. (1986). Functional properties of protein-polysaccharide mixtures. In *Functional Properties of Food Macromolecules*, edited by Mitchell, J.R. and Ledward, D.A., London: Elsevier, pp 385–415

Tompa, H. (1956). *Polymer Solutions*, London: Butterworths

Tsuchida, E. and Abe, K. (1982) Interactions between macromolecules in solution and intermacromolecular complexes. *Advances in Polymer Science*, **45**, 1–119.

Zeman, L. and Patterson, D. (1972). Effect of the solvent on polymer incompatibiilty in solution. *Macromolecules*, **5**, 513–516

3

KINETICS OF DEMIXING

A.H. CLARK
Food Research Department, Unilever Research Laboratory, Colworth House, Sharnbrook, Bedford. MK44 1LQ U.K.

Introduction

Biopolymer solution demixing is currently of widespread interest. This interest extends from students of polymer science concerned with academic aspects of the subject, to technologists involved with practical applications. Such applications include separation and purification procedures, the production of fibres, films, membranes and foams, and the generation of novel soft solids ("gels") with useful material properties. In this last connection, recent developments in the food industry provide a good example. In this context, composite aqueous biopolymer gels have found numerous applications as fat replacement materials.

This chapter will focus particularly on the gel aspect of biopolymer demixing, but initially the discussion will concentrate on the demixing phenomenon itself. The form of the presentation will be that of a review, and although biopolymers will be the polymer type of principal interest, it will be necessary to dig deeply into the literature of synthetic polymer science (and indeed into the literatures of other sciences such as the studies of glasses, ceramics, and metal alloys). The study of solution demixing phenomena is more advanced in these areas, and provides a rich source of both theoretical description and experimental investigation. Indeed, it has been a general trend over the years for developments in the physics and physical chemistry of biopolymer-based systems to follow the synthetic polymer example.

The order of presentation of material consists therefore of a preliminary discussion of the equilibrium aspects of polymer solution phase behaviour, a review of the kinetic theory of demixing, a consideration of experimental results for systems based on one and two polymer types and a solvent, and finally an exploration of the difficult, but practically very important, issue of how, where gelation is also involved, demixing kinetics couple with polymer aggregation to determine final gel material properties. A summary will be given of the conclusions that can be drawn from the progress made so far, as will some indication of the direction of future research.

Equilibrium aspects of solution demixing

It has long been known that where solutions of two components are concerned, non-ideality can lead to conditions of temperature and composition where the homogeneous phase that might have formed under these conditions is thermodynamically unstable. Free energy can be lowered if this unstable phase converts spontaneously to appropriate fractions of two new solutions having the same components but differing in composition.

This situation can be understood in terms of the solution molar free energy G_m. An indication of how this quantity could vary with the mole fraction of solute 1 (X_1), for a solution (solvent 0) demixing as temperature falls, is shown in Figure 3.1(a) for cases in which:

(1) No phase separation is expected ($T > T_C$);
(2) A critical point is expected at a particular critical composition X_1^C and temperature T_C;
(3) Phase separation is expected over a given range of composition ($X_1^A \rightarrow X_1^B$), and at temperature $T < T_C$.

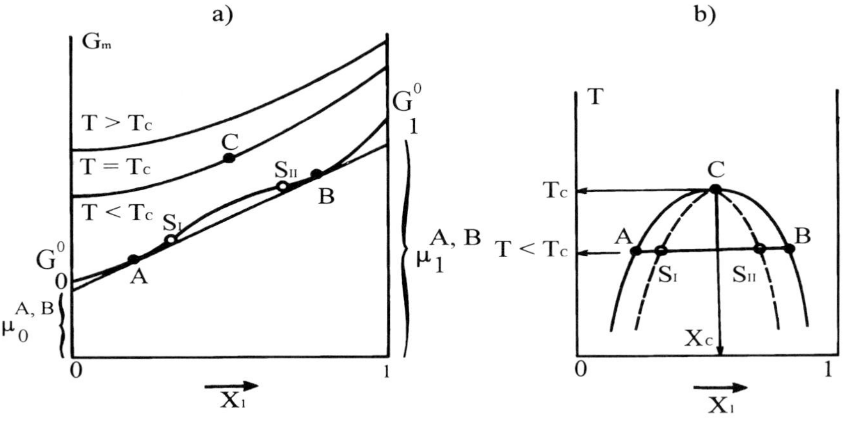

Figure 3.1 (a) Molar free energy (G_m)—solute composition (X_1) curves for a binary solution above, at, and below the critical temperature T_C, for a system demixing on cooling. (b) corresponding temperature—composition phase diagram showing binodal (solid line), spinodal (broken line), tie-line AB at temperature T, and critical point C

Figure 3.1 (b) indicates the kind of temperature–composition phase diagram for the system that would be consistent with Figure 3.1(a).
Figure 3.1(b) shows that above the critical temperature T_C, the solution is stable as a one-phase system at all compositions. However, below the critical temperature,

at any arbitrary temperature, a composition gap ($X_1^A \rightarrow X_1^B$) is found, within which all solutions are unstable with respect to appropriate amounts of phases of composition X_1^A and X_1^B. It is easily shown, of course, that the relative amounts of the phases A and B which are produced, in a particular situation, are given in terms of the tie-line AB by the lever rule.

Overall, as Figure 3.1(b) shows, as temperature is varied continuously from T_C to lower temperatures, the locus of points defining the miscibility gaps at each temperature forms a continuous coexistence curve (including the critical point itself). This is often called the binodal curve. In thermodynamic terms, as Figure 3.1(a) indicates, this coexistence curve can be defined quantitatively at each temperature below T_C, in terms of the solution free energy function. For example, in Figure 3.1(a), the points A and B on the lowest curve, which correspond to the miscibility gap AB in Figure 3.1(b), are shown as points on a common tangent to the free energy function. As is well known from basic physical chemistry, this condition ensures that the chemical potential of either component (μ_1 or μ_0 in Figure 3.1(a)) is the same in solutions of composition X_1^A and X_1^B, and this matching of chemical potentials of all components is of course the condition for equilibrium between phases. When this property is combined with the facts (a) that all solutions lying between points A and B can in principle decompose into appropriate amounts of solutions of composition X_1^A and X_1^B and, (b) that when this happens the total free energy is inevitably lowered, the instability condition operating within composition range X_1^A to X_1^B becomes obvious.

For real systems, phase behaviour is clearly defined by the detailed nature of the solution free energy function (or, equivalently, the free energy of formation of the solution) and its temperature-dependence. For polymer solutions the classical theoretical treatment of this problem is that by Flory and Huggins (see, for example, Kurata, 1982) based on a lattice model. This model will not be examined in detail here. It is noted, however, that according to the Flory–Huggins theory, the condition for the existence of a critical point and a coexistence curve for a single polymer–solvent system, is determined by the value of a parameter χ, which is related to the free energy change involved in replacing polymer segment–polymer segment and solvent—solvent contacts with polymer segment–solvent contacts. When this parameter is greater than 0.5 (for a high molecular weight polymer) demixing occurs. Thus χ takes on the role of temperature in Figure 3.1(b), and indeed for polymer systems, considerable temperature-dependence of χ is expected. Depending on the details of this dependence, demixing can occur either on cooling or on heating solutions, giving possibilities of both upper and lower critical solution (consolute) temperatures (or both). This result will form the basis for future discussion of solution instability of real polymer solutions. It is noted here, however, that for polymer solutions the experimental miscibility gap is far less symmetrical (see, for example, Kurata, 1982) than that shown in Figure 3.1(b), partly because of the large molecular weight difference between polymer and solvent, and partly because of molecular weight polydispersity inherent in real polymer solutions.

Finally in this section, a comment is necessary on the phase behaviour of systems involving two polymer types and a solvent, i.e. ternary systems. Although most of the following discussion will be about demixing of single polymer solutions, some attention will be given to such ternary systems. This is largely because the search for new kinds of gelled systems (particularly in the food industry) has led to the exploitation of the rather more flexible properties of composite gels formed from ternary solutions by demixing and polymer aggregation.

The principles underlying the phase behaviour of ternary systems are very similar to those already discussed, but now two independent composition variables are involved, and phase diagram representation at a single temperature must involve two-dimensional rectangular (or triangular) phase diagrams. A coexistence curve (or binodal) is now defined at a single temperature, and this becomes a three-dimensional surface if temperature is included as a variable. A whole family of tie-lines, and one or more critical points, are now involved at each temperature. The Flory–Huggins theory (three χ values) can still be applied to define the mixing free energy function which is now a two-dimensional surface at any temperature, and the coexistence curve is now defined by a tangent plane to this surface. Tie-lines are also defined in this way, and the lever rule still applies. A graphic representation

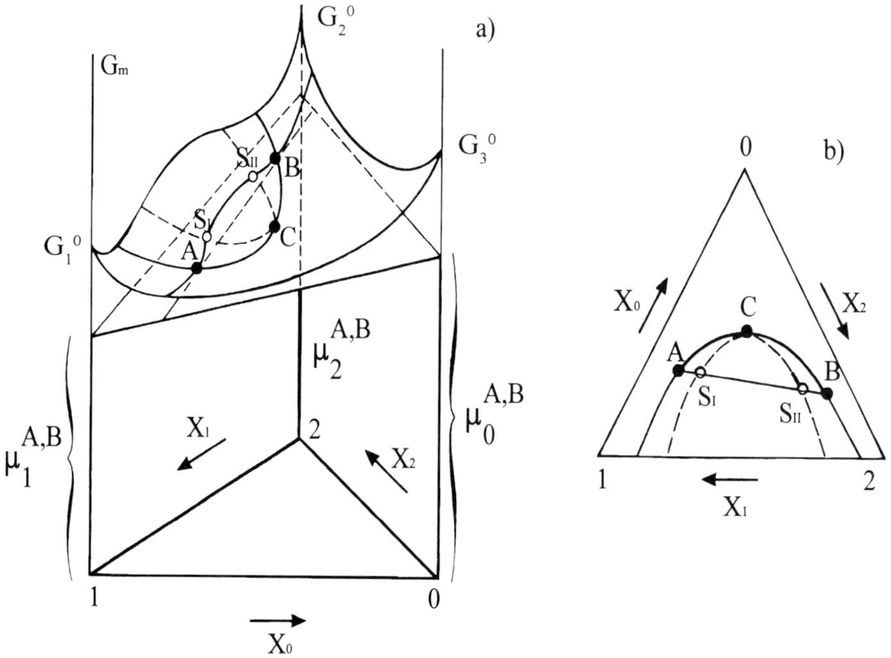

Figure 3.2 Situation analogous to that in Figure 3.1, but generalised to a ternary system consisting of solutes 1 and 2, and solvent (0). Details are given for one temperature only, at which demixing occurs

of this situation appears in Figure 3.2, using a triangular representation for composition, and assuming that the individual polymer components 1 and 2 are soluble in solvent ("component 0") in all proportions. In addition, of course, under certain circumstances more complicated situations can occur for ternary systems, involving more than two liquid phases, mutual solubility of the mixed polymer system itself, and only partial solubility of the individual polymers in the solvent, but these will not be examined here.

Kinetics of solution demixing

The previous section has established the thermodynamic conditions for solution instability with respect to demixing. This information is contained in the functional variation of the solution free energy with respect to composition. This allows a decision to be made about which solution compositions are inherently unstable with respect to demixing and which are not, but there is no indication concerning how fast such events will take place, or by what mechanism. In solving this problem attention must be given to the issue of formulating the free energy function for a system at uniform temperature and pressure but variable composition (i.e. variable from place to place). This quantity, the total free energy of a heterogeneous system, is central to establishing a dynamical model for demixing, as a homogeneous solution must approach the stable two-phase situation through heterogeneous intermediate states.

The problem was first studied not from the polymer solution point of view, but by workers in the field of metallurgy studying demixing of components in metal alloys. In an important series of papers (Cahn and Hilliard, 1958; Cahn and Hilliard, 1959; Cahn, 1965), Cahn and Hilliard formulated the free energy function required, and used it quantitatively to describe two basic mechanisms for solution demixing (nucleation and growth, and spinodal decomposition) and to indicate under what conditions of composition and temperature these mechanisms were to be expected, and how they could be distinguished.

In summary, the argument presented was as follows. The free energy of a heterogeneous system (constant temperature (T), pressure (P) and volume (V), as incompressibility was assumed) can be written as an integral over the system volume using a free energy density function,

$$f'(c) = f(c) + \kappa (\nabla c)^2 \qquad (1)$$

Here f(c) is the free energy of mixing (per unit volume) of a solution of composition c, and the additional term on the right accounts for the contribution of (small) concentration gradients to the free energy density. κ is an energy gradient coefficient which is always positive and is a characteristic of the system.

Eq. (1) is valid only for situations in which the gradient terms are small, i.e. a small deviation from uniform composition. If, under these circumstances, the system considered has an overall average composition c_0, f(c) can be expanded about c_0 in a conventional way, i.e.

$$f(c) = f(c_0) + (c-c_0)\frac{\partial f}{\partial c} + \frac{1}{2}(c-c_0)^2\frac{\partial^2 f}{\partial c^2} + \ldots \tag{2}$$

where the partial derivatives are evaluated at concentration C_0.

Using the fact that the integral of $c-c_0$ throughout the system volume must sum to zero, it is possible to formulate the change in free energy accompanying the shift from an initially homogeneous system to a slightly inhomogeneous system as

$$\Delta F = \int\left[\frac{1}{2}\left(\frac{\partial^2 f}{\partial c^2}\right)(c-c_0)^2 + \kappa\,(\nabla c)^2\right]dV \tag{3}$$

From this result it is immediately obvious that the system is stable to all small fluctuations if the second derivative of the free energy (at c_0) is positive. If this derivative is negative, however, the system can be unstable to such fluctuations, provided that the first term dominates. For random fluctuations of sufficiently long wavelength this is always true, and hence in general the system is always unstable to small fluctuations in this case.

Systems for which the second derivative is positive are clearly in a condition of stable equilibrium (though metastable if composition lies within the coexistence curve), whilst when this derivative is negative, unstable equilibrium is found. The situation where the second derivative is zero provides a point of demarcation between these conditions, and states corresponding to this value have particular significance. Indeed, before continuing the discussion of decomposition mechanisms it will be valuable to comment further on this with the aid of Figures 3.1 and 3.2.

For a two component system, at a particular temperature below T_C, the solution molar free energy (Figure 3.1(a)) shows two composition points at which the second derivative can be zero. These are the points of inflection S_I and S_{II}. As temperature varies, the locus of such points is a second curve, the spinodal curve, lying within the binodal, and coinciding with it at the critical point (Figure 3.1(b)). Evidently the second derivative of the mixing free energy is always negative for compositions lying within the spinodal, and hence from the arguments presented earlier, systems established uniformly within this region (by sudden quenching, for example) will always be unstable with respect to random fluctuations of appropriate wavelength, and will decompose spontaneously into separate phases without the need for an activation energy.

Systems lying in composition between the spinodal and binodal, on the other hand, are metastable, and will decompose only if large local fluctuations lead to nucleation. Thus, if systems are quenched rapidly into the spinodal region, or are quenched at compositions close to the critical composition, a spinodal mechanism is likely to operate, whilst systems quenched into the metastable region (or very slowly through the metastable region) will eventually decompose by nucleation and growth when temperature becomes low enough for stable nuclei to be formed. These various types of quenching (critical, off-critical, deep, shallow, etc.) are shown schematically in Figure 3.3, the metastable region being shown as a shaded area. All systems outside the binodal are of course globally stable (at least to demixing).

For ternary systems the situation is just a generalisation of the above (Kurata, 1982), with a spinodal curve lying within the binodal at any temperature, and defined by an analogous mathematical condition on the matrix of second derivatives of the mixing free energy (Figure 3.2). The concept of a temperature quench must also be generalised for ternary systems, i.e. considered in terms of a movement along a line in a three-dimensional temperature–composition diagram composed of fixed temperature "slices" of the type shown in Figure 3.2(b).

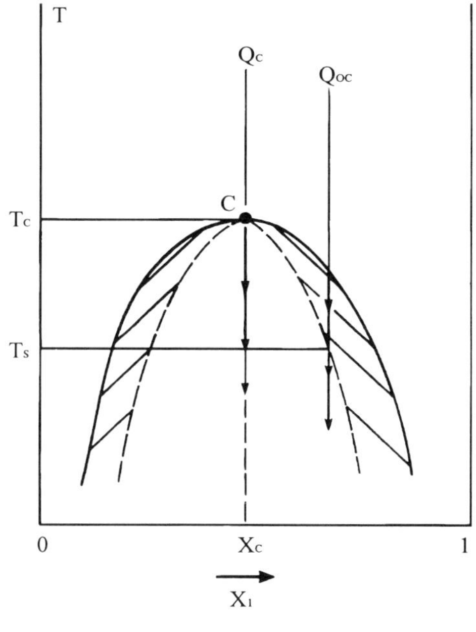

Figure 3.3 Temperature quench experiments for a binary system. Q_C and Q_{OC} indicate critical and off-critical quenches of varying depth, whose spinodal temperatures are T_C and T_S, respectively. The shaded area indicates the metastable region

Returning to a more detailed and quantitative description of demixing mechanisms, attention is now focused on the mechanism of demixing within the spinodal region. Nucleation and growth will not be treated in detail here as most experimental studies have focused particularly on the more novel (and possibly more common) spinodal mechanism and its consequences, consequences which appear to be highly significant where mechanisms of gel formation are concerned.

Spinodal decomposition occurs when a system is in the condition where the second derivative of the mixing free energy is negative, i.e. when its composition lies within a spinodal region of a phase diagram. Cahn (1965) pursued a quantitative description of the dynamics of spinodal decomposition in the following way. First, for the heterogeneous system the chemical potential field describing the difference in partial molar free energy between the components was formulated from Eq. (1) for the overall free energy density field: then the diffusion flux of the components was formulated in terms of this, via the gradient of the chemical potential field and a mobility coefficient (M). Lastly, by applying the usual continuity equation (conservation of mass), a final diffusion equation for the initial stages of decomposition was written as

$$\frac{\partial c}{\partial t} = M \frac{\partial^2 f}{\partial c^2} \nabla^2 c - 2M\kappa\nabla^4 c \qquad (4)$$

whose solution could be written (for isotropic systems) as a time-dependent Fourier summation,

$$c(r) - c_o = \sum_{\beta} e^{R(\beta)t} [A(\beta) \cos(\beta r) + B(\beta) \sin(\beta r)] \qquad (5)$$

where r is the magnitude of the position vector in real space, ß is the wavenumber of a component wave ($=2\pi/\Lambda$, where Λ is the wavelength) and A(ß), B(ß) are amplitudes of component waves of the random fluctuations present initially.

Clearly, Eq. (5) suggests that as time progresses, wave components of wavenumber ß grow exponentially, the exponential factor having a variable rate constant R(ß) which can be shown to be given by

$$R(\beta) = -M \frac{\partial^2 f}{\partial c^2} \beta^2 - 2M\kappa\beta^4 \qquad (6)$$

Here M is the diffusive mobility factor referred to earlier and the other symbols have their earlier meaning. Eq. (6) indicates immediately that fluctuations will grow spontaneously only when the second derivative of the mixing free energy is negative, and then only when the wavenumber of random fluctuations exceeds a critical value $\beta_c = [-(\partial^2 f/\partial c^2)/\kappa]^{1/2}/\sqrt{2}$. The wavenumber of the fluctuation which grows at a maximum rate (β_m) is given by $\beta_c/\sqrt{2}$, and the maximum rate itself by $M(\partial^2 f/\partial c^2)^2/8\kappa$.

The implications of the above equations for the mechanism of spinodal decomposition are quite far-reaching. They show that only certain random concentration fluctuations will grow in amplitude and, more importantly, that for any situation there will be a particular wave of wavenumber β_m that will grow fastest and dominate the situation. The expectation is therefore that in the very early stages of the spinodal mechanism, the uniform concentration field of a given component, present initially, will develop into a periodic distribution of higher and lower concentration values with a characteristic periodicity length (Λ), and hence size, of demixed regions. As time goes on, the prediction is that these regions remain constant in size, but change in local concentration. The initial stages of spinodal decomposition in a system should therefore be identifiable, under microscopic examination, by a characteristic "mottled" appearance of density fluctuations (particularly where the separating phases are bicontinuous). Such observation will of course require that systems have been arrested in the earliest stages of demixing by some form of "quenching". In his original paper, Cahn (1965) modelled this effect by computer and gave indications of the microstructural effects likely to be produced (see Figure 3.4, for example).

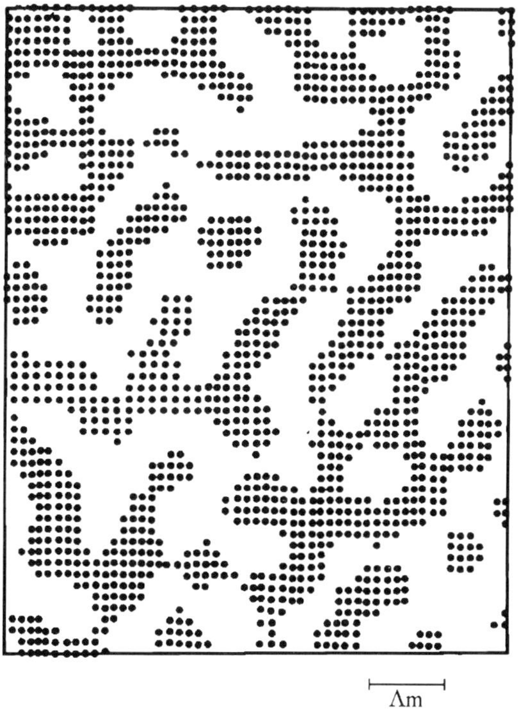

Figure 3.4 Section through computer-generated simulation of spinodal decomposition using the linear theory of Cahn. The situation modelled corresponds to a binary system at 50:50 volume fraction with characteristic wavelength $\Lambda_m = 2\pi/\beta_m$ determined by input parameters. Points indicate regions of higher solute density. (After Cahn, 1965, with kind permission of the American Institute of Physics.)

The model just described is often called the linear theory of spinodal decomposition because it is based on a particular type of diffusion equation which, in turn, is derived from a truncated version of the free energy density function. It is therefore expected to be able to describe only the very earliest stages in the decomposition mechanism. More complex descriptions of the phenomenon including the effects of random thermal fluctuations (Cook, 1970) and non-linear effects (Langer, Bar-on and Miller, 1975) have been published, and these suggest that in reality the wave of maximum growth rate will progressively decrease in wavenumber (increase in characteristic spacing) as time goes on. Finally, at very much later stages, demixing solutions will suffer coarsening through several possible mechanisms such as hydrodynamic break up of bicontinuous phase distributions (Siggia, 1979), coalescence of droplets (Binder and Stauffer, 1974), diffusive evaporation–condensation (Lifshitz and Slyozov, 1961) etc., and eventually a liquid–liquid emulsion will form with sizeable droplets subject to gravitational forces.

All of the above aspects of the spinodal decomposition mechanism have been studied, systems of interest including metal alloys (Rundman and Hilliard, 1967), glassy mixtures (Nielson, 1969) solutions of small molecules (Wong and Knobler, 1981), polymer solutions (Van Aartsen and Smolders, 1970; Aubert, 1990; Lal and Bansil, 1991) and polymer blends (Hashimoto, Kumaki and Kawai, 1983; Hashimoto, Itakura and Hasegawa, 1986a; Hashimoto, Itakura and Shimidzv, 1986b; Voigt-Martin, Leister, Rosenau and Koningsveld, 1986; Maruta, Ohnaga and Inoue, 1993). In most of this work (which will be discussed at more length below) attempts are made to locate the initial demixing regime, and hence to test the applicability of the Cahn model (and its modifications). Later stages of demixing are also studied, and attempts made to identify coarsening mechanisms by, for example, studying the variation of light scattering intensity distributions with time. Time-resolved light scattering is a natural approach to studying spinodal decomposition in real time, as the intensity at any wave-vector q ($= 4\pi \sin(\theta/2)/\lambda$: θ= scattering angle; λ = wavelength of light used) is a measure of the square of the amplitude of a corresponding Fourier component (wavelength $\Lambda = 2\pi/q$: $q \equiv \beta$) of the concentration density distribution. Thus, as will be discussed in more detail in the following section, light scattering experiments on a spinodally decomposing solution should detect (in the earliest stages) an intensity maximum which grows exponentially with time, but which remains fixed in scattering angle. In reality, for the reasons discussed above, most experiments record the drift of this maximum to lower angles of scatter, as nonlinear effects (including longer term coarsening) become important.

Other techniques for the study of spinodal decomposition include microscopy (optical and electron) on quenched samples, quasi-elastic light scattering, turbidity measurements, spectroscopic approaches, neutron and X-ray scattering (for systems showing unusually small characteristic spacings and for studying other short-range aspects of the demixed materials:see, for example, Hill, Tomlins and Higgins, 1985;

Jinnai, Hasegawa, Hashimoto and Han, 1991), and mechanical measurements (e.g. static and dynamic viscosity). These last are particularly applicable when the process leads to substantial mechanical changes through structuring, e.g. gelation. Most of these approaches will be referred to in the experimental sections which follow.

Experimental studies of spinodal decomposition

NON-GELLING POLYMER SOLUTIONS

Where polymer systems are concerned most experimental work has focused on polymer blends containing two polymer types. Considerable information is available for these systems, and the Cahn mechanism has been verified for the earliest stages of demixing. For polymer blends, the linear theory has required some development, however, to make it applicable to the polymer situation (de Gennes, 1980; Pincus, 1981; Binder, 1983; Onuki, 1986; Binder, 1987). In both the theoretical and experimental work, interest has extended to demixing behaviour beyond the linear region.

Unfortunately, demixing in polymer solutions, which is the principal subject of concern here, has been much less thoroughly studied. Pioneering work by Van Aartsen and Smolders (Van Aartsen, 1970; Van Aartsen and Smolders, 1970) demonstrated by time dependence of light scattering measurements (and from microscopy evidence) that the spinodal mechanism applied to demixing in 15% w/w polymer solutions containing poly(2,6-dimethyl-1,4-phenylene ether) in caprolactam, when these were rapidly quenched. Measurements at various quench temperatures, and over a range of scattering angles and times at each temperature, enabled the critical wavenumber to be calculated as a function of temperature. The temperature dependence of β_c was compared with that predicted from an equation derived by Van Aartsen from earlier work by Debye, and the Flory–Huggins solution theory (Van Aartsen, 1970), i.e.

$$\beta_c^2 = 6l^{-2}\left(1 - \frac{T}{T_s}\right) \tag{7}$$

where T_S is the spinodal temperature of the system for the solution composition concerned (Figure 3.3) and l is a characteristic length for the polymer. This last quantity, which can in some cases be equated to the polymer root mean-square radius of gyration, is likely to be temperature-dependent, a result verified by the authors. An assumed characteristic length of 20 nm for the polymer gave roughly the correct spacing for the spinodally decomposed solution as measured by electron microscopy, i.e. $\approx 1\,\mu$.

Further work on synthetic polymer solutions seems scarce. In 1990, Aubert discussed demixing in polystyrene–cyclohexane solutions (Aubert, 1990), using microscopical methods and solution quenching techniques to measure coarsening. A range of concentrations was studied, and a diffusive mechanism proposed for the coarsening event. It was not possible to investigate the earliest stages of decomposition and hence assess the applicability of the Cahn mechanism.

Later, the polystyrene–cyclohexane system received further study by Lal and Bansil (1991). Samples of polystyrene of two molecular weights were considered, but only one (near-critical) concentration was measured for each sample. Demixing was followed by time-resolved small angle light scattering to determine the scattered intensity function I(q,t) for a series of quench depths, increasing to roughly 1 K

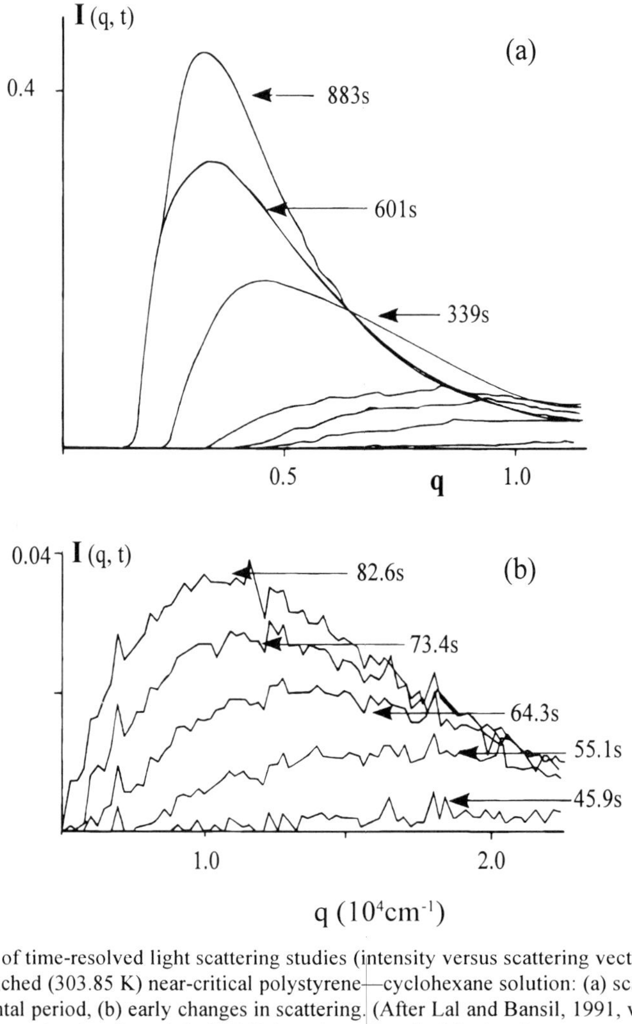

Figure 3.5 Results of time-resolved light scattering studies (intensity versus scattering vector q) of spinodal demixing in a quenched (303.85 K) near-critical polystyrene—cyclohexane solution: (a) scattering profiles over full experimental period, (b) early changes in scattering. (After Lal and Bansil, 1991, with kind permission of *Macromolecules*. Copyright (1991) American Chemical Society.)

below the appropriate spinodal temperatures (as determined by quasi-elastic light scattering). Unlike the work by Aubert, it was possible to study early stages of the decomposition process, at times less than 100 s. In these early stages, exponential growth of intensity was observed, together with a diffuse intensity maximum but, unlike the predictions of the Cahn theory, the wavenumber of this maximum moved to smaller q values as time progressed (Figure 3.5).

In their treatment of the short-time data, Lal and Bansil estimated an ideal value for the wavenumber $q_m(0)$ ($\equiv \beta_m$) of the intensity maximum at zero time using a relationship similar to eq. (7) above, but based on the Ising theory. An average value for the diffusion coefficient, $D = M\partial^2 f/\partial c^2$, was calculated from the structure factor data at each temperature, and used to estimate an initial maximum growth rate appropriate to that temperature. This provided a characteristic time for the system $\tau_o = 1/Dq_m(0)^2$.

In a treatment of the intensity data measured at longer times, Lal and Bansil used the characteristic length scale defined by $q_m(0)$, and timescale τ_o, to examine intermediate and late stages of the decomposition process using scaling arguments. For this purpose data for each sample, and at each quench temperature, which included both the wavenumber of the maximum in intensity q_m, in reduced units $q_m/q_m(0)$, and the maximum intensity value (I_m), were plotted against reduced time $\tau = t/\tau_o$ in log–log form. According to expectation, such data should reduce to a universal power-law plot (irrespective of quench depth) with a power-law exponent indicative of the mechanism of coarsening.

Such a collapse of the wavenumber data was in fact demonstrated, particularly for the lower molecular weight sample (Figure 3.6). For both samples, however, a substantial part of the wavenumber data did not reduce in this way, indicating that intermediate behaviour was being observed over much of the experimental timescale. Interestingly, the power-law exponents obtained from the plots from the linear self-similar sections of the reduced data (0.38 and 0.44) were greater than predicted by various theories of diffusive coarsening (0.33, or less), and also greater than values typical for polymer blends. Power-law exponents calculated from the light scattering behaviour corresponding to time scales shorter than this (i.e. from the non-scaling region) were variable, and higher still. Since a power-law of unity is expected for coarsening processes dominated by hydrodynamic effects, it was concluded that hydrodynamic contributions to coarsening were important for polymer solutions initially in the dilute concentration range, though these effects probably reduced in importance as the solution demixed.

Experimental studies of spinodal decomposition in biopolymer solutions have also been reported, particularly for the polysaccharide agarose (e.g. Feke and Prins, 1974). Since these have always treated a solution which has the potential to gel, discussion of this work is postponed to the following sub-section, which deals specifically with gelling. The present discussion of polymer solution behaviour

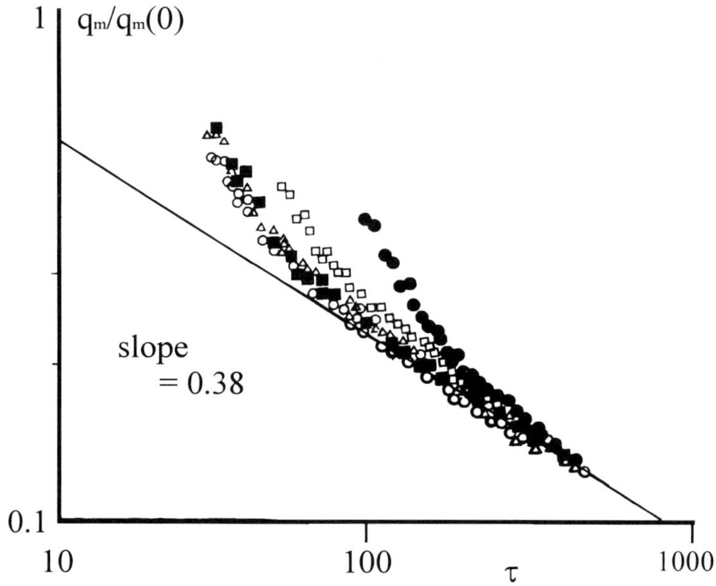

Figure 3.6 Log—log plot of reduced wavenumber of position of maximum scattering, against reduced time, for a polystyrene—cyclohexane system subjected to varying quench temperatures: ○ 24.3°C, △ 24.2°C, ■ 24.1°C, □ 24.0°C, ● 23.8°C. Solid line indicates scaling region. (After Lal and Bansil, 1991, with kind permission of *Macromolecules*. Copyright (1991) American Chemical Society.)

ends with some comment on ternary mixtures, i.e. on solutions containing two polymer types (A and B) in a solvent (S). Although there does not appear to be a lot of kinetic data available for such solutions at present, a ternary solution system has been considered in an interesting paper by Inoue and co-workers on the solution casting of polymer blends (Inoue, Ougizawa, Yasuda and Miyasaka, 1985).

A number of polymer types were considered in the work by Inoue *et al.* including, in particular, the combination poly(vinylchloride)–poly(acrylonitrile-*co*-butadiene) in tetrahydrofuran. Small-angle light scattering studies were performed on solutions of a given fixed starting polymer concentration, and changes in scattering monitored as the solvent was allowed to evaporate (Figure 3.7). Thus, these experiments involved a concentration "jump" not a temperature quench. Experiments were performed to study the effects on the scattering behaviour, and on the final phase morphology of the dried film produced, of varying the rate of evaporative solvent loss (Figures 3.7(a) and 3.7(b)), the polymer type (e.g. molecular weight and Flory–Huggins interactive χ_{AB} values) and the AB blend ratio. Results were interpreted in terms of the Flory–Huggins solution theory applied to ternary systems, and the Cahn model for spinodal decomposition. In most cases scattering experiments revealed the maximum in scattering characteristic of spinodal decomposition, and microscopy work on the final dried films showed the expected "mottled" appearance of the phase separated structure as well as a high level of phase interconnectivity and spatial periodicity. The general trends in behaviour could be explained in terms of a competition between phase separation kinetics as the solution entered the familiar miscibility gap in the ternary phase diagram (e.g. Figure 3.2(b)), and the concentration changes continually introduced during drying.

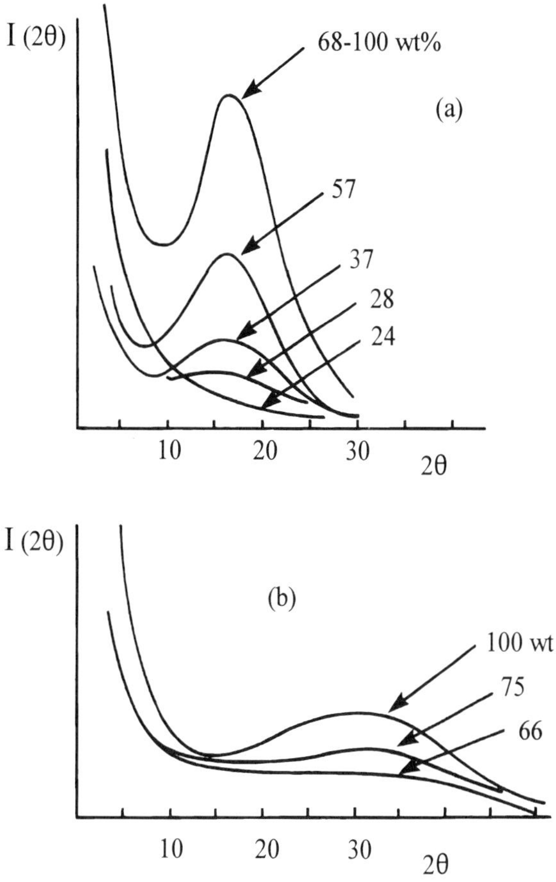

Figure 3.7 Light scattering profiles from poly(vinyl chloride)—poly(acetonitrile-*co*-butadiene)–tetrahydrofuran systems, during evaporative loss of solvent at (a) slow, and (b) fast, evaporation rates. Percentages indicate total polymer concentrations. (After Inoue *et al.*, 1985, with kind permission of *Macromolecules*. Copyright (1985) American Chemical Society.)

This study of a ternary system is of particular interest as it shows some analogy with the process of making composite gels from ternary biopolymer mixtures. There is the difference, of course, that in the composite gel case the solution composition remains constant, and the competition is between the rates of phase separation and of aggregation and gelation, induced (usually) by temperature changes. Gelling polymer systems are now considered.

GELLING POLYMER SOLUTIONS

Gelling, arising from the aggregation of polymer components in solution, is an important structuring process in polymer solutions, and in cases where solvent conditions promote demixing as well, complex demixed gels can result, with properties very much determined by the kinetics of the demixing event, as balanced against the kinetics of aggregation. This contrasts with the case of simple solutions where, unless some quenching process such as rapid freezing and/or freeze drying is performed, demixing ultimately produces total phase separation, and separation into discrete macroscopic phases. Gelling systems, on the other hand, involve an in-built quenching process which usually prevents phase separation on such an obvious macroscopic scale, or even true phase separation at all.

From the thermodynamic point of view, where a fairly dynamic cross-linking process is envisaged, theoretical treatments of aggregating/demixing systems have been published (Coniglio, Stanley and Klein, 1979, 1982; Tanaka, 1990; Tanaka and Stockmayer, 1994) which provide phase diagrams containing both miscibility gaps and sol–gel critical point lines. Thus, in an equilibrium sense, the effect of gelation is to add to the usual temperature–composition diagrams (for a binary system, for example) a line indicating the onset of gel network space continuity (percolation) which cuts across the miscibility gap (Figure 3.8). This usually divides the diagram into regions where either homogeneous sol or gel can be found, or a phase separated system consisting of both sol and gel regions (percolated or otherwise).

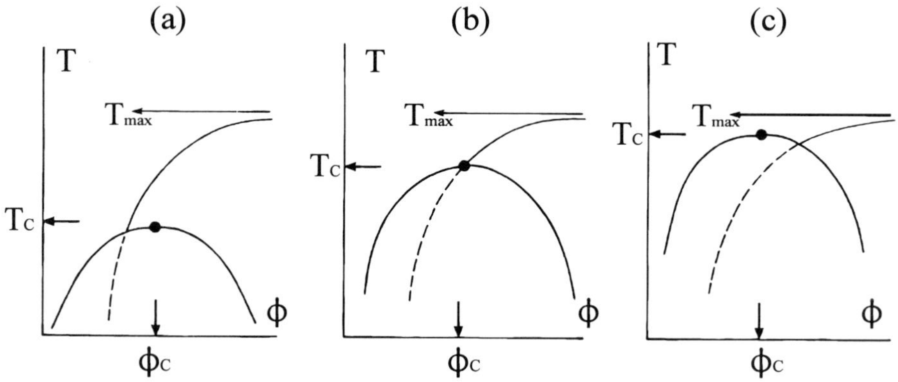

Figure 3.8 General forms of temperature—composition phase diagrams for reversibly gelling binary systems undergoing demixing on cooling, as suggested by both mean-field and percolation theories. T_c and ϕ_c indicate the critical solution point, and T_{max} is a possible limiting gel melting temperature. Different intersections of the sol—gel line and the coexistence curve appear in (a), (b), and (c)

Both the methods of percolation theory (Coniglio *et al.*, 1979, 1982), and of the mean field Flory–Stockmayer approach (Tanaka, 1990; Tanaka and Stockmayer, 1994), have been employed to generate such phase diagrams. Interest attaches to whether the sol–gel line meets the coexistence curve at the critical solution point or to its left or right (see Figure 3.8). Of particular theoretical concern is the nature of the critical point in the case where critical gel and solution points coincide.

The predictions of the equilibrium theory are interesting in that highly turbid microphase-separated gels, and suspensions of microgel particles in solvent, are actually found both for gelling synthetic polymer systems, and for biopolymers. Also, as solvent quality changes (equivalent to temperature), such systems can become homogeneous. This success may only be qualitatively valid, however, as the equilibrium theory is unlikely to apply rigorously to many of these "state diagrams". Strong gels formed by cross-linking events of high affinity (which are commonly encountered) are unlikely to be at equilibrium. They are almost certainly formed under kinetic control.

In the end, to describe such phase-separated systems, it is necessary to return to kinetic arguments and to ask to what extent the mechanism of spinodal decomposition (or indeed, in certain cases, nucleation and growth), when set up in competition with aggregation, can explain the observed results. This is a difficult theoretical problem, and in the first instance it is necessary to appeal to experiment. Experimental studies of kinetic demixing and gelation in polymer solutions are not much more abundant than solution studies, but at least it is possible to find results for biopolymer systems. The latter will be focused on here, but it is worth noting that interesting experimental data can be found in the literature for physical gels based on synthetic polymers (e.g. Matsuo, Kawase, Sugiura, Takematsu and Hara, 1993), and for certain "hybrid" systems (stiff-chain polyamino acids in non-aqueous solvents) investigated by Miller and co-workers (Miller, Kou, Tohyama and Voltaggio, 1978).

That a gelling biopolymer system might involve spinodal decomposition as a mechanism of its formation, was suggested in the early seventies, as a result of a light scattering investigation of agarose solutions (Pines and Prins, 1973; Feke and Prins, 1974). Thus, in a detailed study of the 1% w/w agarose system, buffered to constant pH (4.9), and quenched to several temperatures from 82°C, Feke and Prins (1974) claimed clear evidence for a spinodal decomposition mechanism during the earliest stages of change after the quench. Only at the highest quench temperature (41°C) did it appear that nucleation and growth (characterised by non-exponential growth of scattered intensity) was occurring. Below this temperature (32.5°C) a distinctive maximum in light scattering was observed, which reduced in intensity, broadened, and moved to higher angles of scatter, at lower quench temperatures. Plots of intensity versus time for data at 25.5°C confirmed exponential growth, and a linear plot was also obtained on appropriate plotting of the growth rate constant against scattering vector (Figure 3.9). No discussion was given of the later stages of the process where, presumably, the gelling event was of major importance, and no microscopic evidence was available to provide direct microstructural corroboration of the spinodal nature of the gels.

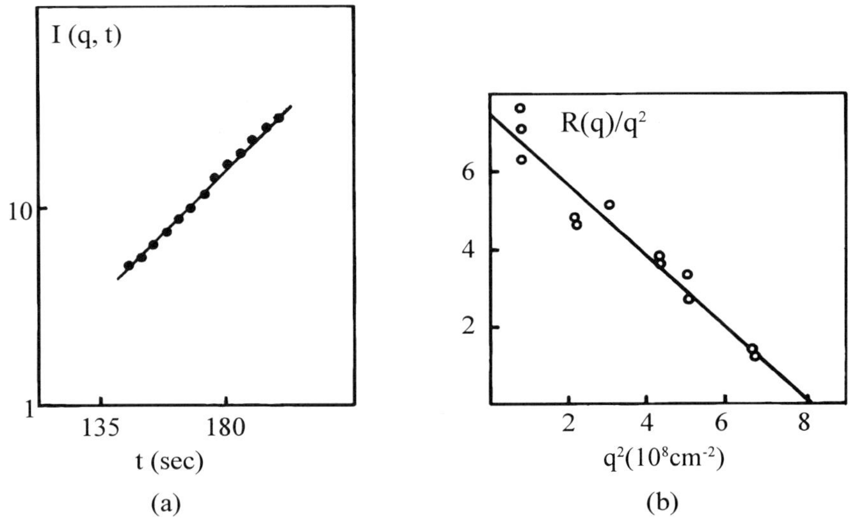

Figure 3.9 Evidence of spinodal decomposition in a quenched (25.5°C) 1% w/w agarose solution. (a) exponential growth of scattered light intensity (fixed q) during early stages, and (b) linear $R(q)/q^2$ versus q^2 plot of exponential growth rate constant as predicted by Cahn. (After Feke and Prins, 1974; with kind permission of *Macromolecules*. Copyright (1974) American Chemical Society.)

The agarose system has subsequently been studied in more detail in a series of investigations by Palma and Palma-Vitorelli and co-workers (e.g. San Biagio, Madonia, Newman and Palma, 1986; Leone, Sciortino, Migliore, Fornili and Palma-Vitorelli, 1987; Palma-Vitorelli and Palma, 1987; Bulone and San Biagio, 1991; Emanuele, Di Stefano, Giacomazza, Trapanese, Palma-Vitorelli and Palma, 1991; Emanuele and Palma-Vitorelli, 1992) using light scattering techniques (static, dynamic and turbidity measurements), chiroptical methods (optical rotatory dispersion (ORD)) and viscoelastic measurements (viscosity and dynamic viscosity). Like Feke and Prins, these authors concluded that spinodal decomposition was fundamentally important as a mechanism in the formation of agarose gels but they observed that, at the temperatures used in their experiments (above 40°C), the gelling event was extremely slow. As a consequence, under these conditions, it was possible to effectively separate decomposition from gelling (see Figure 3.10, for example), and indeed to describe the gelling event as being driven by the local concentration enhancement achieved by spinodal demixing. Only at concentrations higher than perhaps 2.0% w/v was it likely that gel formation occurred without such demixing.

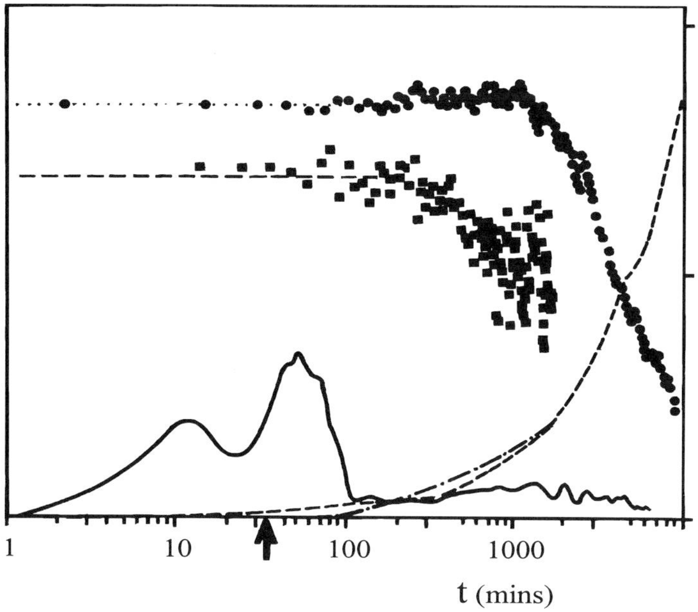

Figure 3.10 Multi-technique monitoring of kinetic processes in a 0.5% w/v agarose solution quenched to 47.5°C (optical rotation -- ●; turbidity -- broken line; onset of optical diffraction -- vertical arrow; number of free coils -- ■; shear viscosity — continuous line; complex viscosity — · —). Early stages are consistent with preliminary spinodal demixing. Later stages involve coil—helix transition, aggregation, and gelation. For more details see text. (After Emanuele *et al.*, 1991, with kind permission of J. Wiley and Sons, Inc. Copyright (1991)

For example, as Figure 3.10 shows, preliminary spinodal decomposition was demonstrated in a 0.5% w/v agarose system, at the earliest times, by appropriate changes in scattered light intensity and turbidity (and by the kinetics of these changes), and by a rise and fall in viscosity coinciding with the demixing event. In addition, after an appropriate delay time, there was a significant change in the noise characteristics of the light intensity fluctuations in dynamic light scattering experiments, and this delay showed critical behaviour in relation to quench temperature. This enabled the spinodal temperature appropriate to the agarose concentration involved to be estimated, and a phase diagram to be constructed for gelling agarose solutions involving both a coexistence curve and sol–gel line (Emanuele *et al.*, 1991).

As Figure 3.10 also indicates, processes occurring at much longer times, such as aggregation of agarose molecules involving coil–helix transitions, and longer term associations of the rod-like aggregates, produced their own characteristic changes in quasi-elastic light scattering, turbidity, ORD behaviour, and viscoelastic response. In connection with this, Emanuele and co-workers make the interesting suggestion that the long-time curing of agarose gels, through what is generally accepted to be side-to-side associations of rod-like double helical aggregates, may also involve a demixing process, occurring this time according to the predictions of

Flory's calculations of the phase behaviour of solutions of rod-like molecules (Flory, 1956). In this situation demixing can produce a dense ordered phase in equilibrium with polymer solution and, in the agarose case, this will lead to enhancement of the gel modulus (further curing).

The above model for the gelation of agarose solutions is clearly very much one of kinetic control. Although phase diagrams are considered, they refer to the agarose solution properties prior to conformational change and aggregation. When these last processes occur, subsequent gel formation is under kinetic control, as there appears to be no sense in which aggregated ordered molecules are able to repartition in relation to the solvent (unless one counts the proposed second type of demixing of the rod-like structures). Gelation appears to "quench" the system in a microstructural form largely dictated by the earliest demixing processes, though subsequently modulated by further coarsening through rod–rod associations. One would therefore expect such microstructures to be highly sensitive to rates and depths of temperature quench, at least at lower agarose concentrations. This aspect of agarose gel formation does not seem to have been sufficiently investigated so far.

A principal proposal from the work just reviewed is that the local concentration enhancement achieved in biopolymer solutions by spinodal decomposition is likely to induce gelling within these regions by accelerating aggregation. This, in turn, will allow overall gelling to occur in solutions of low average concentration, if percolation of the aggregated phase is also achieved. Spinodal decomposition can therefore, it appears, be an important factor in determining the magnitude of critical gelling concentrations, if this concept is extended to include the idea of percolation of phase-separated microgel regions, rather than just simply referring to a concentration above which molecular weight divergence can occur (a common interpretation of critical concentration for homogeneous gelling systems). According to this more general idea, a solution of microgel particles may be consistent with the idea that a polymer solution has passed through a critical gel point but, unless these particles associate, and percolate space, they do not provide gel character to the total system of which they are a part, and a gel has not formed.

Palma and colleagues have also explored the solvent-induced demixing mechanism of structuring for protein systems including haemoglobin (San Biagio and Palma, 1991), and elastin (Sciortino, Prasad, Urry and Palma, 1993), and have proposed that solvent-induced concentration fluctuations may play a substantial part in the self-assembly of biological structures. Interestingly, this is an idea which was also explored by Miller and co-workers in their earlier studies of rod-like polyamino acids gelling in non-aqueous solvents (Miller *et al.*, 1978). They too proposed a spinodal mechanism of solvent-induced microphase separation and, basing their argument on Flory's phase diagram (Flory, 1956), suggested that a dense ordered network would form (gel network), which would resist further coarsening, due to the development of solid viscoelastic character within the concentrated domains.

At present, there appear to be few other reported studies of biopolymer solutions showing spinodal effects. A mechanism something like this, in that it suggests a preliminary demixing of the polymer constituent, has been proposed for the gelation of amylose (Miles, Morris and Ring, 1984), and indeed something similar may be involved for other starch-based polymers and their associated hydrogels. There is also much microscopic evidence that could support the involvement of demixing mechanisms where heat-set globular protein gels are concerned. Thus, it is well established (e.g. Clark and Lee-Tuffnell, 1986) that changing conditions of protein charge, and the screening of this charge (pH/ionic strength effects), can cause such systems to progress from clear gels to turbid gels as the corresponding microstructures vary from reasonably homogeneous networks (Figure 3.11(a)), through networks showing only periodic fluctuations in network density (Figures 3.11(b) and 3.12(a)), to substantially collapsed, or completely phase-separated, structures (Figures 3.12(b) and 3.13).

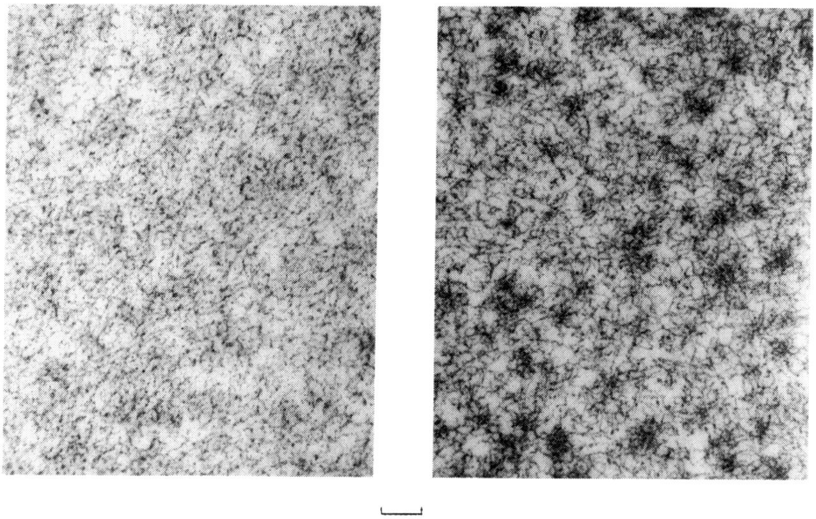

a 200nm b

Figure 3.11 Transmission electron micrographs of heat-set soy protein gels (glycinin, 15% w/w, 0.5 M NaCl, 98°C for 30 minutes) at (a) pH 7.5, and (b) pH 5.4

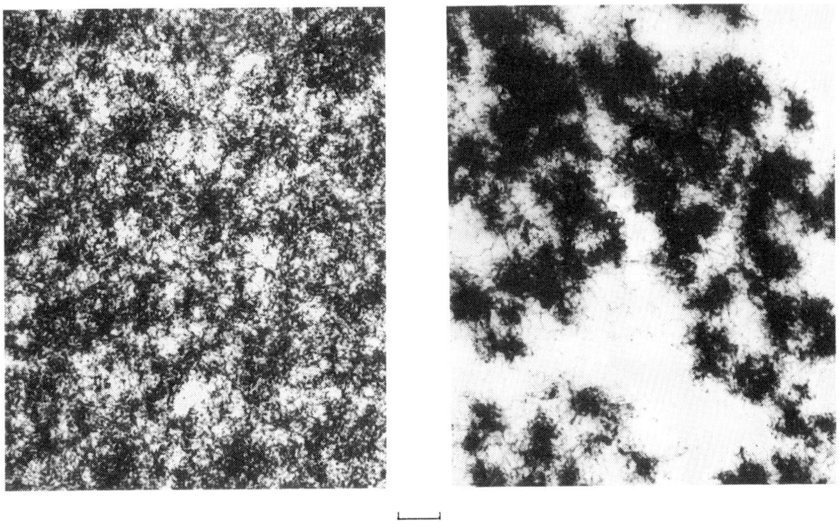

a 200nm b

Figure 3.12 Transmission electron micrographs of heat-set bovine serum albumin (BSA)–ovalbumin co-gels (10% w/w total protein: 6–4 BSA–ovalbumin) at (a) pH 7.0, and (b) pH 6.0

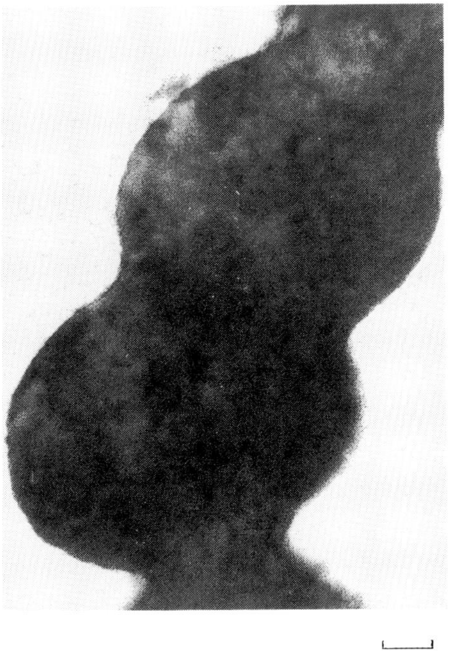

200nm

Figure 3.13 Transmission electron micrograph of heat-set BSA (10% w/w) gel–coagulate at pH 5.1

Recently, Hermansson and co-workers (Langton and Hermansson, 1992; Stading, Langton and Hermansson, 1993) have demonstrated these effects particularly clearly in investigations of whey protein gel systems using electron microscopy. These authors distinguish so-called fine stranded gels from particulate gels, and link mechanical properties of the gels to their microstructures. From the photographs of the particulate gels they present, it is tempting to propose that, in these cases, demixing has dominated gelling (much as is indicated in Figure 3.13), and the microstructure of the gel is that of small uniform spheres of gelled concentrated solution, percolating space.

Where ternary systems are concerned, and particularly mixed gels based on ternary systems, there again appears to be little quantitative evidence for the mechanism of demixing, though the course of gel structuring has been followed kinetically by mechanical spectroscopy (Clark, Richardson, Robinson, Ross-Murphy and Weaver, 1982; Clark, Richardson, Ross-Murphy and Stubbs, 1983). Microscopic evidence (Clark *et al.*, 1982, 1983; Clark, 1987) relating to final gel structures reveals a rich variety of morphologies (see for example, Figure 3.14), including the formation of bicontinuous systems with some of the appearance of photographs of recognised spinodal structures. No doubt the current interest in such systems will stimulate their study by time-resolved light scattering. Hopefully this will provide a less haphazard understanding of the origins of their gel microstructures, and the implications of these structures for material properties.

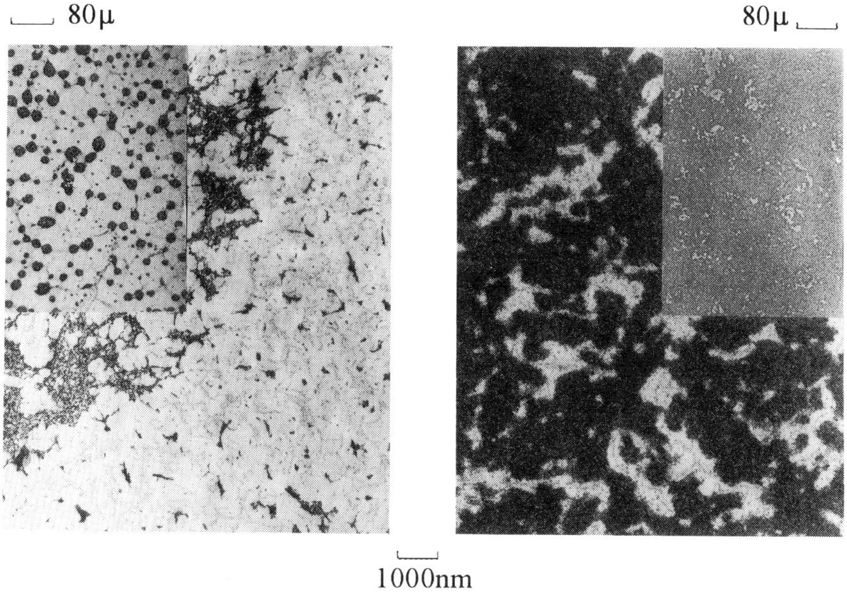

Figure 3.14 Transmission electron (and light) micrographs of (a) agar (1% w/w) and gelatin (1% w/w), and (b) agar (1% w/w) and BSA (10% w/w) thermally prepared co-gels, showing complex microstructures

Conclusions

The present survey has focused on the kinetic aspects of polymer solution demixing and, in particular, on the spinodal decomposition mechanism. Nucleation and growth have also been mentioned, however, and may well be important in some situations. Spinodal decomposition is particularly fascinating in view of the ordered (and sometimes bicontinuous) microstructures it can produce, at least in the earliest stages of demixing. Where gelling is also involved, or the rapid freezing or drying of solutions, this microstructure may persist in the product material and have important implications both for mechanical behaviour and visual appearance.

Where gels are concerned, whether they have one or more polymer types present, a contest between aggregation and demixing will generally be possible (and indeed, inevitable). If this process is adjustable through variation of gelling conditions, potentially it will lead to almost infinite possibilities for gel structure. Thus, if aggregation is much faster than demixing, homogeneous network structures may form which will probably be unstable with respect to longer term processes of solvent loss or structural coarsening. Where the opposite is true, however, and demixing is the dominant effect, a water–water emulsion may arise which, if it gels rapidly enough, could lead to a gel structure based on the percolation of gelled droplets or microgel beads. Between these extremes a whole range of possibilities exists, with the whole picture becoming immensely more complex and flexible as systems progress from only one polymeric component to two or more.

Experimental data, particularly of a quantitative type, as has been principally derived for synthetic polymer blends and some polymer solutions, is clearly not extensive in the area of biopolymer gels, despite the many gelling systems that are known. It is therefore a challenge for workers in the biopolymer gel field to carry out studies analogous to those on blends, and so bring some clarity and order to what appears to be an extremely complicated situation in the materials science of soft aqueous systems.

Finally, it should be noted that in real processing situations, demixing occurs under shear. The work discussed in this review has focused on demixing under quiescent conditions. Clearly, a natural and important progression of both experiment and theory in this area is towards studying shear-induced effects which are likely to influence microstructure (and hence macroscopic properties) quite profoundly.

Acknowledgements

The author is indebted to a former colleague Mr. M.J. Stubbs for provision of electron micrographs, and to Mrs. L.A. Parrott for assistance with figures.

References

Aubert, J.H. (1990) Structural coarsening of demixed polymer solutions. *Macromolecules*, **23**, 1446–1452

Binder, K. and Stauffer, D. (1974) Theory for the slowing down of the relaxation and spinodal decomposition of binary mixtures. *Physical Review Letters*, **33**, 1006–1009

Binder, K. (1983) Collective diffusion, nucleation, and spinodal decomposition in polymer mixtures. *Journal of Chemical Physics*, **79**, 6387–6409

Binder, K. (1987) Dynamics of phase separation and critical phenomena in polymer mixtures. *Colloid and Polymer Science*, **265**, 273–288

Bulone, D. and San Biagio, P.L. (1991) Microgel regions in dilute agarose solutions – The notion of non-gelling concentration, and the role of spinodal demixing. *Chemical Physics Letters*, **179**, 339–343

Cahn, J.W. and Hilliard, J.E. (1958) Free energy of a nonuniform system. I. Interfacial free energy. *Journal of Chemical Physics*, **28**, 258–267

Cahn, J.W. and Hilliard, J.E. (1959) Free energy of a nonuniform system. III. Nucleation in a two-component incompressible fluid. *Journal of Chemical Physics*, **31**, 688–699

Cahn, J.W. (1965) Phase separation by spinodal decomposition in isotropic systems. *Journal of Chemical Physics*, **42**, 93–99

Clark, A.H., Richardson, R.K., Robinson, G., Ross-Murphy, S.B. and Weaver, A.C. (1982) Structure and mechanical properties of agar BSA co-gels. *Progress in Food and Nutrition Science*, **6**, 149–160

Clark, A.H., Richardson, R.K., Ross-Murphy, S.B. and Stubbs, J.M. (1983) Structural and mechanical properties of agar gelatin co-gels – small deformation studies. *Macromolecules*, **16**, 1367–1374

Clark A.H. and Lee-Tuffnell, C.D. (1986). Gelation of globular proteins. In *Functional Properties of Food Macromolecules*, edited by Mitchell, J.R. and Ledward, D.A., Barking, UK: Elsevier Applied Science, pp 273–314

Clark, A.H. (1987) The application of network theory to food systems. In *Food Structure and Behaviour*, edited by Blanshard, J.M.V. and Lillford, P.J., London: Academic Press, pp 13–34

Coniglio A., Stanley, H.E. and Klein, W. (1979) Site–bond correlated-percolation problem: A statistical mechanical model of polymer gelation. *Physical Review Letters*, **42**, 518–522

Coniglio A., Stanley, H.E. and Klein, W. (1982) Solvent effects on polymer gels – a statistical mechanical model. *Physical Review: Part B*, **25**, 6805–6821

Cook, H.E. (1970) Brownian motion in spinodal decomposition. *Acta Metallurgica*, **18**, 297–306

de Gennes, P.G. (1980) Dynamics of fluctuations and spinodal decomposition in polymer blends. *Journal of Chemical Physics*, **72**, 4756–4763

Emanuele, A., Di Stefano, L., Giacomazza, D., Trapanese, M., Palma-Vitorelli, M.B. and Palma, M.U. (1991) Time-resolved study of network self-organization from a biopolymeric solution. *Biopolymers*, **31**, 859–868

Emanuele, A. and Palma-Vittorelli, M.B. (1992) Time-resolved experimental study of shear viscosity in the course of spinodal demixing. *Physical Review Letters*, **69**, 81–84

Feke, G.T. and Prins, W. (1974). Spinodal phase separation in a macromolecular sol-gel transition. *Macromolecules*, **7**, 527–530

Flory, P.J. (1956) Phase equilibria in solutions of rod-like particles. *Proceedings of the Royal Society, Part A*, **234**, 73–89

Hashimoto, T., Kumaki, J. and Kawai, H. (1983) Time-resolved light scattering studies on kinetics of phase separation and phase dissolution of polymer blends. 1. Kinetics of phase separation of a binary mixture of polystyrene and poly(vinyl methyl ether). *Macromolecules*, **16**, 641–648

Hashimoto, T., Itakura, M. and Hasegawa, H. (1986a) Late stage spinodal decomposition of a binary polymer mixture. 1. Critical test of dynamical scaling on scattering function. *Journal of Chemical Physics*, **85**, 6118–6128

Hashimoto, T., Itakura, M. and Shimidzu, N. (1986b) Late stage spinodal decomposition of a binary polymer mixture. 2. Scaling analyses on $Q_m(\tau)$ and $I_m(\tau)$. *Journal of Chemical Physics*, **85**, 6773–6786

Hill, R.G., Tomlins, P.E. and Higgins, J.S. (1985) Preliminary studies of the kinetics of phase separation in high molecular weight poly(methyl methacrylate) / solution-chlorinated polyethylene blends. *Macromolecules*, **18**, 2555–2560

Inoue, T., Ougizawa, T., Yasuda, O. and Miyasaka, K. (1985) Development of modulated structure during solution casting of polymer blends. *Macromolecules*, **18**, 57–63

Jinnai, H., Hasegawa, H., Hashimoto, T. and Han, C.C. (1991) Time-resolved small-angle neutron scattering in intermediate stage and late stage spinodal decomposition of DPB HPI blends. *Macromolecules*, **24**, 282—289

Kurata, M. (1982). *Thermodynamics of Polymer Solutions*. London: Harwood Academic Publishers

Lal, J. and Bansil, R. (1991) Light-scattering study of kinetics of spinodal decomposition in a polymer solution. *Macromolecules*, **24**, 290–297

Langer, J.S., Bar-on, M. and Miller, H.D. (1975) New computational method in the theory of spinodal decomposition. *Physical Review: Part A*, **11**, 1417–1429

Langton, M. and Hermansson, A.-M. (1992) Fine-stranded and particulate gels of β-lactoglobulin and whey protein at varying pH. *Food Hydrocolloids*, **5**, 523–539

Leone, M., Sciortino, F., Migliore, M., Fornili, S.L. and Palma-Vittorelli, M.B. (1987) Order parameters of gels and gelation kinetics of aqueous agarose systems – Relation to the spinodal decomposition of the sol. *Biopolymers*, **26**, 743–761

Lifshitz, I.M. and Slyozov, V.V. (1961). The kinetics of precipitation from supersaturated solid solutions. *Journal of the Physical Chemistry of Solids*, **19**, 35–50

Maruta, J., Ohnaga, T. and Inoue, T. (1993) Demixing kinetics in a binary polymer mixture of poly(methyl methacrylate) and poly(acrylonitrile-co-styrene). *Macromolecules*, **26**, 6386–6390

Matsuo, M., Kawase, M., Sugiura, Y., Takematsu, S. and Hara, C. (1993) Phase separation behavior of poly(vinyl alcohol) solutions in relation to the drawability of films prepared from the solutions. *Macromolecules*, **26**, 4461–4471

Miles, M.J., Morris, V.J. and Ring, S.G. (1984) Some recent observations on the retrogradation of amylose. *Carbohydrate Polymers*, **4**, 73–77

Miller, W.G., Kou, L., Totiyama, K. and Voltaggio, V. (1978). Kinetic aspects of the formation of the orolered phase in stiff-chain helical poly amino acids. *Journal of Polymer Science: Polymer Symposia*, **65**, 91–106

Nielson, G.F. (1969). Spinodal decomposition in soda-silica glass. *Physical Chemistry of Glasses*, **10**, 54–62

Onuki, A. (1986) Late stage spinodal decomposition in polymer mixtures. *Journal of Chemical Physics*, **85**, 1122–1125

Palma-Vitorelli, M.B. and Palma, M.U. (1987) Thermodynamic and microscopic aspects of stability and functional interactions of biomolecules in solution. *Advances in Biosciences*, **65**, 281–288

Pincus, P. (1981) Dynamics of fluctuations and spinodal decomposition in polymer blends – 2. *Journal of Chemical Physics*, **75**, 1996–2000

Pines, E. and Prins, W. (1973). Structure - property relations of thermoreversible macromolecular hydrogels. *Macromolecules*, **6**, 888–895

Rundman, K.B. and Hilliard, J.E. (1967) Early stages of spinodal decomposition in an aluminum–zinc alloy. *Acta Metallurgica*, **15**, 1025–1033

San Biagio, P.L., Madonna, F., Newman, J. and Palma, M.U. (1986) Sol–sol structure transition of aqueous agarose solutions. *Biopolymers*, **25**, 2255–2269

San Biagio, P.L. and Palma, M.U. (1991) Spinodal lines and Flory–Huggins free energies for solutions of human hemoglobins HBS and HBA. *Biophysical Journal*, **60**, 508–512

Sciortino, F., Prasad, K.U., Urry, D.W. and Palma, M.U. (1993) Self-assembly of bioelastomeric structures from solutions – Mean-field and critical behaviour and Flory–Huggins free energy of interactions. *Biopolymers*, **33**, 743–752

Siggia, E.D. (1979) Late stages of spinodal decomposition in binary mixtures. *Physical Review: Part A*, **20**, 595–605

Stading, M., Langton, M. and Hermansson, A-M. (1993) Microstructure and rheological behaviour of particulate -lactoglobulin gels. *Food Hydrocolloids*, **7**, 195–212

Tanaka, F. (1990) Thermodynamic theory of network-forming polymer solutions.

1. *Macromolecules*, **23**, 3784–3795

Tanaka, F. and Stockmayer, W. (1994) Thermoreversible gelation with junctions of variable multiplicity. *Macromolecules*, **27**, 3943–3954

Van Aartsen, J.J. (1970). Theoretical observations on spinodal decomposition of polymer solutions. *European Polymer Journal*, **6**, 919–924

Van Aartsen, J.J. and Smolders, C.A. (1970). Light scattering of polymer solutions during liquid-liquid phase separation. *European Polymer Journal*, **6**, 1105–1112

Voigt-Martin, I.G., Leister, K.H., Rosenau. R. and Koningsveld, R. (1986) Kinetics of phase separation in polymer blends for deep quenches. *Journal of Polymer Science, Part B*, **24**, 723–751

Wong, N.-C. and Knobler, C.M. (1981) Light scattering studies of phase separation in isobutyric acid and water mixtures – hydrodynamic effects. *Physical Review: Part A*, **24**, 3205–3211

4

INFLUENCE OF SHEAR ON THE MICROSTRUCTURE OF MIXED BIOPOLYMER SYSTEMS

C.R.T. BROWN, T.J. FOSTER, I.T. NORTON and J. UNDERDOWN
Unilever Research, Colworth House, Sharnbrook, Bedford. MK44 1LQ U.K.

Introduction

The behaviour of biopolymers when mixed is an area of interest to both the academic scientist and the industrial researcher. Thus there is already fairly extensive literature which is growing rapidly. This has recently been reviewed (Tolstoguzov, 1991). A number of mechanisms seem possible: (a) biopolymers binding together in "synergistic gels" or complex coacervates, (b) co-solubility or (c) they segregate (Piculell, Iliopoulos, Linse, Nilsson, Turquois, Viebke and Zhang, 1994) into distinct phases.

From these previous studies it appears as if the most common behaviour is that the different biopolymers will occupy their own space and thus demix (phase separate). This can happen at all temperature ranges but seems most common when conformational ordering occurs at lower temperatures.

In industry we go one step further, as most hydrocolloids and mixtures of hydrocolloids are processed in order to produce the final product (Tolstoguzov, 1994). The consequence of processing on the conformational ordering or gelation or phase behaviour of mixed systems has received little attention, although the consequences of such changes and failure to control them are well known in industry, resulting in unstable processes and poorly textured products.

If phase separation occurs a number of microstructures can be produced:

(a) Polymer A is continuous and B is dispersed,
(b) Polymer B is continuous and A is dispersed,
(c) Both polymers are continuous (percolated system).

Changes in phase continuity have been observed (Clark, Richardson, Ross-Murphy and Stubbs, 1983; Kasapis, Morris, Norton and Brown/Clark, 1993) and a recent publication (Brown, Madson, Norton and Wesdorp, 1993) suggests that shear can be used to modify this behaviour for biopolymers.

We have been investigating whether these phase separated systems behave like an oil–water emulsion, with the larger phase becoming continuous and a phase inversion occurring over a critical range of phase volume ratios. It is clear that although there is literature covering synthetic polymer mixtures (Heinrich and Wolf, 1990; Larson, 1992) we will in the future need more information on this behaviour and the physical forces that lead to such biopolymer microstructures.

Methodology

PHASE SEPARATION

Gelatin–Maltodextrin

Well resolved peaks in the proton high resolution NMR spectra of gelatin (acid pre-treated pig skin, 250 Bloom, Croda Colloids) and maltodextrin (Paselli SA2, Avebe) in D_2O were calibrated and used to analyse the composition of mixed solutions.

Gelatin, Paselli SA2 or mixtures of the two were prepared by dissolving various weight ratios of the polymers in D_2O or D_2O–NaCl solutions (0.21, 0.53 Molal), forming bulk phase separated fractions from them and recording their proton HR–NMR spectra. Concentrations of gelatin and Paselli SA2 were calculated from calibration plots of component polymer peak areas, referred to pyrazine as 100 units.

The concentration data for each bulk phase of successive weight ratio polymer mixtures were used to construct tie-lines connecting weight ratios of constant effective composition. Binodal lines were constructed from the locus of tie-line ends.

SHEAR EXPERIMENTS

Gelatin–Paselli SA2

Paselli SA2 (6% to 20% w/w) was dissolved with stirring (Silverson lab mixer) in cold de-ionised water and heated to 95°C. After cooling to 70°C solid gelatin (3% w/w) was added and stirring continued at 70°C until all the gelatin had dissolved. Solutions were quenched from 60°C to 15°C in a jacketed shear device fitted with a two vane scraper blade attached to a motor incorporating a torque measuring device (Figure 4.1). The output from probes measuring rotor speed, motor torque, sample temperature and conductivity was recorded continuously with time, using a computer based data capture program (LABTECH NOTEBOOK).

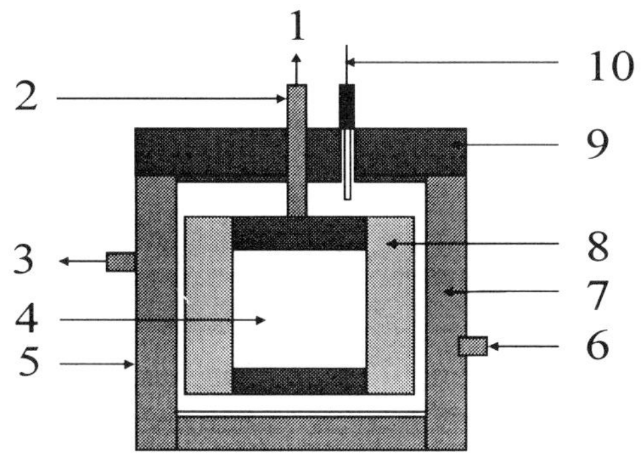

1. drive motor
2. paddle stirer
3. water flow
4. mixture
5. glass cell (300ml volume)
6. water flow
7. temperature jacket
8. teflon baffle
9. sealed lid
10. temperature prob

Figure 4.1 Stirred pot apparatus

Samples of un-sheared and shear-cooled, mixed dispersions were taken for analysis by rheological methods and microscopy.

Gelatin–agar

Agar (Luxara 1253) was dissolved in >95°C hot water wells with stirring (Silverson). Solid gelatin (Type A, acid pigskin 250 Bloom, Croda Colloids) was added and stirring continued until dissolution was complete. The hot solution was transferred to the stirred pot device in exactly the same way as described for gelatin–Paselli mixtures.

κ-Carrageenan–agar

The same procedure as above was used to prepare mixed solutions of κ-carrageenan (X0909, Hercules) and agar in KCl solutions. KCl concentrations were chosen to cause κ-carrageenan to form a gel at either a higher or lower temperature than the gelling temperature of agar.

Methods

MODULUS MEASUREMENTS

The gel formation and subsequent temperature-dependence of gel strength was monitored by small angle oscillatory measurements of bulk modulus (G^*) performed on a Carrimed CSL 500 stress-controlled rheometer using parallel plate geometry (20 mm radius, 1 mm gap) coated with emery paper (600 grade) to prevent slippage and fitted with a bottom plate modified to allow external temperature control. A stress of 1.0 N/m^2 was applied at a frequency of 1 Hz.

Gel formation was monitored by taking repeated measurements of bulk modulus (G^*) with the sample temperature controlled at 10°C.

The same samples were used to follow gel melting by taking measurements of bulk modulus (G^*) at temperature intervals between 10°C and 90°C.

FALLING BALL METHOD

Initial and final samples taken from the stirred pot device (see above) were collected in melting point tubes (Quickfit MF15/2). Lead spheres (10mg) were placed in the samples and the tubes stored in a horizontal position at 5°C for 24 hours or 5 days to allow completion of network formation. The melting point tube was placed in a water jacket (5°C) and the lead ball located in the view finder of a travelling microscope. A thermocouple placed next to the lead sphere was used to monitor sample temperatures. The jacket was raised at 10°C/min and the temperature at which the lead sphere started to flow was recorded and compared to that for gels of each single polymer component.

Phase continuity was determined by comparing the melting temperatures of polymer mixtures to those of single polymer components and assigning the continuous phase according to their correspondence.

MICROSCOPY

Samples taken from the stirred pot experiment were allowed to gel fully during storage at 5°C for 24 hours. Fixation in buffered aldehyde solutions, alcohol dehydration and embedding in acrylic resin (LR Gold) were followed by sectioning and staining with colloidal gold. Finally the dried sections were post stained with 2% uranyl acetate followed by zinc citrate and examined by electron microscopy.

CALORIMETRY

Temperature-dependent conformation transitions of agar and κ-carrageenan were measured by differential scanning calorimetry (DSC) using a Micro DSC II (Setaram) operated at 1°C/min over the range 5°C to 90°C, in order to establish the relationship between gel formation and melting temperatures of agar–κ-carrageenan mixtures.

Results and discussion

GELATIN–MALTODEXTRIN

Gelatin–Paselli SA2 (a potato maltodextrin with dextrose equivalence (DE) 2) phase separates on cooling into two distinct bulk phases. A phase diagram can be produced (Figure 4.2). As highlighted by other papers in this book, salt, pH and temperature can influence the phase diagram. Therefore we studied the effect that salt concentration, pH and temperature have on the phase behaviour of our materials. We found that neither salt or pH have a significant effect on the binodal line position or on the tie-lines.

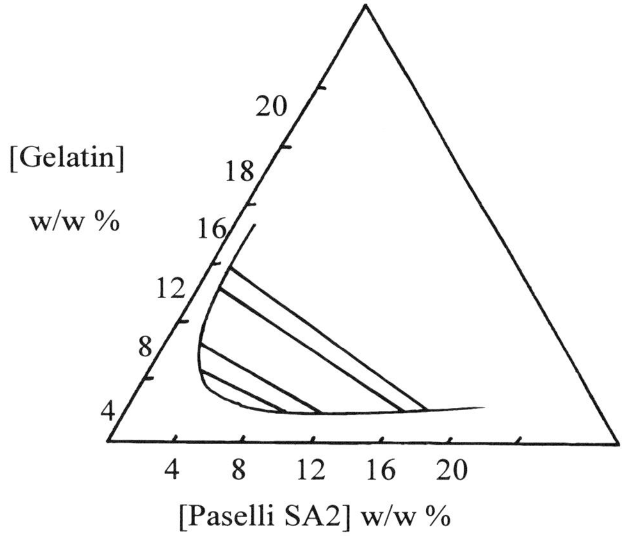

Figure 4.2 Phase diagram for gelatin and Paselli SA in water (pH 6.7)

At a high temperature it appears as if the two polymers are compatible. Phase separation then occurs on cooling. This is of importance when we study the effect of shear. We are probably cooling from a wholly miscible region, and slight variations in salt and pH (although controlled and measured) will have little effect. Recent work (Kasapis *et al.*, 1993) investigated the phase sense of the two polymers and to some extent therefore, the microstructure. They showed that for a fixed level of gelatin (4% LO-2) and increasing levels of Paselli SA2 added, a gelatin-continuous system inverted to become maltodextrin-continuous at about 15%. Equally, for 25% gelatin (LO-2) with SA6 (a maltodextrin with a DE 6) added again the phase inversion still occurred at about 15% of maltodextrin.

These effects were found by microscopy, gelation time and by rheological measurements. The change in phase sense occurred over a small concentration range (about 1–2%) and had a significant effect on the rheological properties of the mixture. More recently it has been published (Brown *et al.*, 1993) that for a mixture where the phase change occurred at about 50% phase volumes, shear during cooling resulted in a maltodextrin-continuous phase when under quiescent conditions it would have been gelatin-continuous.

We have extended this shearing work to look at its effect systematically in terms of shear rate, time and temperature profile. For our materials, and taking the approach of a set gelatin level and adding maltodextrin (Paselli SA2) to it, the microstructure inversion occurs at a constant maltodextrin level for gelatin levels between 3% and 4.5% (Figure 4.3), with the transition again being very sharp and at about 14.5% Paselli addition. Using a stirred pot apparatus (Figure 4.1), which allows us to shear systems continuously while changing temperature, we have studied the influence of shaft speed, polymer concentration and temperature profile on the physical properties of gelatin–Paselli SA2 mixtures. The stirred pot allows

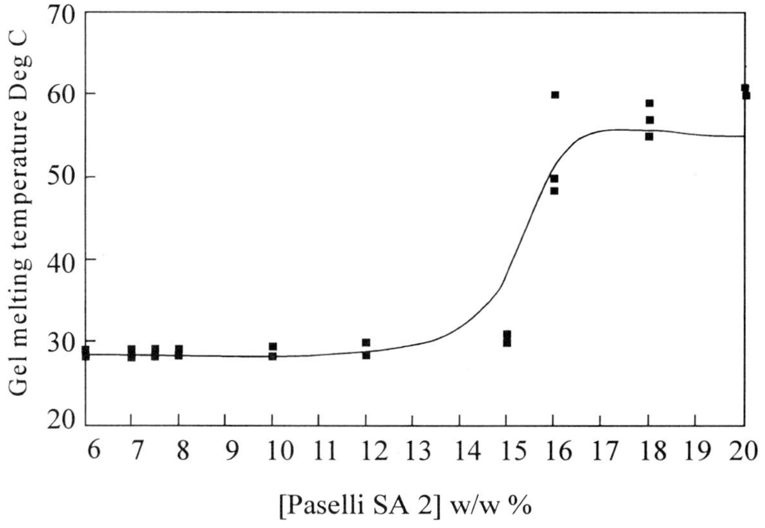

Figure 4.3 Un-sheared gelatin—Paselli composite gel melting temperatures

continuous measurement of torque, so allowing us to estimate the temperature at which material properties change (i.e. gelation of one of the phases, etc.) and obtain an apparent viscosity for the system.

Gelatin (3% w/w) mixtures have been sheared with a range of Paselli SA2 concentrations (6–20%). Samples were extracted after 40 minutes shearing. Taking the method of Kasapis *et al.* (1993) the modulus–temperature profiles were measured (Figure 4.4). Very similar results were found to those reported by Kasapis. At low maltodextrin levels a gelatin-continuous mixture is observed as the maltodextrin level is increased, then a phase change occurs to give a maltodextrin-continuous mixture.

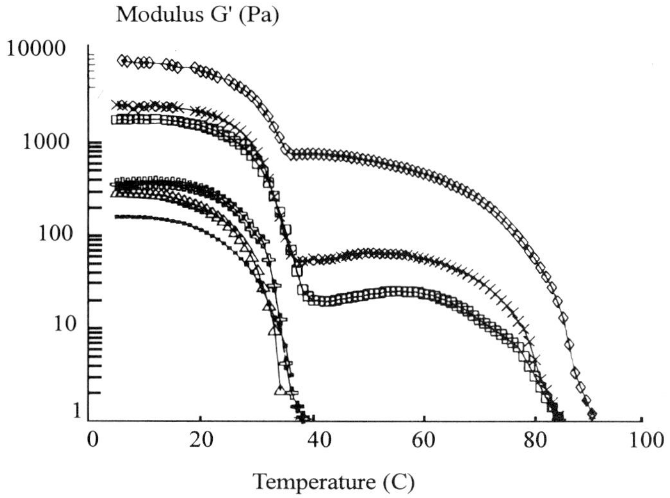

Figure 4.4 Elastic modulus—temperature profiles for sheared composite gels of gelatin (3%) with varying amounts of Paselli SA2 (▪ 10, ▵ 12, ✚ 14, □ 14.5, × 15, ◇ 16 w/w%)

As can be seen in Figure 4.5 the phase transition occurred in sheared systems at about 14.5% maltodextrin as measured by both modulus determination and falling ball experiments. Thus it appears that the melting temperatures obtained from both the quiescent and sheared systems are the same. For these data it appears as if the phase sense of gelatin–Paselli SA2 cannot be modified by applying shear while cooling.

Taking the modulus measurements at 10°C we can obtain a modulus versus concentration plot as shown (Figure 4.6). Initially the modulus is reduced by addition of the maltodextrin as expected for the addition of a soft filler to the gelatin gel. On further addition of maltodextrin a rapid increase in modulus is observed.

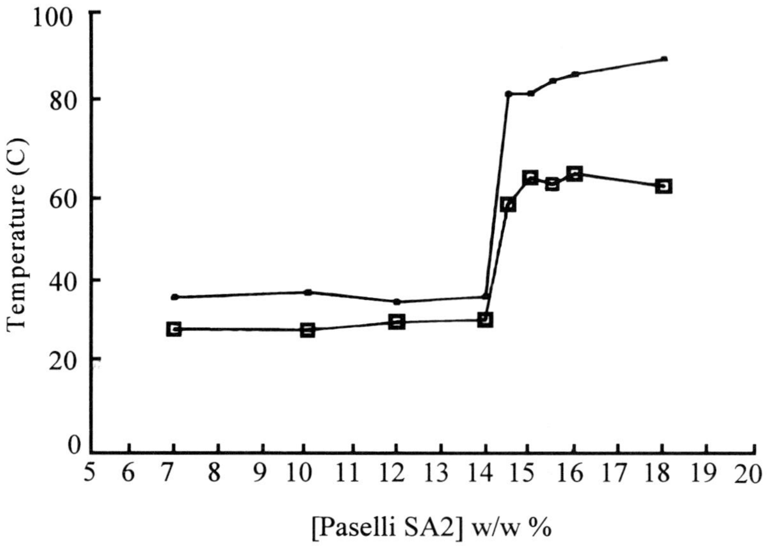

Figure 4.5 Phase inversion for sheared gelatin–Paselli mixtures by modulus–temperature and falling ball methods (▪ Modulus/Temperature, ▫ Falling Ball)

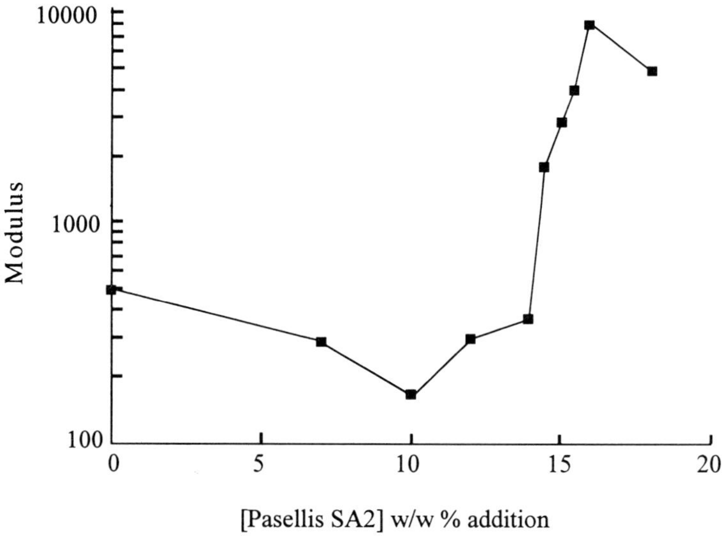

Figure 4.6 Concentration dependence of a 3% gelatin–Paselli SA2 composite gel modulus

Inversion in this experiment occurs at about 90% phase volume of the Paselli-rich phase. However, in this case it is probably not the phase volume which determines the effect, but rather the fact that the effective Paselli concentration must be above its critical concentration for gelatin within the time limit of the experiment. The problem with all this previous work is that by holding a nominal concentration constant (in all experiments it was the gelatin level), a number of parameters of the system are changed between maltodextrin levels:

(a) Effective concentrations of both gelatin and maltodextrin,
(b) The phase volume of the two phases (i.e. both ends of the tie-lines have moved).

To overcome these shortcomings of the previous work we have opted to study the phase behaviour located on two tie-lines. Thus the effective concentrations are constant and it is only the phase volumes of the two phases that change. This allows us to study the microstructure of the system as a function of phase volume, and determine how applied shear influences the microstructure.

Microscopy has shown (Figure 4.7) that at the gelatin-rich end of the tie-line the material is gelatin-continuous with small "droplets" of maltodextrin included. At the maltodextrin-rich end of the tie-line the maltodextrin is a continuous phase with "droplets" of gelatin included. In the middle of the tie-line various degrees of percolated structure exist, but normally with one dominant phase.

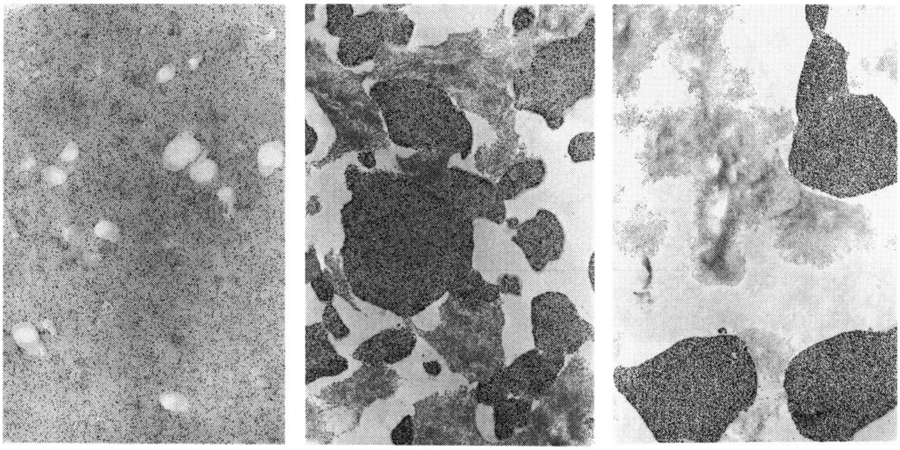

a b c

Figure 4.7 Electron micrographs of sheared composite gels of gelatin—Paselli SA2 mixtures: a) 1.4% Paselli SA2, 12% gelatin; b) 9.0% Paselli SA2, 7.0% gelatin; c) 17.6% Paselli, 2.2% gelatin

Following the procedure of Kasapis *et al.* (1993) the modulus–temperature profiles were obtained. For this tie-line, the gelatin is continuous and dominant until about 50% phase volume. Above 50% phase volume for the maltodextrin-rich phase the final melting occurs at 80–90°C which is typical of a mixture in which the maltodextrin phase dominates. Thus the rheological findings and microscopy pictures are consistent. A very similar result was found from the falling ball experiment.

The second tie-line was at a lower concentration. Similar results were found. A comparison of the modulus versus temperature curves reveals that gelatin is more dominant at phase volumes of 50:50 than was observed for the higher concentration tie-line. The corresponding modulus–temperature profile showed a melting temperature at approximately 35°C with no modulus tail caused by the maltodextrin phase. This profile is probably as a result of the lower effective concentration of maltodextrin for a tie-line at low overall polymer concentrations, which means that the effective concentration of maltodextrin in the Paselli-rich phase is closer to the critical concentration for gelation. Thus we might expect it to have fewer bridges between droplets. A comparison of modulus versus phase volume (Table 4.1) reveals the change in phase behaviour at about 50% phase volume for both tie-lines.

TABLE 4.1 ELASTIC MODULUS AS A FUNCTION OF PHASE VOLUME FOR TWO TIE-LINES FROM THE GELATIN–PASELLI PHASE DIAGRAM (FIGURE 4.2)

Phase volume Paselli-rich phase (%)	*Tie-line 1* G' (Pa)*	*Tie-line 2* G' (Pa)*
0	25,000	20,000
25	20,000	15,000
50	10,000	5,000
63	7,000	5,000
75	7,000	5,000
100	5,000	5,000

*Effective concentrations for gelatin and Paselli SA2 in both phases:

	Gelatin-rich phase		*Paselli-rich phase*	
Tie-line	*Gelatin*	*Paselli SA2*	*Gelatin*	*Paselli SA2*
1	12.0%	1.4%	2.2%	17.6%
2	10.8%	1.4%	2.2%	16.3%

The effect of shear on the microstructure of the 50:50 phase volume mixture of tie-line 1 was studied (Figure 4.8). If sheared for about 30 min we find that the mixture is gelatin-continuous with an elastic modulus at 10°C of about 10,000 Pa. On further shearing of the mixture (for up to 96 hours) inversion to a continuous maltodextrin system and an included gelatin phase occurs which shows a high melting temperature (>90°C) and an elastic modulus typical of the maltodextrin dominated system (i.e. 5,000 Pa).

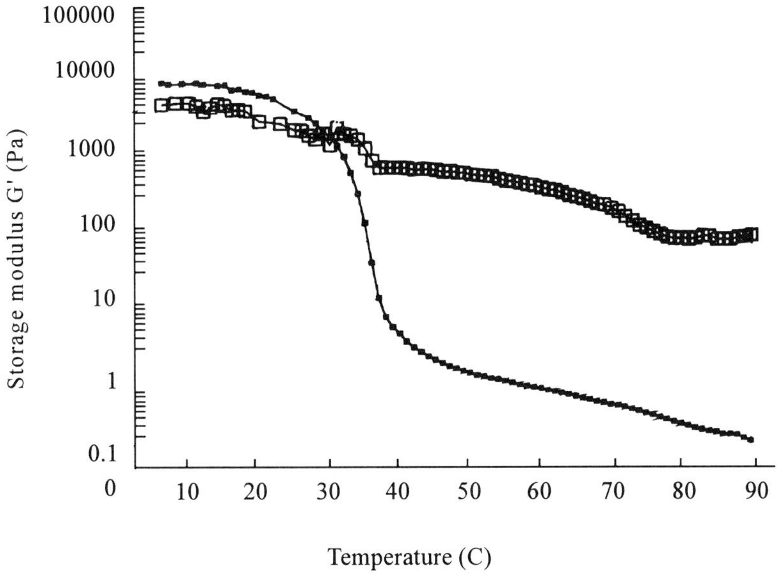

Figure 4.8 Elastic modulus—temperature profiles for a sheared 7% gelatin—Paselli SA2 (9.5%) mixture cured for 0.5 hour and 96 hours (▪ 0.5 hours, ▫ 96 hours)

Thus it would appear that the microstructure of gelatin-maltodextrin mixtures can be significantly influenced by applying shear for much of the time when the gelatin gel is forming. In addition, the maltodextrin has been given enough time to structure so as to reduce the chance of gelatin droplet interaction. In future experiments the time-dependence of this effect will be investigated, as will the range of phase volume ratios where the structure is affected by shear.

GELATIN–AGAR MIXTURES

The microstructures of gelatin–agar composite gels follow a similar pattern to those discussed above for gelatin–maltodextrin. It has been shown by adding gelatin to a 1% agar gel (Figure 4.9) that initially the material is agar-continuous with droplets of gelatin included (Clark *et al.*, 1983). At about 2.5% gelatin the microstructure appears to become somewhat percolated (i.e. both gelatin- and agar-continuous). At 3% gelatin a microstructure comprising agar droplets dispersed within a continuous gelatin matrix is clearly visible. On further addition of gelatin the agar droplets become smaller, probably due to the increased osmotic pressure exerted by the higher levels of gelatin.

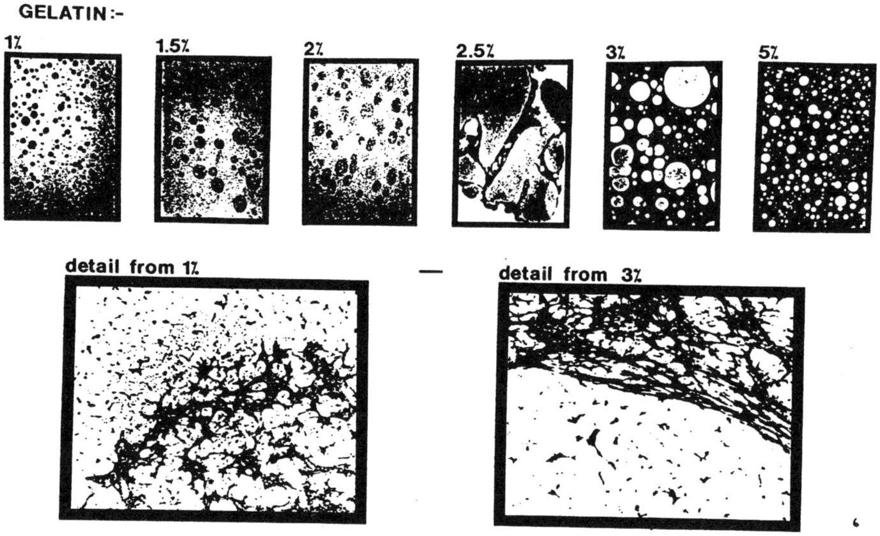

Figure 4.9 Microstructure of un-sheared gelatin—agar gels (Clark *et al.*, 1983)

Using gelatin–agar mixtures we have carried out a similar experiment to Clark *et al.* (1983), but with 3% agar. In our experiment the microscope pictures show an agar-continuous gel with included gelatin up to and including 5% gelatin.

Using the stirred pot, a range of mixtures of agar and gelatin have been sheared at rotor speeds between 100 rpm and 1000 rpm while cooling from 90°C. The torque plot obtained from the sheared pot at 1000 rpm shows (Figure 4.10) that the initial viscosity of the mixture (temperatures 70–50°C) is consistent with an agar-continuous phase. If the mixture was gelatin-continuous a torque value of close to 1 would have been observed. As the agar starts to gel the torque reading increases, then decreases as phase inversion occurs and the mixture becomes gelatin-continuous. By the time most of the agar gel has set (i.e. a temperature of 20°C) the torque increases, probably due to the gelation of the gelatin. However, the mixture is already trapped as a gelatin-continuous/agar included microstructure. The microstructure and rheology of mixtures taken from the pot have been measured.

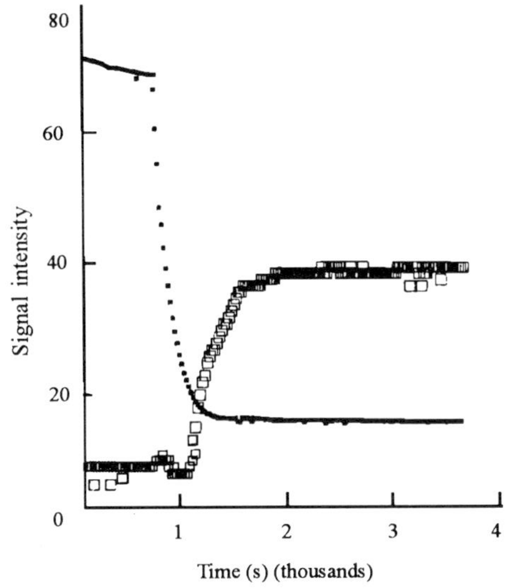

Figure 4.10 Stirred pot for 5% gelatin–agar (3%) using a shaft speed of 1000 rpm (▪ Temperature (C), ▫ Torque x 10)

Microscopy pictures (Figure 4.11) show that if the mixtures are stirred while being cooled through the gelation temperatures of both the agar (45°C) and gelatin (15°C), they become gelatin-continuous with agar included. If sheared at 100 rpm the agar forms fairly large irregular particles. If the stirrer is run at a higher speed

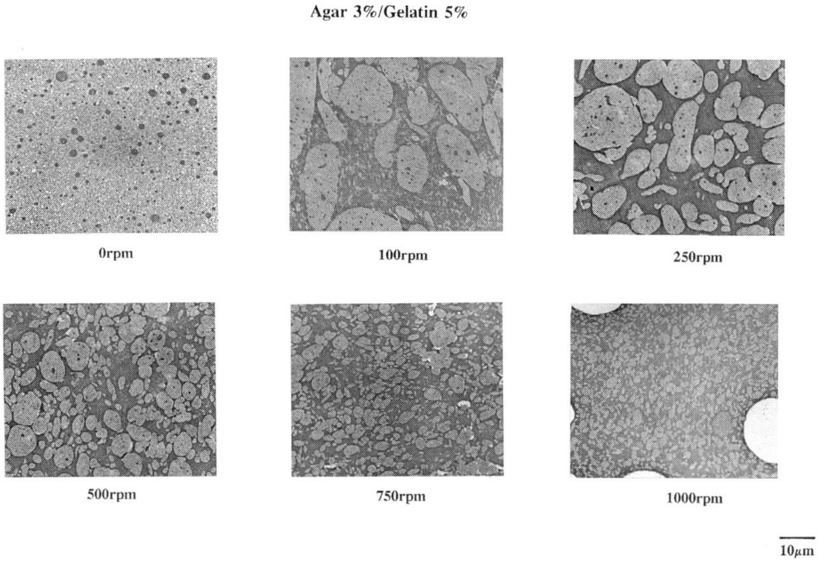

Figure 4.11 Photomicrographs of sheared gelatin–agar gels (magnification 1 cm = 10 μm)

(500–1000 rpm) then the "droplets" of agar become much finer and a narrower distribution is measured (Figure 4.12), e.g. by a Quantimet analysis of the photomicrographs.

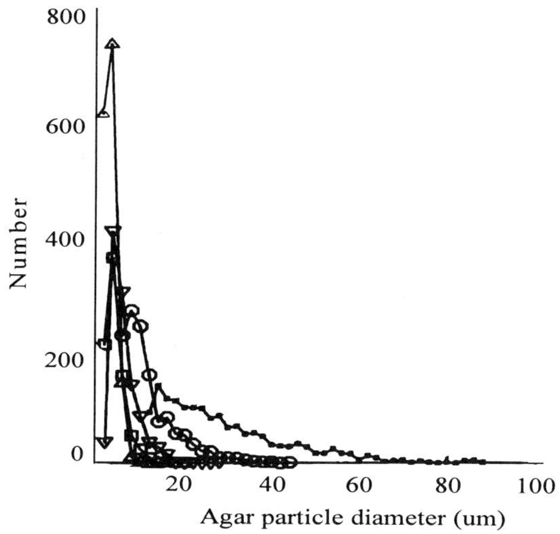

Figure 4.12 Size distributions for agar particles determined by image (Quantimet) analysis of 3% gelatin–agar (3%) mixtures. Results obtained after 1 hour of shearing in the stirred pot (Figure 4.1) ▪ 100 rpm, ○ 250 rpm, ▽ 500 rpm, □ 750 rpm, △ 1000 rpm

This behaviour is the same for gelatin levels between 1% and 5%.

Thus:

(a) Phase inversion of the gelatin–agar mixture is brought about by shearing (even gentle agitation) through the gelling temperatures,
(b) Increasing the shear forces leads to smaller and more spherical particles.

COLLOID THEORIES

Colloid theories have been developed for simple oil–water mixtures which describe phase behaviour and droplet break-up under shear (Chesters, 1975; de Bruijn, 1989). These theories and experiments show that in order for an emulsion to lower its energy while under shear, the phase with the highest viscosity becomes included. From the torque measurements taken from the stirred pot, the agar has a higher viscosity than the gelatin initially (i.e. at high temperatures) and so would be expected to become the included phase immediately after phase separation occurs. Samples taken at temperatures up to the point at which gelation of the agar occurred showed that the agar was continuous. If the simple viscosity rule is valid for "biopolymer emulsions", this suggests that either:

(a) Phase separation only occurs once gelation of agar is under way,
(b) The system is kinetically trapped with no mechanism for the interfacial layer to be disrupted.

Further work involving the study of phase behaviour at different concentrations and for different materials is necessary to distinguish between these interpretations.

We do find that, as we observed for the gelatin–maltodextrin mixtures, the biopolymer which gelled first becomes an included phase.

The other colloid theory of interest to us is about the prediction of droplet break-up. The theory for laminar flow suggests that whether a droplet will break up or not depends on the ratio of the applied shear stress and the Laplace Pressure. This ratio is known as the capillary number.

$$\text{Capillary number } \Omega = \frac{G\mu_c}{\sigma/R} \qquad (1)$$

where:
$G\mu_c$ = shear stress
G = velocity gradient (shear rate)
μ_c = viscosity of the system
σ/R = Laplace Pressure
σ = interfacial tension
R = radius of the particle

Work on simple oil–water emulsions (Shamlou and Titchener-Hooker, 1993) has shown that for droplet break-up the critical value for the capillary number depends only on the viscosity ratio of the two fluids. For mixed biopolymer emulsions it is expected that the "droplets" will break up when the applied shear stress exceeds the sum of the elastic modulus and the Laplace pressure of the particle. Under these conditions the gelled particle will start to "flow" and under the continuous shear field break up into pieces. The surface tension for mixed biopolymer systems is unlikely to have any effect as it is expected to be very small. Thus by increasing the rotor speed in the stirred pot experiment from 100 rpm to 1000 rpm we would expect the particles to be reduced in size, reaching a limiting value once the optimum conditions for break-up are reached. This appears to be observed (Figure 4.12). The change in droplet size shows a smooth trend, and the particles are monotonic, which implies that "flow" of the gel and droplet break-up are dominant; it is not brittle fracture of gel structures.

From modelling of the flow in the stirred pot and measuring the power dissipation as a function of shaft speed it appears as if for our mixed biopolymers the flow is laminar.

From the capillary number argument it would appear as if the droplet breakup should occur most efficiently in the liquid state or very early stages of gelation (i.e. low particle modulus), the two phases then have the closest match in terms of "viscosities" and the flow is right for breakup. This should also be true if gelation occurred at the walls as the gelled material should be scraped off the surfaces and broken up in the pot.

Taking this approach, it should be possible to predict break-up conditions and particle size. For mixed biopolymer systems there are still a number of unknowns (e.g. interfacial tensions, flow regions in which break-up occurs, etc.). The insight and predictive capabilities that could result from such studies will be highly valuable.

Considering the modulus measurements for the sheared mixtures of agar and gelatin, we find that the measurements are consistent with expected trends from the microstructure analysis, i.e. on applying shear the modulus is decreased, particularly for the 1% gelatin case (Table 4.2), indicating the switch to a gelatin-continuous mixture. The difference observed between 100 rpm and 1000 rpm reflects the somewhat percolated nature of the material formed at the lower shaft speed.

TABLE 4.2 SHEAR EFFECT ON THE ELASTIC MODULUS OF GELATIN–AGAR MIXTURES AT 10°C

Composition (%)			*Modulus (Pa)*	
Agar	*Gelatin*	*0 (rpm)*	*100 (rpm)*	*1000 (rpm)*
3.0	0	60,000	1,500	500
3.0	1.0	60,000	4,000	900
3.0	3.0	40,000	6,000	4,000
3.0	5.0	20,000	9,000	8,800

At higher gelatin levels, once the system has been inverted to a gelatin-continuous structure, the modulus shows the effects of shear less, due to the higher effective gelatin concentrations in the continuous phase.

Plotting the elastic modulus against nominal concentration of gelatin shows how the quiescent gel has a slight reduction in modulus as expected for the inclusion of softer particles into the agar gel. With the sheared gel we see first (Figure 4.13) that the agar alone has a reduced modulus. This is due to the formation of particles in the shear field. Then, on addition of gelatin, we see increases in the modulus.

We attempted to fit the isostress (stress uniform, strain discontinuous) and isostrain (strain field uniform, stress discontinuous) models (Clark, 1987).The appropriate equations were:

$$\text{Isostrain} \quad G_c = \phi_x G_x + \phi_y G_y \tag{2}$$

$$\text{Isostress} \quad \frac{1}{G_c} = \frac{\phi_x}{G_x} + \frac{\phi_y}{G_y} \tag{3}$$

where:
G_c = shear modulus of the composite gel
G_X = modulus of gelled polymer X
G_Y = modulus of gelled polymer Y
ϕ = phase volume fraction

The quiescent gel initially fits to the hard matrix–soft filler (isostrain) predictions of this approach (Figure 4.13). The modulus then starts to dip towards the soft matrix–hard filler (isostress) limit. This probably indicates that phase inversion is

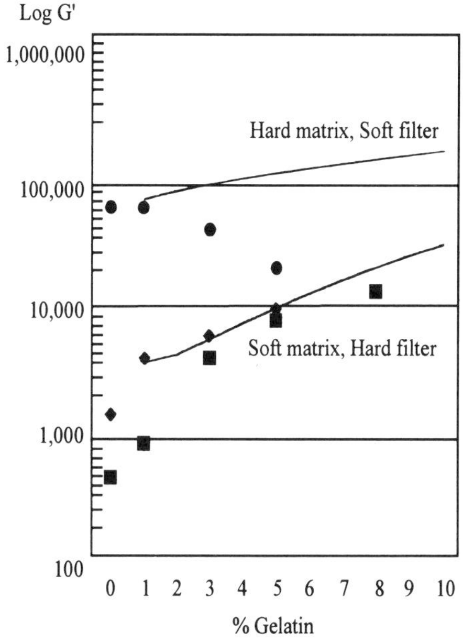

Figure 4.13 Composite gel modulus fits to the isostress—isostrain model (Clark *et al.*, 1983) for gelatin–agar mixtures (— Qu (p=1.0), ● Qu (exp), ■ Sh (1000), ♦ Sh (100))

starting to occur, as a consequence of the higher phase volumes of the gelatin. For the sheared gels the data obtained for shaft speeds of 100 rpm and 1000 rpm, fits the behaviour of the soft matrix–hard filler (isostress) analysis and follows the lower bound up to 8% gelatin concentrations. This supports the microstructure data showing a soft gelatin-continuous matrix with hard agar particles included.

If the data obtained for agar sheared alone (i.e. in the absence of gelatin) are used for the analysis, the data fit very well to the soft matrix–hard filler theory with a p-factor of 1. Thus it appears that the sheared mixed biopolymer gels fit the theories in line with predicted behaviour from the microstructure analysis.

AGAR–κ-CARRAGEENAN MIXTURES

In order to check these ideas we have designed an experiment where we can change which polymer gels first. Thus we have taken κ-carrageenan with various KCl concentrations. These were selected so that the carrageenan would gel either before, at the same time or after the agar. We are at present studying mixtures with 0.15 M KCl in which the carrageenan gels before the agar, 0.07 M KCl in which the carrageenan gels at about the same temperature as the agar, and with 0.02 M KCl in which the carrageenan gels at a temperature well below the agar.

Initial results show that the carrageenan is continuous when it gels after the agar but it is included when it is gelled first. This supports the proposed mechanism. However, we need to obtain more details on the melting temperature and microstructure of the mixed system.

Conclusions

It seems a good analogy to treat phase separated biopolymer mixtures in water as if they were simple two-phase classical emulsions. The biopolymer which has the highest viscosity (first to gel) is forced by shear to become an included phase. Thus phase inversion and microstructure of the system can be significantly influenced by processing parameters, as well as by the ingredients used and their concentrations. Particle break-up and coalescence in these "emulsions" appear to be understandable from the Emulsion theories. In the future they will hopefully be predictable as well.

Acknowledgements

The authors extend their thanks to Alex Aldred and Karen Buxton for their technical assistance in preparing this paper.

References

Brown, C.R.T., Madsen, R.A., Norton, I.T. and Wesdorp, L.H. (1993). European Patent Number 574973

de Bruijn, R.A., (1989). Ph.D. thesis, Technical University of Eindhoven, pp 5–31

Chesters, A.C. (1975). *International Journal of Multiphase Flow*, **2**, 191–212

Clark, A.H., Richardson, R.K., Ross-Murphy, S.B. and Stubbs, J.M. (1983) Structural and mechanical properties of agar gelatin co-gels – small deformation studies. *Macromolecules*, **16**, 1367–1374

Clark A.H., (1987) The application of network theory to food systems. In *Food Structure and Behaviour*, edited by Blanshard, J.M.V. and Lillford, P.J., London: Academic Press, pp 13–34

Heinrich, M., and Wolf, B.A., (1990) Kinetics of phase separation – Trapping of molecules in non-equilibrium phases. *Macromolecules*, **23**, 590–596

Kasapis, S., Morris, E.R., Norton, I.T. and Brown, C.R.T. (1993) Phase equilibria and gelation in gelatin maltodextrin systems. 3. Phase separation in mixed gels. *Carbohydrate Polymers*, **21**, 261–268

Larson, R.G., (1992) Flow-induced mixing, demixing, and phase transitions in polymeric fluids. *Rheologica Acta*, **31**, 497–520

Piculell, L., Iliopoulous, I., Linse, P., Nilsson, S., Turquois, T., Viebke, C. and Zhang, W. (1994) Association and segregation in ternary polymer solutions and gels. *Gums and Stabilisers for the Food Industry*, **7**, 309–322

Shamlou, P.A. and Titchener-Hooker, N. (1993). *Processing of Solid–Liquid Suspensions*, Oxford: Butterworth–Heinemann, pp 1–25

Tolstoguzov, V.B. (1991) Functional properties of food proteins and role of protein–polysaccharide interaction. *Food Hydrocolloids*, **4**, 429–468

Tolstoguzov, V.B. (1994) Some physio-chemical aspects of protein processing in foods. *Gums and Stabilisers for the Food Industry 7*, edited by Phillips, G.O., Williams, P.A. and Wedlock, D.J., Oxford: IRL Press, pp 115–124

5

SMALL DEFORMATION RHEOLOGICAL BEHAVIOUR OF BIOPOLYMER MIXTURES

S.B. ROSS-MURPHY
Biopolymers Group, Division of Life Sciences, King's College London, Campden Hill Road, Kensington, London. W8 7AH U.K.

Introduction

Over the last decade or so there has been an impressive growth of activity and interest in the structure and rheology of biopolymer mixtures. The reasons for this are several, in the biomedical area there is renewed activity in extra-cellular processes, particularly in collagen–proteoglycan interactions, and also in controlled drug release (which can usefully employ gel technology). Of particular interest in the present context is work on food gels and stabilizers. The latter is driven by two main thoughts, to reduce the dependence on "unnatural" additives, and, with the growth in perception of healthy eating, in the design of low fat structured products such as creams, yoghurts, fat spreads, etc. The latter problem is simply put. If we take all the fat out of a fat spread, we need to add back structuring ingredients to give an acceptable, predominantly solid, spreadable material.

Although this problem is quite trivial, it is rather a long way from the traditional bounds of materials science. In the latter case it is usually a requirement to design for strength and or (lack of) wear. Here we need to design in controllable (plastic or viscoplastic) failure, so the spread still looks like a solid in the refrigerator or container and can be easily spread, but then remains on the bread without flowing off (or through!) it. In this case the obvious design approach (and an effective solution) is to incorporate a soft particulate filler (to encourage failure by the introduction of heterogeneities) mixed with a gel matrix phase to provide the static structure. This chapter will address some of the topics in structure/rheology relationships of biopolymer mixtures, but in a non-system specific manner. Since the area is already semi-mature, it will also adopt a more philosophic and future looking approach.

Biopolymer mixtures

In the present article we will assume that a biopolymer mixture consists of (at least) two biopolymer systems plus, in volume terms by far the largest component, an aqueous "solvent". This might itself contain salts and other low molecular weight water soluble components. In this case, depending upon the nature of the biopolymers, we have several potential blends. If both biopolymers form thickening solutions, we have a fluid–fluid system. If one of the biopolymers is largely insoluble or unswellable (e.g. starch at low temperatures), we can have a solid dispersed in a fluid (or depending upon phase volumes the converse system) and if both biopolymers separately give gels, we can have a mixed gel system (see Figure 5.1). The mixed gel system can itself be subdivided into mixed networks, interpenetrating networks, etc., as pointed out by Brownsey and Morris (1988). Much of this chapter will concentrate on mixed solid–solid systems. In the next sections we briefly discuss mixed fluid and filled fluid systems.

Liquid/liquid *Liquid parameters*	*Liquid/solid* *Liquid phase*	*Solid/solid* *"Filled systems"*
Newtonian Non-Newtonian	As left	As left
Power law "Yield stress"	Solid phase	Mixed systems
	Soft/hard filler Size distribution "Shape" Volume fraction	Composition Phase continuity Heterogeneity

Figure 5.1 Different combinations of fluid and solid two-phase systems.

There are a number of studies of mixed fluid biopolymer systems in the literature, particularly polysaccharide co-gel systems, including carrageenan (normally gelling) and galactomannan (non-gelling) systems at low concentration (see for example Turquois, Doublier, Taravel and Rochas, 1994). The viscosity of a single biopolymer fluid system depends upon polymer concentration (dilute or semi-dilute), shear rate, and polymer size and shape. As the concentration increases it can be said that the latter effect becomes less pertinent, since then the solution becomes more segment homogeneous. However what is not clearly established (nor as far as the author is concerned can it be) is that the viscosities, say at a low shear rate, are either additive or that the mixture fluid viscosity can be simply predicted from the properties of the components. Chemical engineers use a number of empirical blending rules for predicting mixed fluid viscosities, but these have little theoretical validity. The important consideration is that attempts to deduce synergism (or anti-synergism) from an apparent increase (or decrease) in viscosity for mixed fluid systems are based upon a misconception. After all we should not expect additivity of "simple" thermodynamic parameters (vapour pressure, osmotic pressure). This does not mean that there is no practical worth in measuring such synergism. If the viscosity of a mixture is substantially higher than either component we may be able to use less expensive biopolymers and so save money. However, it does suggest that an early appeal to "molecular" interpretations, for example based upon a viscosity measured at a single (or zero) shear rate, is somewhat risky.

Solid in fluid systems

The vast majority of work in the rheological literature on such systems has concentrated upon investigations of small (usually spherical) particles dispersed in a Newtonian fluid. For hard sphere systems there is a clear maximum disperse phase packing volume (ϕ_{max}) above which the system will look solid (or become glassy). Many (at the last count more than 50) equations (see, for example Ross-Murphy, 1984) have been suggested to relate the viscosity of the disperse system , to that of the solvent η_s and the packing fraction (ϕ) or packing fraction ratio (ϕ/ϕ_{max}), where ϕ_{max} is the maximum packing fraction. This depends upon the particle shape. For hexagonally close-packed hard spheres ϕ_{max} is around 0.74, decreasing to about 0.1 for hard rod-shaped particles.

$$\eta = \eta_s (1 + f(\phi/\phi_{max})) \qquad (1)$$

The simplest equation is, of course, that due to Einstein in which $f(\phi)$ is just 2.5 ϕ. Other expressions introduce a series of coefficients in ϕ^2,.., but all realistic equations introduce some singularity (where $\eta \rightarrow \infty$) as $\phi/\phi_{max} \rightarrow 1$.

To extend this to non-spherical particles (e.g. by Simha), it is necessary to replace the Einstein 2.5 by higher coefficients (say 15) as the particles become increasingly anisotropic (rod-like). More significant in the present context is that as $\phi \rightarrow \phi_{max}$ the behaviour becomes less and less Newtonian. Indeed it may develop a power-law type shear rate—viscosity profile, with an apparent yield stress at low shear rates. The other important consideration is that soft particles are often found in practice. For these, ϕ_{max} can be greater than 100%, since the particles compress, compact, deform and interpenetrate. This means that rather than the (logarithmic) singularity at $\phi \to \phi_{max}$ we get a more gradual transition in apparent viscosity, at least at finite shear rates.

Small and large deformation measurements

Since this chapter concentrates on small deformation measurements, it is important to clarify precisely what these terms imply. For simple "elastic" solids the distinction is quite clear, in small strain uniaxial deformation a solid is ideal and Hookean, with stress proportional to strain, and consequently with a strain-independent shear (G) and Young's modulus (E). Large deformation implies fracture or failure of the sample (particularly in tensile measurements). The widespread growth in oscillatory measurements has enabled both steady and oscillatory measurements to be performed on viscoelastic fluids and solids.

For most polymeric fluids a full characterisation (Ferry, 1980; Ross-Murphy, 1994) would involve measurements of shear viscosity, η, versus shear rate $\dot{\gamma}$ and, using oscillatory (sinusoidal) measurements, of G', G" and η^* (= G^*/ω), where G′ is the shear storage modulus, G" the loss modulus (G"/G' = tan δ, with δ the phase angle), G^* is the complex modulus, η^* the complex viscosity and ω the oscillatory frequency (in radians per second). Comparison of the (η, $\dot{\gamma}$) and (η^*, ω) profiles usually show a great similarity, and this commonality is known as the Cox—Merz rule.

In these terms, such a fluid shows the same behaviour in small (oscillatory) and large strain (steady shear rate) measurements. A viscoelastic solid such as a gel cannot be measured in steady shear since, like the elastic solid, it will fracture or rupture. Nevertheless in oscillatory measurements G^* is independent of shear strain γ only over a range of γ (linear viscoelastic behaviour). At higher strains G^* becomes strain-dependent (normally G^* decreases with γ but for some systems e.g. fibrin gels, strain hardening is seen). Figure 5.2 shows the strain dependence of G^* (or its normalised form). For some structured fluids ("weak gels") the small deformation frequency behaviour (mechanical spectrum) is solid-like (G' > G" and both are rather frequency-independent) but measurements can be made in steady shear. However, in this case the system does not usually obey the Cox—Merz superposition, and indeed the (η, $\dot{\gamma}$) < (η^*, ω) profiles are power-law fluid-like just as we saw earlier for the high volume fraction filled systems.

The implication here is that "structure" or "heterogeneity" is correlated with small values of the linear viscoelastic strain, γ_{lin}. Such correlations are implicit in classical colloid science, and there would be related to the particle—particle correlation function (the Fourier transform of the radial distribution function). In the mixed and weak gel area this suggests a study in which image analysis (particularly transformed to Fourier space), to provide spatial averaging, might be related to γ_{lin} with changes in composition.

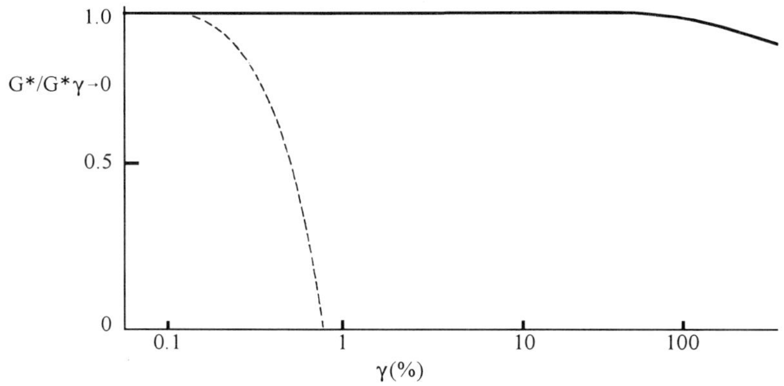

Figure 5.2 Typical modulus versus strain response for a "weak gel" (- - - -) and a strong gel or viscoelastic fluid (——). (From *Physical Techniques for the Study of Food Biopolymers*, 1994, edited by S.B. Ross-Murphy, with kind permission of Chapman & Hall, London.)

Prediction of mixed gel modulus

As other chapters discuss in more detail, when two biopolymer systems are mixed at low concentrations they may co-exist within a single aqueous phase. At higher concentrations, and in the absence of preferential ("specific") interactions, two phases separate, each containing a predominance of one biopolymer component and a smaller amount of the other. In systems and conditions where both biopolymers can gel the end result is usually a biphasic, composite gel.

Upper and lower bound models

For bulk mixed networks or melts, the modulus of the composite can be related by so-called blending laws. If a high modulus system (filler) is dispersed within a low modulus phase we can, by exact analogy with eq. (1) for the viscosity of filled fluid phases discussed earlier, write

$$G_c = G_x\ (1 + f\ (\phi)) \tag{2}$$

where ϕ is, as before, the volume fraction of the filler phase, and G_c and G_x are the composite modulus and the modulus of the X-phase material respectively. Such an approach has been applied to both small and large deformation properties of glass filled gels by several workers (Richardson, Robinson, Ross-Murphy and Todd, 1981; Ross-Murphy and Todd, 1983; Langley and Green, 1989). In these treatments it is assumed that G_y, the modulus of the Y or filler phase, is effectively infinite. When G_y is of the same order as G_x, the approach pioneered by Takayanagi (Manson and Sperling, 1976) is to apply series and parallel models. For $G_y > G_x$ this gives

$$\frac{1}{G_c} = \frac{(1 - \phi)}{G_x} + \frac{\phi}{G_y} \tag{3}$$

and for $G_x > G_y$

$$G_c = (1-\phi)\ G_x + \phi G_y \tag{4}$$

The justification for these two so-called upper and lower bound models are that in the first case the filler is deformed proportionally less than the matrix, so both the components experience the same stress. Conversely when the matrix is stronger the total deformation is controlled by that of the matrix. Consequently the upper bound model is sometimes called the isostress model, and the lower bound model the isostrain (as illustrated in Figure 5.3). Figure 5.4 shows the corresponding modulus bounds calculated from eqs. (3) and (4).

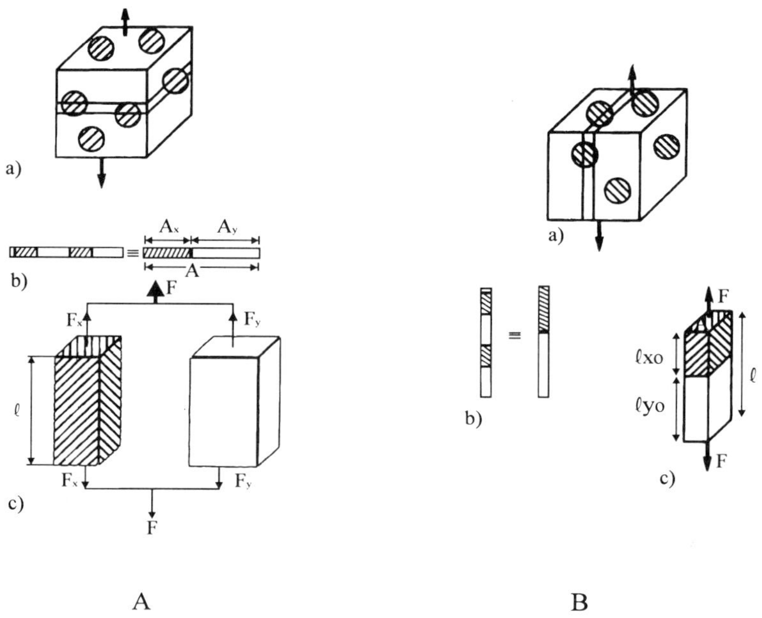

Figure 5.3 Schematic diagram of the isostrain (a) and isostress (b) models. (From McEvoy, Ross-Murphy and Clark, 1985, with kind permission of Butterworth Heinemann Ltd.)

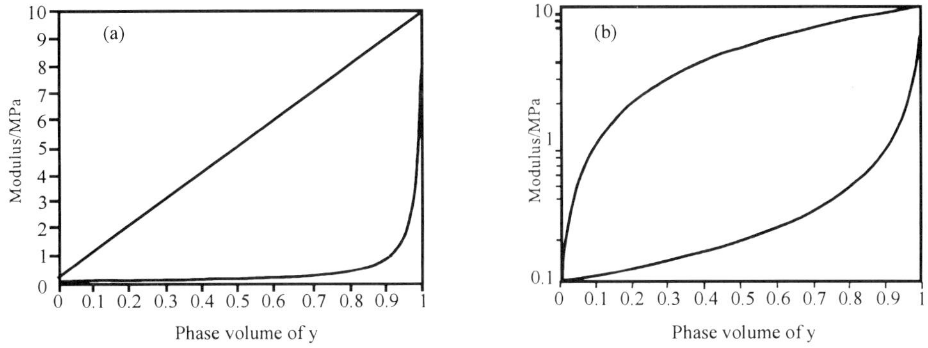

Figure 5.4 Modulus versus phase composition for (top) isostrain and (bottom) isostress models plotted (a) linearly and (b) logarithmically.

The problem with this approach is implicit in these terms "upper" and "lower" bounds. The properties of any particular system will, in all likelihood, be far more complex than such simple mechanical analogues suggest. Indeed the widely accepted Van der Poel model for the second case actually assumes a formal three-phase structure of filler, matrix and a filler–matrix interface mesophase (Nielsen, 1974; Theocaris, 1990). This means that, just as for electrical networks, it is usual to combine series and parallel models, and so produce a specific model which falls between the upper and lower bounds. Indeed, for polymer blends one could assume that a cross-over from near one (say lower) to the other bound occurs around the phase inversion point. The important point is, however, that a materials scientist, who would be dealing with pure phase components (for example a blend of two high molecular weight polymers) would only rarely employ the absolute upper and lower bound Takayanagi models. Instead he would probably adopt the approach of adjusting the proportion of isostress and isostrain components *a posteriori* in order to obtain an appropriate fit to the observed behaviour.

Modified Takayanagi model

The problem with applying even the simplest bounds to predicting the properties of a mixed gel system is that the nominal concentrations of polymer will be changed when they are mixed and phase separation takes place. In other words the phase volume is determined not only by the amount of each (bio)polymer but by how the (aqueous) solvent is partitioned. This also means that adjusting the bounds to fit, as discussed above, is an inappropriate strategy.

In the modified Takayanagi (MT) model of Clark and co-workers (Clark, Richardson, Ross-Murphy and Stubbs, 1983; Clark, 1987), a parameter, p, originally assumed and now more clearly established to be related to the cross-term χ_s in the Flory–Huggins lattice polymer mixing theory (Durrani, Prystupa, Donald, and Clark, 1993), is introduced. As explained elsewhere in this volume, the parameter χ (Flory, 1953) is a measure of interaction between polymer and solvent or polymer and polymer. For a three component (two polymers and one solvent) solution there will be three χ values; a χ_{XS} between component (phase) X and solvent s, a corresponding χ_{YS}, and a third χ_{XY} cross-term, usually small and which can be neglected when the solvent volume fraction is large. The value of p is given by:

$$p = \frac{\alpha y}{x(1-\alpha)} \tag{5}$$

and α is a parameter such that αw of the solvent is associated with the X phase and $(1 - \alpha)$w with Y, where $w = 1 - x - y$ and x and y are the weight fraction of X and Y phases.

Under these circumstances the effective concentration of the x-phases (C_x) is given by

$$C_x = \frac{x}{x + \alpha w}$$
$$C_y = \frac{y}{y + (1+\alpha)w} \tag{6}$$

which can be related directly to p (we deliberately avoid repeating the algebra). When the density is close to unity, mass fraction and volume fraction are the same, and the volume fractions of x and y phases are

$$\phi_x = \frac{x(p + (1-p)\ y)}{px + y}, \qquad \phi_y = 1 - \phi_x \tag{7}$$

Provided some relationship can be established between modulus and concentration (or volume fraction) for each of the original single component gels, the modulus of the mixed gel can be evaluated.

A number of approaches have been made to relate G and C, the simplest of which simply assumes $G \propto C^2$, an expression which is probably quite reasonable for $C > 10\ C_0$, where C_0 is the critical gel concentration. On the basis of the C^2 law Morris (1992) has elaborated more detailed conditions and applied his modified approach to gelatine—maltodextrin gels (Kasapis, Morris, Norton and Clark, 1993). Perhaps a more realistic approach (although algebraically less tractable) is to assume, as Walkenström and Hermansson (1994) did in their work on gelatin—whey protein mixed gels that the modulus $G \propto (C - C_0)^b$.

In the original model of Clark and co-workers (1983), G was related to C using a cascade model. Although more complicated, this approach tends to look like the previous equation, but G is implicitly dependent upon $(C - C_0)$, so that the exponent b varies (from around 2 at high concentrations to much higher values as $C \rightarrow C_0$). This is probably the least empirical method. However, it has to be said that the detailed form of the G(C) function is almost irrelevant to the method.

Limitations of the MT method

The real limitations of the method lie more with the underlying assumptions. Firstly, the model assumes that the two biopolymers are confined entirely to their respective phases. This is never true in practice, and indeed analysis of the phase separation behaviour would predict that there will be concentrations, albeit low under most conditions, of the second component. Recent work by Durrani and co-workers (Durrani *et al.*, 1993; Durrani and Donald, 1994), reported elsewhere in this volume, has established non-destructive methods for analysing phase compositions. Moreover, micrographs (for example Clark *et al.*, 1983) confirm that each "pure phase" has inclusions of the other, which in turn have sub inclusions of the first, apparently *ad infinitum*. A second assumption is that the volume of polymer is negligible compared to the solvent. On the surface this is reasonable, but the behaviour of polymer plus solvent together makes the definition of polymer separability rather obscure.

The third assumption is, practically, the most limiting. The introduction of a solvent avidity p parameter implies that the solvent can actually repartition after gelation. In practice this (equilibrium) assumption may be impossible to follow and, indeed, the system may be under kinetic control. Finally, as we have already mentioned, it may be that the subsequent uses of absolute upper and lower bounds is quite inappropriate. In view of the real structural complexity of mixed gels (particularly those now being investigated using "real" food ingredients) I suggest that this may be the case.

The gel de-swelling approach

A further technique which may still be useful in this case is that suggested by McEvoy, Ross-Murphy and Clark (1985). This does address the question of equilibrium solvent redistribution. In this work gelatin gels were allowed to dry in air. Denoting the initial modulus and concentration by G° and C° respectively we can write

$$\frac{G^o}{G} = \left(\frac{C^o}{C}\right)^q \qquad (8)$$

Clearly if a real equilibrium existed we would expect, for example following the G ∞ C^2 law discussed previously, that q = 2. However an alternative interpretation is that any loss of solvent gives rise to an increase in modulus governed simply by de-swelling rather than by a change in the number of effective cross-links (the underlying assumption of the C^2 dependence). In this case classical Flory–Rehner de-swelling (Flory, 1953) would predict q = 2/3 . Experimentally, q was found to be approximately 1.7, i.e. between the two values. When one phase gels before the other, say agar in an agar—gelation co-gel, we can generalise eq. (8) to give

$$\frac{G^{nom}}{G} = \left(\frac{C^{nom}}{C^{eff}}\right)^q \qquad (9)$$

where the superscripts ^{nom} and ^{eff} imply nominal and effective concentrations (moduli) respectively. It is important to emphasise that, in principle at least, this de-swelling contribution acts in addition to (rather than instead of) the contribution p. Indeed, the observed minimum in mechanical properties of the gelatin–agar system is well reproduced even if p is set to 1, i.e. there is no preferential solvent affinity factor (Figure 5.5). Indeed we might expect aqueous gelling biopolymers to have

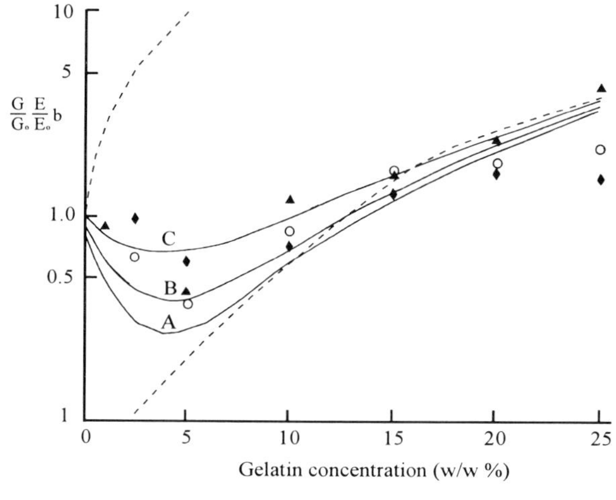

Figure 5.5 Shear modulus (▴), Young's modulus (○) and large deformation strain concentration factor (♦) for mixed gelatin and agar (1%) gels. Moduli are plotted relative to the value for pure agar gels. Solid curves give the isostrain predictions for p = 1 and q = (A) 0, (B) 0.33 and (C) 0.67. Heavy curves give the isostrain (upper) and isostress (lower) predictions for a system with fully labile cross-links. (From McEvoy, Ross-Murphy and Clark, 1985, with kind permission of Butterworth Heinemann Ltd.)

values of the Flory—Huggins χ_{ps} of ≈ 0.5, where ps denotes either a polymer X or Y plus solvent , and then, as stated previously, for χ_{XY} to be ≈ 0.

In the recent work of Kasapis *et al.* (1993) a hybrid approach was adopted, in which the de-swelling and/or the solvent distribution (MT) approach was adopted as appropriate, depending upon the precise mechanism of mixed gel formation. The argument proposed by McEvoy, Ross-Murphy and Clark (1985) suggests that when some co-gels are formed, the first gelling species will form, effectively without interference from the other, at its original nominal concentration. The second species gelling within the confines of the first gel network will tend to create a separate phase, but can only do so by taking solvent from the first gel species. This in turn will indeed tend to de-swell the first network.

The problem is that the experimental q value for the agar–gelatin system is much greater than expected from de-swelling arguments alone (about 5/3 compared to 2/3), suggesting that the real behaviour is far more complex. In their work on gelatin–maltodextrin Kasapis and co-workers (1993) calculated absolute bounds using the Morris (1992) approximation of the MT method. They showed that for low concentrations of the maltodextrin (gelatin-continuous) the data agreed well with the de-swelling method, whereas at higher levels of added maltodextrin there was an apparent transition towards the lower bounds of the MT method. The p value obtained for this system was around 1.8, a value which, as discussed below, could be deduced from the relevant values.

Unpublished work by Clark, cited in Kasapis *et al.* (1993), states that fits to the gelatin—maltodextrin phase diagram are consistent with χ values of 0.46 for gelatin (marginally better than a θ system where χ is 0.5) and 0.55 for starch (consistent with a poor polymer solvent system). This would suggest that gelatin is more solvent (water) loving than starch or starch derivatives, and indeed the cited values of p corresponding to this apparently lie in the range $1.3 < p < 2.2$

Although the agreement of experimental and predicted p values appears to be good, the weakness of this approach may be the assumption that χ is constant. Much work by Köningsveld (1989) and co-workers has shown how χ (in modern parlance g) is molecular weight-, concentration- and MW distribution-dependent. Indeed, even accepting the constancy of thermodynamic parameters, the assumption of thermodynamic equilibrium may be inappropriate.

Extension to large deformation

Using the de-swelling approach, and by employing a constitutive equation of the form

$$\sigma = \frac{2E}{3n}\left(\lambda^{n-1} - \lambda^{-(n+2)/2}\right) \qquad (10)$$

originally due to Blatz, Sharda and Tschoegl (1974), where σ represents the nominal tensile stress, E is Young's modulus and λ is the extension ratio, it was possible to relate small strain shear and tensile modulus, and *failure properties* to composition for agarose–gelatin gels.

Eq. (10) is an elastomer type equation (indeed with the exponent $n = 2$ it reduces to the stress–strain equation for a rubber network) but experimentally n does not depend on concentration. By using this together with calculations of appropriate values of E_x, E_y n_x and n_y, the complete stress–strain profile of a composite gel can be predicted.

Conclusion

Small deformation measurements on mixed biopolymer systems will continue to be of great importance in the next few years. However, it may be the time is right for a greater emphasis on large (and intermediate) deformation regimes. We also suggest that direct determination of phase composition as discussed by Durrani and co-workers (1993, 1994) and mixed gel structure (for example microscopy as discussed in Chapter 12 by Hermansson) should be of increasing emphasis. Although models such as the MT method and the de-swelling model have proved of value, it is altogether too easy to regard these as parametric delights, since there is much flexibility for successful data fitting. In this author's view now is perhaps not the best time for more models, but it is perhaps appropriate to appraise the present state of the art. It is presumably the purpose of this book to furnish just such a service.

Acknowledgements

The author is delighted to acknowledge past (and future) collaborations and discussions with Prof. A.H. Clark (Unilever Research Colworth) and Prof. E.R. Morris (Cranfield University).

References

Blatz, P.J., Sharda, S.C. and Tschoegl, N.W. (1974) Strain energy function for rubber-like materials based upon a generalised measure of strain. *Transactions of the Society of Rheology*, **18**, 145–161

Brownsey, G.J. and Morris, V.J. (1988) Mixed and filled gels – Models for foods. In *Food Structure – Its Creation and Evaluation*, edited by Blanshard, J.M.V. and Mitchell, J.R., London: Butterworths, pp 7–23

Clark, A.H., Richardson, R.K., Ross-Murphy, S.B. and Stubbs, J.M. (1983) Structure and mechanical properties of agar–gelatin co-gels – small-deformation studies. *Macromolecules*, **16**, 1367–1374

Clark, A.H. (1987) The application of network theory to food systems. In *Food Structure and Behaviour*, edited by Blanshard, J.M.V. and Lillford, P.J., London: Academic Press, pp 13–34

Durrani, C.M., Prystupa, D.A., Donald, A.M. and Clark, A.H. (1993) Phase-diagram of mixtures of polymers in aqueous-solution using Fourier-transform infrared-spectroscopy. *Macromolecules*, **26**, 981–987

Durrani, C.M and Donald, A.M. (1994) Fourier-transform infrared microspectroscopy of phase-separated mixed biopolymer gels. *Macromolecules*, **27**, 110–119

Ferry, J.D. (1980). *Viscoelastic Properties of Polymers*, 3rd edition, New York: John Wiley

Flory, P.J. (1953). *Principles of Polymer Chemistry*, Ithaca, New York: Cornell University Press

Kasapis, S.A., Morris, E.R., Norton, I.T. and Clark, A.H. (1993) Phase equilibria and gelation in gelatin/maltodextrin systems. Part IV: Composition-dependence of mixed gel moduli. *Carbohydrate Polymers*, **21**, 269–276

Köningsveld, R. (1989) Thermodynamics of macromolecular systems. *Pure and Applied Chemistry*, **61**, 1051–1064

Langley, K.R. and Green, M.L. (1989) Compression strength and fracture properties of model particulate food composites in relation to their microstructure and particle–matrix interaction. *Journal of Texture Studies*, **20**, 191–207

McEvoy, H., Ross-Murphy, S.B. and Clark, A.H. (1985) Large deformation and ultimate properties of biopolymer gels. II: Mixed gel systems. *Polymer*, **26**, 1493–1500

Manson, J.A. and Sperling, L.H. (1976). *Polymer Blends and Composites*, London: Heyden

Morris, E.R. (1992) The effect of solvent partition on the mechanical properties of biphasic biopolymer gels: an approximate theoretical treatment. *Carbohydrate Polymers*, **17**, 65–70

Nielsen, L.E. (1974). *Mechanical Properties of Polymers and Composites*, New York: Marcel Dekker

Richardson, R.K., Robinson, G., Ross-Murphy, S.B. and Todd, S. (1981) Mechanical spectroscopy of filled gelatin gels. *Polymer Bulletin*, **4**, 541–548

Ross-Murphy, S.B. and Todd, S. (1983) Ultimate tensile measurements of filled gelatin gels. *Polymer*, **24**, 481-486

Ross-Murphy, S.B. (1984) Rheological methods. In *Biophysical Methods in Food Research* (*Critical Reports in Applied Chemistry*, Vol. 5), edited by Chan, H.W.-S., London: Society of Chemical Industry, pp 138–199

Ross-Murphy, S.B. (1994) Rheological methods. In *Physical Methods for the Study of Food Biopolymers*, edited by Ross-Murphy, S.B., Glasgow: Blackie Academic and Professional, pp 343–392

Theocaris, P.S. (1990) The exact shear modulus of particulates based on the concept of mesophase. *Colloid and Polymer Science*, **268**, 1118–1130

Turquois, T., Doublier, J.L., Taravel, F.R. and Rochas, C. (1994) Synergy of the -carrageenan galactomannan blend inferred from rheological studies. *International Journal of Biological Macromolecules*, **16**, 105–107

Walkenström, P. and Hermansson, A.-M. (1994) Mixed gels of fine-stranded and particulate networks of gelatin and whey proteins. *Food Hydrocolloids*, **8**, 589–607

6

PHYSICAL METHODS TO STUDY PHASE SEPARATION IN PROTEIN–POLYSACCHARIDE MIXTURES

A.M. DONALD, C.M. DURRANI, R.A.L. JONES, A.R. RENNIE and R.H. TROMP
Polymers and Colloids Group, Cavendish Laboratory, University of Cambridge, Madingley Road, Cambridge. CB3 0HE U.K.

Introduction

Many food systems include as major components complex mixtures of biopolymers in solution in water. There is a direct link between the morphology of such mixtures and their macroscopic and rheological properties, which, in turn, underlie the subjective appreciation of texture by the consumer. Thus the phase behaviour of such mixtures has understandably aroused much interest. However the equilibrium thermodynamics of such mixtures, and in particular their phase diagrams, while of central importance, are not the whole story. Observed morphologies are almost never equilibrium morphologies; mechanisms of transport in systems containing polymers are very slow and usually insufficient time is available for equilibrium to be reached during processing. Thus in order to understand observed morphologies, and to be able to design process routes to control structure, one needs to understand both the equilibrium phase diagram of the mixture and the kinetics of the phase separation process. In many cases of practical importance transport mechanisms actually get slower as phase separation proceeds, often seizing up completely. This may occur when one phase becomes glassy, or when one or both components can form a physical gel. Thus in these cases the observed morphology is the result of a competition between phase separation processes and these processes of kinetic arrest.

A simple example of such biopolymer mixtures comprises two polymer species in a single solvent, water. Usually water is a good solvent for both biopolymers, and in such a situation the water serves mainly to dilute unfavourable interactions between the two biopolymers. This can lead to phase separation for some ranges of temperature and concentration. Such systems offer the hope of a relatively simple starting point for the understanding of phase separation kinetics, in that one might hope that straightforward extensions of the ideas applied to binary systems such as metallic alloys and polymer blends might be applicable (see Binder (1991) and Hashimoto (1993) for recent reviews). However one must be aware that even in these supposedly simple systems our understanding of phase separation mechanisms is far from complete, and reliable theoretical results are rare. Biopolymer systems exhibit a number of serious complications and it is still too early to say how important these complications are likely to be in practice. First, biopolymer mixtures are never true ternary mixtures; the biopolymers themselves are usually polydisperse, and other components such as lipids, surfactants and inorganic ions will usually be present. Secondly, in addition to phase separation other slow time-dependent processes may well be active. Perhaps the most important of these processes is, as mentioned above, physical gelation, which can be expected to have a profound effect both on the thermodynamics and the kinetics of phase separation. This complication may in fact offer great potential for the control of texture, and for this reason this aspect forms a central part of the current study. It is also worth noting that such systems may have other technological applications, for example as membranes for separation processes (Petrak, 1984).

In this work we report results on the phase behaviour of blends of two polysaccharides with gelatin. We have studied both the equilibrium phase diagrams of the mixtures and the kinetics of the phase separation process. Our method of mapping out phase diagrams, based on Fourier transform infra-red spectrometry (Durrani, Prystupa, Donald and Clark, 1993), allows us to determine not only the partitioning of water between the phases, but also the concentrations of the two biopolymers in the coexisting phases; it is usually a bad approximation to assume that the coexisting phases are in fact solutions of a single biopolymer. In order to follow the kinetics of phase separation we use time-resolved small angle light scattering; by varying the temperature at which phase separation takes place we are able to observe the kinetics of phase separation both in the presence and the absence of gelation. In addition to the statistical information about the decomposition pattern obtained from light scattering, we have used phase contrast microscopy to make direct observations of the phase separated morphologies.

Background

The equilibrium thermodynamics of any mixture is fully determined by the free energy of mixing as a function of composition and temperature. From this free energy function the phase diagram can be derived. Here we will briefly review the features of the phase diagram for a two component mixture (Figure 6.1), before going on to discuss ternary solutions. Points defined by a double tangent construction on the free energy diagram define the limits of coexistence at a given temperature; the locus of such points is the coexistence curve or binodal. Concentrations within this curve can lower their overall free energy by phase separating. Thus points within this curve are unstable in a global sense; however they may not necessarily be unstable to small local fluctuations in concentration. This local stability limit is defined by the points of inflection of the free energy curve; that is, the condition for local instability is that the second derivative of free energy with respect to concentration is negative. The set of points defining this stability limit is called the spinodal. The spinodal and coexistence curve meet at a single point, the critical point, which defines the extreme temperature at which phase separation takes place. For ternary solutions the phase diagram is three-dimensional, with the coexistence and stability limits defined by surfaces.

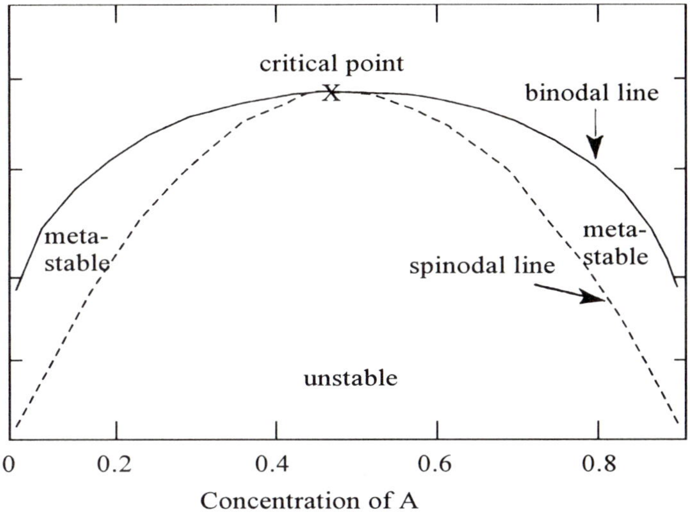

Figure 6.1 Schematic phase diagram for a two component mixture, showing the coexistence curve, the spinodal line and the critical point

When one takes a sample which is in the one-phase region of the phase diagram at a particular temperature, and cools it down into the two-phase region, one observes that at some temperature the initially clear solution becomes cloudy due to the onset of phase separation, manifested by the appearance of concentration inhomogeneities bigger than the wavelength of light. Making such observations for a variety of different starting compositions allows one to construct a cloud-point curve. This curve will fall somewhere between the coexistence curve and the spinodal, its precise location depending on the details of the way the experiment was carried out, and in particular the cooling rate used.

Let us now turn to a more detailed consideration of the mechanisms and kinetics of phase separation. A crucial distinction is between those parts of the phase diagram within the spinodal, and those parts between the coexistence curve and the spinodal line. In the second case, a small concentration fluctuation does not lower the system's free energy, and will normally die away again, because the saving of free energy caused by having a different composition to the bulk composition is outweighed by the free energy cost associated with the interface created. This means that a droplet of a certain size has to be nucleated before it can grow. This mechanism of phase separation is thus known as homogeneous nucleation and growth, and is characterised by Arrhenius kinetics. However the activation barrier may be hugely reduced or even eliminated by the presence of interfaces with third phase particles (heterogeneous nucleation). Such "seed" particles are almost always likely to be present; thus in practical systems homogeneous nucleation is not likely to be important.

When a system is rapidly cooled into the unstable part of the phase diagram, within the spinodal line, any small fluctuations in concentration of the type that inevitably exist due to thermal fluctuations will grow rather than shrink (Cahn, 1965). Formally, the mutual diffusion coefficient is negative, so material is transported from regions of low concentration to high concentration. However, not all concentration fluctuations grow at the same rate; a strong size selection effect is at work which means that one particular size of fluctuations grows the fastest. This size selection mechanism is explained by the linear theory due to Cahn and Hilliard. The physical mechanism is simple to explain; short length-scale fluctuations have associated with them an energy penalty associated with the steep concentration gradients they imply, and are suppressed. Long length-scale concentration fluctuations, on the other hand, grow slowly, simply because material has to be transported longer distances. Thus there exists one length scale which grows the fastest, and therefore dominates the phase separation morphology at very early times. The characteristic phase separation pattern one thus expects to see is a random structure with a given length. If one does a scattering experiment the characteristic length scale gives rise to a broad scattering peak at a wave-vector q_{max} corresponding to the fastest growing length scale.

According to the very simplest theory of spinodal decomposition one expects one single length scale to dominate the morphology, and for concentration fluctuations at this length scale to grow exponentially. Thus in a scattering pattern one would expect to see a broad peak grow in intensity without the peak position moving. In fact, what one typically sees is that the peak moves in to lower wave-vectors — corresponding to a larger length scale dominating the morphology — as time goes on. The reason for this can be mathematically stated as being due to those non-linear terms in the Cahn–Hilliard equation that are linearised in the simple theory. In fact, this is illustrated in Figure 6.2; the size of the concentration fluctuations cannot keep on growing indefinitely, because at some point they actually reach the coexisting compositions. The system is still not at equilibrium, as there are a lot of interfaces present. After a complicated "intermediate" regime, the system then continues to evolve, driven by the tendency to reduce the total surface energy. This is the so-called late stage of phase separation.

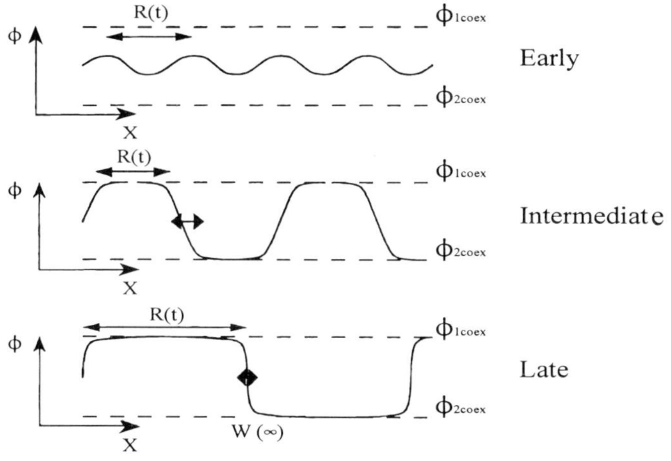

Figure 6.2 The different stages of phase separation by spinodal decomposition. In the early stage, the characteristic length scale R(t) is constant and the amplitude changes. In the intermediate stage, R(t) is growing, while the interface width w(t) between the incipient domains is getting smaller. Finally, in the late stage, the interface width has achieved its equilibrium value w(∞), while the characteristic size of the domains R(t) is still increasing

Recently attention has focused on the possibility that this late stage is characterised by a dynamic scaling property (Binder and Stauffer, 1974). The argument is that because there is now only one relevant length scale in the problem (the interface widths having achieved their equilibrium values) the morphology at any later time is simply (in a statistical sense) equivalent to a magnified version of the morphology at an earlier time. Thus one reduces the problem of characterising the morphology at any time to two simpler problems; that of finding the single time-independent scaled morphology, and that of finding the growth law of the single length scale that rescales the universal morphology. Such dynamic scaling has been observed in computer simulations and in some experiments, including experiments on alloys (Komura, Takeda, Osamura and Okuda, 1985), small molecule liquids (Chou and Goldburg, 1981) and binary polymer mixtures (Bates and Wiltzius, 1989).

In order for the morphology to coarsen, material has to be transported through the system; the type of growth law one observes will depend on the mechanisms of transport which are active. Such mechanisms may include simple molecular diffusion, which should lead to coarsening as the cube root of time — the Lifshitz–Slyozov law (Lifshitz and Slyozov, 1961) — or hydrodynamic flows, which are expected to lead to much faster growth, possibly linear with time (Siggia, 1979). Different mechanisms are expected to dominate at different stages of the coarsening process, so in practice one may see different regimes of power-law behaviour, or no clear power-law regimes at all.

A key assumption of the dynamic scaling hypothesis is that no qualitative changes in morphology take place during late stage coarsening. In fact one important change is sometimes observed (Hasegawa, Shiwaku, Nakai and Hashimoto, 1988); a change from an interconnected morphology to one featuring discrete droplets of the minority phase in the majority phase. In elementary discussions of morphologies of phase separated systems the assertion is often made that droplet morphologies result from a nucleation and growth mechanism, while interconnected morphologies result from a spinodal mechanism. This is not in general true; very often the transition to a droplet morphology occurs late in the phase separation process and thus the presence of this morphology at late times offers no clue to the mechanism of phase separation at early times.

The aim of this paper is to elucidate both the nature of the equilibrium phase separation, and the kinetics with which this is effected. Two different physical techniques are used to accomplish this aim: Fourier transform infra-red spectrometry is used to analyse the phase diagram, and time-resolved small angle light scattering to follow the kinetics.

Experimental

SAMPLES

Commercial food-grade gelatin and amylopectin samples were obtained from Unilever and D_2O (99.9% isotopic purity) was bought from Aldrich Chemical Company, Dorset. Two fractions of amylopectin of different molecular weight were used to plot two separate gelatin–amylopectin–D_2O phase diagrams. The amylopectin fractions were prepared by pyrolysis of waxy maize starch under dry acid conditions for different periods of time. Both fractions dissolved easily in water at temperatures above ~70°C. The biopolymers were used without further treatment.

For the determination of the phase diagram at 51°C using Fourier transform infra-red spectroscopy (FTIR), samples were made by weighing out the appropriate masses of amylopectin, gelatin and D_2O into a sealed bottle. D_2O is used in preference to H_2O so that the amide bands arising from the gelatin are not obscured by bands due to the OH stretch from the water. All samples were calculated on a weight percentage basis. Samples were left overnight at room temperature to allow the gelatin to swell, and were then heated to 85°C for 10 minutes while being mixed with a magnetic stirring bar.

For the light scattering experiments we used mixtures of gelatin with dextran rather than amylopectin; dextran, unlike amylopectin, does not crystallise or gel, allowing us to study phase separation in the absence of any gelling component. Dextran (molecular weight 160,000) and gelatin (molecular weight 200,000) were purchased from Sigma, and were dissolved in a 0.5 M solution of NaCl, to rule out complexation of the dextran and gelatin molecules, and also to bring the phase separation temperature of a wider range of solute compositions within the temperature window between the gelation temperature of the gelatin and the boiling point of the solvent.

PHASE DIAGRAM DETERMINATION

The method for determining the equilibrium phase diagram using FTIR relies on an algorithm for analysing the spectra known as "partial least squares" (PLS). The PLS method is based on taking the spectra of a series of standards of known concentration to form a matrix of calibration spectra — whose elements are the infra-red absorbance at each experimentally sampled frequency for each spectrum — and then using this matrix to predict the concentrations in samples of unknown concentration. Full details of this approach for these systems are given elsewhere (Durrani *et al.*, 1993).

Phase separation was allowed to proceed to equilibrium at 51°C for each composition, so that bulk phase separation occurred. Samples of the top and bottom phases so formed were extracted and injected into Specac semi-permanent liquid cells which were made up of two 4 mm CaF_2 windows separated by a 12.5 μm tin spacer. Cells were pre-heated to 51°C before injecting the samples. Infra-red spectra were recorded with a Mattson 4020 FTIR spectrometer equipped with a deuterated triglycine sulphate (DTGS) detector. The spectrometer was well purged with dry nitrogen gas. For each sample, 100 interferograms recorded at 4 cm^{-1} resolution were co-added and Fourier-transformed using Beer–Norton strong apodization.

To compensate for differences in sample thickness between samples, the spectral data of both the standards and the unknowns were normalized by setting the lowest absorption in the range 980 cm^{-1} to 2000 cm^{-1} to zero and then scaling the D_2O peak at 1120 cm^{-1} to an absorption of unity. The PLS algorithm was then used to predict the ratios of gelatin to D_2O and amylopectin to D_2O. We note that in this way the D_2O, together with the CaF_2 windows and any HDO and H_2O, were modelled implicitly as the background. The concentrations of amylopectin, gelatin and D_2O were then calculated from the two ratios and the conservation of mass.

REAL TIME LIGHT SCATTERING TO STUDY KINETICS

To study the phase separation process in real time we use time-resolved small angle light scattering (Figure 6.3). A helium–neon laser beam is led through a pinhole, lens and diaphragm to a mirror which reflects it through the temperature-controlled sample cell. The resulting scattering pattern is imaged onto the light sensitive array of a charge-coupled device (CCD) camera; the camera can capture all light scattered by an angle of less than 17°. A beam stop protects the CCD from the intense, unscattered, beam; the size of this beam stop determines the lower limit of wave-vectors that can be observed. In practice, for the instrument configuration we use, reliable data can be obtained over a range of wave-vectors from 0.2×10^{-4} Å^{-1} to 2×10^{-4} Å^{-1}, corresponding to distance scales from about 6 μm to 60 μm.

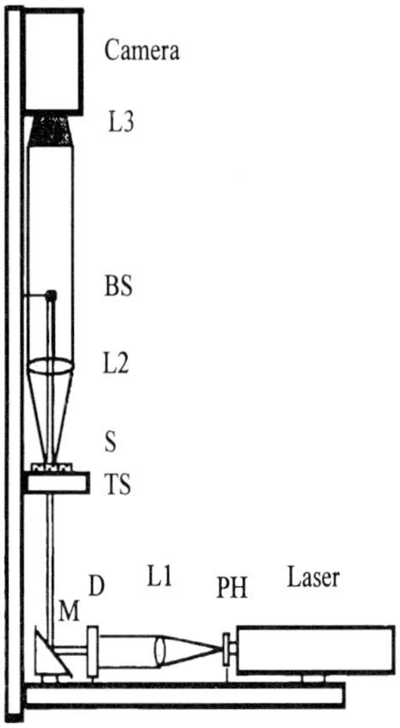

Figure 6.3 The small angle light scattering apparatus used to study the kinetics of phase separation. A He–Ne laser beam passes through the pinhole (PH), lens (L1) and diaphragm (D) before being reflected by a mirror (M) to pass through the temperature controlled sample stage (TS) in which the sample crucible (S) is mounted. The scattering pattern is imaged by lenses (L2 and L3) onto a CCD camera, with the unscattered light being removed by a beamstop (BS)

In order to carry out an experiment, hot sample solution is pipetted into a 1 mm deep sapphire crucible which is preheated on the sample stage. After about 20 minutes to equilibrate, the sample is quenched to the destination temperature. The sample stabilises at this temperature after about 40 seconds; this lag sets an upper limit on the speed of decomposition processes that we can study.

The map of intensity on the CCD camera is isotropic, and can thus be radially averaged to give a curve of intensity versus scattering vector. The scattering vector itself is calibrated by measuring the diffraction pattern from a pinhole of known diameter. A typical result for a two-dimensional scattering map is shown in Figure 6.4. This shows the typical "spinodal ring" expected for the spinodal decomposition, as discussed in detail below.

Complementary direct microscopic observations were carried out using a Zeiss Axioplan microscope in phase contrast mode, using the same sample container and temperature controlled stage that were used for the scattering experiments.

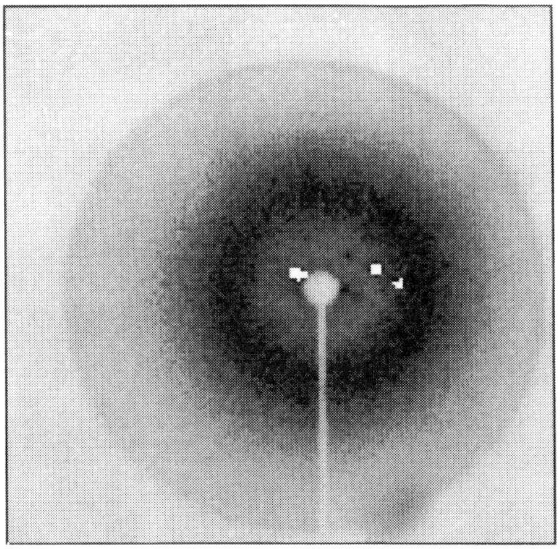

Figure 6.4 A typical intensity map of the scattering pattern produced by a phase separating sample. Note that the scattering is isotropic, and that the scattering intensity is peaked to form a ring around the straight through beam position

Results and discussion

The phase diagrams of mixed systems containing amylopectin and gelatin in D_2O at 51°C are given in Figures 6.5 and 6.6. Figure 6.5 was made with the higher molecular weight amylopectin sample and Figure 6.6 with the lower molecular weight amylopectin. In both mixed systems phase separation into a gelatin-rich upper phase and an amylopectin-rich lower phase took place. The points on the curves represent the weight concentrations in the top and bottom phases as determined by our PLS method. Both phase diagrams also include the composition of the starting samples. The tie-line drawn through each set of three compositions (representing the starting composition, the top phase and bottom phase) is calculated by a least-squares best fit to the three data points.

From the shape of the curve separating the one- and two-phase regions in Figures 6.5 and 6.6, it can be seen that this method determines a phase boundary essentially similar to phase diagrams determined by standard techniques for other systems. A clear indication that the method is very successful is the fact that the best fit straight line through each set of three points passes closely through each set; a straight tie-line connecting the points is expected by the lever rule.

Both phase diagrams are asymmetric and are skewed towards the branched amylopectin component. The effect of a decrease in molecular weight of the amylopectin is to shift the apex of the curve towards the amylopectin corner of the phase diagram and to increase the area of the one-phase region on the phase diagram. The gelatin-rich top phase is capable of containing more of the lower molecular

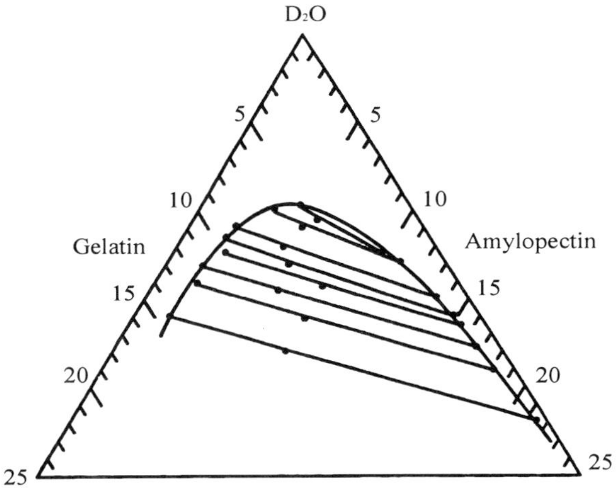

Figure 6.5 Phase diagram of the high molecular weight amylopectin–gelatin mixture, as determined by the FTIR method

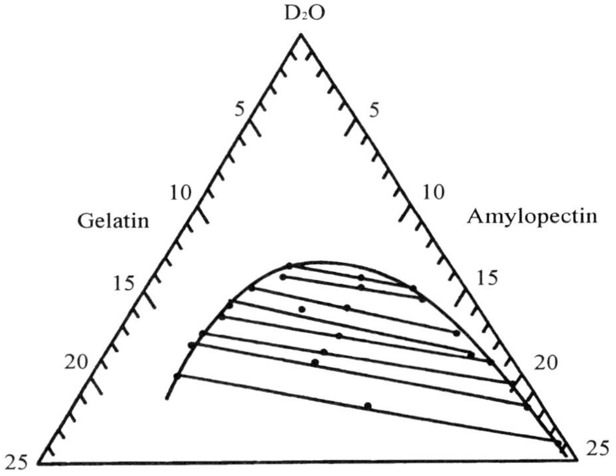

Figure 6.6 Phase diagram of the low molecular weight amylopectin–gelatin mixture, as determined by the FTIR method

weight amylopectin than it can of the higher molecular weight amylopectin, which is to be expected on thermodynamic grounds. The lower amylopectin-rich phase appears to be similar in both phase diagrams in that it is capable of supporting less of the other polymer than the top phase can. However it is clear that complete demixing has not occurred in either phase, so that each phase contains both polymers.

On the phase diagram containing the lower molecular weight amylopectin (Figure 6.6), the tie-lines are approximately parallel to one another with a root-mean-square deviation about the mean slope of the lines of only 1.2°. However on the phase diagram containing the higher molecular weight amylopectin, the slopes of the tie-lines increase as the concentrations of the end-points of the tie-lines move up the phase boundary, such that the lines appear to "rotate". This effect has been observed in other protein–polysaccharide systems (Tolstoguzov, 1986).

The quality of these results permits an analysis of them by the Flory–Huggins model of polymer solutions. This analysis, described in more detail for this system elsewhere (Durrani *et al.*, 1993), performs tie-line calculations based on the iterative minimisation procedure described previously by Hsu and Prausnitz (1974), using an experimental estimate for the molecular weight of the amylopectin and then allowing the χ values and the gelatin molecular weight to vary. The fit for the system containing the low molecular weight amylopectin produces χ values for amylopectin and gelatin in water (0.51 and 0.46 respectively) which agree well with literature values, and a χ_{AG} between the two polymers of 0.082. In order to fit the data of the phase diagram containing the higher molecular weight amylopectin, apart from the need to use the different experimental amylopectin molecular weight estimate, only χ_{AG} needed to be altered, with a best fit value of 0.059. That this should be different in the two cases is not surprising, given the crudity of the Flory–Huggins theory, particularly in its treatment of branched polymers such as amylopectin, which, in a further complication, varies with molecular weight in degree of branching.

In discussing the kinetics of phase separation, let us first describe the situation in which the biopolymer mixture is quenched to a temperature below the phase separation temperature, but above the gelation temperature. This simpler situation may be expected to be more directly comparable to other studies on polymer mixtures. Figure 6.7 shows a number of scattering curves at various times after the quench, obtained by radially averaging intensity maps such as that shown as Figure 6.4. What can be seen clearly is the characteristic broad peak which results from the spinodal decomposition mechanism, which grows in intensity but moves to lower wave-vectors as time goes on. This indicates that the situation cannot be accurately described by the simple linear theory (this is confirmed by more extensive data analysis (Tromp, Rennie and Jones)); non-linear effects are always important at the times we can probe experimentally. It may be that there is some time up to which the linear theory works, but in this system the contrast is too weak and the wave-vector range too restricted to observe this, even if we were able to quench at a faster rate.

We can investigate the validity of the dynamic scaling hypothesis by attempting to superimpose the normalised scattering functions (Figure 6.8). The wave-vector is divided by the peak wave-vector, and the scattered intensity, multiplied by the

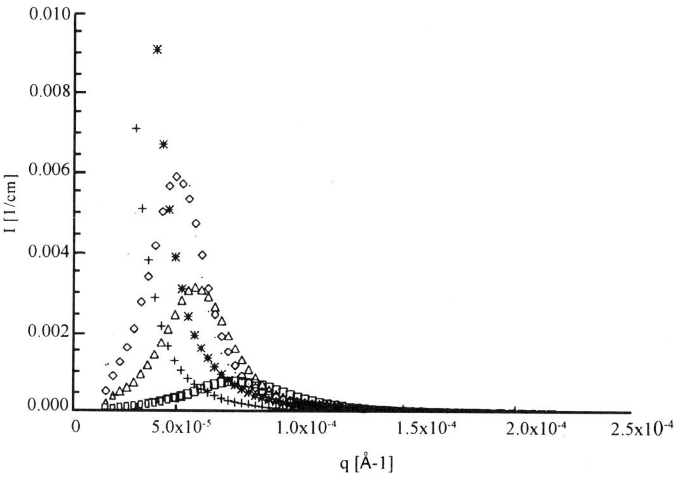

Figure 6.7 Light scattering pattern after a quench to 30°C of a solution of 4.2% dextran and 4.2% gelatin (phase transition at 38°C, gelation temperature 25°C). □ : 314 s, ▵ : 470 s, ◇ : 628 s, ● : 786 s, * : 1268 s, + : 4018 s

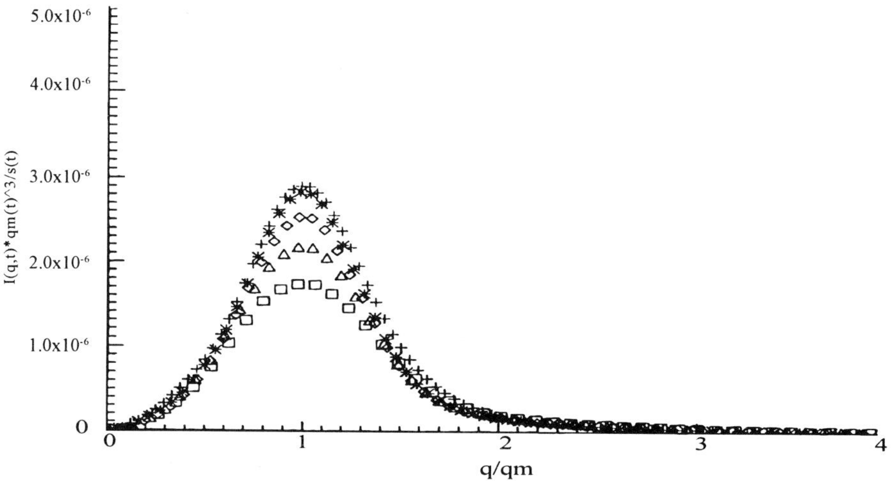

Figure 6.8 Scaled scattering curves for a sample quenched to 30°C. + : 314 s, * : 470 s, ◇ : 628 s, ▵ : 786 s, □ : 952 s

cube of the peak wave-vector, is divided by the time dependent scattering power s(t), which accounts for the change in refractive index difference between the phases as their composition changes. Even though s(t) is still changing, indicating that the phase compositions are not yet constant, the phase separation pattern is indeed self-similar in time until about 700 seconds after the quench, after which the shape of the scattering curves starts to change qualitatively. A similar transition appears if we look at the variation of the peak wave-vector with time (Figure 6.9); up to about 700 seconds the peak wave-vector decreases with a power law with an exponent of about -0.6, whereas after that time the rate of coarsening speeds up, with an apparent power of about -1.2. A clue to the origin of this transition is obtained if we examine light micrographs of the phase separating material; at about 700 seconds we see a transition from the type of morphology shown in Figure 6.10(a), with a complex, interconnected phase separation pattern, to the morphology shown in Figure 6.10(b), with discrete spherical particles of the gelatin-rich phase in a continuous dextran-rich matrix. What is the origin of the two growth laws? We cannot answer this with certainty. However, it is clear that the Ostwald ripening mechanism in which material transport is entirely diffusive, with a predicted exponent of -1/3, cannot account for the observed ripening behaviour; bulk, hydrodynamic flow of material must also be taking place to account for these larger exponents.

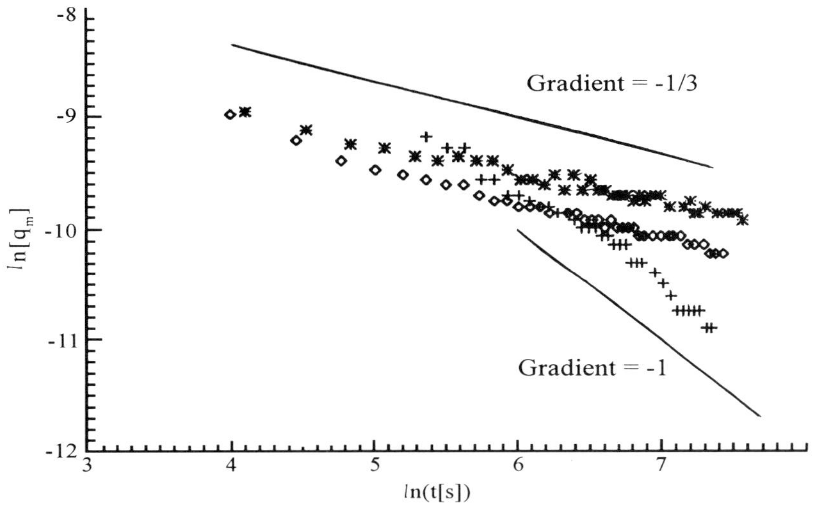

Figure 6.9 Time dependence of the position q_m of the scattering maximum at 30°C (+), 25°C (◇) and 21°C (*)

The kinetics when the sample is quenched to below the gelation temperature are quite different. Figure 6.11 shows the scattering curves as a function of time for a quench to 23°C. Initially the patterns are very similar to those obtained for the quench to 28°C; a broad peak appears indicative of spinodal decomposition. The peak appears at a larger value of the wave-vector; this can be understood qualitatively in terms of the linear theory, in that we expect the dominant length scale of spinodal decomposition to be smaller for a deeper quench. However even at the earliest

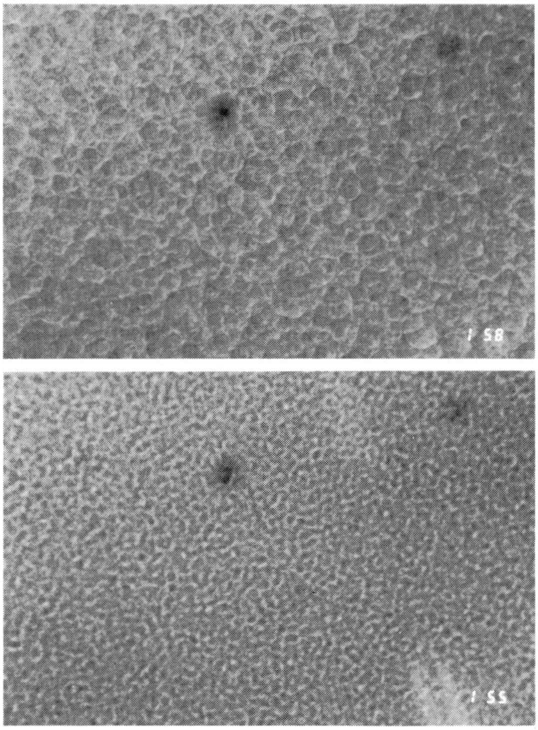

Figure 6.10 Phase contrast microscope images of the phase separating gelatin—dextran solution at 30°C at two times after the quench: (a) 4 minutes; (b) 24 minutes. The short side of the photographs is 450 μm

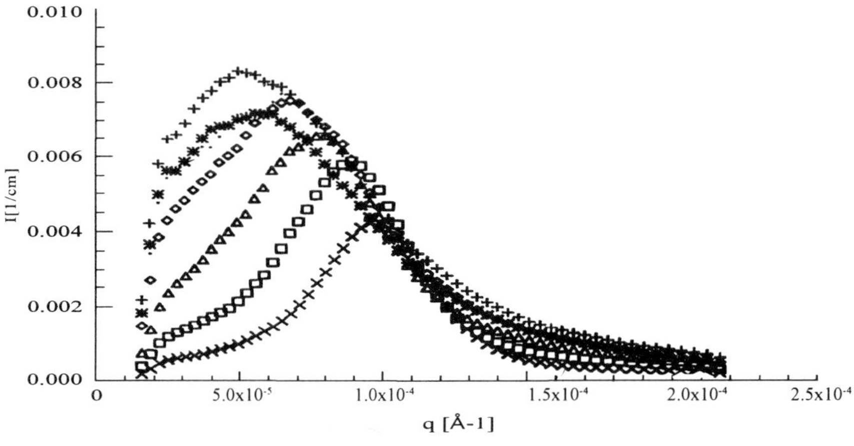

Figure 6.11 Light scattering pattern after a quench to 21°C of a solution of 4.2% dextran and 4.2% gelatin. ×:126 s, □ : 198 s, △ : 378 s, ◇ : 560 s, ● : 744 s, *:1160 s, +:1382 s

times the shape of the curve is qualitatively changing, indicating the failure of dynamical scaling. The rate of coarsening, too, is quite different to the non-gelling case; Figure 6.9 shows that the position of the peak wave-vector accurately obeys a power law of -0.3 over the whole time that a peak can be unambiguously defined. Thus we can conclude that the hydrodynamic flows that allowed the pattern to coarsen so quickly at late times in the non-gelling case are completely suppressed by gelation and material can only move by diffusion of individual chains.

At this stage the interpretation of the change of shape of the scattering curves can only be tentative. Such behaviour could be accounted for by the initiation at later times of further phase separation in pre-existing domains. This could be because as the gelation proceeds the phase boundaries move; as the effective average molecular weight increases the system in effect becomes more immiscible, and the phase-separated domains that were originally at one of the coexisting compositions now find themselves within the phase boundary and therefore susceptible to further phase separation. A further factor arises from the network elasticity of the gelled domains; this elasticity will have the effect of suppressing long wavelength concentration fluctuations (Binder and Frisch, 1984), promoting what is in effect microphase separation rather like that observed in interpenetrating networks. It is anticipated that FTIR microscopy experiments, currently in progress, will shed light on exactly how the composition of phase separated domains changes with time for samples quenched below the gelation temperature.

A tentative summary of the course of phase separation in the two cases is illustrated in Figure 6.12. The major differences between the cases with and without gelation are as follows. With gelation one observes no changeover to a droplet morphology — the system appears to remain interconnected at all times. The coarsening process is slower with gelation than without, and appears to proceed entirely by diffusion — a $t^{1/3}$ growth law is obeyed well at all times. In contrast, without gelation there is no simple power law behaviour, and we conclude that hydrodynamic transport must be important. Finally, in the presence of gelation, phase separation results in a structure with a very wide range of length scales present, whereas without gelation at very late stages the distribution of droplet sizes remains relatively narrow.

Conclusion and outlook

This paper has described how two different physical techniques can be used to build up a picture both of the nature of the equilibrium phase separation in an aqueous biopolymer mixture, and the way in which this separation occurs above and below the gelation temperature. Fourier transform infra-red spectrometry has been used to determine the compositions of the two bulk phases into which a

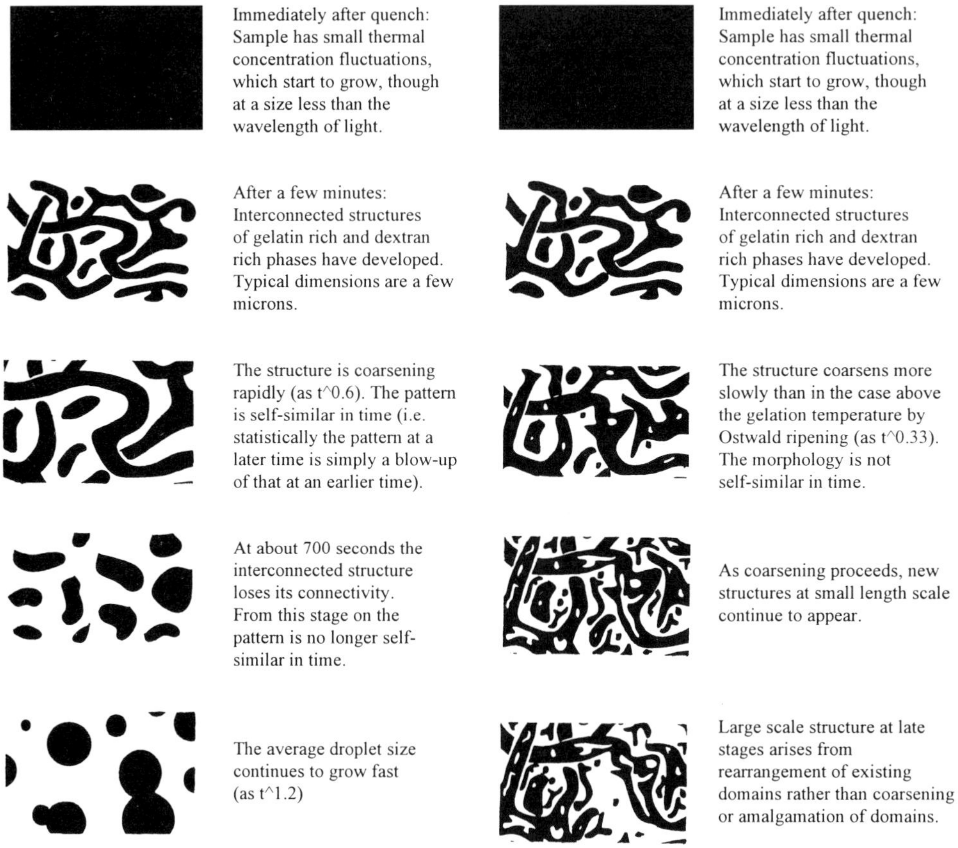

Figure 6.12 Schematic sketches showing the hypothesised stages of phase separation in the case where the quench temperature is above the gelation temperature (a); and below the gelation temperature (b)

gelatin–amylopectin mixture splits above the gelation temperature. This analysis can be carried out without making the assumption that each phase only contains one of the polymers. It is found that as the molecular weight of the amylopectin content is increased, the amylopectin-rich phase can hold less of the gelatin. The nature of the phase separation can be analysed within the framework of the Flory–Huggins theory, although this theory does not explicitly allow for branching of the molecules.

The FTIR analysis described here applies to systems which have bulk phase separated. By combining this method based on "partial least squares" with the use of an FTIR microscope, it is possible to analyse regions of samples of the order of 40 μm across. This therefore allows compositional analysis to be carried out at a much earlier stage of the phase separation. Current work is directed towards mapping out local compositions in both amylopectin–gelatin and dextran–gelatin systems with the use of a motorised stage.

In order to probe the phase separation at earlier times when the domains are even smaller, the technique of time-resolved small angle light scattering is particularly suitable. This has allowed us to characterise the growth laws of the characteristic size of the phase separated structure, as well as to characterise the morphology statistically and to test the dynamic scaling hypothesis. The characterisation of the morphology is complemented by real space studies using phase contrast light microscopy.

The achievements of our work include the development of an accurate way of determining phase diagrams and thus obtaining thermodynamic information, as well as the development of techniques for the quantitative study of the kinetics. What is just as important as these instrumental developments is the realisation that, despite the complications inherent in biopolymer systems, the framework for understanding their phase separation behaviour is exactly the same as has been developed for synthetic polymer mixtures, small molecule liquids, and alloys. There is, however, still much work to be done. First, using the phase diagram information, thermodynamic models need to be developed until they have real predictive power. Secondly, some issues in the phase separation kinetics that are poorly understood even for simpler systems need to be addressed. These include the range of validity of the dynamic scaling hypothesis, the nature of the apparent percolating—non-percolating morphology transition, and the origin of the growth laws. Thirdly, there needs to be more theoretical and experimental work developing a theory of phase separation in the presence of gelation, perhaps incorporating computer simulation. This work needs to include consideration of situations in which one has an interplay between phase separation and aggregation. Finally, systematic work is needed to relate this growing understanding of the origins of morphology to macroscopic properties such as rheology.

Acknowledgements

Some of the work described here was carried out by C.M. Durrani during an A.F.R.C. cooperative studentship with Unilever plc, and the rest was supported by A.F.R.C. and B.B.S.R.C. We are grateful to these funding agencies for their support, and to Dr. A. Clark for many useful discussions.

References

Bates, F.S. and Wiltzius, P. (1989) Spinodal decomposition of a symmetric critical mixture of deuterated and protonated polymer. *Journal of Chemical Physics*, **91**, 3258–3274

Binder, K. and Stauffer, D. (1974) Theory for the slowing down of the relaxation and spinodal decomposition of binary mixtures. *Physical Review Letters*, **33**, 1006–1009

Binder, K. and Frisch, H.L. (1984) Phase stability of weakly cross-linked interpenetrating polymer networks. *Journal of Chemical Physics*, **81**, 2126–2136

Binder, K. (1991). In *Phase Transformations in Materials*, edited by Haasen, P., Weinheim: VCH, pp 405–472

Cahn, J.W. (1965) Phase separation by spinodal decomposition in isotropic systems. *Journal of Chemical Physics*, **42**, 93–99

Chou, Y.C. and Goldburg, W.I. (1981) Angular distribution of light scattered from critically quenched liquid mixtures. *Physical Review A*, **23**, 858–864

Durrani, C.M., Prystupa, D.A., Donald, A.M. and Clark, A.H. (1993) Phase-diagram of mixtures of polymers in aqueous solution using Fourier-transform infrared-spectroscopy. *Macromolecules*, 981–987

Hasegawa, H., Shiwaku, T., Nakai, A. and Hashimoto, T. (1988). In *Dynamics of Ordering Processes in Condensed Matter*, edited by Komura, S. and Furukawa, H., New York: Plenum, pp 457–462

Hashimoto, T. (1993) In *Structure and Properties of Polymers*, edited by E.L. Thomas, E.L., Weinheim: VCH, pp 252–300

Hsu, C.C. and Prausnitz, J.M. (1974). *Macromolecules*, **7**, 320–324

Komura, S., Takeda, T., Osamura, K. and Okuda, K. (1985) Time evolution of the structure function of quenched Al–Zn and Al–Zn–Mg alloys. *Physical Review B*, **31**, 1278–1301

Lifshitz, I.M. and Slyozov, V.V. (1961). *Journal of Physics and Chemistry of Solids*, **19**, 35–50

Petrak, K.L. (1984) The effect of phase separation on the structure and permeability of gelatin/dextran films. *Journal of Applied Polymer Science*, **29**, 555–565

Siggia, E. (1979) Late stages of spinodal decomposition in binary mixtures. *Physical Review A*, **20**, 595–605

Tolstoguzov, V.B. (1986). In *Functional Properties of Food Macromolecules*, edited by Mitchell, J.R. and Ledward, D.A., London: Elsevier, pp 385–415

Tromp, R.H., Rennie, A.R. and Jones, R.A.L. (1995) *Macromolecules*, **28**, 4129–4138

7

THE USE OF NMR IN CHARACTERIZING BIOPOLYMER MIXTURES

M.A.K. WILLIAMS, R.D. KEENAN and T.K. HALSTEAD
Department of Chemistry, University of York, York. YO1 5DD U.K.

Introduction

A full understanding of the properties of mixed biopolymer systems will require detailed information about their composition, structure, dynamics, and morphology at several distance scales, ranging from the molecular to the macroscopic. At our current level of understanding even basic information, such as the composition, size and topology of the various phases, is lacking for many mixed biopolymer systems, and this information may be difficult to obtain. In this context, the term "phase" should be used with caution and should not be taken to imply that the system is in a state of thermal equilibrium. Over the last 15 years, a number of solutions and gels containing a single biopolymer component, e.g. a protein or polysaccharide, have been studied by NMR, and applications of NMR to food science have been reviewed (Belton, Colquhoun and Hills, 1992). On the other hand, mixed systems, containing more than one biopolymer component, have received little attention, partly because interpretation of the results is complex.

This review first considers a model that has been developed to describe water proton relaxation in an aqueous system containing a single biopolymer component, giving particular attention to the variation of important parameters with water content. Secondly, NMR techniques used successfully to study synthetic polymer blends are considered with regard to their potential application to mixed biopolymer systems.

Model for proton relaxation in aqueous biopolymer solutions and gels containing a single biopolymer component

In an NMR experiment, the probe nuclei can be labelled by a preparation pulse sequence, consisting of one or more pulses separated by time delays. In this way, nuclei, can be distinguished (a) by their resonance frequency (chemical shift), which reflects the chemical environment; (b) by their time-averaged dipolar coupling, which is determined by the length of the internuclear vector, its orientation with respect to the applied magnetic field, and by the rate at which these fluctuate; and (c) in the case of deuterium (and other nuclei with $I > \frac{1}{2}$), by the time-averaged quadrupolar coupling. The dipolar coupling is modulated as a result of molecular diffusion, chemical exchange, and reorientation of molecules or groups of atoms. Similarly, the time-averaged quadrupolar coupling of deuterium depends on which atom (C, O, or N) it is bonded to, and on the rate at which the orientation of the bond fluctuates as a result of molecular reorientation or chemical exchange. Thus magnetic labels can be associated with different parts of the biopolymer system, for example, biopolymer CH protons can be distinguished from water protons because they have different chemical shifts, or because isotropic tumbling reduces the dipolar coupling of the water molecules compared with that of the more rigid CH protons. The labelling can persist for times up to that of the spin-lattice (longitudinal) relaxation time, T_1, which is typically of the order 1 s, and the observation of more rapid blurring of the labelling can be used to obtain information about the rate at which a proton (or its magnetization) diffuses from one part of the system to another. NMR has a large dynamic range and experiments can be devised to detect fluctuations having correlation times from 10 s to 10^{-11} s, where the longer time is governed by T_1, and the shorter by the inverse of the resonance frequency, $1/\omega_0$.

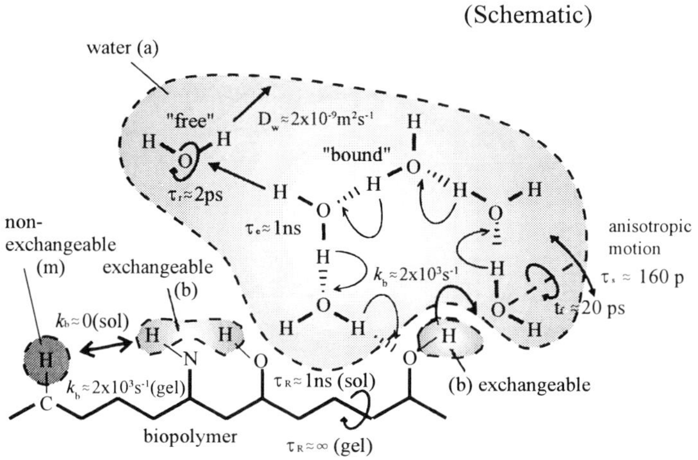

Figure 7.1 A schematic diagram showing the varied proton environments that exist in a typical aqueous biopolymer system. Exchange rates are denoted by k with an appropriate subscript. The correlation times τ_r and τ_f refer to the reorientation of the free and bound water respectively; τ_s refers to further anisotropic reorientation of bound water, and τ_e to the exchange between bound and bulk water. D_w is the diffusion coefficient of bulk water

Even systems containing only a single biopolymer component are found to be complex, with protons in many different environments, as shown schematically in Figure 7.1. In water-rich systems, some water molecules exist as "bulk" or "free" water and are free to tumble rapidly and isotropically. The reorientational correlation time is $\tau_r \approx 2$ ps and the translational diffusion constant for water is $D_w \approx 2\times10^{-9}$ m^{2}s^{-1} In all except the driest systems, some water is "bound" to the biopolymer, probably via hydrogen bonds to the OH or NH groups. The motion of these molecules is anisotropic and more restricted and is characterized by two correlation times: $\tau_f \approx$ 20 ps and $\tau_s \approx$ 160 ps, which are discussed again later. The diffusion constant of this bound water is only about 5–10 times smaller than in bulk water. For isotropic diffusion in three dimensions, the mean square displacement, *a*, of a water molecule in a time interval *t* is given by

$$a = (6Dt)^{1/2} \qquad (1)$$

corresponding to a characteristic diffusion time constant $a^2/(6D_w)$. There is some uncertainty about the value of the numerical constant as it varies according to which model is used, e.g. 4/3 for diffusion from an extended source, 2 for diffusion perpendicular to lamellae, etc. (Henrichs, Tribone, Massa and Hewitt, 1988). Here we are mainly interested in an order of magnitude estimate, therefore we assume the value 6, keeping in mind that the actual average distance moved by a water molecule may be as much as 2–3 times the value calculated with eq. (1). A graph of $a^2/6D_w$ versus *a* for $D_w = 2\times10^{-9}$ m^{2}s^{-1}, the value for water, shown in Figure 7.2, indicates that a water molecule diffuses a distance of the order of 1 nm in 1 ns. This suggests that, as the lifetime for the bound water $\tau_e \approx 1$ ns, no reasonable distinction can be made between free and bound water.

The protons in OH and NH groups on the biopolymers are able to exchange with bound water protons. The exchange rate, k_b, is found to be 10^3–10^4 s^{-1} in water-rich systems, but is much smaller in water-poor systems. The remaining protons on the biopolymers, e.g. CH, CH_2, methyl, phenyl, etc., are non-exchangeable and form a relatively isolated pool. In water-poor systems, and probably in the junction zones of water-rich gels, the motions of the biopolymers are slow and of restricted amplitude, therefore the non-exchangeable protons will be coupled together and communicate with each other via spin-diffusion or cross-relaxation as a result of strong dipolar interactions. Molecular motions tend to average out the dipolar interactions, and therefore the non-exchanging protons in flexible parts of the biopolymers, e.g. in interjunction zones or side chains of gels, or in tumbling molecules in solutions, will not be coupled together. Because of their small gyromagnetic ratio, any deuterons introduced will be much more weakly coupled together.

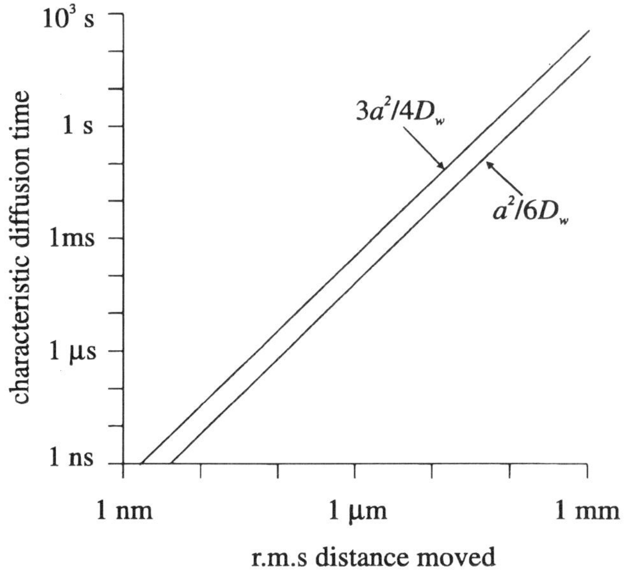

Figure 7.2 A graph showing the distance a water molecule can diffuse in a given time,shown for two extreme models; that is, diffusion on a sphere (1/6),and from an extended source (3/4). 'a' is the r.m.s distance travelled and D_w is the diffusion coefficient of water

A model describing aqueous solutions of proteins and carbohydrates has been shown to account for the NMR behaviour of water-rich systems and to provide a starting point for discussing water-poor systems. (Hills, 1991a).

As Figure 7.3 illustrates, there are four major populations of protons to be considered: free (bulk) water, bound (possibly non-freezable) water, exchangeable biopolymer protons, and non-exchangeable biopolymer protons. For reasons given above, at room temperature the free and bound waters can be treated as a single water population (denoted (a) in Figure 7.3). The proton fraction of water is denoted by P_a. A ^{17}O relaxation study has shown that the reorientation of bound water is anisotropic (Halle and Wennerstrom, 1981), and that a fraction ($1-S^2$) of the intramolecular water dipolar interaction is modulated by a fast reorientational motion with a correlation time $\tau_f \approx 20$ ps, where $S^2 \approx 10^{-4}$ is an order parameter. The value of τ_f is only 5–10 times longer than τ_r in dilute solutions but is several orders of magnitude longer in water-poor systems. The residual dipolar interaction proportional to S^2 is modulated by a slower motion with correlation time $\tau_s \approx 160$ ps. The intrinsic longitudinal and transverse relaxation times of the water protons, denoted by T_{1a} and T_{2a} respectively, are in the range 2–3 s.

The exchangeable biopolymer protons (C—OH and/or C—NH), are denoted by the subscript 'b' in Figure 7.3 and the non-exchangeable ones (C—H) by the subscript 'm'. The intrinsic relaxation times of the biopolymer protons T_{1b}, T_{1m}, T_{2b}

and T_{2m}, depend on the mobility of the biopolymer chains, which ranges from almost isotropic tumbling in dilute sols to rigid solid-like behaviour in the junction zones of water-poor systems. The resonance frequencies of the biopolymer protons are found to be shifted by about 1 ppm relative to the water protons, giving $\delta\omega \approx 300$ Hz.

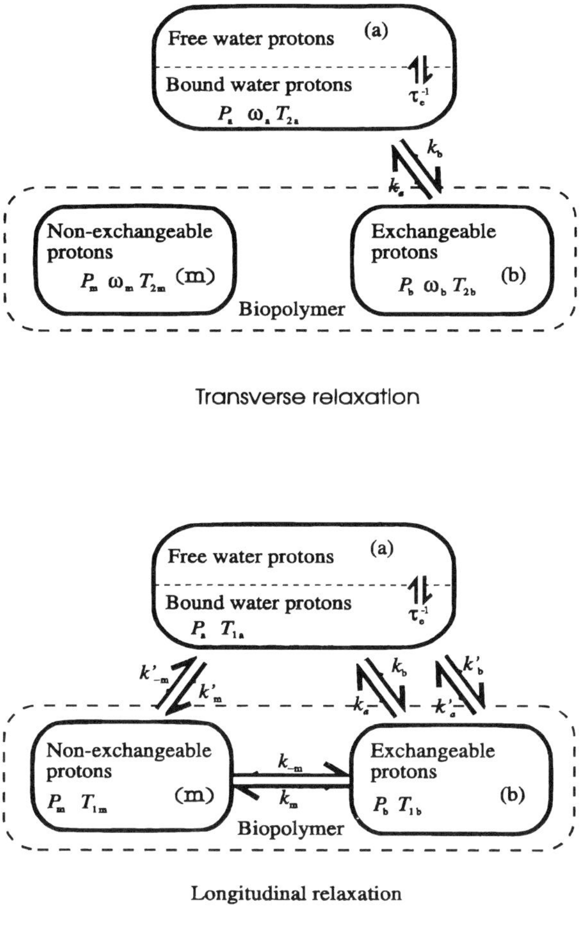

Figure 7.3 An illustration of the four populations of protons considered to exist in a typical aqueous biopolymer system and their involvement in the basic scheme used to model the NMR behaviour of transverse and longitudinal relaxation. The population of each proton pool is denoted by P with an appropriate subscript. Relaxation parameters T_1 and T_2, and resonance frequencies ω of each pool are also labelled in this way. Exchange rates are denoted by k and correlation times by τ

Transverse magnetization is only transferred between the different populations by the chemical exchange of protons (see Figure 7.3), but longitudinal magnetization, in the laboratory and rotating frames, can be transferred either by the chemical exchange, or by quantum mechanical exchange of the magnetization, often referred to as cross-relaxation (see Figure 7.3(b)). Hills (1991a) suggests that the cross-relaxation rate constant, k_m, between exchangeable and non-exchangeable biopolymer protons, is related to the secular dipolar cross-relaxation rate between a single pair of protons, σ, by the equation $k_m = n\sigma$, where n is the average number of pairwise interactions (Hills, 1991a). σ is a function of the correlation time, τ_c, of processes modulating the dipolar interaction. σ can be evaluated by a modification of the theory describing the nuclear Overhauser effect (NOE). Since the interaction is modulated by both molecular reorientation and chemical exchange, it follows that $\tau_c^{-1} = \tau_R^{-1} + k_b$, where τ_R is the reorientation correlation time for the biopolymer and k_b is the chemical exchange rate constant, i.e. the reciprocal lifetime of the exchangeable biopolymer protons. For a free tumbling globular protein, $\tau_R \approx 1$ ns and $k_b \approx 10^3$ s^{-1}, therefore $\tau_c = \tau_R$ and k_m is negligible. In gels, $\tau_R \rightarrow \infty$, therefore chemical exchange becomes the dominant process in the modulation of the dipolar interaction. Thus $\tau_c = k_b^{-1}$, and consequently k_m is increased by several orders of magnitude on gelation. In water-poor systems $k_b << k_m$, and when the motion of the biopolymer is sufficiently slow for the rigid lattice criterion to be satisfied, cross-relaxation will be limited by the rate of flip-flop transitions and the rate constant $k_m = (1/120)(M_2)^{1/2}$, where M_2 is the biopolymer proton dipolar second moment (Tanner, Hills and Parker, 1991).

Table 7.1 TYPICAL PARAMETERS FOR PROTON RELAXATION IN AQUEOUS BIOPOLYMER SYSTEMS

Parameter	*Sol*	*Water-rich gel*	*Water-poor gel*
τ_s	160 ps		60 μs
τ_f	20 ps		1 ns
τ_e	1 ns		100 ns
P_m	0.05	0.05	0.35
T_{1m}	30 ms	360 ms	1.6 s
T_{2m}	100 ms	33 μs - 25 ms	20 μs
P_b	0.01	0.005 (rigid) /0.001 (mobile)	0.35
T_{1b}	30 ms	500 ms / 25 ms	1.6 s
T_{2b}	30 ms	33 μs / 25 ms	20 μs
P_a	0.96	0.96	0.3
T_{1a}	1 s	1 s	130 ms
T_{2a}	1 s	1 s	300 μs
k_b	2x10^3 s^{-1}	2x10^3 s^{-1}	0
k_m		2x10^3 s^{-1}	2.5x10^3 s^{-1}
$k'_b + k'_{-m}$			320 s^{-1}
$\delta\omega$	2πx1.25 ppm	2πx1.25 ppm	

The bound-water cross-relaxation rate constants k'_m and k'_b are negligible in solutions because the lifetime of the bound water is only of the order 1 ns. The situation will be different in the junction zones of gels and in water-poor systems where k'_m and k'_b may become significant. In dilute solutions of proteins and carbohydrates, exchange between the water protons and the exchangeable biopolymer protons is fast on the timescale of T_2 with $k_b > 10^3$ s^{-1}. Proton exchange is slower in the water-poor regime. Only k_b, k'_b, k_m and k'_m need be treated as independent parameters as the principle of microscopic reversibility ensures that $P_b k_m = P_m k_{-m}$, etc. Values for these, and other, parameters are given in Table 7.1 for typical systems: a water-rich solution (e.g. 20% w/w gelatine sol), a water-rich gel (e.g. 20% w/w gelatine gel) (Hills, 1991b) and a water-poor system (e.g. starch containing 10% w/w water) (Tanner *et al.*, 1991).

There has been considerable controversy about the interpretation of proton relaxation in aqueous biopolymers. In the opinion of Hills, the neglect of the anisotropy of bound water, reorientation and the effects of proton chemical exchange means that much of the published work on water relaxation in biopolymer systems, especially in the food literature prior to 1990, needs reinterpretation (Hills, 1990).

ANALYSIS OF TRANSVERSE RELAXATION DATA

Transverse relaxation is usually observed with a Carr-Purcell-Meiboom-Gill (CPMG) sequence (Figure 7.4(a)) $90_x — (t_1 — 180_y — t_1 — \text{echo} — t_1)_n$ and an effective transverse relaxation time T_2^{CPMG} is obtained by fitting the echo envelope. T_2^{CPMG} often exhibits dispersion, i.e. its value decreases as t_1 increases. This dispersion is a consequence of proton exchange, which dominates the transverse relaxation of water protons in dilute solutions of proteins and carbohydrates. Because cross-relaxation is ineffective in transferring transverse magnetization, the non-exchangeable biopolymer protons relax independently and a two-site model is adequate to describe the time evolution of the exchangeable biopolymer—water protons during the CPMG sequence. Hills has shown that corrected Carver-Richards expressions are able to describe the characteristic t_1-dependence of T_2^{CPMG} in terms of the parameters k_a, k_b, p_a, p_b, $\delta\omega$, T_{2a} and T_{2b}, where $\delta\omega = \omega_b - \omega_a$. Not all of these are independent, although it is possible that some of them can be determined independently (Hills, Takacs and Belton, 1989a, 1989b). At very large pulse spacing, when $P_b \ll 1$ and $t_1 \gg (2k_b)^{-1}$, the relaxation is given by the Swift-Connick expression (Hills, Wright and Belton, 1989; Hills, Cano and Belton, 1991).

$$\frac{1}{T_2^{\text{CPMG}}}(\text{long } t_1) + \frac{1}{T_{2a}} + \frac{P_b k_b}{P_a}\left\{\frac{T_{2b}^{-2} + T_{2b}^{-1}k_b + (\delta\omega)^2}{(T_{2b}^{-1} + k_b)^2 + (\delta\omega)^2}\right\} \tag{2}$$

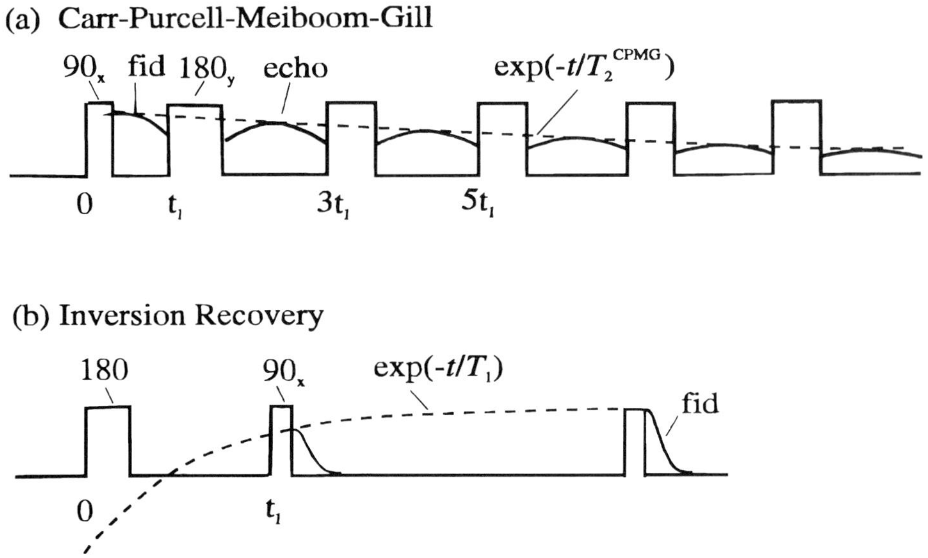

Figure 7.4 Typical pulse sequences used to measure (a) T_2 and (b) T_1

If T_{2b} is short so that the terms in $(\delta\omega)^2$ are negligible, or in the limit of very short t_1, the expression reduces to

$$\frac{1}{T_2^{\text{CPMG}}}(\text{short } t_1) = \frac{1}{T_{2a}} + \frac{P_b}{P_a T_{2b}} \tag{3}$$

The chemical shift effect is thus responsible for the t_1- and ω_0-dependence of T_2^{CPMG}. Simulated dispersion curves for some typical systems are shown in Figure 7.5. An estimate of the chemical exchange rate can be obtained from the characteristic frequency at the inflexion of the CPMG dispersion curve where it can be shown that $(t_1)^{-1} \approx k_b$. Because of the rigidity of their gel networks, starch, amylose and amylopectin gels do not show any proton exchange CPMG dispersion (Hills *et al.*, 1991).

It is difficult to make reliable measurements of water proton relaxation times for biopolymer concentrations much higher than about 20% w/w because of the signal from non-exchangeable protons. One way of overcoming this problem is to make measurements of deuterium relaxation with samples dissolved in ^2H_2O. The ^1H transverse relaxation dispersion reported for glucose solutions has been analysed with a model analogous to that described above (Hills, 1990).

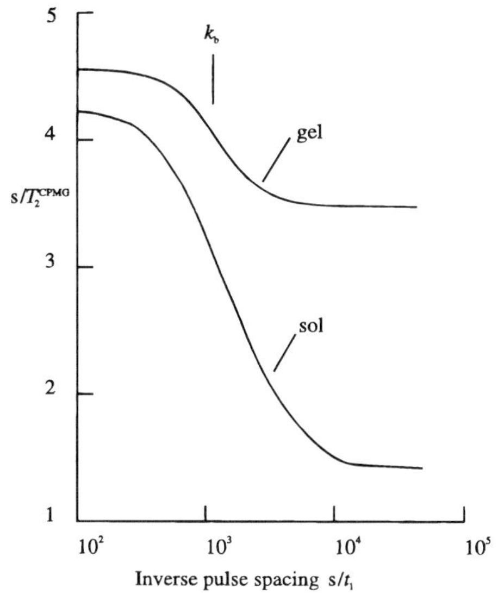

Figure 7.5 Typical simulated dispersion curves for a gel and a sol. The chemical exchange rate, k_b, can be obtained from the characteristic frequency of the inflexion of the curve

ANALYSIS OF LONGITUDINAL RELAXATION DATA

The analysis of longitudinal relaxation (see Figure 7.4(b)) differs from that of transverse relaxation in two major respects. Firstly, because longitudinal magnetization does not evolve under the Zeeman interaction, resonance frequency differences between protons in different pools have no effect. Secondly, longitudinal (but not transverse) magnetization can be transferred by quantum mechanical cross-relaxation so that the non-exchangeable biopolymer protons can, in principle, couple both with the exchangeable protons and water protons. For dilute solutions, we have stated previously that on account of rapid tumbling of the biopolymer, k_m is negligible, and because of the short lifetime of water molecules near the biopolymer, k'_m is also negligible. Consequently, the non-exchangeable protons are uncoupled from the water and appear only as a more rapidly relaxing component in non-selective (hard pulse) T_1 experiments, i.e. where all protons are excited by the rf pulse. The relaxation of the slower relaxing component is dominated by the relaxation of the exchangeable biopolymer protons. This accounts for the observed dispersive dependence on ω_0.

The major effect of gelation on longitudinal relaxation is the increase in T_{1m} and T_{1b} and in the cross-relaxation rate constant k_m. Hills has predicted the behaviour of selective (soft pulse) and non-selective (hard pulse) T_1 experiments by numerically solving the coupled Bloch equations for a three-site model that includes cross-relaxation and chemical exchange effects (Hills, 1991a). Direct dipolar cross-relaxation between water and non-exchangeable biopolymer protons is negligible compared with an indirect proton exchange cross-relaxation pathway. This proceeds via proton exchange between water and exchangeable biopolymer protons followed by quantum mechanical cross-relaxation between exchangeable and non-exchangeable biopolymer protons. The cross-relaxation rate constant in gels, k_m, can be measured by creating a magnetization gradient between the non-exchangeable and exchangeable biopolymer protons and monitoring its disappearance with time. The gradient can be created by exploiting differences in transverse relaxation rates, as in a Goldman-Shen experiment (Hills, 1991a; Hills and Favret, 1994) or differences in linewidths, as in experiments involving selective off-resonance saturation of the non-exchangeable protons (Grad, Mendelson, Hyder and Bryant, 1990; Koenig, Brown, Pande and Ugolini, 1993).

The Goldman-Shen pulse sequence (Figure 7.6) consists of

$$90_x\text{-}t_1\text{-}90_x\text{-}t_2\text{-}90_x\text{-}t_3\ (\text{fid}) \tag{4}$$

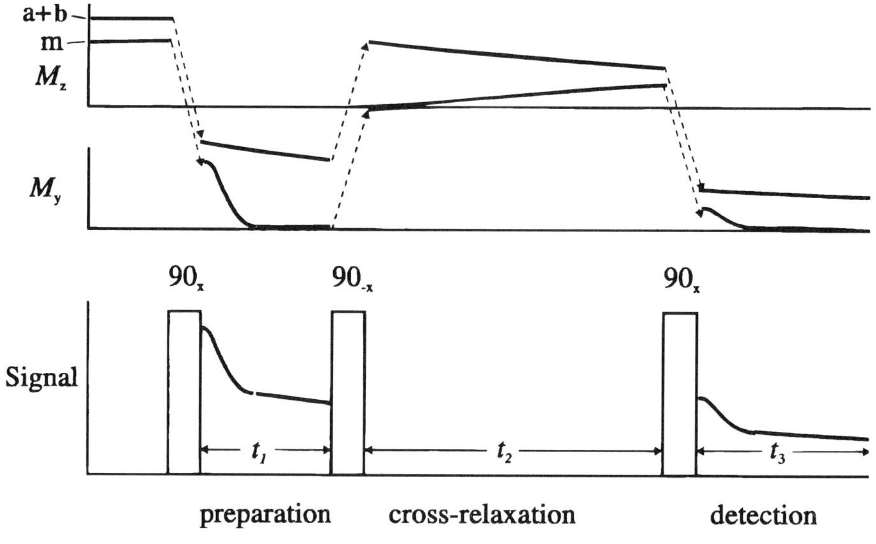

Figure 7.6 A typical Goldman-Shen pulse sequence and a schematic representation of how this could be utilised in an experiment to measure cross-relaxation rates. The non-exchangeable protons are denoted by 'm', and the exchangeable and water protons by 'b' and 'a' respectively

If the preparation period t_1 is set to be about $5T_{2m}$, the transverse magnetization of the non-exchanging proton pool is completely removed, while the transverse magnetization of the water and exchanging biopolymer protons are merely attenuated. The 90_x pulse converts the transverse magnetization into longitudinal magnetization, where it can be transferred by cross-relaxation to the non-exchanging biopolymer pool during the contact time t_2. The recovery of non-exchangeable biopolymer magnetization is monitored by the free induction decay (Hills, 1991b; Hills and Favret, 1994). In a modification of the sequence, during the periods t_1 and t_3 the spins are subjected to a CPMG sequence of pulses, which helps in separating the signals for the two pools of protons.

The three-site cross-relaxation model also describes relaxation in the rotating frame (Hills, 1991a). In this case the magnetization of all three proton pools must be considered as well as their chemical shift differences. Useful information can be obtained from the ω_1 dependence of the spin-lattice relaxation time in the rotating frame $T_{1\rho}$ (Hills, 1990). It has been shown that the CPMG experiment is equivalent to a spin-locking experiment in an effective locking field of frequency $\frac{1}{2}t_1$, where $T_2^{CPMG} = T_{1\rho}$ (Santyr, Henkelman and Bronskill, 1988).

MULTIEXPONENTIAL RELAXATION IN SINGLE BIOPOLYMER SYSTEMS

For a system containing only a single biopolymer component, single exponential relaxation is expected only when molecular diffusion or chemical exchange enables a given proton to sample every environment in the system during the timescale corresponding to the resulting average relaxation time for the system. In very dry, rigid systems, strong dipolar coupling would ensure single exponential relaxation by allowing a given proton to exchange magnetization with protons in all other environments by spin-diffusion. On the T_2 timescale, as has been discussed above, the non-exchangeable protons are isolated and they manifest themselves as a more rapid component in the free-induction decay or CPMG envelope. T_1 experiments give much more time for averaging to occur. In solutions, the rapid motion of the biopolymer molecules ensures that the non-exchangeable protons are isolated and therefore relax independently. In gels and water-poor systems, even if the non-exchangeable protons, the exchangeable protons and the bound water protons are coupled together, multiexponential behaviour will be observed unless the biopolymer regions are separated by "bulk" water regions that are sufficiently small for diffusion to ensure that a given water molecule sees an average environment. Assuming that the biopolymer regions are on average a distance $2a$ apart, then single exponential transverse relaxation will be observed only when the condition $a^2/6D_w << T_2$. Taking $T_2 \approx 2$ s and $D_w \approx 2\times10^{-9}$ m^{2}s^{-1} for pure water, it follows that micro-heterogeneity on a scale larger than 2×150 μm cannot be diffusionally averaged and will give rise to multiexponential behaviour. The relaxation behaviour of water in frozen-thawed agarose gels (see Figure 7.7) has been interpreted in this way by assuming that spaces occupied by ice crystals in the frozen gels remain as water-filled holes in the gel network after thawing (Belton, Hills and Raimbaud, 1988).

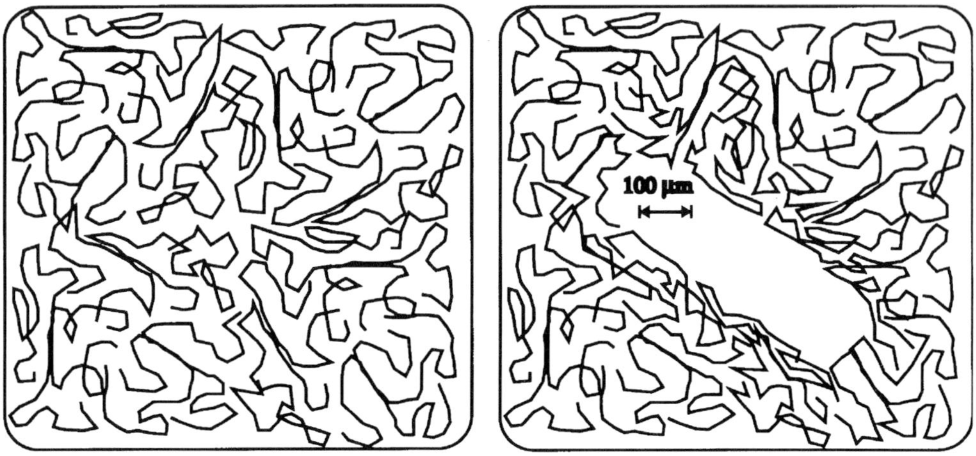

Figure 7.7 A schematic diagram showing morphology in a frozen gel which would give rise to multi-exponential water proton relaxation behaviour

Deuterium relaxation studies of ß-cyclodextrin hydrate provide evidence for deuteron exchange between exchangeable hydroxyl groups and water molecules as well as for mobile non-exchanging water molecules even at temperatures well below the freezing point of water (Kitchin, 1993).

RELAXATION IN COMPARTMENTALIZED SYSTEMS

In many systems of current interest, such as foods, cellular tissue and porous solids, aqueous solutions are compartmentalized in pores and/or cellular structures. In these systems, multiexponential relaxation behaviour can be used to obtain information about the scale of heterogeneity. Figure 7.8 illustrates a simple model for systems consisting of two kinds of compartment, each of size $2a$, containing solutions A and B, separated by barriers of permeability κ. When the scale of heterogeneity within each compartment is less than about 200 µm, each will be characterized by the parameters: T_1^A, T_2^A, D^A, ω_0^A, and T_1^B, T_2^B, D^A and ω_0^A, respectively. When $\kappa = 0$, the compartments are isolated and biexponential relaxation is expected. In phase-separated biopolymer systems, however, the barriers are assumed to be infinitely permeable, i.e. $\kappa = \infty$, and there will be diffusive exchange between the compartments. A description of the relaxation behaviour for the general case requires a numerical solution (Belton and Hills, 1988). It should be noted that it is inappropriate to treat diffusive exchange by equations describing chemical exchange between two sites, as has sometimes been done in the past.

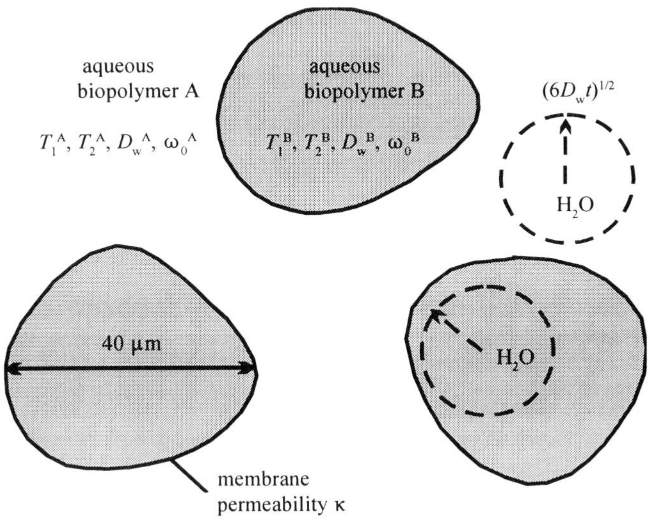

Figure 7.8 A simple model for systems consisting of two kinds of compartment containing biopolymer solutions A and B, characterized by relaxation parameters T_1 and T_2, resonance frequencies T and water diffusivities D with appropriate superscripts, separated by barriers of permeability 6. When 6 = 0 the compartments are isolated and biexponential relaxation is expected; however, in general a description of the relaxation behaviour requires a numerical solution

A graph of $a^2/6D_w$ versus a for $D_w = 2 \times 10^{-9}$ m^{2}s^{-1}, the value for water, is shown in Figure 7.2.

When the resonance frequencies, diffusion coefficients and compartment sizes (2a) are equal, but there is relaxation dispersion, i.e. $|1/T_2^A - 1/T_2^B| = |\Delta R_2| \neq 0$, the criterion for the observation of single exponential decay is $(a^2/6D_w)\,|\Delta R_2| << 1$. The exact condition depends on the particular geometry of the system. In the slow diffusive exchange regime, this criterion does not hold, and complex multiexponential transverse relaxation is observed. This kind of behaviour has been observed for water-saturated beds of Sephadex beads (Hills *et al.*, 1989).

When the resonance frequencies in the two compartments are different, it has been shown that transverse relaxation dispersion is expected when the CPMG pulse spacing t_1 becomes comparable to the characteristic diffusion time constant $a^2/6D_w$, i.e. $6D_w t_1/a^2 \approx 1$, analogous to the chemical exchange dispersions found when $k_b t_1 \approx 1$ (Hills *et al.*, 1989). Double dispersions have been observed in systems where both diffusive and chemical exchange occur in a heterogeneous system. A complete dispersion curve will only be observed for $a \approx 1$ µm, corresponding to $a^2/6D_w \approx 10^{-4}$ s^{-1}. For heterogeneity scales larger than 10 µm, $a^2/6D_w$ becomes so large that the single component transverse relaxation rate, $1/T_2^{CPMG}$, will increase monotonically with pulse spacing. This has been observed in carrageenan gels (Hills *et al.*, 1991) and for packed Sephadex 20 µm beads (Hills *et al.*, 1989c).

The temperature-dependence of T_2 of water in frozen, randomly packed beds of Sephadex beads, which exhibit many of the features of complex food systems, has provided useful "finger-print" information about the amount and microscopic distribution of ice and non-freezing water (Hills and Floc'h, 1994).

The effects of magnetization transfer have considerable potential for providing new information in NMR imaging by allowing some control of image contrast. In these experiments, the immobile tissue protons are irradiated with an rf field that is approximately 5–10 kHz off-resonance from the water protons. The water protons respond to the perturbation of the solid-component spin system and these protons can be detected through observation of the water protons by a technique known as Z-spectroscopy. Theoretical models considered so far appear to neglect chemical exchange (Bryant and Lester, 1992).

Sensitivity of proton dispersions, arising from diffusive exchange, to the distance scale characterizing the heterogeneity suggest that it may be a useful probe of complex morphology changes in heterogeneous systems generally.

Techniques used in the investigation of synthetic polymer blends

NMR experiments in polymer blends can generally be split into two categories, namely those which seek to obtain information on the degree of miscibility of the two polymers with a view to characterizing the scale and extent of mixing, and those that investigate the spatial dimensions of specific domains that are rich in one of the polymer components and are referred to as phase separated regions. Both of these topics are also of relevance in the field of mixed biopolymer systems. The first can be related to the investigation of the fractal nature of phase separated mixed biopolymer systems — that is, the degree of heterogeneity within phase separated regions — and to the possibility of the study of interfacial regions. The second is also clearly of interest in mixed biopolymer systems, where phase separation and the consequent formation of a tangible microstructure determines textural properties. Hence, the applicability of techniques used to investigate these aspects of polymer blends to mixed biopolymer systems is worthy of consideration. There is, however, an extra proton signal observed in mixed biopolymer systems compared with polymer blends, namely that provided by the large amounts of mobile water. Thus another key issue NMR techniques might hope to address is the partitioning of the water between regions in a phase separated system. It is not apparent in such systems that this distribution necessarily needs to be identical to that exhibited in systems which have undergone bulk phase separation. NMR techniques that have been used successfully in the study of polymer blends, their limitations, and comments on their applicability to mixed biopolymer systems, will be discussed. Some ways in which the proton water signal might be exploited are also suggested.

MISCIBILITY IN BLENDS

The essence of a miscibility study is to examine the intimacy of mixing; i.e. the proximity of the different polymer types that constitute the blend. An NMR experiment is required that discriminates swiftly on the basis of distance, and, as such, the exploitation of the dipolar interaction, which is a function of r^{-6} (where r is the relevant internuclear distance), is ideal. This feature is common to all measurements carried out on polymer blend miscibility, although techniques vary. Ironically, it is the dipolar interaction which is the main cause of the experimental difficulties, in that both intra- and inter-molecular dipolar coupling between protons cause resonance line broadening, resulting in the indistinguishability of the separate polymer signals. One approach to circumventing this problem is to dilute the proton spins on one of the polymers, so that the resonances sharpen and become resolvable. This resolution advantage exhibited by dilute spins is routinely manifest in ^{13}C observation.

Proton dilution can be achieved by incomplete deuteration of one of the polymers. The principle of using this approach to study miscibility is that when the proton-diluted polymer is near to the other blend constituent, the normally resolved proton resonance lines will be broadened, and correspondingly their intensity will drop. Thus, in a simple experiment the proton spectrum reflects the degree of intimate mixing. The distance scale of the mixing probed is of the order of nanometres. This method involves direct proton observation in the presence of magic angle spinning (MAS). Macroscopic reorientation of the sample at the magic angle to the applied field causes lines to narrow when the spinning rate is larger than the static linewidth, as a consequence of the angle-dependent scaling of (anisotropic) dipolar and chemical shift interactions. If the residual protons on the deuterated polymer account for only 2% of those in the undeuterated state, moderate spinning speeds of the order of 2 kHz would be required to narrow the lines. This method has been successfully applied to a polymer blend containing atactic polystyrenes (Vander Hart and Manders, 1986).

This kind of experiment can be used to study mixed biopolymer systems. The morphology of these systems, however, can be perturbed by the centripetal forces generated by spinning, e.g. spinning rates of greater than 3 kHz have been found to produce instabilities in 5% amylose gels (Gidley, 1989). Therefore low spinning rates are desirable. Attempts to deuterate complex natural products such as gelatin and pectin pose considerable, but not intractable, synthetic problems, and, for most purposes, the deuteration of one biopolymer is sufficient. Deuterated collagen fibrils have been produced with the use of cell cultures (Jelinski, 1980). On the other hand, specifically deuterated pectin derivatives have been prepared by esterification with deuterated methanol (Cros, Penhoat, Bouchemal, Ohassan, Imbuty and Perez, 1992), and methods for deuterium-labelling carbohydrates with simple monomer units such as poly-α-D-glucans have been reported in the literature. The effect, however, of these chemical modifications on structural parameters such as molecular weight distribution and degree of polymerization should not be overlooked.

An alternative way of studying polymer-polymer miscibility is to examine intermolecular polarization transfer directly. Figure 7.9 illustrates a ^{1}H-^{13}C cross-polarization procedure in which only one of the polymer constituents contains a substantial proton pool and the ^{13}C signal of the other perdeuterated polymer is detected. Consequently, if the two polymers are miscible, the dilute proton polymer will be visible in the ^{13}C CP/MAS of the blend only as a result of intermolecular polarization transfer during the contact time. This method has been used successfully to study a polymer blend of PVC and PMMA-d_8 (Parmer, Dickenson, Chien and Porter, 1987). Advantages of this technique include the ability to study blends with less than 10% of one component and blends in which the relaxation times of the

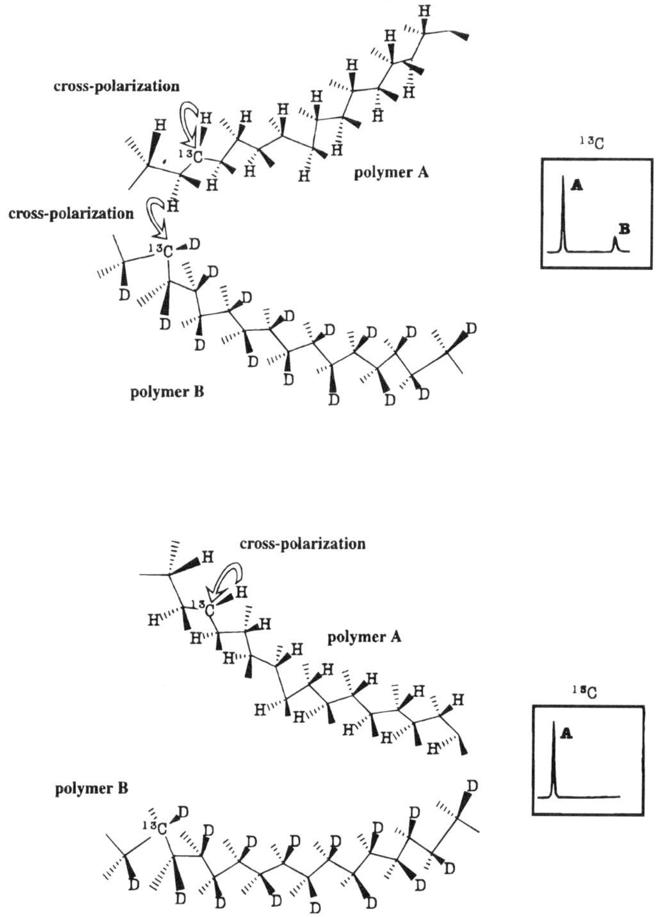

Figure 7.9 An illustration of the use of a ^{1}H-^{13}C cross-polarization procedure for studying polymer—polymer miscibility. This scheme relies on the deuteration of one of the polymer components

two constituents are not distinct. The technique can also be applied to mixtures of immiscible polymers in which the investigation of interfacial mixing is of interest (Gobbi, Silvestri, Russel, Lyerla and Fleming, 1987). In some blends where one component contains ^{19}F exclusively, ^{19}F-^{13}C intermolecular polarization transfer has also been used in this respect (Eijkelenboom, Maas, Veeman, Buning and Vankan, 1992). For techniques that depend on the observation of ^{13}C to be useful, the component polymers must give distinct peaks in the ^{13}C spectrum of the mixture. In terms of protein-polysaccharide mixed biopolymer systems this poses no problem as proteins such as gelatin contain aromatic residues that have ^{13}C chemical shifts of greater than 110 ppm, whereas the carbons on polysaccharides have shifts of less than 110 ppm. Mixed polysaccharide systems pose more of a problem in this respect.

DOMAIN SIZE

It has proved possible in polymer blends to make use of the measurement of spin diffusion to obtain useful information on homogeneity and microstructure scales, in a completely analogous way to the use of molecular diffusion of water in aqueous biopolymer systems already discussed. Spin diffusion leads to communication between spins without involving molecular motion. It is the transfer of magnetization via the "flip-flop" term of the dipolar Hamiltonian, and as such it will homogenize the magnetization profile and average relaxation parameters. The spin diffusion coefficient, related to the dipolar second moment, scales as the cube root of the proton density, and is found to be of the order of 10^{-16} m^2s^{-1}. If the timescale is defined by T_1, the observation of a single relaxation parameter demonstrates that the system is homogeneous over the corresponding distance scale, $2\times(6D_{sd}T_1)^{1/2}$. Figure 7.10 illustrates this method. It has been successfully applied to the study of a polymer blend of PC-PMMA (Asano, Takegoshi and Hikichi, 1992), where the observation of a single $T_1 \approx 300$ ms corresponds to homogeneity on the 20 nm scale, and the complementary measurement of $T_{1\rho} \approx 8$ ms indicates heterogeneity on the 4 nm scale. Once again the problem of spectral resolution of the distinctive polymer resonances is tackled by cross polarization and subsequent observation of carbon signal (see Figure 7.11).

It is worthwhile at this point, before discussing the use of the spin diffusion phenomenon to probe the domain size of regions rich in a specific polymer, to comment on the application of this type of experiment to mixed biopolymer systems. Some insight may be gained from the consideration of elastomers which, while still lacking large amounts of mobile water, will be a closer dynamical analogy to mixed biopolymer systems. At room temperature the proton linewidth, under MAS conditions, for polybutyl acrylate is only about 150 Hz (Voelkel, 1988). The

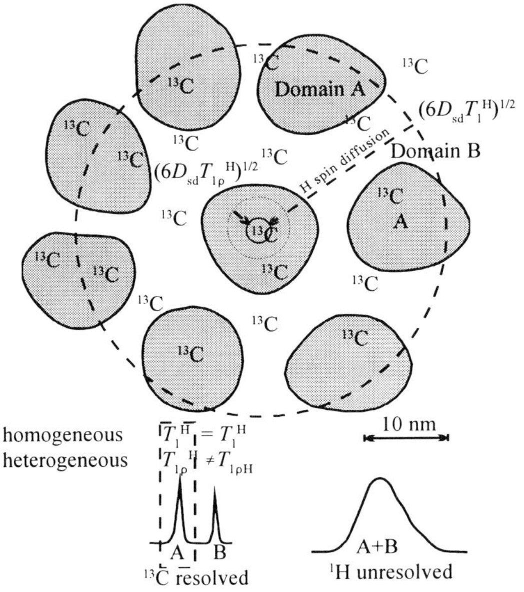

Figure 7.10 A schematic diagram showing how the process of spin diffusion can be used to probe heterogeneity over different distance scales by examining the averaging or otherwise of relaxation parameters, for example T_1 and $T_{1\rho}$. D_{sd} is the appropriate diffusion coefficient

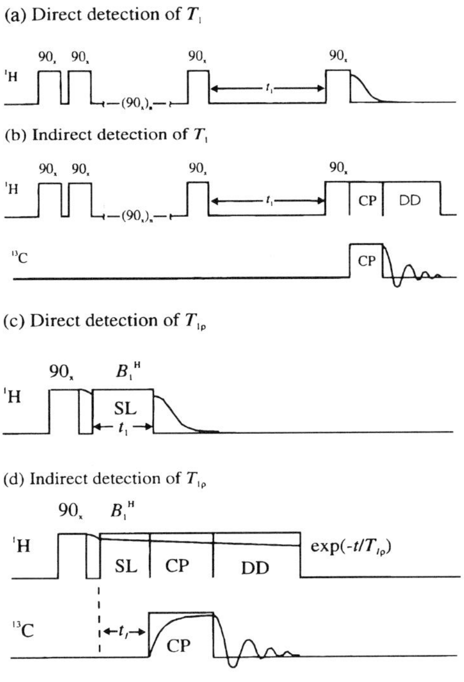

Figure 7.11 Typical pulse sequences used in the indirect detection of T_1 and $T_{1\rho}$, utilizing cross-polarization (CP) and dipolar decoupling (DD). (SL) denotes spin-locking

molecular mobility averages the proton-proton dipolar interaction and reduces the spin diffusion rate to such an extent that an individual $T_{1\rho}$ is found for each proton. When the system is cooled to 248 K, however, the proton linewidth increases to approximately 13 kHz because reduced molecular motion is insufficient to average the dipolar second moment, leading to faster spin diffusion and to a common $T_{1\rho}$. It is likely that in mixed biopolymer systems there will be a distribution of mobilities. Flexible regions, such as cross-linking chains and side-chains may act as bottlenecks preventing the transfer of magnetization by spin diffusion.

The implementation of these types of experiments for the study of mixed biopolymer systems will not be straightforward, although difficulties may not be completely prohibitive. In an analogy to polymer blends, the methodologies discussed rely on the NMR observation of biopolymer protons or carbons, and as such the water proton signal in these cases is extraneous. This can easily be eliminated by the use of ^2H_2O, although it should be noted that exchangeable protons on the biopolymers will themselves be depleted. The signal to noise ratio will be poor as commonly only about 30% of the sample will consist of the biopolymers, and a common $T_{1\rho}$ is unlikely at room temperature because of high mobility. It is possible to overcome these difficulties either by concentration of the sample by freeze-drying or by lowering the temperature, although one must be aware of changes in morphology resulting from these bulk manipulations. In this respect, more promising results using this kind of technique are likely to be obtained from water-poor systems, i.e. less than 50% ^2H_2O content.

For systems in which these types of experiments may be viable, an elegant method for observing spin diffusion is that first suggested by Goldman and Shen (1966). This experiment has been discussed previously in the context of measuring cross-relaxation rate constants, k_m. The method has been used in several studies of spin diffusion in polymer systems (Tanaka and Nishi, 1986). The basis of the experiment (see Figures 7.6 and 7.12) is, as before, that different spatial regions are discriminated between on the grounds of a relaxation parameter. This is classically T_2, as in the first case examined, so that after an initial system preparation, a magnetization map is created in which certain regions (those with short T_2) consist wholly of dephased spin isochromats. The magnetization is transferred back to the *z* direction, and, after a variable time delay, it is observed with a read-pulse. The re-emergence of a signal from the short T_2 region demonstrates that the spin diffusion is active in homogenising the magnetization profile. Variants of the experiment discriminate on the basis of $T_{1\rho}$, or even the system response to multiple pulse sequences such as MREV-8 (Haeberlen, 1976). These have the advantage that spin diffusion in the preparation and detection periods is eliminated. The application of these ideas to mixed biopolymer systems is of course subject to the limitations discussed in the last section, along with the restriction that the timescale of the experiment, and hence the spatial resolution obtainable, rely on T_1 being long, so as not to contribute to the evolution of the magnetization profile during the observation of spin diffusion. It is unlikely that an initial discrimination on the basis of a relaxation parameter accompanied by sufficiently long spin-lattice relaxation times will be a general feature of mixed biopolymer systems. Even if this method is successful, the slow rate of spin diffusion limits the domain size that can be investigated to substantially lower than the included phase region dimensions and this might be thought more conventionally to be the province of microscopists. In this respect, the use of the more rapid water diffusion ($D_w \approx 10^{-9}$ m^{2}s^{-1}) in equilibrating the magnetization profile created between two water "pools" warrants investigation.

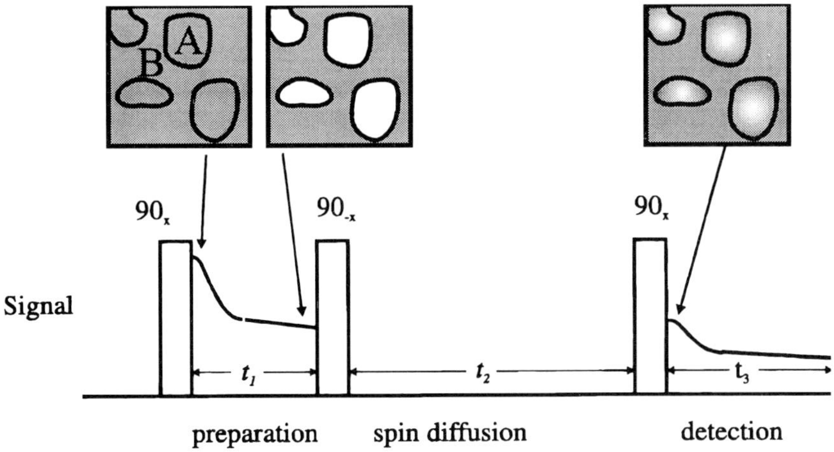

Figure 7.12 An illustration of how a Goldman-Shen type experiment can be used to probe distance scales. Spin diffusion can be seen to be active in homogenizing the magnetization profile created in the preparation period. The rate at which this occurs reflects the dimensions of the magnetization 'map', and as such indicates the extent of different regions

Methods exploiting the presence of mobile water

To be able to use the water (H_2O) signal to obtain useful information about mixed biopolymer systems, a method of discrimination is required which allows differentiation of water signals originating from different regions of the system. In this respect, the previous consideration of compartmentalized water is particularly useful. The sizes ($2a$) of included phase regions are generally distributed between about 10 μm and 200 μm, peaking at ~ 40 μm (Abeysekera, 1994), corresponding to characteristic diffusion time constants, $a^2/6D_w$ of 2 ms to 1 s. Therefore, if discrimination is to be possible on the basis of a difference in relaxation rates between the included and supporting regions, ΔR, the condition $(a^2/6D_w) \mid \Delta R \mid >> 1$ must be satisfied, therefore $|\Delta R|$ must be >> 500 s^{-1} for the smaller regions and >> 1 s^{-1} for the larger regions. Consequently, all samples are expected to show single exponential longitudinal relaxation, whereas those with smaller included regions should show biexponential transverse relaxation. When the transverse relation times of each region can be measured, the values should, according to the model previously described, provide information about the composition of the two regions.

The usefulness of this method could be improved if ΔR could be put under the control of the experimenter. One approach that has been used in compartmentalized systems with impermeable walls, e.g. plant tissues, is to reduce the water relaxation times by the introduction of low concentrations of paramagnetic ions, e.g. Mn^{2+} (Lauffer, 1987). The relaxation time of water protons in the coordination sphere of the paramagnetic complex ions is short as a result of the modulation of the strong dipolar coupling by the reorientational motion of the aquo-ion. The relaxation times of the water associated with the paramagnetic ion can be dramatically reduced by increasing the reorientational correlation time of the complex by binding it the biopolymer. If there is still rapid exchange of the water between the bound complex and the bulk water, the average relaxation time of the water is also greatly reduced;

$$(T_i^{-1})_{obs} = (T_i^{-1})_{water} + R_i[\mathrm{M}] \quad i = 1,2 \tag{5}$$

where R_i is the relaxivity and [M] the concentration of the paramagnetic species.

It is suggested that in a phase separated mixed biopolymer system paramagnetic ions will be associated more with regions rich in biopolymers with charged sites, for example gelatin and pectin, in preference to, for example, neutral starch. If this is the case then doping with paramagnetic ions will reward the experimenter in a number of ways. Firstly, by specifically binding to regions rich in one biopolymer, water molecules in that domain will clearly be distinguishable by their short T_2 values, i.e. the paramagnetic ions are more efficient relaxation sinks in regions where they are involved in specific interactions (Belton and Hills, 1987). A hypothetical doped mixed biopolymer system is shown schematically in Figure 7.13. (It should be noted that a corresponding reduction in T_1 may be problematic from the point of view of some experiments and this is discussed again in due course.) As well as being distinguishable, the numerical values of relaxation times will be reduced, so that there is the possibility of distinctly labelling included "phases", rich in one biopolymer type, of smaller dimensions. Perhaps most importantly from the point of view of designing a procedure which will reveal information on separate "phase" dimensions as well as on concentrations, this methodology gives the experimenter the opportunity to tune the relaxation time simply by varying the concentration of paramagnetic ions in the system; i.e. to select the distance scale which will apply to the distinct resolvable components.

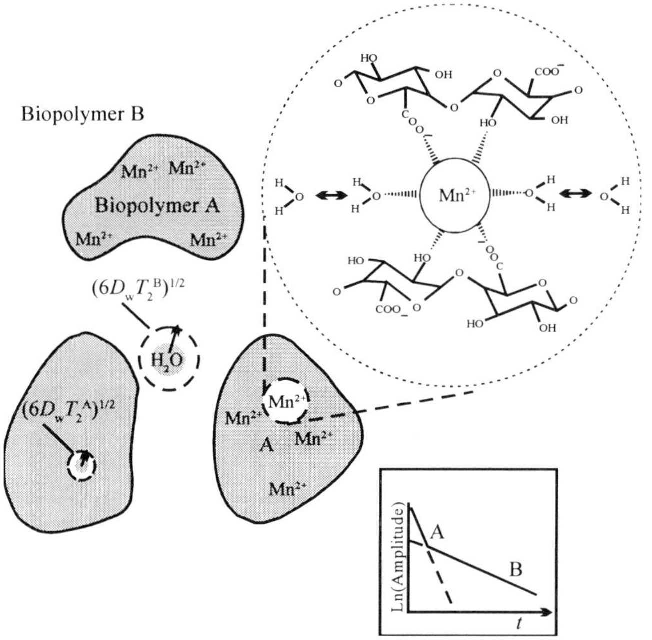

Figure 7.13 A hypothetical mixed biopolymer system doped with paramagnetic Mn^{2+} ions. Specific binding to biopolymer A creates a discrimination in the water proton relaxation times, for example T_2, between regions of the two biopolymers. This will be observable when the water diffusion (diffusion coefficient D_w) is unable to average the distinct times on the timescale of the experiment

Hence, the variation of the composite measurable T_2 from a function closely representing an average (longest T_2 possible) to a function closely representing two distinct signals (T_2 small enough to distinctly label the vast majority of included region) can communicate information about the distribution of included region dimensions. In the latter case the relative sizes of the two components of the measurable T_2 will represent the partition of water between the regions.

This hypothetical system has two components distinguishable by T_2, and as such warrants comment on the applicability of Goldman-Shen type experiments. In this case the components with short T_2 would be allowed to dephase completely before the magnetization was returned to the *z* direction. After a variable time a read-pulse would be applied and magnetization characteristic of the short T_2 region inferred to have diffused into that domain, as mobile water, during the variable time delay. Thus, the extent of the short T_2 regions could conceivably be investigated in this manner. However, it is essential that T_1 is long compared with the timescale of this experiment; i.e. magnetization returning to the *z*-direction of the short T_2 region non-diffusively, via T_1 processes, must be negligible. This is problematic as the paramagnetic ions reducing T_2 also reduce T_1. It is unlikely, even if the ion exhibits a large scalar interaction such as manganese, that the T_1 of the water in

Table 7.2 SUMMARY OF THE APPLICABILITY AND LIMITATIONS OF NMR TECHNIQUES

Information / *Technique*	*Scale of domain Microstructure*	*Partition of water between regions*	*Degree of heterogeneity within domains*	*Concentration of biopolymers in domains*	*Investigation of interfacial regions*	*Bipolymer molecular dynamics*	*Invesigation of bound water*
Proton MAS (in D_2O)			By diluting proton spins (relies on deuteration)		By diluting proton spins		
Proton 2D NOE (in D_2O)					Observation of cross peaks, CP		
Deuterium						Via relaxation and lineshape analysis	Via multi-quantum filtering
Proton relaxation		Need appropriate discriminatory relaxation parameter	Single relaxation time implies homogeniety at corresponding distance	^{13}C observed. Relies on different relaxation between regions			
Paramagnetic tuning	Paramagnetic ion concentration determines relaxation times and so the distance scale of the experiment	Specific interaction with one biopolymer discriminates regions by relaxation times	Possible that relaxation times can be tuned to look inside as well as outside specific regions	Possible from actual values of relaxation times from degree of binding			
Goldman-Shen (Spin diffusion)	Possible but regions of high chain mobility will be bottlenecks for spin diffusion		Requires adequate spin diffusion and long T_1 c.f timescale of experiment				
Goldman-Shen (Water diffusion)	Applicable to larger dimensions than spin diffusion but need T_1 long c.f discriminatory relaxation						

the paramagnetic ions reducing T_2 also reduce T_1. It is unlikely, even if the ion exhibits a large scalar interaction such as manganese, that the T_1 of the water in these systems would be more than one order of magnitude larger than T_2. The longitudinal magnetization of such a system will have recovered 9.5% of its equilibrium value simply during the preparation period, and it is not obvious how this difficulty could be overcome. This does not limit the applicability of the techniques discussed in the last paragraph, which appear promising from the point of view of obtaining information on both water partition and also the distribution of spatial dimensions.

A summary of the applicability and limitations of NMR techniques discussed is given in Table 7.2.

Acknowledgements

This work was supported by a grant under the LINK Food Processing Sciences programme (FSP 102) sponsored by M.A.F.F., involving Unilever Research, Cerestar U.K. Ltd., Nestec York Ltd., Hercules Ltd. and Sanofi Bioindustries. R.D. Keenan wishes to thank EPSRC and Nestec York Ltd. for a CASE studentship.

References

Abeysekera, R., (1994). Personal communication

Asano, A., Takegoshi, K. and Hikichi, K. (1992) Solid state NMR study of miscibility and phase separation of polymer blend: polycarbonate/polymethylmethacrylate. *Polymer Journal*, **24**, 555–562

Belton, P.S. and Hills, B.P. (1987) The effects of diffusive exchange in heterogeneous systems on NMR line shapes and relaxation processes. *Molecular Physics*, **61**, 999–1018

Belton, P.S., Hills, B.P. and Raimbaud, E.R. (1988) The effects of morphology and exchange on proton NMR relaxation in agarose gels. *Molecular Physics*, **63**, 825–842

Belton, P.S., Colquhoun, I.J. and Hills, B.P. (1992) Applications of NMR to food science. *Advances in Optical and Magnetic Resonance*, New York: Academic Press

Bryant, R.G. and Lester, C.C. (1992) Magnetic relaxation coupling in heterogeneous systems. *Journal of Magnetic Resonance*, **B101**, 121–125

Cros, S., Penhoat, C.H.D., Bouchemal, N., Ohassan, H., Imbuty, A. and Perez, S. (1992) Solution conformation of a pectin fragment disaccharide using molecular modelling and magnetic resonance. *International Journal of Biological Macromolecules*, **14**, 313–320

Eijkelenboom, A.P.A.M., Maas, W.E.J.R., Veeman, W.S., Buning, G.H.W. and Vankan, J.M.J. (1992) Triple resonance ^{19}F, ^{1}H, ^{13}C CPMAS NMR study of the influence of PMMA tacticity on the miscibility in PMMA/PVF_2 blends. *Macromolecules*, **25**, 4511–4518

Gidley, M.J. (1989) Molecular mechanisms underlying amylose aggregation and gelation. *Macromolecules*, **22**, 351–358

Gobbi, G.C., Silvestri, R., Russel, T.P., Lyerla, J.R. and Fleming, W.F. (1987) Intermolecular polarization transfer study of polymer blend compatibility. *Journal of Polymer Science: Part C: Polymer Letters*, **25**, 61–65

Goldman, M. and Shen, L. (1966) Spin relaxation in LaF_3. *Physical Review*, **144**, 321–331

Grad, J., Mendelson, D., Hyder, F. and Bryant, R.G. (1990) Direct measurement of longitudinal relaxation and magnetization transfer in heterogeneous systems. *Journal of Magnetic Resonance*, **86**, 416–419

Haeberlen, U. (1976). *High Resolution NMR in Solids*, New York: Academic Press

Henrichs, P.M., Tribone, J., Massa, D.J. and Hewitt, J.M. (1988) Blend miscibility of bisphenol A polycarbonate and poly(ethylene terephthalate) as studied by solid-state high-resolution ^{13}C NMR spectroscopy. *Macromolecules*, **21**, 1282–1291

Halle, B. and Wennerstrom, H. (1981) Interpretation of magnetic resonance data from water nuclei in heterogeneous systems. *Journal of Chemical Physics*, **75**, 1928–1943

Hills, B.P., Takacs, S.F. and Belton, P.S. (1989a) The effects of proteins on the proton NMR transverse relaxation time of water. I. Native bovine serum albumin. *Molecular Physics*, **67**, 903–918

Hills, B.P., Takacs, S.F. and Belton, P.S. (1989b) The effects of proteins on the proton NMR transverse relaxation time of water. II. Protein aggregation. *Molecular Physics*, **67**, 919–937

Hills, B.P., Wright, K.M. and Belton, P.S. (1989) Proton NMR studies of chemical and diffusive exchange in carbohydrate systems. *Molecular Physics*, **67**, 1309–1326

Hills, B.P. (1990) Multinuclear NMR studies of water in solutions of simple carbohydrates. I. Proton and deuterium relaxation. *Molecular Physics*, **72**, 1099–1121

Hills, B.P. (1991a) The proton exchange cross-relaxation model of water relaxation in biopolymer systems. *Molecular Physics*, **76**, 489–508

Hills, B.P. (1991b) The proton exchange cross-relaxation model of water relaxation in biopolymer systems. II. The sol and gel states of gelatine. *Molecular Physics*, **76**, 509–523

Hills, B.P., Cano, C. and Belton, P.S. (1991) Proton NMR relaxation studies of aqueous polysaccharide systems. *Macromolecules*, **24**, 2944–2950

Hills, B.P. and Favret, F.A. (1994) A comparative multi-nuclear relaxation study of protein–DMSO and protein–water interactions. *Journal of Magnetic Resonance*, **B103**, 142–151

Hills, B.P. and Le Floc'h, G. (1994) NMR studies of non-freezing water in randomly packed beds of porous particles. *Molecular Physics*, **82**, 751–763

Jelinski, L.W. (1980) ^{2}H NMR study of molecular motion in collagen fibrils. *Nature,* **284**, 531–534

Kitchin, S.J. (1993). *Deuterium NMR Studies of β-cyclodextrin complexes*, thesis, University of York

Koenig, S.H., Brown, III, R.D., Pande, A. and Ugolini, R. (1993) Rotational inhibition and magnetization transfer of -crystallin solutions. *Journal of Magnetic Resonance*, 172–177

Lauffer, B.L. (1987) Paramagnetic metal complexes as water proton relaxation agents for NMR imaging: Theory and design. *Chemical Reviews*, **87**, 901–927

Parmer, J.F., Dickenson, L.C., Chien, J.C.W. and Porter, R.S. (1987) Polymer–polymer miscibility determination via CPMAS in blends of deuterated and protonated polymers. *Macromolecules*, **2**, 2308–2310

Santyr, G.E., Henkelman, R.M. and Bronskill, M.J. (1988) Variation in measured transverse relaxation in tissue resulting from spin locking with the CPMG sequence. *Journal of Magnetic Resonance*, **79**, 28–44

Tanaka, H. and Nishi, T. (1986) Spin diffusion in block copolymers as studied by pulsed NMR. *Physical Review B*, **33**, 32–42

Tanner, S.F., Hills, B.P. and Parker, R. (1991) Interactions of sorbed water with starch studied using NMR spectroscopy. *Journal of the Chemical Society, Faraday Transactions*, **87**, 2613–2621

Vander Hart, D.L and Manders, W.F. (1987) Proton magic-angle spinning NMR method for determining intimate mixing in polymer blends. *Macromolecules*, **20**, 1725–1728

Voelkel, R. (1988) High-resolution solid state ^{13}C NMR spectroscopy of polymers. *Angewante Chemie – International Edition in English*, **27**, 1468–1483

8

MICROSCOPY AS AN ANALYTICAL TOOL IN THE STUDY OF PHASE SEPARATION OF STARCH–GELATIN BINARY MIXTURES

R.M. ABEYSEKERA and A.W. ROBARDS
Institute for Applied Biology, University of York, P.O. Box 373, York. YO1 5YW U.K.

Introduction

Modern microscopy offers a range of techniques to study food structure, from gross morphology to atomic resolution. Using a simple phase-separated carbohydrate–protein binary mixture as a model, we have illustrated that microscopy is a powerful analytical tool that can be used in the study of phase separation of polymer mixtures. Some of the most useful techniques are illustrated with a discussion of their contributions and limitations. Finally, new techniques are mentioned in relation to probable future developments.

Biopolymer mixtures form the basis of the majority of processed foods, yet our understanding of how polymers interact, separate into microphases or bulk separate into layers is incompletely understood. Different techniques are used to study the behaviour of biopolymer mixtures but microscopy is one of the most useful and direct of all currently available techniques.

For our model system we have selected gelatin and starch. The polymeric component of gelatin, collagen, is a flexible rod containing predominantly the amino acids proline, hydroxyproline and glycine. Starch consists of two polymer types. Amylose is a straight chained α1–4 linked glucan chain. Amylopectin, a much larger molecule, contains the same glucan backbone with α1–6 linked branches. A mixture of these polymers is thermodynamically incompatible and phase separates when mixed above their gelling temperatures.

Methodology

MATERIAL

Modified maize starch, C*Set 06598 Batch SH 1338 provided by Cerestar U.K. Ltd.

Potato maltodextrin, C*Pur 01906 Batch HQ 1265 provided by Cerestar U.K. Ltd. Pig skin gelatin PC1 (lime hide) provided by Systems Bio-Industries.

PREPARATION OF GELS

The gelatin (10% w/w) was heated to 60°C. The starch (35% w/w) and maltodextrin (20% w/w) were heated to 100°C and 85°C respectively and cooled to 60°C before mixing with gelatin. The mixed gels were stirred for 30 minutes and cooled on ice and stored at 5°C for approximately 18 hours prior to processing for microscopy.

Results and Discussion

LIGHT MICROSCOPY

Phase contrast and differential interference contrast microscopy

The ideal manner in which to examine structure is in a way that avoids interference. Collagen, amylopectin and amylose are unpigmented and lack contrast when examined with direct light. Both the phase contrast and the differential interference contrast techniques provide enhancement of unstained structures with different refractive indices. They allow the path of the light to be modified by shifting phase and subsequently recombining to increase contrast between different refractive indices. Figure 8.1 is a phase-separated unstained sample of starch and gelatin mixture examined by phase contrast optics. The same sample examined by bright field optics cannot distinguish the excluded starch phase. Both viscous solutions of binary mixtures spread on glass slides or sections of solidified binary gels can be examined in this manner without any chemical treatment or staining. The technique is rapid and can be used to provide information on dimensions of different phases and therefore the extent of phase separation. Further, because dyes are unnecessary for providing contrast, dynamic activities such as the stages of phase separation can be monitored and recorded by video detection.

Bright field microscopy

This is the most commonly used of the light microscopy techniques (Yiu, 1993). White light is transmitted through the specimen, where histochemical dyes are used to provide contrast to specific chemical components.

The best manner to examine a specimen is without any interference or processing at all. However, if an unstable sample is not stabilised (chemically fixed or

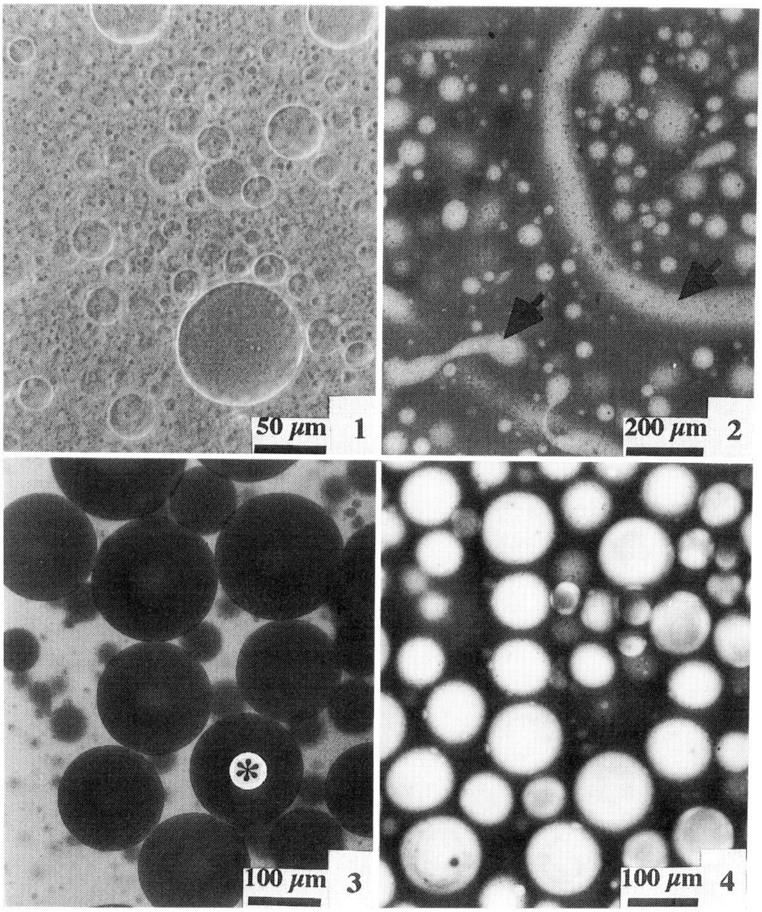

Figure 8.1 A phase contrast image of an unstained phase separated starch–gelatin mixture. The gelatin is in the continuous and the starch is in the excluded phase

Figure 8.2 A bright field image of a phase separated starch–gelatin mixture stained with iodine vapour. The arrows point to distortion of the gelatin phase as a result of cover slip pressure

Figure 8.3 A bright field image of a Sirius red stained phase separated starch–gelatin mixture. The (*) marks the swollen excluded gelatin phase

Figure 8.4 A bright field image of a proportion of the same sample as in Figure 8.3, but stained with iodine vapour. The continuous starch phase is stained by the iodine

embedded), structural deformation can be easily induced. For example, if one or both of the polymers in a binary mixture is not gelled, physical deformation can be introduced by placing a cover slip on the sample. Figure 8.2 illustrates shearing in the excluded gelatin phase caused by a cover slip. Use of small props or spacers between the cover slip, the sample and the glass slide, for instance, can prevent such deformation effects.

In bright field microscopy, it is necessary to use a dye and therefore it is important to bring the dye to its target by diffusion without modifying structure along the way. Most histochemical dyes used in food research are prepared in an aqueous medium. If the sample is not chemically fixed and embedded in a resin, then diffusion of an aqueous dye can have detrimental effects on the native structure. In such samples, movement, displacement and swelling are common phenomena. Therefore, the use of an aqueous dye for quantitative analysis should be carried out with extreme caution. The example in Figure 8.3 illustrates swelling of the excluded gelatin phase due to staining with aqueous Sirius red. Another portion of the same sample examined after staining with iodine vapour is illustrated in Figure 8.4. The non-invasive staining by vapour leaves the structure undisturbed. Iodine is commonly used in its aqueous state (Lugol's iodine solution) to identify amylose, which forms inclusion complexes with iodine (stained blue) and amylopectin (stained red) in starches (Flint, 1988; Svegmark, Kidman and Hermansson, 1993). Iodine vapour (Flint, 1988) is, however, less commonly used but provides a very useful non-invasive approach to identify starch components in un-embedded samples. Images obtained by the vapour staining technique in combination with image analysis can confidently be used to study the phenomenon of phase separation of starch with other biopolymer mixtures.

Both un-gelled viscous mixtures and very thin sections of gelled mixtures can be used where only single layers of the excluded particles are present. We have determined, by examining sections of gels end-on after embedding in a resin that, where sections are cut in which gelatin is in the excluded phase, the plane of the cut does not go through the gelatin spheres but along their interface. Therefore, in an image such as Figure 8.4, the diameter of the excluded phase is its true diameter, allowing quantitative analysis. In a similar way, all gels should be examined carefully to determine if they are suitable for quantitative analysis by bright field microscopy.

Starch and gelatin binary mixtures can also be chemically fixed and embedded in resin to carry out histochemical staining for qualitative and quantitative analysis. This is time consuming and can introduce shrinkage (Figure 8.5) or loss of material. Therefore unless chemical fixation and embedding is absolutely necessary, e.g. for techniques such as colloidal gold probing (see below), use of un-embedded starch–gelatin is recommended for studies of their interactions by light microscopy, because the same information can be obtained more rapidly by iodine vapour staining of fresh gels (see Figure 8.6 for comparison of a periodic acid/Shiffs' stained resin embedded starch–gelatin binary mixture).

The specificity of histochemical staining should be interpreted with caution. As commonly known, iodine reacts with starches, but it is less well recognised that it also binds to gelatin and lipids, giving a pale yellow coloration (Flint, 1988). The periodic acid/Shiffs' reaction stains some cell wall polysaccharides as well as starch, giving a brilliant pink coloration. A histochemical stain can easily distinguish a

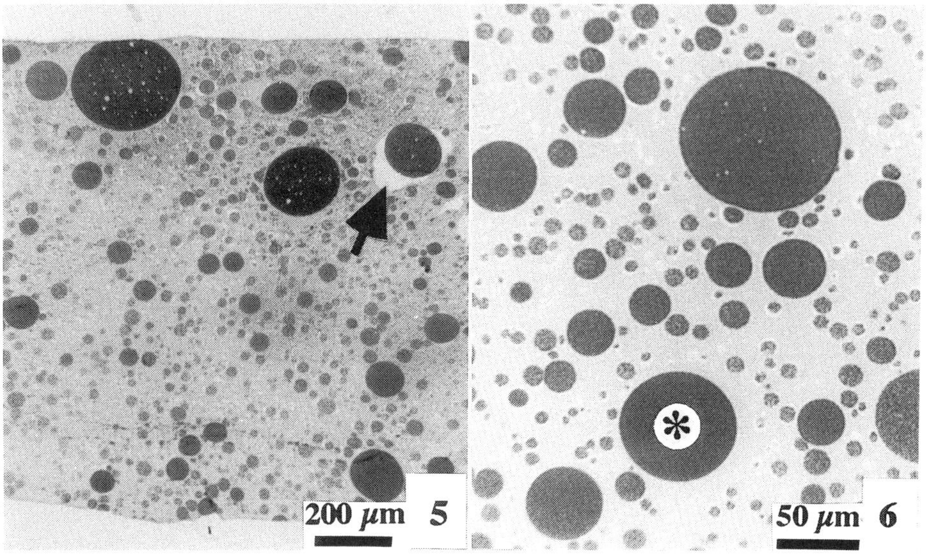

Figure 8.5 A section of a resin embedded starch–gelatin gel. The arrow points to areas of shrinkage of the continuous gelatin phase during processing of the material

Figure 8.6 A bright field image of a resin embedded starch–gelatin gel stained with periodic acid / Shiffs' stainstain. The excluded starch phase is intensely stained (*)

protein from a carbohydrate. However, chemical identification is more difficult between two different types of proteins or carbohydrates. These stains are therefore very helpful for chemical analysis if there is an initial knowledge of the components making up a polymer mixture. In the absence of that understanding, it is important to use other chemical identification techniques (see below) alongside histochemistry.

Epifluorescence microscopy

In this technique, the specimen is illuminated with ultraviolet light (light of short wavelength), usually from a high pressure mercury source. Re-emitted light (longer wavelength) is detected as fluorescence (see Rost, 1991). The technique requires either the component to be studied to be autofluorescent or, if not, to be tagged with an external fluorochrome. If a gel is to be studied un-embedded, the same problems described for the use of histochemical stains by bright field microscopy will be encountered. Some workers prefer to add solid crystals of the fluorochrome directly onto the specimen to allow diffusion with minimum disturbance (Brooker, 1991a). Here the crystal size must be small to ensure that there will be no artificial clustering of the dye crystals along the diffusion path to its target.

Fluorescein isothiocyanate (FITC) is a fluorochrome commonly used to identify gelatin. The plant lectins, Concanavalin A and *Lens culinaris* agglutin labelled with fluorescein can be used as markers for starches (Miller, Yiu, Fulcher and Altosaar, 1984; Yiu, 1993).

An observation that is widely ignored by food scientists is the autofluorescence of gelatin (see Aguilera and Stanley, 1990 for references). This autofluorescence is fairly stable (see the section on confocal microscopy) and can be used as a qualitative and an analytical tool in the study of phase separation of binary mixtures of starches with gelatin. External fluorochromes can also be used in quantitative microscopy (see Rost, 1991) to determine the concentration of the gelatin and starches during phase separation.

Confocal laser scanning microscopy

The confocal laser scanning microscope (CLSM) is a significant development in the field of light microscopy. The technique allows the use of very large samples whereby the non-invasive penetration of the laser permits serial optical sectioning of material focused only in a single focal plane. This allows very rapid three-dimensional reconstructions and stereo imaging of structure. Binary images thus produced can be rapidly processed for image analysis. For example, particle size, distribution and variance are processed as a part of the data collection directly. Most commercial CLSMs can be equipped with image analysis facilities.

Today, the scope of confocal microscopy in the food industry is well established (Brooker, 1991a). Dynamic activity can be measured by taking a time-lapse series of confocal images. For example, by using a temperature-controlled stage attached to the CLSM the dynamic activities of phase separation and coalescence can be monitored. Figure 8.7 is a time-lapse series of confocal images illustrating the coalescence of a phase-separated starch and gelatin mixture examined at 40°C after 30 minutes of separation and at 60°C after 40, 50 and 60 minutes of separation. Although faded to a certain extent, the autofluorescence of gelatin is detected and still apparent after a 60 minutes of monitoring.

As with fluorescence microscopy, this technique is also open for quantitative analysis by use of specific fluorochromes (Rost, 1991).

Colloidal gold probes (see below) are also used in the reflection contrast mode of the CLSM to give a resolution of 40 nm (Shotton, 1989), allowing the CLSM to be used as a powerful analytical tool.

Colloidal gold probes as molecular markers

A colloidal gold probe is a gold sphere coated with a selected protein (usually an

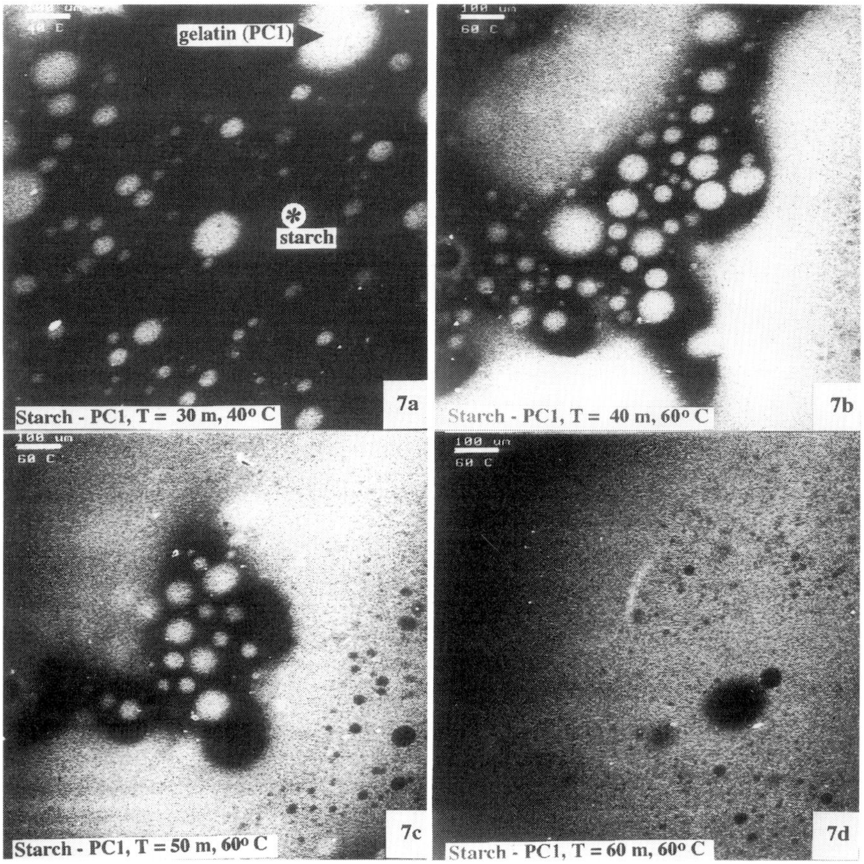

Figure 8.7 A series of time-lapse CLSM images of the process of coalescence of a phase separated starch–gelatin mixture. The light areas are the autofluorescence of the gelatin

antibody) marker. The gold is prepared by reducing tetrachloroauric acid to form small spheres of selected diameter, usually ranging between 2—50 nm. The technique, since its inception in 1971 (Faulk and Taylor), has been used extensively in immunocytochemistry (Bullock and Petrusz, 1985; Beesley, 1989). The technique principally involves the adsorption of a selected protein to the colloidal gold. The specific mechanism of this adsorption is not known (Beesley, 1989) but is assumed to be electrostatic in nature where the negatively charged gold is believed to interact with positively charged proteins by Van der Waals attractive forces (Roth, 1983). Despite the lack of understanding of the mechanism of colloidal gold binding, under controlled conditions, colloidal gold probing provides one of the most powerful analytical tools in microscopy.

Although lengthy and time consuming, quantitative analysis has also been carried out using colloidal gold probing (Bendayan, 1984; Griffith and Hoppeler, 1986). It requires strict controls, keeping polymer concentrations, fixation and embedding schedules all constant. The gold probe to substrate ratio is not one to one. Only the very small, 2 nm gold particles may approach this ratio.

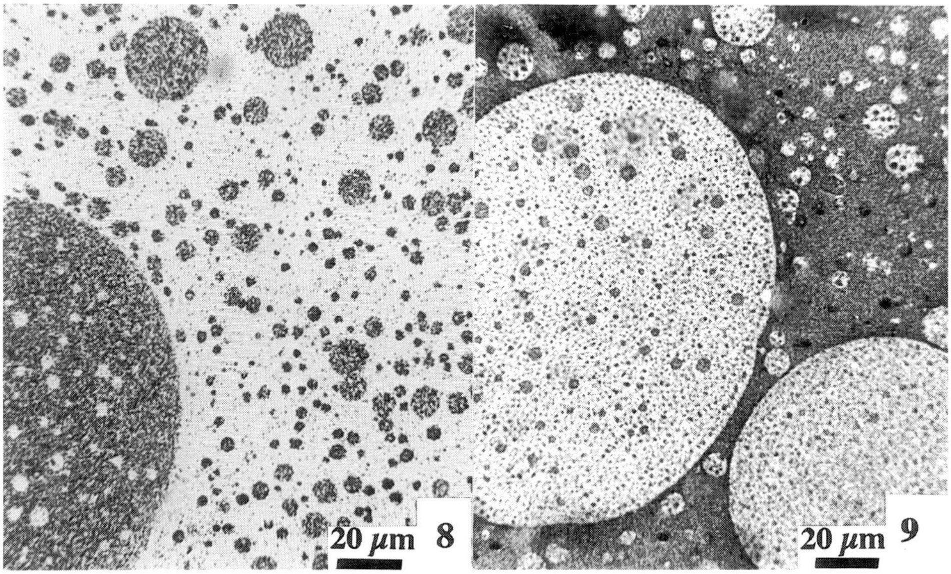

Figure 8.8 A bright field image of a resin embedded section of starch–gelatin labelled with pullulinase–colloidal gold followed by silver enhancement. The excluded starch phase is labelled

Figure 8.9 A bright field image of a resin embedded section of starch–gelatin gel labelled with pure colloidal gold followed by silver enhancement. The continuous gelatin phase is labelled

In principal, pure colloidal gold can be used to localise proteins directly in sections of embedded material. Figure 8.9 is a section of a resin embedded starch–gelatin gel labelled with pure colloidal gold. The gold is enhanced by the silver enhancement technique (Holgate, Jackson, Cowen and Bird, 1983) to visualise the bound sites by bright field light microscopy. The section in Figure 8.9 is taken from the immediate vicinity to that illustrated in Figure 8.6, where the excluded phase is clearly identified as starch by the periodic acid/Schiffs' reaction which labels 1,2-glycols (Pearse, 1972). Therefore, in Figure 8.9, the pure colloidal gold, as expected, has bound to the continuous phase of the charged protein, gelatin, as distinguished from the presumably uncharged excluded polysaccharide phase. However, one must be extremely cautious in distinguishing proteins from carbohydrates in mixtures in this manner. Horisberger and Rosset (1977) observed that neutral polysaccharides such as mannose and polyethylene glycol also bind pure colloidal gold.

For colloidal gold probing, the gels have to be embedded in a suitable non-epoxy resin that will not alter the molecular conformation of the probe binding sites (the substrates). The reader is referred to Beesley (1989) for a selection of appropriate methodologies for colloidal gold probing. The pH and the concentration of the gold and the probe protein are critical to ensure optimal adsorption that will lead to specific binding (Geoghegan and Ackerman, 1977). This is not a technique that is recommended to the casual microscopist.

Bendayan (1985) showed that enzymes can also be adsorbed onto colloidal gold and the enzyme–gold probe was used in turn to locate substrate sites. This is a useful way to locate specific carbohydrates in mixtures. Figure 8.8 is a section of a region of the starch–gelatin gel in the vicinity of that shown in Figure 8.9 but labelled with a pullulinase–gold probe with specific binding at the α1–6 sites on the glucan chains. The gold was enhanced by silver staining. Note the binding of the label to the excluded starch phase (compare the contrasting images of Figures 8.6 and 8.8 with Figure 8.9).

SCANNING ELECTRON MICROSCOPY

This is perhaps the most commonly used of the microscopy techniques by food scientists (see examples in Lewis, 1988; Schmidt and Buchheim, 1992). Just about every food type has been examined by scanning electron microscopy (SEM). According to Aguila and Stanley (1990), since the scanning electron microscope became commercially available in the mid-1960s, the instrument has been recognised by food scientists as “the primary source of microstructural information.” Very large samples can be used that will provide three-dimensional topographic detail with a great depth of field, far more than ever obtainable by light microscopic methods. Although, theoretically, an actual magnification of approximately 100,000× can be achieved by SEM, in practice at magnifications above 20,000× the resolution becomes limiting.

Specimen preparation by direct sputter coating, critical-point drying or by cryo-preservation is less extensive than for transmission electron microscopy (TEM). However, any preparation step can give rise to artifacts if not monitored carefully. Material can be extracted during critical point drying while ice crystal growth during low temperature processing can damage delicate molecular organisation.

In our experience in the study of phase separation of biopolymer mixtures, especially that of starches and gelatin, we have found that the topographical information obtained by SEM techniques was not different from that obtained from light microscopy or CLSM. Further, the limited operational resolution of SEM precludes the acquisition of substantial molecular information. Therefore we have not included any SEM data obtained on phase-separated gels in this publication.

Cathodoluminescence and X-ray microanalysis in combination with cryo-SEM can be powerful analytical tools. In X-ray microanalysis, emitted X-rays from elements bombarded with electrons reflect characteristics of the elemental composition of the sample. This analysis is widely used in the food industry (Brooker, 1991b). Diffusion and localisation of ions in binary gels can be detected in this manner. The technique is, however, usually limited to the detection of elements bearing an atomic weight higher than or equal to sodium.

Mention should be made of newly developed environmental SEM. This will allow the examination of hydrated samples at ambient temperatures and pressures (see Roziewski and Russel, 1992; McDonough, Gomez, Lee, Waniska and Rooney, 1993). However, this is still at a very early stage of development, although when completed it will provide a very useful means of examining dynamic processes at a relatively high resolution.

TRANSMISSION ELECTRON MICROSCOPY

Transmission electron microscopy (TEM) can provide an actual molecular resolution up to 0.2 nm. However, processing of material for TEM becomes even more critical because the smallest introduced deformation is detected at sub-nm resolution.

An un-gelled viscous solution or liquid polymer mixture can be chemically fixed or fixed by freezing and substituting the ice with a chemical fixative at low temperatures before embedding in a resin. A simple and an effective design for the encapsulation of liquids for TEM processing is described by Alleyne, McMahon, Youssef and Hekmat (1993). A solid gel, on the other hand, is easier to process.

Thin sections of 40 nm thickness can be sectioned from the embedded material for conventional TEM. Figures 8.10, 8.11, 8.12 and 8.13 are TEM images of starch or maltodextrin phase-separated with gelatin viewed at various magnifications. Information such as the distribution, interaction of the polymers and intermolecular porosity only become visible at the resolution obtainable with the TEM. Macromolecular aggregates are visible in the SEM but individual molecules are difficult to resolve. Note the difference between the molecular network structures formed by the oxidised waxy maize starch (Figures 8.11 and 8.12) and the potato maltodextrin (Figure 8.13). Although waxy maize starch has a higher amylopectin content than the potato maltodextrin, the network structural differences depicted above can be due to the differences in the chain lengths of the two carbohydrates. The network structure of gelatin formed after phase separation, on the other hand, appears unaffected whether it was initially mixed with starches or maltodextrins (Figures 8.12 and 8.13).

Figure 8.10 depicts a gradient of excluded gelatin spheres approaching the starch–gelatin interface, increasing their total volume as they approach the gelatin-continuous phase. Although the larger 2 μm gelatin spheres will be visible by light microscopy, the smaller particles and therefore the gradient will probably not be detected by this microscopical method.

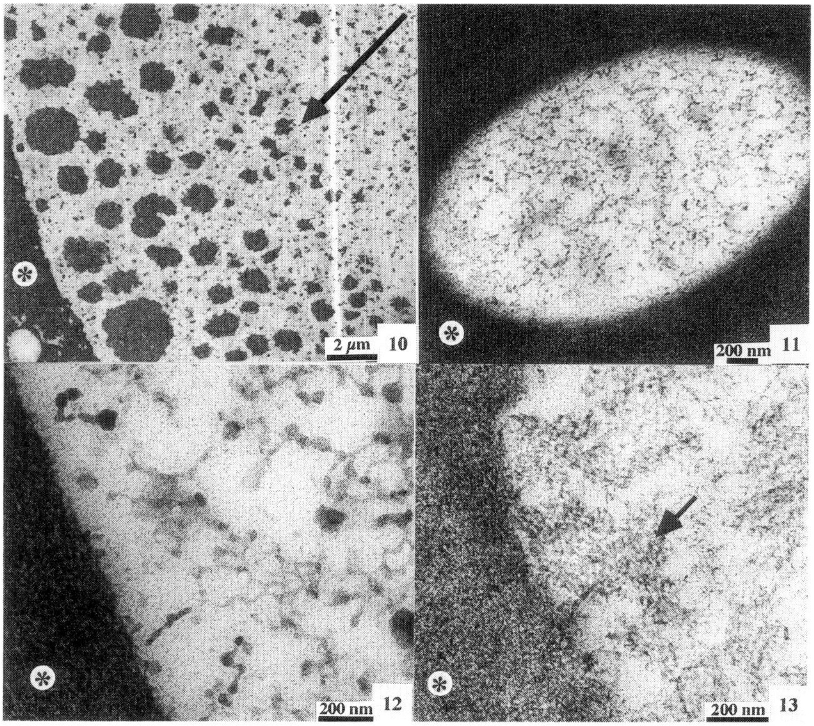

Figure 8.10 A TEM image of a thin section of phase separating starch–gelatin stained with uranyl acetate and lead citrate. The osmiophilic gelatin (*) is stained dark and the arrow points in the direction of the increasing gelatin particle size approaching the starch–gelatin interface

Figure 8.11 A high magnification TEM image of phase separated starch–gelatin mixture. The gelatin phase is indicated by (*)

Figure 8.12 Details of the starch and gelatin network illustrated in Figure 8.11. The gelatin phase is indicated by (*)

Figure 8.13 A high magnification TEM image of the network of phase separated maltodextrin–gelatin mixture. The gelatin phase is indicated by (*) and the maltodextrin network is indicated by the arrow

The thin-sectioning technique can only provide two-dimensional information unless extensive time is spent on serial sectioning that can provide information for three-dimensional image reconstruction. Alternatively, metal replication techniques (Robards and Abeysekera, 1992; Abeysekera, Bergström, Goodall, Norton and Robards, 1993; Djabourov, Bonnet, Kaplan, Favard, Favard, Lechaire and Maillard, 1993; Stokke and Elgsaeter, 1994) of polymers and gels can provide a three-dimensional image of the polymers (Figure 8.14) and a clearer view of polymer–polymer interactions.

By TEM, one type of polymer may not appear visibly significantly different from another. For example, a collagen chain in gelatin (Djabourov *et al.*, 1993) does not appear significantly different from the amylose chains in amylose gels (Leloup, Colonna, Ring, Roberts and Wells, 1992). Here, the more specific probe labelling of the polymers described above using colloidal gold markers is used. Figure 8.15 is a thin section of a maltodextrin network labelled with α-amylase tagged to colloidal gold. At this magnification, the α-amylase adsorbed gold particles (10 nm) are seen aligned over the maltodextrin network.

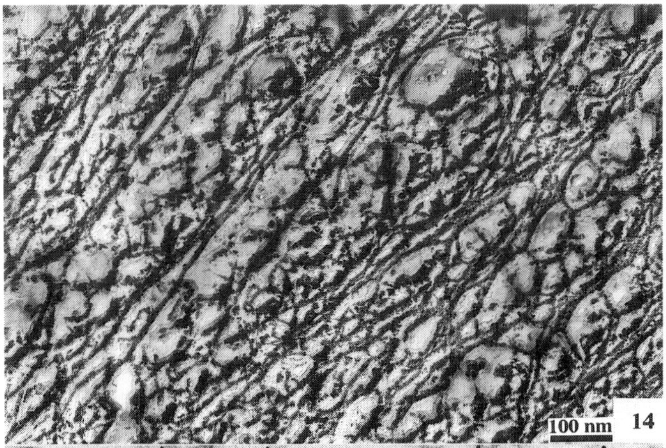

14

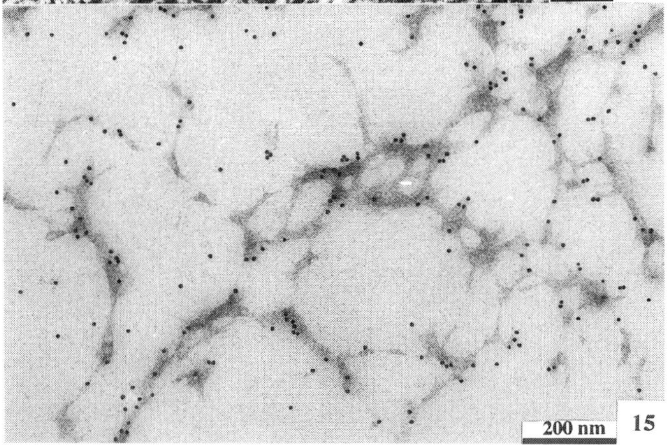

15

Figure 8.14 A TEM image of a platinum–carbon metal replica of a gelatin network illustrating the collagen fibrils

Figure 8.15 A TEM image of a thin section of maltodextrin network labelled with α-amylase–colloidal gold

FUTURE DEVELOPMENTS

The newly developed family of scanning probe microscopes (SPMs) provides the potential to examine the topography of molecules with atomic resolution at ambient pressures in the hydrated state. Roberts, Williams, Davies, Jackson and Tendler (1994) give a good informative description of the technique. Although extensively used in the surface sciences in studies of semiconductors, SPM techniques are still new to biologists. Many of the early studies on the biological systems have been done on DNA (e.g. Arscott, Lee, Bloomfield and Evans, 1989). The work of V.J. Morris and co-workers over the last five years has shed hope for food scientists to examine the structure and interactions of food-related polymers. The structures of single polymer types, e.g. glucoamylases (Kramer, Gunning, Morris, Belshaw and Williamson, 1993), carrageenan (Lee, Atkins and Miles, 1992), xanthan gum (Gunning, McMaster and Morris, 1993), poly(γ-benzyl-*L*-glutamate) (McMaster, Carr, Miles, Cairns and Morris, 1990), wheat gluten proteins (Miles, Carr, McMaster, Kenneth, Belton, Morris, Field, Shewry and Tatham, 1991) and vicilin (Welland, Miles, Lambert, Morris, Coombs and Pethica, 1989) have all been studied by scanning tunnelling microscopy (STM). The samples were prepared by depositing them on highly oriented pyrolitic graphite and air drying. Miles and co-workers (Miles, McMaster, Carr, Tatham, Shewry, Field, Belton, Jeenes, Hanley, Whittam, Cairns, Morris and Lambert, 1990) also reported images of the drug, ethidium bromide, binding to DNA. This is the only SPM study that we are aware of where the study of the interaction of two different types of biologically relevant molecules has been attempted by STM. Cellulose microfibrils have been examined by atomic force microscopy (AFM) (Hanley, Giasson, Revol and Gray, 1992).

Wilkins and co-workers (Wilkins, Davies, Jackson, Mitchell, Roberts, Stokke and Tendler, 1993) have studied the structure of xanthan gum after coating with platinum and carbon. Although the coating gives better resolution and stabilisation of the molecules than the technique of V.J. Morris, the images obtained did not provide new information compared to the same metal replicas examined by TEM. The report by Meyer, Rouquet, Lecourtier and Toulhoat (1992), who were actually able to produce images of mobile microgels of xanthan in a fluid cell, is quite promising.

Cryogenic force microscopy of collagen was reported recently by Shattuck, Gustafsson, Fisher, Yanagimoto, Veis, Bhatnagar and Clark (1994). Unlike in the research described above, Shattuck and co-workers designed a force microscope operating at -130°C. Promising preliminary observations of inter- and intra-molecular interactions of collagen molecules were reported.

Artifacts also abound this newly explored technology. Artifacts resembling periodic molecular structure have been reported to be commonly created by the technique (Lee *et al.*, 1992). Therefore interpretation plays a significant role in this new technology. As a technique, the SPM technology has great potential for the food scientist, although it is still at a very early stage of development.

Conclusions

Microscopy is a powerful tool. The above description clearly illustrates the significance of microscopy in analysing structures from gross morphology to atomic resolution of biopolymer mixtures. It is imperative, however, to identify the microscopical technique that will most rapidly and most effectively provide us with the answer to the specific question asked, and then to use this method to observe the specimen with as little structural alteration as possible. In Table 8.1 we have summarised the relevant techniques, their capabilities and limits, in the model system of phase-separated starch and gelatin studied. By carefully selecting the technique to suit a specific problem, unlike any other available technique, we can use microscopy as a tool to study a range of events of biopolymer phase separation from the morphologies of the phases down to the potential for analysing the placement and chemical nature of atoms.

Table 8.1 SUMMARY OF THE CAPABILITIES AND LIMITATIONS OF MICROSCOPY TECHNIQUES WHEN INVESTIGATING STARCH—GELATIN BINARY MIXTURES

	Qualitative	*Quantitative*	*Chemical*	*Kinetic*	*Resolution*
Phase & interference contrast	+	-	-	+	Phases
Bright field	+	+	+	-	Phases
Epi-fluorescence	+	+	+	+	Phases
CSLM	+	+	+	+	Phases
Colloidal gold probing	+	+	+	-	Phases
SEM	+	+	+	+	Macro-molecular
TEM	+	+	+	-	Molecular
SPM	+	+	+	+	Atomic

Acknowledgements

We thank Cerestar U.K. Ltd. for providing the starch samples and Systems Bio-Industries for providing the gelatin. We thank Dr. Eric Kolodziejczyk at the Nestlé Central Research Laboratory for assistance with the colloidal gold probe work, Dr. John North at the Nestec York Ltd. Research Laboratory for assistance with confocal microscopy and Mrs. Sophia Lihong Cheng for technical assistance. This work is funded by a grant from a MAFF DTI LINK project.

References

Abeysekera, R.M., Bergström, E.T., Goodall, D.M., Norton, I.T. and Robards, A.W. (1993) Ultrastructural evidence for intramolecular double stranding in -carrageenan. *Carbohydrate Research*, **248**, 225–231

Aguilera, J.M. and Stanley, D.W. (1990). *Microstructural Principles of Food Processing and Engineering*, Amsterdam: Elsevier Science Publishers Ltd., pp 1–57

Alleyne, M.C., McMahon, D.J., Youssef, N.N. and Hekmat, S. (1993) An apparatus for a new microcube encapsulation of fluid milk in preparation for transmission electron microscopy. *Food Structure*, **12**, 21–30

Arscott, P.G., Bloomfield, V.A. and Evans, D.F. (1989) Scanning tunnelling microscopy of Z-DNA. *Nature*, **339**, 484–486

Beesley, J.E. (1989). In *Colloidal Gold: A New Perspective for Cytochemical Marking*. Royal Microscopical Society Microscopy Handbooks, **17**, Oxford: Oxford University Press

Bendayan, M. (1984) Protein-A gold electron microscopic immunocytochemistry – Methods, applications, and limitations. *Journal of Electron Microscope Technology*, **1**, 243–270

Bendayan, M. (1985). In *Techniques in Immunocytochemistry*, **3**, edited by Bullock, G.R. and Petrusz, P., London: Academic Press Inc., pp 179–201

Brooker, B.E. (1991a). *Microscopy and Analysis*, November 1991, 13–15

Brooker, B.E. (1991b) Recent developments in the application of X-ray microanalysis to the study of food systems. *Food Structure*, **10**, 145–152

Bullock, G.R. and Petrusz, P. (Editors) (1985). *Techniques in Immunocytochemistry*, **3**. London: Academic Press

Djabourov, M., Bonnet, N., Kaplan., H., Favard, N., Favard, P., Lechaire, J.P. and Maillard, M. (1993) 3D analysis of gelatin gel networks from transmission electron microscopy imaging. *Journal de Physique II*, **3**, 611–624

Faulk, W.P. and Taylor, G.M. (1971). *Immunocytochemistry*, **8**, 1081–1083

Flint, F.O. (1988) The evaluation of food structure by light microscopy. In *Food Structure – Its Creation and Evaluation*, edited by Blanshard, J.M.V. and Mitchell, J.R., London: Butterworths, pp 351–365

Geoghegan, W.D. and Ackerman, G.A. (1977). *Journal of Histochemistry and Cytochemistry*, **25**, 1182–1200

Griffith, G. and Hoppeler, H. (1986) Quantitation in immunocytochemistry – Correlation of immunogold labelling to absolute number of membrane antigens. *Journal of Histochemistry and Cytochemistry*, **34**, 1389–1398

Gunning, A.P., McMaster, T.J. and Morris, V.J. (1993) Scanning tunnelling microscopy of xanthan gum. *Carbohydrate Polymers*, **21**, 47–51

Hanley, S.J., Giasson, J., Revol, J.-F. and Gray, D.G. (1992) Atomic force microscopy of microfibrils – Comparison with transmission electron microscopy. *Polymer*, **33**, 4639–4642

Holgate, C.S., Jackson, P., Cowen, P.N. and Bird, C.C. (1983) Immunogold silver staining – New method of immunostaining with enhanced sensitivity. *Journal of Histochemistry and Cytochemistry*, **31**, 938–944

Horisberger, M. and Rosset, J. (1977). *Journal of Histochemistry and Cytochemistry*, **25**, 295–305

Kramer, G.F.H., Gunning, A.P., Morris, V.J., Belshaw, N.J. and Williamson, G. (1993) Scanning tunnelling microscopy of *aspergillus niger* glucoamylases. *Journal of Chemical Society, Faraday Transactions*, **89**, 2595–2602

Lee, I., Atkins, E.D.T. and Miles, M.J. (1992) Visualization of the algal polysaccharide carrageenan by scanning tunnelling microscopy. *Ultramicroscopy*, **42–44**, 1107–1112

Leloup, V.M., Colonna, P., Ring, S.G., Roberts, K. and Wells, B. (1992) Microstructure of amylose gels. *Carbohydrate Polymers*, **18**, 189–197

Lewis, D.L. (1988) An electron microscopist's view of foods. In *Food Structure – Its Creation and Evaluation*, edited by Blanshard, J.M.V. and Mitchell, J.R., London: Butterworths, pp 367–385

McMaster, T.J., Carr, H., Miles, M.J., Cairns, M.P. and Morris, V.J. (1990) Polypeptide structures imaged by the scanning tunnelling microscope. *Journal of Vacuum Science Technology*, **A8**, 648–651

McDonough, C., Gomez, M.H., Lee, J.K., Waniska, R.D. and Rooney, L.W. (1993) Environmental scanning electron microscopy evaluation of tortilla chip microstructure during deep fat frying. *Journal of Food Science*, **58**, 199–213

Meyer, A., Rouquet, G., Lecourtier, J. and Toulhoat, H. (1992). *Physical Chemistry of Colloids and Interfaces in Oil Production*, **27**, 275–279

Miles, M.J., McMaster, T., Carr, H.J., Tatham, A.S., Shewry, P.R., Field, J.M., Belton, P.S., Jeenes, D., Hanley, B., Whittam, M., Cairns, P., Morris, V.J. and Lambert, N. (1990) Scanning tunnelling microscopy of biomolecules. *Journal of Vacuum Science Technology*, **A8**, 698–702

Miles, M.J., Carr, H.J., McMaster, T.C., Kenneth, I.J., Belton, P.S., Morris, V.J., Field, J.M., Shewry, P.R. and Tatham, A.S. (1991) Scanning tunnelling microscopy of a wheat seed storage protein reveals details of an unusual supersecondary structure. *Proceedings of the National Academy of Sciences*, **88**, 68–71

Miller, S.S., Yiu, S.H., Fulcher, R.G. and Altosaar, I. (1984) Primary evaluation of lectins as fluorescent probes of seed structure and composition. *Food Microstructure*, **3**, 133–139

Pearse, A.G.E. (1972). In *Histochemistry: Theoretical and Applied*, 3rd edition, London: Churchill–Livingston

Robards, A.W. and Abeysekera, R.M. (1992). In *Electron Microscopy 92*, **3**, edited by Megias-Megias, L., Rodriiguez-Gracia, M.L. and Rios, A., Granada: Secretario de Publicaciones de la Universidad de Granada, pp 77–78

Roberts, C., Williams, P.M., Davies, M.C., Jackson, D.E. and Tendler, S.J.B. (1994) Atomic force microscopy and scanning tunnelling microscopy – Refining techniques for studying biomolecules. *Trends in Biotechnology*, **12**, 127–132

Rost, F.W.D. (1991). In *Quantitative Fluorescence Microscopy*, Cambridge: Cambridge University Press

Roth, J. (1983). In *Techniques in Immunocytochemistry*, **2**, edited by Bullock, G.R. and Petrusz, P., London: Academic Press Inc., pp 216–284

Roziewski, K. and Russell, A.J. (1992) Effect of hydration on the morphology of enzyme powder. *Biotechnology and Bioengineering*, **39**, 1171–1175

Schmidt, D.G. and Buchheim, W. (1992) The application of electron microscopy in dairy research. *Journal of Microscopy*, **167**, 105–121

Shattuck, M.B., Gustafsson, M.G.L., Fisher, K.A., Yanagimoto, K.C., Veis, A., Bhatnagar, R.S. and Clarke, J. (1994) Monomeric collagen imaged by cryogenic force microscopy. *Journal of Microscopy*, **174**, RP1–RP2

Shotton, D. (1989). *Microscopy and Analysis*, July 1989, 7–12

Stokke, B.T. and Elgsaeter, A. (1994) Conformation, order–disorder conformation transitions and gelation of noncrystalline polysaccharides studied using electron microscopy. *Microscopy*, **25**, 469–491

Svegmark, K., Kidman, S. and Hermansson, A.-M. (1993) Molecular structures obtained from mixed amylose and potato starch dispersions and their biological behaviour. *Carbohydrate Polymers*, **22**, 19–29

Welland, M.E., Miles, M.J., Lambert, N., Morris, V.J., Coombs, J.H. and Pethica, J.B. (1989) Structure of the globular protein vicilin revealed by scanning tunnelling microscopy. *International Journal of Biological Macromolecules*, **11**, 29–33

Wilkins, M.J., Davies, M.C., Jackson, D.E., Mitchell, J.R., Roberts, C.J., Stokke, B.T. and Tendler, S.J.B. (1993) Comparison of scanning tunnelling microscopy and transmission electron microscopy image data of a microbial polysaccharide. *Ultramicroscopy*, **48**, 197–201

Yiu, S.H. (1993) Food microscopy and the nutritional quality of cereal foods. *Food Structure*, **12**, 123–133

9

DEPLETION FLOCCULATION

P.A. WILLIAMS and N.J. SMITH
Centre for Water Soluble Polymers, The North East Wales Institute, Plas Coch, Wrexham, Clwyd. LL11 2AW U.K.

Introduction

When particles are dispersed in a polar medium they will acquire a surface charge due, for example, to ionisation of surface groups, adsorption of ions from solution or differential dissolution of ions from the surface. The charges are neutralised by counterions present in solution which, due to thermal motion, form a diffuse cloud around the particles. This is referred to as the electrical double layer. At sufficiently high surface charges particle aggregation is prevented by the osmotic repulsive forces that occur when electrical double layers overlap. The particles are said to be charge (or electrostatically) stabilised and a quantitative theory (Derjaguin and Landau, 1941; Verwey and Overbeek, 1948) — known as DLVO theory — was developed to predict this phenomenon (Hunter, 1987). The thickness of the electrical double layer is a function of the solution ionic strength and in the presence of sufficient electrolyte it is compressed to such an extent that the particles are able to approach each other closely enough for van der Waals attractive forces to predominate and cause the particles to aggregate. This process results in the formation of a densely packed sediment or "cake" which is often very difficult to redisperse and hence is undesirable in many industrial applications.

The stability of particulate dispersions is strongly influenced by the addition of polymers (Napper, 1983). At low levels of polymer addition, bridging flocculation occurs due to polymer molecules adsorbing simultaneously onto two or more particles. The flocs produced are more voluminous than particle aggregates formed in the absence of polymer and result in a loosely packed sediment. When sufficient polymer is present to completely coat each particle then aggregation is prevented by the elastic and osmotic repulsions which arise from the overlap of the adsorbed polymer layers. This phenomenon is termed steric stabilisation (Heller and Pugh, 1954) and has been quantitatively described by Hesselink, Vrij and Overbeek (1971). Of more recent interest is the aggregation of particles brought about by the presence of non-adsorbed polymer (Fleer, Scheutjens and Vincent, 1984). This is often referred to as depletion flocculation and was first predicted by Asakura and Oosawa (1954, 1958). The centre of mass of a polymer close to a surface cannot approach the surface to a distance less than the polymer radius (Figure 9.1(a)). Hence the polymer segment concentration at the interface is less than in the bulk solution and results in a loss of configurational entropy of the polymer. This gives rise to a so-called depletion layer. On the close approach of two particles the depletion layers will overlap and polymer molecules will be excluded from the space between the particles (Figure 9.1(b)). Since the osmotic pressure of the bulk polymer solution is greater than in the excluded zone containing pure solvent the particles are pushed together. The net effect is that an attractive force is set up, the magnitude of which depends on the osmotic pressure of the polymer solution and on the size and shape of the polymer molecules. The osmotic pressure and radius of gyration, R_g, of polyelectrolytes are very much greater than for non-ionic polymers of the same molecular mass and hence they are expected to be far more effective at inducing depletion flocculation. Suprisingly very little work has been reported in the literature concerning this aspect (Snowden, Clegg, Williams and Robb, 1991; Nashimba, Sudo and Furusawa, 1992).

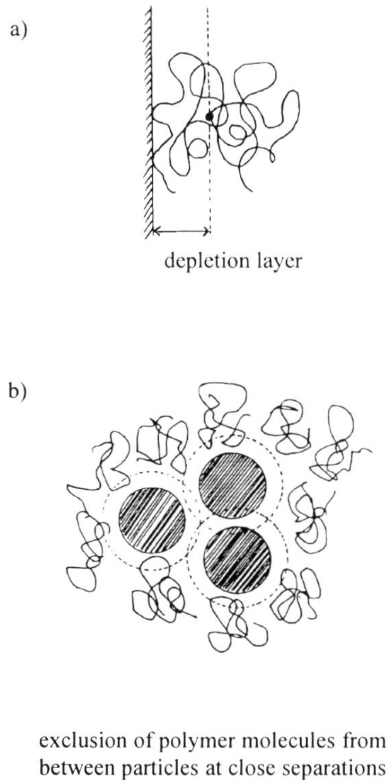

Figure 9.1 Schematic representation of: (a) Polymer depletion layer; and (b) Exclusion of polymer molecules from between particles at close separations leading to depletion flocculation

This paper describes work that we have undertaken to investigate this phenomenon. We have used a model system, namely polystyrene latex in the presence of hydroxyethyl cellulose (HEC) and sodium carboxymethyl cellulose (SCMC). HEC is non-ionic and adsorbs onto the negatively charged latex, whereas SCMC is a polyelectrolyte and will only adsorb at high ionic strengths.

Materials

Monodisperse polystyrene latices were prepared by the surfactant-free emulsion polymerisation method of Goodwin, Hearn, Ho and Ottewill (1974) using potassium persulphate as initiator. The latices were found to have a diameter of 789±3 nm by electron microscopy and a zeta potential of -57 mV by particle microelectrophoresis.

The HEC and SCMC were provided by Aqualon (U.K.) Ltd. and details of their physico-chemical characteristics are given in Table 9.1.

Table 9.1 PHYSICO CHEMICAL CHARACTERISTICS OF HEC AND SCMC

Code name	*D.S.*[a]	*M.S.*[b]	M_w[c]
Natrosol 250GR		2.5	250.000
Blanose	0.9		250.000

[a] degree of substitution quoted by manufacturer
[b] molar substitution quoted by manufacturer
[c] molecular weight quoted by manufacturer

The equivalent sphere hydrodynamic radius, R_h, was determined by photon correlation spectroscopy in 0.5 M NaCl at 25°C using the Oros 801 (Biotage). Measurements were made at a range of concentrations and R_h was obtained by extrapolation to infinite dilution. The values obtained were 36 nm and 63 nm for HEC and SCMC respectively. R_h for SCMC in the absence of electrolyte was estimated as 93 nm using an expansion factor of 1.59 (Schneider and Doty, 1954).

The fact that in water SCMC has an R_h almost three times that of HEC, even though they have similar molecular masses may be a consequence of coil expansion, which occurs as a consequence of intramolecular repulsion between the charged carboxylate groups along the SCMC chain.

The osmotic pressures of SCMC and HEC solutions were determined as a function of concentration using the Knauer membrane osmometer and the results are shown in Figure 9.2. The very much larger values obtained for SCMC compared to HEC are due to the presence of the sodium counterions. The Flory–Huggins interaction parameter, χ, was calculated from the second virial coefficients obtained from the osmotic pressure measurements, the values were found to be 0.463 and 0.468 for HEC and SCMC respectively. The adsorption capacity for HEC on polystyrene latex was taken as 1.2 mg/m^2 and the adsorbed layer thickness as 9 nm (Tanaka, Meadows, Phillips and Williams, 1989).

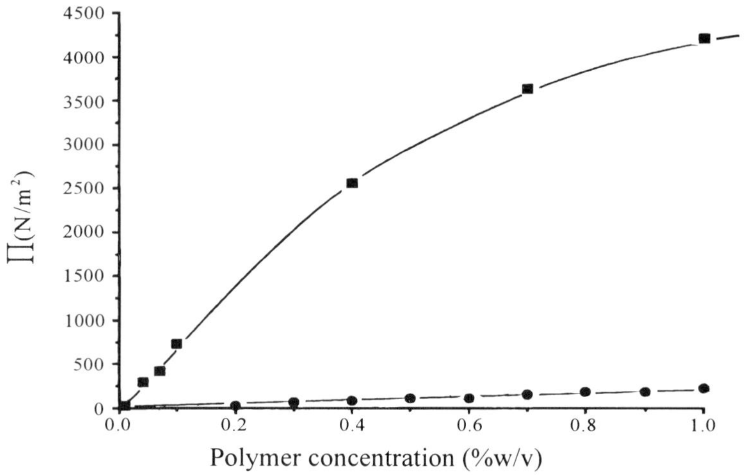

Figure 9.2 Osmotic pressure of solutions of: ■, sodium carboxymethylcellulose, and ●, hydroxyethylcellulose at varying concentrations

Methods

The stability of the latex dispersions was monitored from measurements of the wavelength dependence of the dispersion turbidity, τ. It has been shown by others (Long, Osmond and Vincent, 1973) that

$$\frac{\tau}{C_{c\to 0}} = K\lambda^{n} \qquad (1)$$

where C is the concentration of particles, λ is the wavelength of light and n and K are constants; n is related to particle size and if λ is much larger than the particle dimensions has a value of -4. As the particle size increases n tends towards zero.

0.1cm^3 of the polystyrene latex dispersion (3.5% w/w) was added to a series of polymer solutions at varying concentrations. The samples were sonicated for 2 minutes using an MSE Soniprep 150 sonicator in order to disrupt any aggregates and then left to stand for 18 hours. The sample tubes were then gently inverted to resuspend any sedimented particles and the absorbance of each sample was measured between 500—700 nm at 50 nm intervals. The plot of log absorbance against log wavelength was found to be linear over this wavelength range: n was obtained from the slope of the line.

Theory

The interaction between particles can be represented by a potential energy diagram in which the total potential energy of interaction (V_T) is plotted as a function of particle separation. V_T is obtained by summation of the individual interaction potentials arising from van der Waals attraction (V_A), electrostatic repulsion (V_R), steric repulsion (V_S) and the depletion force (V_D), i.e.

$$V_T = V_A + V_R + V_S + V_D \qquad (2)$$

The resultant potential energy curve often displays a primary and secondary minimum. This is illustrated in Figure 9.3 which shows how V_T, V_A and V_R vary with interparticle separation. If the height of the primary maximum (V_{max}) is of the order of 15—20 kT then aggregation in the primary minimum is prevented since the particles are not able to approach each other closely enough. At lower values of V_{max} strong irreversible aggregation will result. If the depth of the secondary minimum (V_{min}) is greater than a few kT (i.e. greater than the translational energy of the particles which is of the order of 1.5 kT) then weak reversible particle aggregation will occur. The individual interaction potentials can be calculated as follows:

$$V_A = \frac{-A}{6}\left(\frac{2}{S^2-4} + \frac{2}{S^2} + \ln\left\{\frac{S^2-4}{S^2}\right\}\right) \qquad (3)$$

where A is a composite Hamaker constant and S is the ratio of the separation between the centres of the particles.

$$V_R = 2\varepsilon_0\varepsilon_r a\pi\Psi^2_0 \ln\left(1 + e^{-\tau(s-2)}\right) \qquad (4)$$

where ε_0 and ε_r are the permittivities of free space and the medium respectively, Ψ_0 is the surface potential, a is the particle radius and τ is the ratio of a to the electrical double layer thickness ($^1/_\kappa$).

$$V_s = V_{s_{mix}} + V_{s_{el}} \qquad (5)$$

$$V_{s_{mix}} = \frac{4\pi aRT}{V_1^0} (\phi_a)^2 \left(\frac{1}{2} - \chi\right) \left(\delta - \frac{H}{2}\right)^2 \quad (6)$$

where V_1^0 is the molar volume of the solvent, ϕ_a is the average volume fraction of the polymer in the adsorbed layer, χ is the Flory–Huggins interaction parameter, δ is the adsorbed layer thickness and H is the particle surface to surface separation.

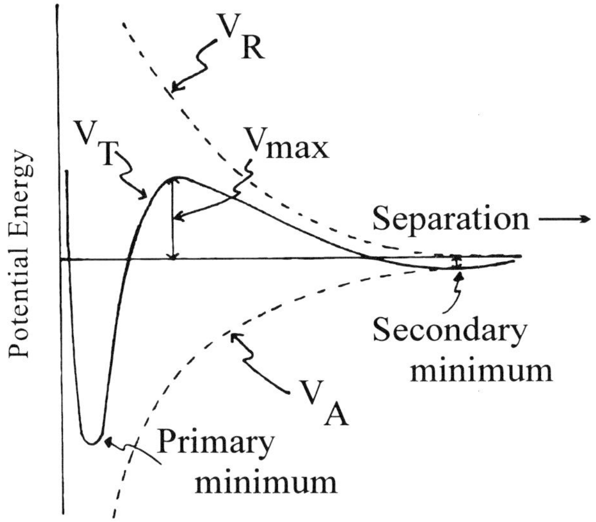

Figure 9.3 Potential energy diagram for two interacting particles showing V_A, V_R and V_T as a function of particle separation

$V_{s\,el}$ only operates at separations < δ and can be neglected since $V_{s\,mix}$ itself is normally >> V_A and V_D and the particles are prevented from approaching each other close enough to experience $V_{s\,el}$.

For bare particles,

$$V_D = 2a\,\pi P_{osm} \left(\Delta - \frac{H}{2}\right)^2 \left(1 + \frac{2}{3a} + \frac{H}{6a}\right) \quad (7)$$

where P_{osm} is the osmotic pressure of the polymer solution and Δ is the depletion layer thickness. Δ equals R_g below the critical coil overlap concentration, $\Phi_{overlap}$. At higher concentrations Δ decreases according to the relationship (Vincent, 1990)

$$\Delta = R_g \left(\frac{\Phi_b}{\Phi_{overlap}} \right)^m \tag{8}$$

where Φ_b is the bulk polymer concentration and m has values of -0.75 and -1.0 in athermal and θ-solvents respectively.

If the particles have an adsorbed polymer layer then

$$V_D = 2a\,\pi P_{osm} \left(\Delta + \delta - p - \frac{H}{2} \right)^2 \left(1 + \frac{2(\Delta + \delta - p)}{3a} + \frac{H}{6a} \right) \tag{7}$$

p is a function of the adsorbed layer characteristics and for a linear profile is given by

$$p = \frac{(\mu_1 - \mu_1^0)\,\Delta}{RT(1-2\chi)\phi_a \phi_b} \tag{10}$$

where μ_1 and μ_0^1 are the chemical potentials of the polymer solution and pure solvent respectively. It is readily noted from eqs. (6) and (8) that V_D will increase with increasing osmotic pressure of the polymer solution and Δ. When V_D increases it gives rise to an increase in the depth of the secondary minimum in the total potential energy curve and, therefore, induces weak aggregation. The influence of these parameters on the depth of V_{min} is illustrated in Figure 9.4. Figure 9.4(a) shows that Δ has a significant effect on V_{min} as expected, since the depletion potential is proportional to the square of the depletion layer thickness (eqs. (6) and (8)). V_{min} increases from about 1 kT to ~ 25 kT as Δ increases from 14 nm to 100 nm. Figure 9.4(b) shows the effect of polymer osmotic pressure on the total potential energy curve and V_{min} is seen to increase from ~1 kT to ~22 kT as the osmotic pressure of the solution increases from zero to 250 N/m^2.

Results and discussion

The effect of HEC and SCMC on the stability of the polystyrene latex is illustrated in Figure 9.5. The plot shows that n becomes less negative as the polymer concentration increases, indicating that particle aggregation is occurring. The critical polymer concentrations at which aggregation begins to occur (C_{floc}) are 0.4% and 0.02% for HEC and SCMC respectively and are attributed to a depletion flocculation

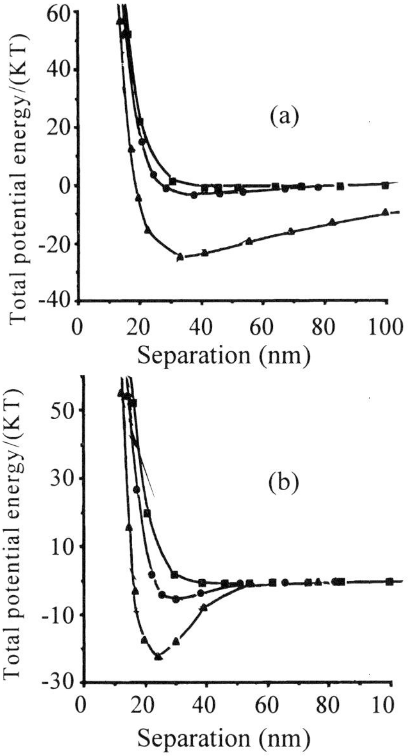

Figure 9.4 (a) Potential energy diagrams for particles in the presence of polymer with increasing molecular mass. ■ 14 nm; ● 48 nm; ▲ 100 nm. (b) Potential energy diagram for particles in the presence of increasing polymer concentration. P_{osm} ■ is 0; ● 100 N/m^2; ▲ 250 N/m^2

mechanism. We have reported similar findings previously in studies on the behaviour of silica particles in the presence of HEC and SCMC (Snowden *et al.*, 1991). Furthermore, Nashima and Furusawa (1991) studied the effect of hydroxypropyl cellulose (HPC) on the rheological properties of polystyrene latex. They pointed out that the HPC adsorbed onto the latices and found that at low shear rates the viscosity of 15 wt% latex dispersions increased at HPC concentrations in the range 0.3% (for 925,000 molecular mass HPC) to 5% (for 53,000 molecular mass HPC) which they attributed to depletion phenomena. In a further study using non-adsorbing polystyrene sulphonate and styrene—sodium-vinylphenylsulphonate copolymer latices they found aggregation to occur at much lower polymer concentrations i.e. 0.1%, which they again attributed to a depletion mechanism.

As noted earlier, HEC adsorbs onto polystyrene latices and it could be argued that in our experiments flocculation is brought about by polymer bridging. However, this should occur only at very low surface coverages corresponding to polymer concentrations about four orders of magnitude lower than the concentration at which

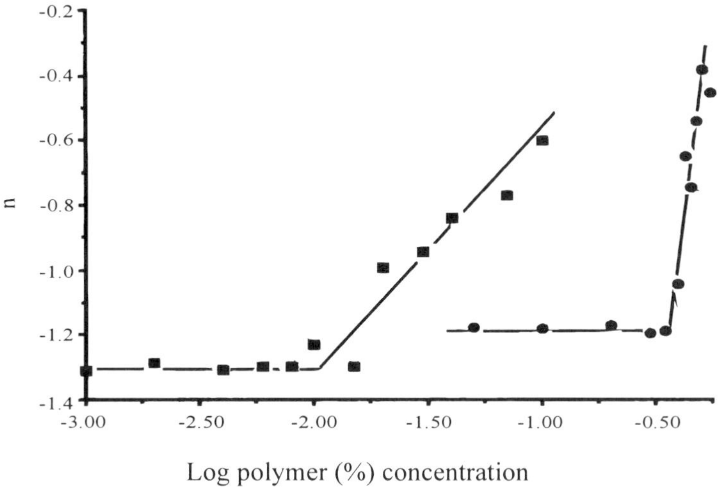

Figure 9.5 Plot of n [d (log absorbance) / d (log wavelength)] for polystyrene latex dispersions in the presence of ■, sodium carboxymethylcellulose and ●, hydroxyethylcellulose

flocculation was observed. Furthermore, aggregation was found to be reversible on dilution. This is expected for particles aggregated due to polymer depletion but not for particles aggregated by polymer bridging.

For systems containing SCMC it is possible that aggregation is induced simply by the fact that the solution ionic strength increases due to the polymer itself; however, following stability experiments in the presence of simple electrolyte it was found that the critical electrolyte concentration required to cause particle aggregation was 0.01 M. This corresponds to an ionic strength about ten times higher than due to the SCMC at C_{floc} and hence depletion flocculation is clearly the most likely cause of aggregation.

As discussed above, it is possible to predict particle flocculation from the depth of the secondary minimum. In experiments involving addition of simple electrolyte to the latices in the absence of polymer we found that V_{min} at the electrolyte concentration required to induce particle aggregation was of the order of ~5 kT and hence this value has been used as an indication of stability. The potential energy curves for the polystyrene particles in the presence of various concentrations of HEC and SCMC are given in Figures 9.6 and 9.7 respectively. The concentrations of HEC and SCMC required to give a value of ~5 kT for V_{min} are seen to be in the ranges 0.3—0.4% and 0.005—0.007% respectively. These values agree well with the experimental values for C_{floc} reported above, supporting the view that flocculation is due to a depletion mechanism. The correlation is perhaps far better than might be expected since the two polymers used are polydisperse and, therefore, the values used for Δ can only be an approximation. Furthermore, in polydisperse systems

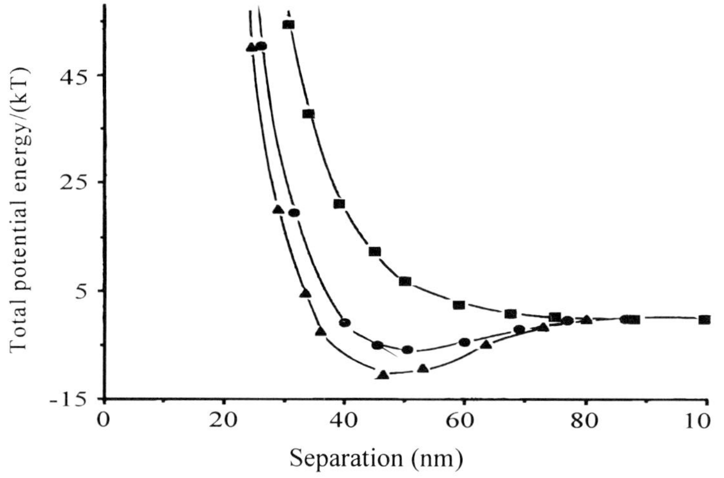

Figure 9.6 Potential energy diagrams for polystyrene particles in the presence of increasing concentrations of hydroxyethylcellulose: ■ none; ● 0.3%; ▲ 0.4%

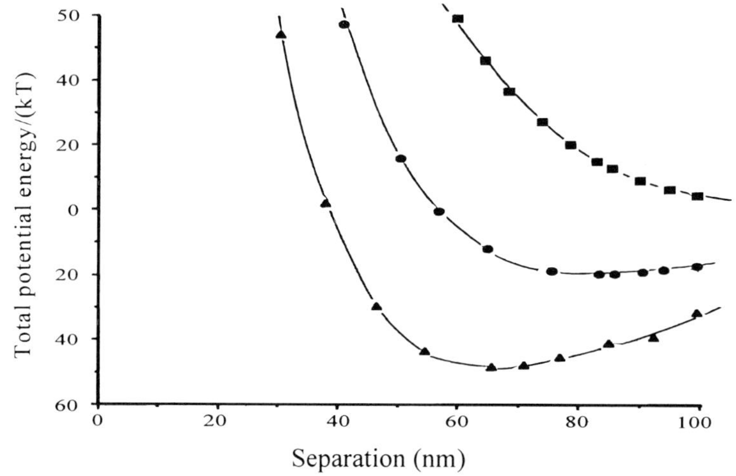

Figure 9.7 Potential energy diagrams for polystyrene particles in the presence of increasing concentrations of sodium carboxymethylcellulose: ■ 0.005%; ● 0.0075%; ▲ 0.01%

the larger molecules will be excluded from the depletion zone more readily than smaller molecules thus reducing the osmotic pressure differential between this region and the bulk solution. In addition, in the case of SCMC, sodium counterions can migrate into the depletion zone which is inaccessible to the polymer chains themselves, again resulting in a reduced osmotic pressure differential. Despite these considerations the theoretical treatment described is clearly adequate to predict the occurrence of depletion flocculation for dilute systems. More sophisticated treatments have now been developed to describe the stability and phase behaviour of concentrated particulate dispersions (Gast, Hall and Russel, 1983; Gast, Russel and Hall, 1986; Vincent, Edwards, Emmett and Croot, 1988; Lekkerkerker, Poon, Pusey, Stroobants and Warren, 1992). Gast *et al.* (1983,1986) calculated the total particle interaction potential and then applied thermodynamic perturbation theory to calculate the phase boundaries at varying particle volume fractions. One of the concerns with this approach is that it does not allow for polymer partitioning in the co-existing phases. Lekkerkerker *et al.* (1992) overcame this problem using a statistical thermodynamical approach based on the depletion layer concept. The model assumes that the only effect of the interaction between the colloid and the polymer is to restrict the volume in which the polymer molecules can move. More recently Clegg, Williams, Warren and Robb (1994) have extended this theory to predict the phase behaviour in concentrated polymer–surfactant mixtures and the current aim is to develop it further to study polymer–protein (globular) systems.

References

Asakura, A. and Oosawa, F. (1954) On interaction between two bodies immersed in a solution of macromolecules. *Journal of Chemical Physics*, **22**, 1255–1256

Asakura, A. and Oosawa, F. (1958) Interaction between two particles suspended in solutions of macromolecules. *Journal of Polymer Science*, **33**, 183–192

Clegg, S.M., Williams, P.A., Warren, P.B. and Robb, I.D. (1995) Phase behavior of polymers with concentrated dispersions of surfactants. *Langmuir*, **10**, 3390–3394

Derjaguin, B.V.and Landau, L. (1941). *Acta Physica-Chimica USSR*, **14**, 633–662

Fleer, G.J., Scheutjens, J.H.M.H. and Vincent, B. (1984) The stability of dispersions of hard spherical particles in the presence of nonadsorbing polymer. *American Chemical Society Symposium Series*, **240**, 245–263

Gast, A.P., Hall, C.K. and Russel, W.B. (1983) Polymer-induced phase separations in non-aqueous colloidal suspensions. *Journal of Colloid and Interface Science*, **96**, 251–267

Gast, A.P., Russel, W.B. and Hall, C.K. (1986) An experimental and theoretical study of phase transitions in the polystyrene latex and hydroxyethylcellulose system. *Journal of Colloid and Interface Science*, **109**, 161–171

Goodwin, J.W., Hearn, J., Ho, C.C. and Ottewill, R.H. (1974) Studies on the preparation and characterisation of mondisperse polystyrene latices. *Colloid and Polymer Science*, **252**, 464–471

Heller, W. and Pugh, T.L. (1954) "Steric protection" of hydrophobic colloidal particles by adsorption of flexible macromolecules. *Journal of Chemical Physics*, **22**, 1778

Hesselink, F.Th., Vrij, A. and Overbeek J.Th.G. (1971) On the theory of stabilization of dispersions by adsorbed macromolecules. II. Interaction between two flat plates. *Journal of Physical Chemistry*, **75**, 2094–103

Hunter, R.J. (1987). *Foundations of Colloid Science*, Oxford: Oxford Science Publications

Lekkerkerker, H.N.W., Poon, W.C.K., Pusey, P.N., Stroobants, A. and Warren, P.B. (1992) Phase behavior of colloid plus polymer mixtures. *Europhysics Letters*, **20**, 559–564

Long, J.A., Osmond, D.M.J. and Vincent, B. (1973) The equilibrium aspects of weak flocculation. *Journal of Colloid and Interface Science*, **42**, 545–553

Napper, D.H. (1983). *Polymeric Stabilisation of Colloidal Dispersions*, London: Academic Press

Nashima, T. and Furusawa, K. (1991) The rheological behaviour of a latex dispersion including a nonanionic polymer. *Colloids and Surfaces*, **55**, 149–161

Nashima, T., Sudo, H. and Furusawa, K. (1992) Flocculation of polymer latex dispersion induced by a polyelectrolyte. *Colloids and Surfaces*, **67**, 95–100

Snowden, M.J., Clegg, S.M., Williams, P.A. and Robb, I.D. (1991) Flocculation of silica particles by adsorbing and nonadsorbing polymers. *Journal of the Chemical Society, Faraday Transactions*, **87**, 2201–2207

Schneider, N.S. and Doty, P. (1954) Macro-ions. IV. The ionic strength dependence of the molecular properties of sodium carboxymethylcellulose. *Journal of Physical Chemistry*, **58**, 762–769

Tanaka, R., Meadows, J., Phillips, G.O. and Williams, P.A. (1989). In *Cellulose: Structural and Functional Aspects*, edited by Kennedy, J.F., Phillips, G.O. and Williams, P.A., Chichester (U.K.): Ellis Horwood Publishers, pp 329–335

Verwey, E.J.W. and Overbeek, J.Th.G. (1948). *Theory of Stability of Lyophobic Colloids*, Amsterdam: Elsevier

Vincent, B. (1990) The calculation of depletion layer thickness as a function of bulk polymer concentration. *Colloids and Surfaces*, **50**, 241–249

Vincent, B., Edwards, J., Emmett, S. and Croot, R. (1988) Phase separation in dispersions of weakly-interacting particles in solutions of non-adsorbing polymer. *Colloids and Surfaces*, **31**, 267–298

10

PECTIN–GELATIN PHASE SEPARATION: THE INFLUENCE OF POLYDISPERSITY

A.C. CLEWLOW and A.J. ROWE
National Centre for Macromolecular Hydrodynamics, Department of Biochemistry, Leicester University, Adrian Building, University Road, Leicester. LE1 7RH U.K.

M.P. TOMBS
National Centre for Macromolecular Hydrodynamics, Department of Applied Biochemistry and Food Science, Nottingham University, Sutton Bonington Campus, Sutton Bonington, Loughborough. LE12 5RD U.K.

Introduction

The first mention of immiscibility shown by mixtures of two or more polymers in solution goes back to 1896, when Beijernick observed that mixing gelatin and agar, or gelatin and starch, led to turbidity and spontaneous separation into two liquid layers. Bungenberg de Jong and co-workers (see, for example, Bungenberg de Jong, 1949) also found that mixtures of gelatin and gum arabic led to phase separation and invented the term "complex coacervation" to describe it. However, the realisation that phase separation could be associated with uncharged molecules, and a definitive description of incompatible phase separation is due to Dobry and Boyer-Kawenoki (1947) for numerous mixtures of synthetic polymers in organic solvents and who correctly ascribed it to coexclusion phenomena. Tompa (1956) then provided a far reaching theoretical account of polymer solutions, including phase separation, which contains all the ideas current today and which still repays study. The main theoretical description of synthetic polymers in solution (Flory, 1953), based on Flory–Huggins lattice theory, has been applied and is still being developed (Guan and Lilley, 1993; Guan, Lilley and Treffry, 1993). Theoretical work on biological macromolecules, for the most part charged and soluble in water, took a rather different route.

The work of Albertsson (1960) led to general recognition of the interest in and possible utility of a number of water soluble polymer mixtures which show phase separation.

The idea of thermodynamic incompatibility of proteins and polysaccharides was exploited by Ogston (1962), building on earlier work on the interpretation of virial coefficients when looking at mixtures of hyaluronic acid and serum albumin. Further work on this system (Laurent and Ogston, 1963; Preston, Davies and Ogston,

1965) led to a full theoretical formulation based on osmotic virial expansions (Ogston and Edmond, 1968), suggesting that under suitable conditions incompatible phase separation could be obtained with many protein–polymer and protein–polysaccharide mixtures. The "precipitation" of proteins by polyethylene glycol (PEG), which came into use for blood protein fractionation, and phase separation phenomena involving groundnut and soy proteins, with dextrans and corn syrups (Tombs 1972; Tombs, Newsom and Wilding 1974), could all be understood in these terms. Recent studies of protein–polysaccharide mixtures include gelatin–amylopectin (Durrani, Prystupa, Donald and Clark, 1993), soy 11S globulin–dextran (Dickinson and Semenova, 1992), soy 11S globulin–κ-carrageenan (Semenova, Bolotina, Dmitrochenko, Leontiev, Polyakov, Braudo and Tolstoguzov, 1991) and gelatin–maltodextrin (Kasapis, Morris, Norton, Clark, Gidley and Brown, 1993).

Although there have been some attempts to take account of the effect of polydispersity in synthetic polymers this problem has largely been ignored in work on biological polymers.

In the gelatin–pectin system we are interested in, both macromolecules will be to some extent polydisperse. (See Tolstoguzov (1990) for a review of gelatin–pectin incompatibility; the review is largely confined to delineating the conditions for phase separation in a variety of solvent pH and ionic strengths.)

Mixtures of gelatin and pectin are now being considered more widely for industrial use and are sometimes producing unexpected and unwanted phase separations as well as those that are actually intended. The effect of the polydispersity of either component and some knowledge of which aspects might be important is needed, if only to lead to better specification of raw materials.

The work described here has so far suggested that two of the most important properties of commercial polymers from the point of view of predicting their behaviour in incompatible mixtures are the number average molecular weight, and the breadth of the distribution. Further work will be needed to demonstrate this experimentally.

Experimental

PECTIN AND GELATIN

An industrial low-methoxy pectin (Hercules) and a gelatin (Extraco, 250 Bloom) were used (both pectin and gelatin were gifts from Dr. A.H. Clark, Unilever Research). Concentration of pectin solutions for partial specific volume, sedimentation coefficient, molecular weight determination and intrinsic viscosity was determined either by using a differential refractometer or by weight and corrected for moisture content. The pectin was dissolved in 0.5 M NaCl and adjusted to pH 5.5±0.05 with a few drops of NaOH solution. Concentrations of gelatin solutions were determined by weight corrected for moisture content. The gelatin was dissolved in 0.5 M NaCl and adjusted to pH 5.5±0.05.

PARTIAL SPECIFIC VOLUME

The partial specific volume, $\bar{v}$, was determined by precision densimetry using an oscillating mechanical density-meter (Anton-Paar Instruments, Austria) of the Kratky type (Kratky, Leopold and Stabinger, 1973), thermostated to 20°C ± 0.01°C.

LOW SPEED SEDIMENTATION EQUILIBRIUM

Determination of weight-average molecular weights

A Beckman XL-A analytical ultracentrifuge was used employing scanning absorbance optics at wavelengths 220 nm (pectin) and 265 nm (gelatin). All experiments were performed at 20°C in 0.5 M NaCl (pH 5.5). Six-sector 10-mm optical path length cells with charcoal-filled epon centre pieces and quartz windows were used. Data sets were collected after reaching equilibrium (12 to 16 h) at rotor speeds of 8000–12000 r.p.m. Evidence for equilibrium was obtained by superimposing data sets taken 4 h apart. At sedimentation equilibrium, monodisperse polymers in a centrifugal field distribute as an exponential (Cantor and Schimmel, 1980):

$$c_r = c_o \exp\left[\frac{M(1-\bar{v}\rho)\,\omega^2\,(r^2-r^2_m)}{2RT}\right] + \text{base} \qquad (1)$$

where c_r is the concentration of the species at a distance, in centimetres from the centre of rotation, c_o is the concentration of the species at a reference radial position. M is the molecular mass, $\bar{v}$ is the partial specific volume, ρ is the solvent density, ω is the angular velocity, r_m is the radial position of the meniscus, R is the universal gas constant, T is the absolute temperature, and *base* is a baseline term to account for non-sedimenting absorbance. Sedimentation equilibrium data were analysed using the non-linear least squares method found on the XL-A data analysis software Microcal Origin.

Determination of Z-average molecular weights

An MSE MK II analytical ultracentrifuge was used employing a scanning Schlieren optical system. All experiments were performed at 20°C at rotor speeds of 8000—12000 rev/min with standard MSE 20-mm optical path length cells. The concentration gradient at equilibrium for a single component in an ideal two-component solution is

$$\frac{1}{n}\frac{dn}{dr} = \frac{\omega^2 rM(1-\bar{v}\rho)}{RT} \quad (2)$$

Integrating the above equation between the meniscus (a) and a point r gives

$$\frac{d\ln n}{d(r^2)} = \frac{\omega^2 rM(1-\bar{v}\rho)}{2RT} \quad (3)$$

A plot of ln C versus r^2 should give a straight line with a slope of $\omega^2 M(1-\bar{v}\rho)/2RT$. A computer digitising tablet was used to capture the data off-line.

VISCOSITY MEASUREMENTS

The viscosity of the polymer solutions was measured with a Schott Geräte AVS 400 apparatus at (20.0±0.1)°C. The intrinsic viscosity was calculated using the Huggins equation (1942)

$$\frac{\eta_{sp}}{c} = [\eta] + k'[\eta]^2 c \quad (4)$$

Therefore by plotting η_{sp}/c against the concentration, the intercept is the intrinsic viscosity, where η_{sp} is the specific viscosity, η is the intrinsic viscosity and k' is a solute–solvent interaction coefficient.

BINODAL CURVE

The solutions of pectin and gelatin (pH 5.5, 0.5 M NaCl) were stirred for 2 h at 80°C and then centrifuged for 2 h at 40°C at 30,000 r.p.m; this was followed by re-equilibration at 80°C and the concentration of gelatin in each phase was then found (Bradford, 1976). The content of pectin in each phase was measured using a phenol-sulphuric acid method (Dubois, Gilles, Hamilton, Rebers and Smith, 1956). Standard curves were determined using the appropriate assays on known concentrations of the pectin and gelatin samples. All analysis was in triplicate and the degree of standard error in determining pectin and gelatin concentrations was found to be approximately 5%.

Results

THE BINODAL

The phase diagram (see Figure 10.1) clearly indicates that higher concentrations of gelatin than pectin are required for phase separation to occur. Mixtures of pectin and gelatin at similar concentrations but this time with 0.4 M NaCl did not result in phase separation, showing that at sufficiently high ionic strength electrostatic complexing of the gelatin and pectin is inhibited (see also Tolstoguzov, 1990). Measurements were made at 80°C to ensure that no significant gelatin-gelatin interaction of the kind that leads to gelation might occur. Similarly the ionic strength was chosen so as to suppress charge-charge interactions.

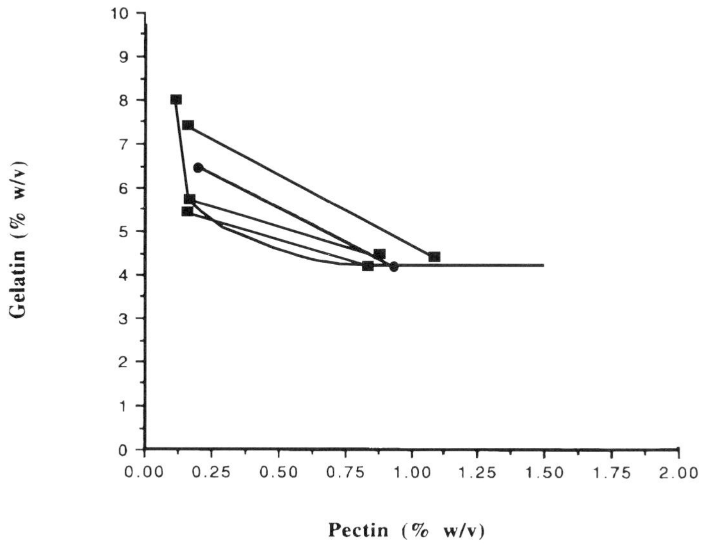

Figure 10.1 The phase diagram of the system pectin-gelatin-water (80°C, 0.5 M NaCl, pH 5.5)

DETERMINATION OF THE MOLECULAR WEIGHTS AND SECOND VIRIAL COEFFICIENTS

The reciprocal of apparent molecular weights determined from the analytical ultracentrifuge plotted against concentration for both the weight and Z-averages are shown (Figure 10.2), revealing the dependence of the apparent molecular weight on the concentration for the pectin and gelatin. The true weight and Z-average molecular weights were determined by extrapolation to infinite dilution and taking the reciprocal of the intercept with the y-axis to yield the infinite dilution molecular weight.

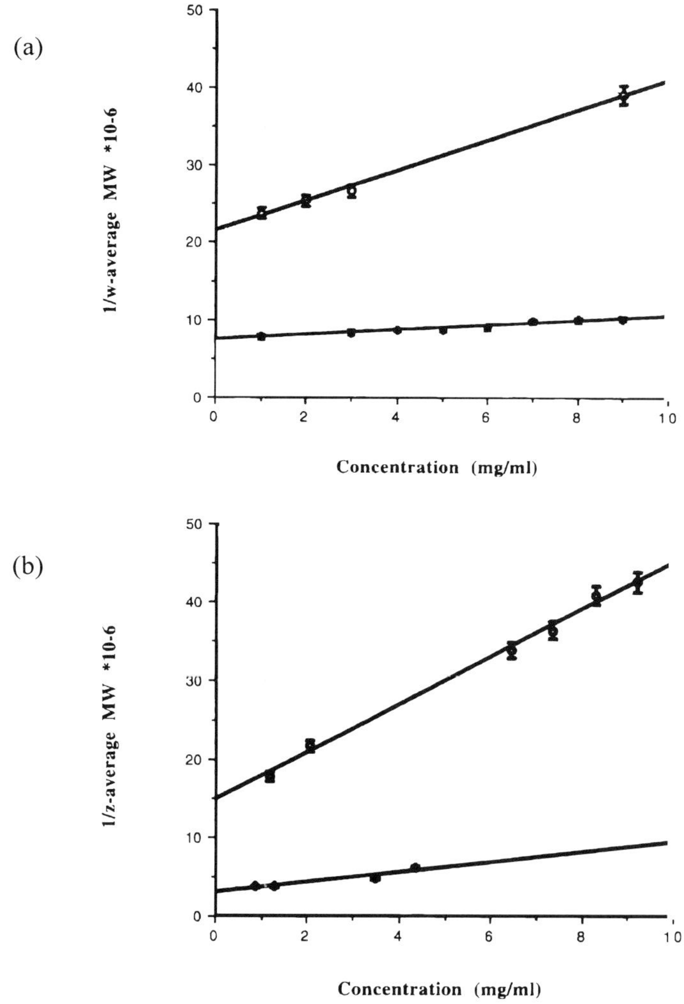

Figure 10.2 Plots of the reciprocal of the (a) weight average and (b) Z-average molecular weight for the pectin and gelatin samples in 0.5 M NaCl at pH 5.5 (○ Pectin, ● Gelatin)

The second virial coefficient was also calculated from the apparent molecular weight reciprocal plots. The equation

$$\frac{1}{M_{app}} \approx 2Bc_2 + \frac{1}{M} \qquad (5)$$

was used and gave a good linear fit showing that only a single coefficient is needed to describe the data, as expected charge-charge interactions have been suppressed. M_{app} is the apparent molecular weight, M the molecular weight at infinite dilution and c_2 is the concentration. 2B is the second thermodynamic or "osmotic pressure" virial coefficient and is taken from the slope of the plot. Measurements were also made for mixtures of gelatin and pectin (Figure 10.3) and second virial coefficients calculated. Values for the molecular weights and second virial coefficients for the pectin and gelatin are given in Table 10.1. Self interaction (c and d) and cross interaction coefficients (a) were calculated from the second virial coefficients and molecular weights using (Edmond and Ogston, 1968)

$$c = \frac{2M_2^2B}{1000} \qquad d = \frac{2M_3^2B}{1000} \qquad a = \frac{2M_2M_3B}{1000} \qquad (6)$$

where M_2 and M_3 are the molecular weights of component 2 (gelatin) and component 3 (pectin) respectively.

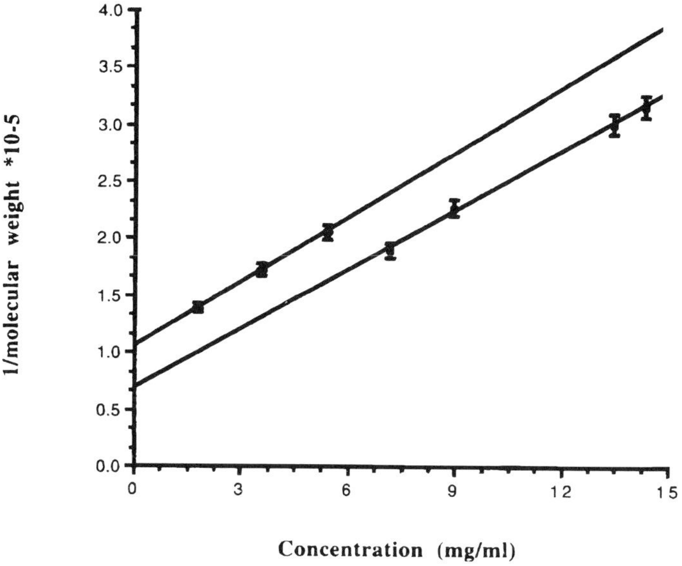

Figure 10.3 Plots of the reciprocal of the weight and Z-average molecular weights for pectin and gelatin mixtures (at a 50:50 ratio by weight) in 0.5 M NaCl at pH 5.5 (■ Z-MW, ● w-MW)

Table 10.1 PROPERTIES OF GELATIN AND PECTIN

	Gelatin	*Pectin*	*Gelatin/Pectin*
M_w	136000±4000	47000±1500	95000±2850
M_z	339000±9000	70000±2100	146200±4000
M_z/M_w	2.45	1.49	1.54
B (ml mol/g^2)	0.00035	0.00211	0.00060
c or d	13050	9320	
a			14470

INTRINSIC VISCOSITY

The intrinsic viscosity is strongly dependent on ionic strength (Figure 10.4), and the changes indicate that the molecule becomes more compact and spherical at high salt concentration. This was a further reason for choosing to make measurements in 0.5 M NaCl.

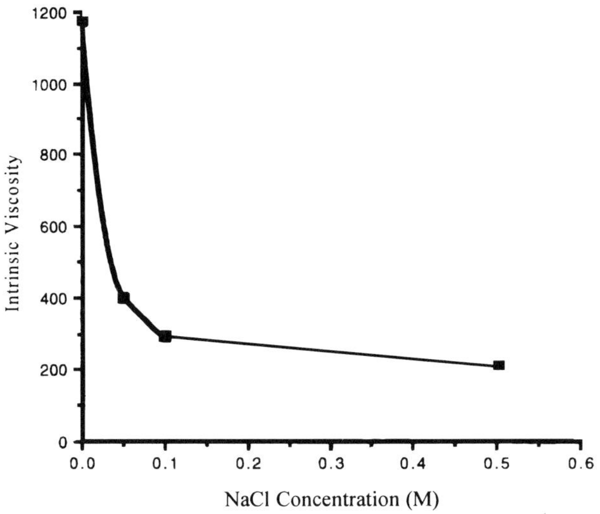

Figure 10.4 Intrinsic viscosity of pectin at varying NaCl concentrations

Discussion

In interpreting our data we have used the simplest possible approach, based mainly on descriptions developed by Edmond and Ogston (1968), Ogston (1962) and Ogston (personal communication).

Writing the standard, and precise, relationship for the change in the chemical potential of solute component 2, as a function of the concentration x_2 and an activity coefficient γ_2,

$$\frac{\mu_2 - \mu_2^0}{RT} = \ln x_2\gamma_2 \qquad (7)$$

we are faced as usual with the problem of how to eliminate or otherwise deal with the very inconvenient activity coefficient. We can define it by

$$\frac{d\ln\gamma_2}{dx_2} = b + ax_2 \qquad (8)$$

where the coefficients a and b may eventually be interpreted as representing attractions (a) and repulsions (b) between molecules although there is no need, strictly speaking, to introduce the idea of molecules at all.

It is difficult to conceive of a simpler formulation, and to some extent it is optional. It would certainly be possible to postulate more complex equations, and the only effective test is whether the formulation chosen can account adequately for the data. This is not to say that by using extra information about molecular properties and different theoretical approaches it is not possible to develop much more complex treatments, (see for example recent work by Lilley and co-workers, Guan and Lilley, 1993; Guan *et al.*, 1993). However, in this work we have more limited objectives and believe that, for two- and three-component systems we can adequately account for all our data so far by using equations based on eq. (8). Application of eq. (8), together with the Gibbs-Duhem equation for a two component system, and the approximation for $\ln x_2$

$$\ln x_2 = (x_2 - 1) - \frac{1}{2}(x_2 - 1)^2 + \frac{1}{3}(x_2 - 1)^3 \ldots\ldots\ldots\ldots\ldots$$

leads directly to the osmotic pressure formulation, in terms of the chemical potential of the solvent

$$\frac{\mu_1^0 - \mu_1}{RT} = x_2 + (1 + b)\frac{x_2^2}{2} + (1+a+b)\frac{x_2^3}{3} \tag{9}$$

and since at equilibrium all phases must be in osmotic equilibrium this can be applied to phase separations. Different treatments mostly differ in the interpretation placed on the virial coefficients, and the extent to which they are related to molecular properties and the way in which the species (including the solvent) might interact.

Once eq. (8) has been taken as the simplest formulation, the next step is to adopt the simplest treatment of the second and third virial coefficients. From eq. (9) the second virial coefficient contains only repulsions, and the simplest interpretation of these is an excluded volume. The third virial contains both repulsions and attractions but, in the great majority of cases, work with biological macromolecules is not sufficiently precise to require a third coefficient for a statistically valid treatment of the data. No doubt this is also due to the usual practice of choosing conditions of ionic strength and pH which tend to minimise charge interactions. This is because most work is aimed at determining a molecular weight, and a simple linear extrapolation to zero concentration is desirable (Tombs and Peacocke, 1974, who also describe systems where a third coefficient **is** needed). Finally, measurements are almost invariably made in salt solutions, and therefore regarding the solvent as a single component is not accurate.

We thus regard the relationships that we have used as semi-empirical, though of the correct form, and have adhered to the simplest approach, only introducing further terms where this cannot account for the data. A further advantage of this approach is that it can also be used to describe phase separations where one component is present in the form of a cross linked gel matrix (see for example Ogston and Silpananta, 1970). The excluded volume due to gelatin, for example, is not likely to change very much when it goes from solution to a gel and we hope to extend these studies to include the case where the gelatin has gelled.

PHASE SEPARATIONS

Extending this treatment to three components, with two solutes, 2 and 3, of molal concentrations m_2 and m_3, the osmotic equation is (Edmond and Ogston, 1968)

$$\mu_1 = \mu_1^0 - \frac{RTM_1}{1000}\left(m_2 + m_3 + \frac{c}{3}m_2^2 + \frac{d}{2}m_3^2 + am_2m_3\right) \tag{10}$$

and now contains three coefficients; a, which represents repulsive interactions between components 2 and 3, while c represents those of component 3 with itself and d those of 2 with itself. We make no assumptions at this stage about the nature of the interactions, which might be binary, ternary or of even higher order.

Ogston obtained

$$\left(\frac{1}{m_2} + c\right)\left(\frac{1}{m_3} + d\right) - a^2 = 0 \qquad (11)$$

as the equation for the spinodal, and much more complicated equations for the binodal

$$\frac{Q}{a} + (m_3' - m_3'') + \frac{cQ^2}{2a^2}\left[\frac{e^p+1}{e^p-1}\right] + \frac{d}{2}(m_3'^2 - m_3''^2) + \frac{Q}{e^p-1}(m_3'^p - m_3'') = 0 \qquad (12)$$

$$m_2' = m_2'' e^p = \frac{Qe^p}{a\,(e^p-1)}$$

where

$$p = \frac{c}{a}\ln\left(\frac{m_3'}{m_3''}\right) + \frac{cd-a^2}{a}(m_3' - m_3'') \qquad Q = -\left[\ln\left(\frac{m_3'}{m_3''}\right) + d(m_3' - m_3'')\right]$$

Using these equations it is possible to calculate the binodal curve if the values for c, d and a are known by taking values of m_3', then computing the corresponding value of m_3''. The corresponding values of m_2' and m_3'' can then be found using eq. (12). We have written a procedure for these calculations on an Apple Macintosh computer using Pascal programming language. This was validated by checking the calculations given in Edmond and Ogston (1968).

As is shown in Figure 10.5, the difference between the spinodal and binodal is not great. (It is not surprising that in the days before calculators were generally available most work was done on the spinodal, which fortunately coincides with the binodal at the critical point.) We can also note that unless

$$cd - a^2 \leq 0 \tag{13}$$

phase separation cannot occur. Some other useful relationships are

$$\left(\frac{1}{m_2^{crit}} + c\right) = a\left(\frac{m_3^{crit}}{m_2^{crit}}\right) \qquad \left(\frac{1}{m_3^{crit}} + c\right) = a\left(\frac{m_2^{crit}}{m_3^{crit}}\right)^{2/3} \tag{14}$$

for the critical point compositions.

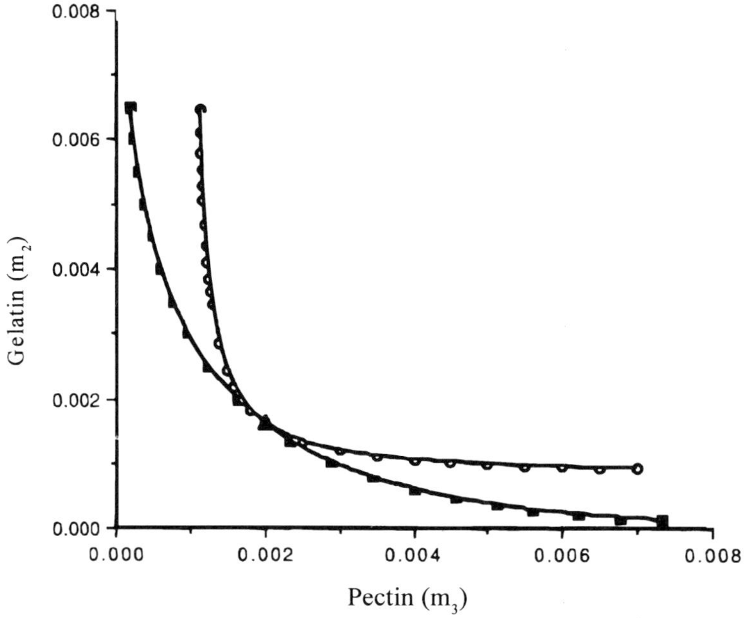

Figure 10.5 Comparison of the theoretical spinodal calculated from eq. (12) and the theoretical binodal curve calculated from the eqs. (13) for the pectin—gelatin system using coefficients measured from the molecular weight non-ideality plots (■ Binodal, ○ Spinodal, ▲ Critical point)

EVALUATION OF VIRIAL COEFFICIENTS IN POLYDISPERSE SYSTEMS

Values for c and d can be obtained in three ways. The most direct is by measuring the variation of the apparent molecular weight as a function of concentration. Table 10.1 contains values for c and d measured for our pectin and gelatin. A value for a has been estimated from measurements made on a mixture of the two.

The values obtained from M_w are much as expected but it is alarming that those derived from M_z should be so numerically different. This affects the calculations in two ways. The molalities apparently change, and since the coefficients have units of ml/mole this must also affect the numerical values.

Clearly when dealing with mixtures the question of which average to use is not a trivial one. The various averages are defined by

$$\overline{M}_n = \frac{\sum_i^j M_i m_i \, .\,.\,.\,.\, M_j n_j}{\sum_i^j m_i \, .\,.\,.\,.\, m_j}$$

$$\overline{M}_w = \frac{\sum_i^j M_i^2 n_i \, .\,.\,.\,.\, M_j^2 n_j}{\sum_i^j m_i n_i \, .\,.\,.\,.\, m_j n_j} \qquad (15)$$

$$\overline{M}_z = \frac{\sum_i^j M_i^3 n_i \, .\,.\,.\,.\, M_j^3 n_j}{\sum_i^j m_i^2 n_i \, .\,.\,.\,.\, m_j^2 n_j}$$

and choosing the simplest approach, the number average $\overline{M}_n$ would be appropriate. There is a further reason for choosing $\overline{M}_n$ (Ogston, 1968; Ogston and Edmond, 1968). Eqs. (3) and (4) are osmotic pressure formulations, and in this case the molecular weight is always a number average (Tombs and Peacocke, 1974). Although we intend to measure $\overline{M}_n$ for these samples it is unfortunately much less accessible than $\overline{M}_w$. $\overline{M}_z$ however appears to be even less appropriate than $\overline{M}_w$ as might be expected.

The other way in which we can obtain experimental values for a, c and d is by measuring the binodal for the two macromolecules, in particularly the critical point compositions, and then use eqs. (14) to obtain values for any two of the coefficients, with one value needed from other data. Estimates can also be obtained by directly calculating the excluded volumes and hence the coefficients (see below) from molecular parameters, but all these calculations require values for the molalities and hence a choice of the molecular weight average. Edmond and Ogston (1968) used coefficients found from $\overline{M}_w$ together with those based on $\overline{M}_n$ but considered that their dextran and PEG had relatively low polydispersity. (It is not clear whether there is much to be gained by consistently using the $\overline{M}_w$, though this has probably been usual). The overall agreement they obtained between coefficients obtained from the binodal and those from direct M_{app} measurement was held to justify the use of the approach outlined above.

In fact a comparison of coefficients and calculated binodal obtained by the consistent use of either $\overline{M}_n$ or $\overline{M}_w$ with measured phase separation data, also employing consistent $\overline{M}_n$ or $\overline{M}_w$, could also be used as an experimental test of the question of the most appropriate molecular weight average. With our much more polydisperse systems this question becomes more important and we hope to do this.

Finally, by using data obtained from a variety of techniques for the size and shape of the molecules it is possible to calculate molecular dimensions, though assumptions about shape will be necessary. Presumably molecular parameters derived from $\overline{M}_n$ will yield number average dimensions.

EFFECT OF POLYDISPERSITY ON THE BINODAL

In the absence of detailed information about molecular weight distribution (the ratio of $\overline{M}_w$ to $\overline{M}_z$ shows that it must be fairly wide, but with commercial samples it may not even be unimodal) the best way of testing the effect of molecular weight variation on the binodal seems to be to calculate values for the coefficients directly, making assumptions about shape. Thus variation in the molecular weight of gelatin, which is rod-shaped, would be reflected in a change in length of the rod, but not its diameter. For spherical molecules, and we treat pectin as such in our conditions, it would appear in the radius of the molecule.

CALCULATION OF BINODALS

Using eq. (12) for the binodal, and the virial coefficients derived from the M versus concentration plots for pectin, gelatin and a mixture of the two, we calculated the binodal. However from Figure 10.6 it can be seen that the calculated binodal is distant from the measured binodal. The reason for this discrepancy is not obvious, but a choice of $\overline{M}_n$ rather than $\overline{M}_w$ would almost certainly show better agreement.

We then calculated the self interaction and cross interaction coefficients based on molecular dimensions (see Tables 10.2 and 10.3). Then using the following relationships (Ogston, 1970) we calculated the self interaction coefficient of pectin, where r_2 is the radius of the sphere.

$$10_3 c = \frac{4\pi N}{3} (2r_2)^3 \qquad (16)$$

For the self interaction coefficient of gelatin, where r_2 is the radius of the rod and L is the half-length of the rod,

$$10^3 d = 4\pi N r \left[L^2 + \frac{2}{3}(2r_3)^3 + 4Lr_3 \right] \qquad (17)$$

The cross interaction coefficients were calculated using the rod–sphere model,

$$10^3a = 2\pi N \; L\,(r_2 + r_3)^2 + \frac{2}{3}(r_2 + r_3)^3 \qquad (18)$$

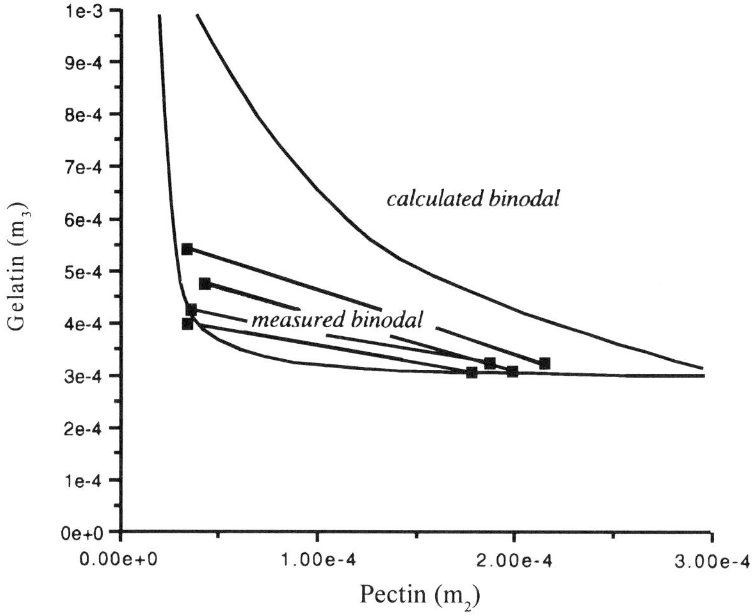

Figure 10.6 A comparison of the measured binodal and the calculated binodal using coefficients determined from analytical ultracentrifuge data

Table 10.2 EFFECT OF 10% VARIATION IN MOLECULAR WEIGHT OF GELATIN ON THE SELF AND CROSS INTERACTION COEFFICIENTS (ASSUMING A CONSTANT RADIUS OF 15 X 10^{-8} cm.

	$\overline{M_w}$	*L(Å)*	*d*	*a*
Gelatin	122,400	220.5	7008	7577
	136,000	245.0	8470	8247
	149,600	269.5	10058	8917

We then varied the molecular weight of either the gelatin or pectin to see the effect this would have on the binodal. In the case of the gelatin we kept the radius constant but in changing the molecular weight changed the length of the gelatin molecule.

Table 10.3 EFFECT OF 10% VARIATION IN MOLECULAR WEIGHT OF PECTIN ON THE SELF AND CROSS INTERACTION COEFFICIENTS

	M_w	$R(Å)$	c	a
Pectin	42,300	68	6050	7828
	47,000	70	6722	8247
	51,700	72	7394	8678

We have not as yet attempted calculations based on the critical point composition, and measured c or d, to obtain values for a, though these are clearly possible.

The results of these calculations are shown in Figures 10.7 and 10.8. They show relatively little effect of changing the pectin molecular weight, but a larger effect for gelatin, as expected from the rod shape of the molecule.

Taking the extreme values for both pectin and gelatin gave larger shifts in the binodal, but the indications are that relatively narrow distributions are not likely to greatly affect the position of the binodal. Preliminary indications are that the number average molecular weight might be an important part of the specifications for industrial samples of pectin and gelatin where phase separations are involved.

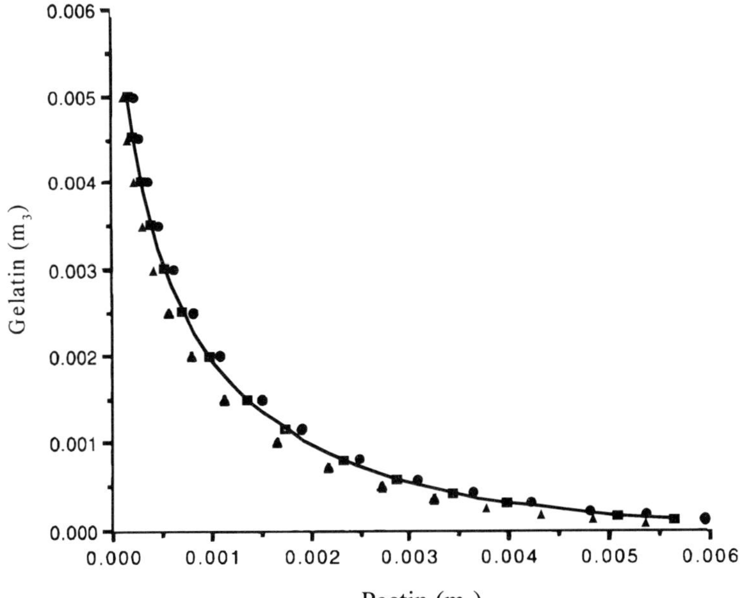

Figure 10.7 The effect on the binodal when varying the molecular weight of the pectin (■ Median, ● 10% lower molecular weight, ▲ 10% higher molecular weight)

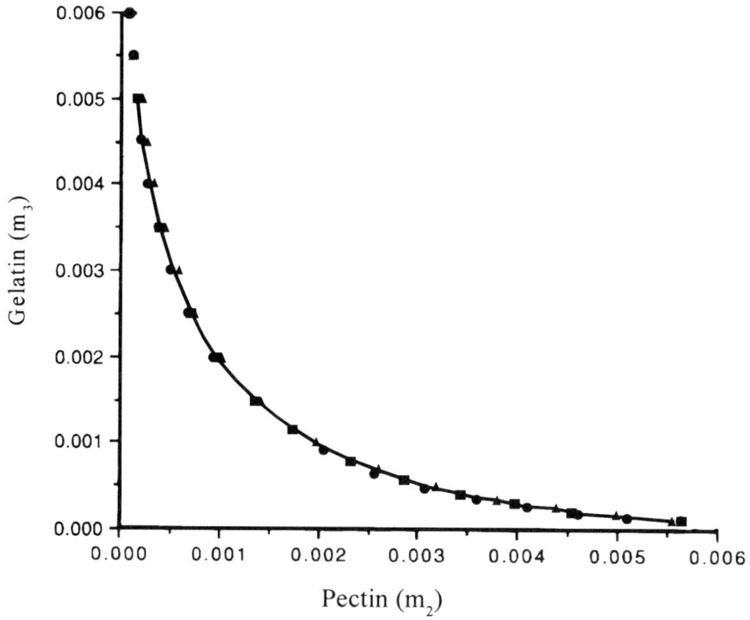

Figure 10.8 The effect on the binodal when varying the molecular weight of the gelatin (■ Median, ● 10% lower molecular weight, ▲ 10% higher molecular weight)

References

Albertsson, P.-Å. (1960). *Partition of Cell Particles and Macromolecules*, 1st edition, New York: Wiley

Beijernick, M.W. (1896). *Centralblatt für Bakteriologie und Parasitenkunde*, **2**, 627–698

Bradford, M.M. (1976) A rapid and sensitive method for the quantitation of microgram quantities of protein utilizing the principle of protein–dye binding. *Analytical Biochemistry*, **72**, 248

Bungenberg de Jong, H.G. (1949). *Les Coacervats et leur importance en Biologie*, Amsterdam: Elsevier

Cantor, C.R. and Schimmel, P.R. (1980). *Biophysical Chemistry, Part 2: Techniques for the Study of Biological Structure and Function*, San Francisco: W.H. Freeman and Co.

Dickinson, E. and Semenova, M.G. (1992) Emulsifying behaviour of protein in the presence of polysaccharide under conditions of thermodynamic incompatability. *Journal of the Chemical Society, Faraday Transactions*, **88**, 849–854

Dobry, K. and Boyer-Kawenoki, H.B. (1947) Phase separation in polymer solution. *Journal of Polymer Science*, **2**, 90–100

Dubois, M., Gilles, K.A., Hamilton, J.K., Rebers, P.A. and Smith, F. (1956) Colorimetric method for determination of sugars and related substances. *Analytical Chemistry*, **28**, 350–356

Durrani, C.M., Prystupa, D.A., Donald, A.M. and Clark, A.H. (1993) Phase diagram of mixtures of polymers in aqueous solution using Fourier-transform infrared-spectroscopy. *Macromolecules*, **26**, 981–987

Edmond, E. and Ogston, A.G. (1968) An approach to the study of phase separation in ternary aqueous solutions. *Biochemical Journal*, **109**, 569–576

Flory, P.J. (1953). *Principles of Polymer Chemistry*, Ithaca (New York): Cornell University Press

Guan, Y., Lilley, T.H. and Treffry, T.E. (1993) Theory of phase equilibria for multicomponent aqueous solutions: Applications to aqueous polymer two-phase systems. *Journal of the Chemical Society, Faraday Transactions*, **89**, 4283–4298

Guan, Y., Lilley, T.H., and Treffry, T.E. (1993) A new excluded volume theory and its application to the coexistence curves of aqueous polymer two-phase systems. *Macromolecules*, **26**, 3971–3979

Huggins, M.L. (1942) The viscosity of dilute solutions of long-chain molecules. IV. Dependence on concentration. *Journal of the American Chemistry Society*, **64**, 2716

Kasapis, S., Morris, E.R., Norton, I.T., Clark, A.H., Gidley, M.J. and Brown, R.T. (1993) Phase equilibria and gelation in gelatin/maltodextrin systems. I. Gelation of individual components. *Carbohydrate Polymers*, **21** 243–276

Kratky, O., Leopold, H. and Stabinger, H. (1973) The determination of the partial specific volume of proteins by the mechanical oscillator technique. *Methods in Enzymology*, **27**, 98

Laurent, T.C. and Ogston, A.G. (1963) The interaction between polysaccharides and other macromolecules. 4. The osmotic pressure of mixtures of serum albumin and hyaluronic acid. *Biochemical Journal*, **89**, 249–257

Ogston, A.G. (1962) Some thermodynamic relationships in ternary solutions, with special reference to the properties of systems containing hyaluronic acid and protein. *Archives of Biochemistry and Biophysics*, **1**, 39–51

Ogston, A.G. (1970) On the interaction of solute molecules with porous networks. *Journal of Physical Chemistry*, **74**, 668–669

Ogston, A.G. and Silpananta, P. (1970) The thermodynamics of interaction between sephadex and penetrating solutes. *Biochemical Journal*, **116**, 171–175

Preston, B.N., Davies, M. and Ogston, A.G. (1965) The composition and physicochemical properties of hyaluronic acids prepared from ox synovial fluid and from a case of mesothelioma. *Biochemical Journal*, **96**, 449–470

Semenova, M.G., Bolotina, V.S., Dmitrochenko, A.P., Leontiev, A.L., Polyakov, V.I., Braudo, E.E. and Tolstoguzov, V.B. (1991) The factors affecting the compatability of serum albumin and pectinate in aqueous medium. *Carbohydrate Polymers*, **15**, 367–385

Tolstoguzov, V.B. (1990) Interaction of gelatin with polysaccharides. In *Gums and Stabilisers for the Food Industry 5*, edited by Phillips, G.O., Williams, P.A. and Wedlock, D.J., Oxford: IRL Press, Oxford

Tombs, M.P. (1972). British Patent, 1,265.661

Tombs, M.P. and Peacocke, A.R. (1974). *The Osmotic Pressure of Biological Macromolecules*, Oxford: Clarendon Press

Tombs, M.P., Newsom, B.G. and Wilding, P. (1974). *Int J Protein Peptide Research*, **6**, 253

Tompa, H. (1956). *Polymer Solutions*, London: Butterworths

11

PHASE SEPARATION IN HYDROCOLLOID GELS

STEFAN KASAPIS
Department of Food Research and Technology, Cranfield University, Silsoe College, Silsoe, Bedford. MK45 4DT U.K.

Introduction

This chapter will provide an outline of the attempts made to rationalise phase separation phenomena in biopolymer gels, on the basis of physical theories from the realm of synthetic polymers, rather than a comprehensive review of all the published data on these complex systems. In this respect, the first published study appeared in the early eighties when a breakthrough was made in the study of agar–gelatin gels by using the Takayanagi model and ideas of solvent partition between the phases of two incompatible biopolymers at equilibrium. Two years later, changes in the mechanical properties as a function of water redistribution between the gelatin and agarose phases were reconsidered on the basis of progressive deswelling of the agarose network which possessed crosslinks (considered permanent under the experimental regime used). Recently, the macromolecular organisation and the phase structure of concentrated mixed gels of several biopolymers have been studied, taking into account density differences between phases and designing an explicit analysis of solvent partition in the binary systems to improve agreement with the experimental data. It was found that mixing of conformationally dissimilar macromolecules (such as maltodextrins with milk proteins) produces biphasic composite gels whose water partition between the two phases is profoundly affected by the phase inversion in the system. This behaviour was rationalised on the basis of mixed gels under kinetic control where formation of a continuous network (caused by a faster gelling component or due to excessive amounts of one polymer) "freezes" the system and prevents diffusion of water between the two phases. All along, the conclusions derived from the theoretical analysis were cross-examined using visual observations, microscopy, differential scanning calorimetry and mechanical spectroscopy. The gelling and melting profiles of composite gels, in combination with the temperature course of individual components, were found to be particularly helpful. Quandaries about the general applicability of this approach to composite gels are also discussed and further research is proposed for their clarification.

Work in this laboratory has also documented the time–temperature superposition principle, using mechanical spectra of dynamic oscillation, for high solute systems of gellan polysaccharide in a manner identical to the procedures of Ferry for acrylic polymers. Application of this method to rheologically interesting polysaccharides might establish the transition from a rubber to glass-like consistency in single and mixed systems of interest to both academia and the food industry. Finally, opportunities for the application of aspects of steric incompatibility to product development are discussed as a tribute to our industrial partners whose moving spirit and finances made much of this research possible.

Mechanical properties of two-phase, synthetic polymer blends

In their classic work Takayanagi, Harima and Iwata (1963) demonstrated that the viscoelastic behaviour of polymer blends formed by bonding together films of polybutadiene (PBD) and styrene–acrylonitrile copolymer (STAN) could be calculated according to the parallel or series model depending on the direction of the force passing through the specimen. The composite modulus (G_c) for the parallel and series arrangements can be expressed as follows:

$$G_c = \phi_x G_x + \phi_y G_y \tag{1}$$

and

$$G_c = \frac{1}{\dfrac{\phi_x}{G_x} + \dfrac{\phi_y}{G_y}} \tag{2}$$

where G_x and G_y are the shear moduli, with ϕ_x and ϕ_y ($\phi_x + \phi_y = 1$) being the phase volumes of the components X and Y respectively. It was also found that macroscopically homogenous dispersions of PBD in STAN copolymer formed a phase-separated mixture due to the immiscibility of the components. The tendency to minimise interfacial area resulted in spherical PBD inclusions. Above the glass transition temperature (T_g) of PBD, but below that of STAN (where $G_{PBD} << G_{STAN}$), the modulus of the composite was estimated from the known properties of the two component polymers on the basis of a parallel model. Therefore, by assuming extreme cases in the distribution of strain and stress within the mixed system eq. (1) applies to isostrain conditions, where the continuous phase is more rigid than the disperse phase and the strain is uniform throughout the material (parallel model), whereas eq. (2) refers to isostress conditions, where the supporting phase is weaker than the discontinuous phase and both phases are now subjected to the same stress (series model). Obviously, when the continuous phase is stronger than the disperse phase the composite modulus achieves values closer to the upper bound limit of the stronger component in isolation whereas phase inversion in the mixture produces viscoelastic functions which are closer to the lower bound limit set by single preparations of the weaker component. However, alignment of the dispersed phase, in layers of aggregated strands, perpendicular to the direction of the applied force as confirmed by optical microscopy, results in a change in the mechanical properties towards the character expected from the series connection in the PBD—STAN system. Therefore the Takayanagi model describes adequately the idealised extremes in the distribution of stress and strain for the case of dispersed spherical particles in a continuous matrix but it fails to predict the intermediate modulus—composition relationships arising from different shapes and perhaps sizes of the discontinuous filler for a given phase volume. In practical applications, the usefulness of the model lies in defining an area of modulus versus composition in which experimental results should be confined if the physical organisation of the material is determined by simple phase separation.

Provided that the phase domains (i.e. ϕ_x and ϕ_y) are well defined, the equilibrium shear modulus (G_x or G_y), and ultimately the modulus of the composite, is derived either from experimental measurements or theoretically (for a well characterised polymerisation process) by the Flory—Stockmayer (Stockmayer, 1943; Flory, 1953) theory of gelation:

$$G = \phi N_e \tag{3}$$

where N_e is the number of elastically effective network chains (EANCs), formed through the covalent interaction of monomers of functionality f. At a temperature T, N_e depends on the degree of cross-linking of functionalities, α. When a network is just able to form, α is given by the following critical value:

$$\alpha_c = \frac{1}{f-1} \tag{4}$$

ϕ is a proportionality constant given by:

$$\phi = \frac{gRT}{V_{mol}} \tag{5}$$

where RT is the entropy gain per mole of strained network chains, V_{mol} is the volume per mole of repeat units and g (front factor) is a measure of the number of RT units contributed to the modulus by each mole of load-bearing chains in the network. In purely entropic networks (ideal rubbers) g takes the value of 1.

Concentration-dependence of modulus in biopolymer gels

Gels of chemically cross-linked macromolecules derive their elasticity from the decrease in entropy of stretched chains due to a perturbation from the statistically most probable end-to-end distance (Manson and Sperling, 1976), but physically cross-linked biopolymer gels develop an elastic component due to the higher energy conformations of long junction zones when occupying portions of the strained macromolecular chains (Ross-Murphy, 1992). The enthalpic nature of non-permanent crosslinks in the Flory—Stockmayer model was accommodated by allowing the front factor to vary in eq. (5), with values of g greater than unity taken as evidence of increasingly non-rubberlike behaviour in gelling biopolymers (Clark and Ross-Murphy, 1985). Furthermore, defects in biopolymer structure as a result of aggregates partially attached to the end of a network (i.e. not contributing to elastic phenomena) or single molecular "dangling ends" that are not capable of supporting stress were considered using the concept of extinction probability u for the recalculation of N_e as follows (Dobson and Gordon, 1965; Gordon and Ross-Murphy, 1975):

$$N_e = \frac{f\alpha(1 - u)^2(1 - b)}{2} \tag{6}$$

where $b = \{(f-1)\alpha u/(1-\alpha+\alpha u)\}$. Substitution of Φ and N_e, as determined by eqs. (5) and (6) respectively, into eq. (3) gives G as a function of α and f:

$$G = gRT\left(\frac{Nf\alpha(1 - u)^2(1 - b)}{2}\right) \tag{7}$$

where N is the ratio of concentration (c) to the molecular weight (M) of the monomer. The above equation constitutes the core of cascade formalism (Gordon, 1962) and it has been used to analyse the mechanical characteristics of elastically active network chains of partially unfolded BSA molecules (Richardson and Ross-Murphy, 1981).

Biopolymer networks not only waste a proportion of the physical crosslinks in terms of enthalpic-elasticity behaviour but they also rearrange the elastically active junction zones as a part of a thermodynamic equilibrium process (cross-link formation is opposed by bond breaking). Assuming that gel formation due to non-covalent interactions between macromolecules is described by a monomer–dimer equilibrium (link ↔ two free sites) and relating the modulus to concentration (Hermans, 1965) the equilibrium constant for association is:

$$K = \frac{\alpha M}{cf(1 - \alpha)^2} \tag{8}$$

As the course of association proceeds, there will be a critical degree of reaction where an effectively infinite network is formed. For a given temperature, this occurs at a minimum critical gelling concentration (c_0) and the proportion of functional groups which has reacted is given by eq. (4). Substitution of α_c into eq. (8) allows the c_0 to be written in terms of f, K and M as:

$$c_o = \frac{M(f - 1)}{Kf(f - 2)^2} \tag{9}$$

By combining eqs. (7) and (9) and substituting c/M for N an expression for the shear modulus G is derived:

$$\frac{GK}{gRT} = \left(\frac{c}{c_o}\right) \left(\frac{(f - 1)\alpha(1 - u)^2(1 - b)}{2(f - 2)^2}\right) \tag{10}$$

The above approach to biopolymer gelation takes into account the relationship between modulus development and extent of the reaction. It also predicts the commonly observed c_0 in gelling polysaccharides, which was not taken into consideration in earlier theories since the individual rheological properties of rubberlike network elements are independent of the macroscopic amount present. For each hypothetical value of functionality a universal "master curve" of G vs. c can be derived and the calculated moduli (at concentrations between or beyond those studied experimentally) may then be inserted in eqs. (1) and (2) to estimate the equilibrium modulus of the composite (G_c).

Solvent distribution in biphasic biopolymer gels

THE ASSUMPTIONS OF PHASE EQUILIBRIUM AND POLYMER DESWELLING USED FOR THE DESCRIPTION OF MECHANICAL PROPERTIES IN COMPOSITE GELS

Having available a theoretically valid procedure that can predict the modulus development as a function of polymer concentration, we are almost ready to apply the blending laws in aqueous three-component systems. However, the inclusion of solvent (water) in the composite introduces one further complication that must be addressed first, namely, variable phase volumes (ϕ_x and ϕ_y), depending on the partition of solvent between the two gelling components. Obviously, effective concentrations after phase separation are higher than the initial, nominal amounts of the two polymer constituents and have to be estimated for use in the modulus–concentration relationship of eq. (10). Clark (1987), tackled the problem of solvent distribution by introducing the p parameter, the ratio of solvent in one phase to the original (nominal) concentration of the appropriate polymer divided by the corresponding ratio for the second polymer:

$$p = \frac{\left(\frac{w_x}{x}\right)}{\left(\frac{w_y}{y}\right)} \tag{11}$$

Therefore, in a system where both the weights (x and y) of two polymers and water ($w = w_x + w_y$) are known the p factor defines the phase volumes and hence the effective concentrations and real moduli of each biopolymer within its own phase. The above analysis was applied on composite agar–gelatin gels as follows: Experimental moduli were recorded for a series of mixed gels (holding the overall concentration of agar constant at 1% or 2% and varying the amount of gelatin in each agar series from 0% to 25%) and trial values of p were used to generate calculated upper and lower bound curves until a convincing fit was obtained (Clark, Richardson, Ross-Murphy and Stubbs 1983). Figure 11.1 shows the upper and lower bounds for the 1% agar gel series, assuming that the p factor equals 1. Upper and lower bound moduli (G_u and G_l) were calculated using the above approach (eqs. (1 to 11)) and making the following two assumptions:

(i) Bulk phase separation to thermodynamic equilibrium takes place first in solution with gelation then occurring subsequently and independently in each phase; and
(ii) The two polymers are confined entirely to their respective phases.

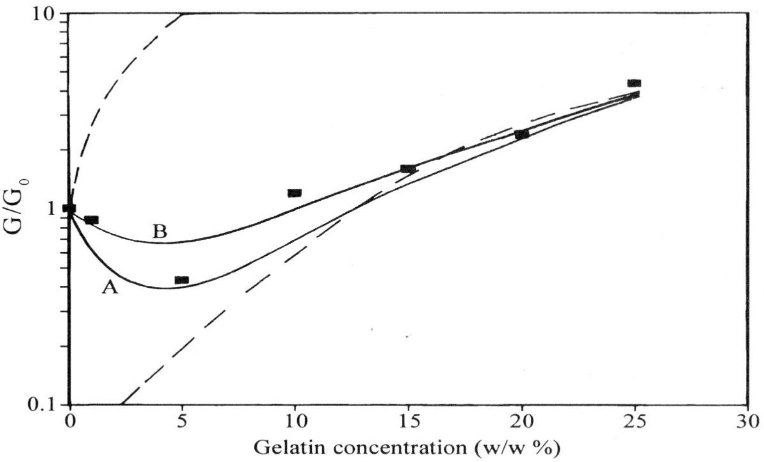

Figure 11.1 Shear modulus for the 1% agar series. Moduli are relative to the value for pure agar (G_0). Dashed lines show the upper and lower bounds for a phase separated system at thermodynamic equilibrium. Solid curves give the isostrain predictions for p=1 and q=0.33 (A) and 0.67 (B), based on deswelling (from McEvoy *et al.*, 1985)

Alternatively, McEvoy, Ross-Murphy and Clark (1985) argued that agar solutions during cooling set at higher temperatures that gelatin, effectively forming gels across the whole system at the nominal concentration (c^{nom}), of modulus G^{nom}. Then the agar network is taken to a higher concentration (c^{eff}), of modulus G^{eff} by removal of water as the gelatin gels. In this process, the agar crosslinks should be sufficiently labile and re-adjust to the new equilibrium condition in order that the cascade fit predicts the concentration dependence of modulus. In view of the fact that agar networks remain effectively permanent within the conditions of the experiment, verified by dynamic oscillatory and tensile evidence (McEvoy, Ross-Murphy and Clark, 1985a), the mechanical nature of a de-swelled agar network was described using the empirical formula:

$$\frac{G^{nom}}{G^{eff}} = \left(\frac{c^{nom}}{c^{eff}}\right)^{\sigma} \tag{12}$$

where the exponent σ was given values derived from the classical Flory (1953a) type swelling theory. The deswelling approach to agar–gelatin phase separation gave upper bounds (A and B in Figure 11.1) that reproduced the minimum in the experimental data at about 5% of gelatin in the mixed system. Calculation of the modulus bounds A and B required a water distribution identical to that for phase separated gels at equilibrium (p factor assumed to be one in both cases), but the assumption that the kinetically controlled deswelling of agarose network will allow re-partition of water and creation of similar volume fractions to those of the phases at equilibrium is rather unlikely. Nevertheless the spirit of both analyses reflects a substantial progress, since it was the first time that the complicated mechanics of composite aqueous gels were rationalised with theories of polymer physics.

EXPLICIT ANALYSIS OF SOLVENT PARTITION AND POLYMER-CHAIN CONTRIBUTION TO PHASE VOLUME IN GELLING BIOPOLYMER BLENDS

Recently the solvent partition approach was re-examined as a result of commercial interest in low fat products where phase separation phenomena between gelling biopolymers in the marketable product are of prime importance (Mageean and Jones, 1989). A new difficulty has arisen in these systems due to high overall polymer concentration. The earlier investigations (Clark *et al.*, 1983; McEvoy *et al.*, 1985) used dilute aqueous gels and thus assumed that the volume of polymer chains was negligible in comparison with the volume of the solvent. In the combinations studied lately, however, polymers constituted about a third of the total sample, so that their direct contribution to phase volume could no longer be ignored. The problem was addressed by calculating the concentration dependence of gel density (D) for each polymer X and Y (Kasapis, Morris, Norton and Clark, 1993). Then, for a particular combination of the two polymers the total weight of a phase (tw), being the sum of polymer and water in each phase, was calculated for each possible distribution of solvent between the component phases. At the end, phase volumes in the composite system were given in terms of the relative volumes (V) of regions X and Y as follows:

$$\phi_x = \frac{V_x}{V_x + V_3}$$

$$\phi_y = \frac{V_y}{V_x + V_y} \qquad (13)$$

where each relative volume was estimated by the corresponding ratio of tw to D. These values of phase volumes could be used directly in eqs. (1) and (2), and for the derivation of shear modulus of the composite gel (G_u and G_l) the moduli of individual components (G_x and G_y) were also obtained by the cascade or de-swelling treatment at the effective polymer concentrations (e.g. the c^{eff} of component X is equal to x over tw_x).

Figure 11.2 demonstrates the computerised output (Morris, 1992), derived for a phase separated mixed gel at which the nominal concentrations of the two components across the whole system would give moduli M_x and M_y, respectively, with $M_y = \Delta M_x$ (in this example, Δ is assumed to have the value of 4). Variation in calculated moduli is plotted against S_x, the fraction of solvent in the polymer X phase. As the value of S_x increases from 0 to 1, the calculated values of G_x decrease whereas those of G_y increase. A point is reached where the solvent partition between the biopolymers is such that the two curves cross, making the rigidity of the phases identical. At the same time the upper bound (X-continuous phase; isostrain conditions) descends from the top-left corner of the diagram, with the lower bound (Y-continuous phase; isostress conditions) rising from the bottom-left one, meeting again at the same "critical" point. Beyond this common point the physical significance of the bounds swaps over, with the Y-continuous phase now representing the upper bound (isostrain conditions) and with converse significance for the X-continuous phase (lower bound; isostress conditions). If the identity of the continuous phase is known from the type of experimental evidence cited in the following section, it should be possible to determine the precise value of S_x required to give best agreement with the observed modulus for a specific binary gel. The experimentally

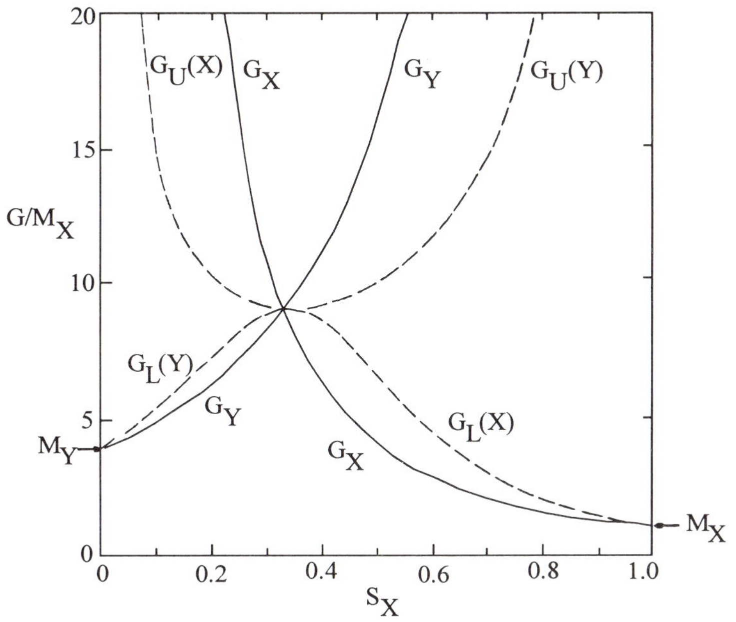

Figure 11.2 Effect of solvent fraction in the polymer X phase (S_x) on the value of calculated moduli (G/M_x). At their nominal concentration, polymer Y is four times stronger than X ($M_y = 4M_x$). Solid lines trace the variation in network strength of the individual components (G_x and G_y), whereas broken lines represent the shear moduli of the composite according to the isostrain and isostress model (G_u and G_l respectively). In both cases the continuous network is shown in brackets (from Morris, 1992)

acceptable value of S_x can then be used to estimate the solvent partition in the mixture from eq. (11), since the weight of water in phase X is the total weight of water in the system multiplied by S_x.

Quantitative analysis of composite moduli in biopolymer gels

COMPUTERISED MODELLING

The theory and working approach, described by means of eqs. (1) to (13) were put to the test initially in mixed gels comprising limed-ossein gelatin (LO-2) and potato maltodextrin (Paselli SA-6 and SA-2). The family of bounds was obtained on the basis that gelation of both components occurs at their final concentrations in their respective phases. These concentrations were used in the cascade formalism of equation (10), to calculate respective moduli. The bounds were plotted in relation to the solvent fraction of the gelatin phase (S_x) and the parameter p, in Figures 11.3a and 11.3b respectively (Kasapis *et al.*, 1993). Overall, the analysis of solvent-partition between the two components suggests that maltodextrin at concentrations above 15% in the system forms a weak, continuous matrix (lower bound; isostress conditions) surrounding the stronger inclusions of gelatin, which however are 80% more hydrophilic than the supporting matrix of the polysaccharide ($\log p \approx 0.26$). However, below 15% maltodextrin in the composite, results remain close to or fall below the gelatin modulus at nominal concentration (5%), rendering unsuitable the aforementioned hypothesis. This pattern of behaviour was treated on the basis of de-swelling ideas (eq. 12), assuming that the faster-gelling species, gelatin, does so at its original concentration throughout the system, followed by ordering of maltodextrin that claims some of the solvent from the gelatin network. The shallow curve in Figure 11.4 demonstrates the standard of agreement achieved between experimental points of the composite gels and expected moduli, calculated on the basis of initial gelation and subsequent de-swelling of gelatin. In doing so, the p value was held constant at 1.8 and the swelling factor, σ , was taken at the theoretical value for a permanent network ($\sigma = 2/3$). The steep line in Figure 11.4 shows the calculated moduli for a maltodextrin continuous network beyond the phase inversion point (15% Paselli in the system), formed at its final concentration (treatment analogous to that in Figure 11.3).

Recent work on another phase separating system of direct practical relevance has shown that the postulate of initial phase separation and subsequent gelation of the two components separately in their own phases describes well the steric exclusion phenomena between maltodextrin (Cerestar 1906) and a commercial milk protein preparation (Promilk 602), in effect replacing gelatin with a globular protein in the

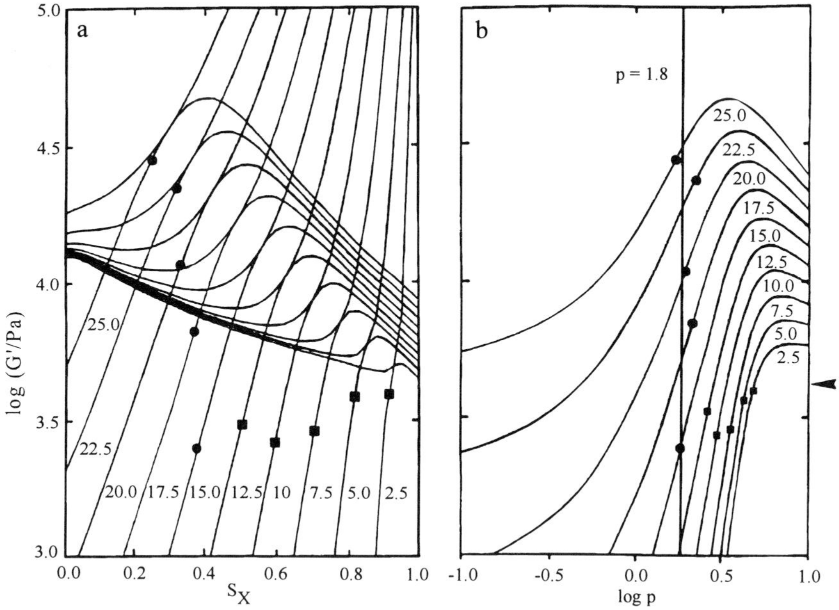

Figure 11.3 Phase separation analysis for 5% w/w LO-2 with SA-6 at the concentrations (% w/w) shown, plotted as a function of solvent fraction in the gelatin phase, S_x (a: both upper and lower bounds are shown) and solvent avidity parameter, p (b: only the lower bounds are plotted). Experimental values are shown on the maltodextrin continuous bounds for physically realistic (●) and physically unrealistic (■) situations. The experimental modulus for 5% LO-2 in isolation is shown by the arrow on the right-hand axis (from Kasapis *et al.*, 1993)

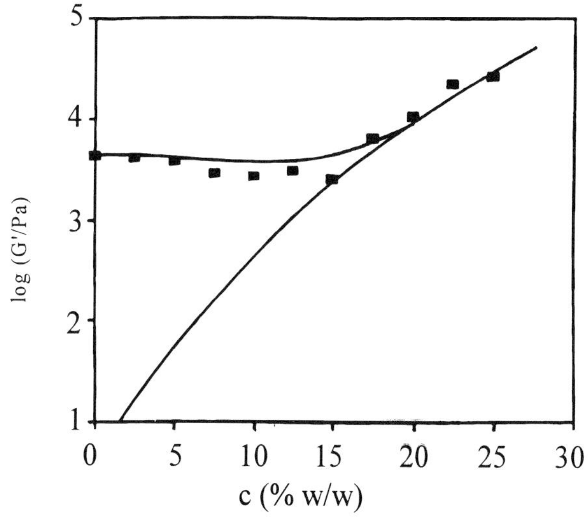

Figure 11.4 Experimental (■) and calculated (—) moduli for 5% LO-2 with varying concentrations (c) of SA-6. The shallow curve corresponds to a gelatin-continuous network formed at its nominal concentration, with subsequent deswelling. The steep line shows the calculated moduli for a maltodextrin-continuous network formed at its final concentration (from Kasapis *et al.*, 1993)

first investigation. In Figure 11.5 the milk protein is regarded as polymer X throughout, so that the parameter S_x refers to the fraction of solvent in the milk protein phase (Chronakis, Kasapis and Richardson, in press). Mixed gels were prepared using a fixed concentration of milk protein (16.5% w/w) with maltodextrin concentration from 2% to 18% w/w. Analysis of the phase separation between the two components (Figure 11.5a) indicates phase inversion from a weak, protein-continuous phase (lower bound) to a strong, maltodextrin matrix (upper bound) at about 13% maltodextrin. Solvent fractions derived from Figure 11.5a were used for analysis of water partition between the two phases (Figure 11.5b), yielding $p \approx 1.7$ ($\log p \approx 0.23$) for the intercepts in the milk protein continuous systems, whereas the data beyond the phase inversion point (maltodextrin continuous systems) are better fitted with a value of $p \approx 1.1$ ($\log p \approx 0.04$).

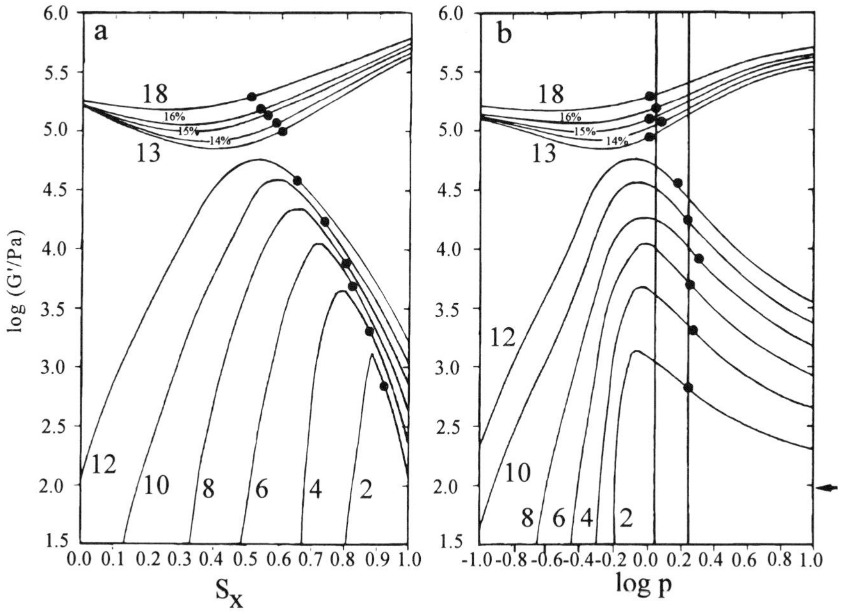

Figure 11.5 Calculated bounds for 16.5% milk protein with maltodextrin concentrations (% w/w) shown, plotted as a function of solvent fraction in the milk protein phase, S_x (a) and solvent avidity parameter, p (b). In the milk protein continuous systems (from 2% to 12% maltodextrin in the composite) only the lower bounds are drawn whereas at maltodextrin concentrations beyond the inversion point (from 13% to 18%) the isostrain predictions are illustrated. Experimental values are shown to intercept the bounds and the experimental modulus for 16.5% milk protein in isolation is noted by the arrow on the right-hand axis (Chronakis, Kasapis and Richardson, in press)

Substitution of maltodextrin, employed in the preceding study, with soya protein (Supro 760) creates a composite system of two cold-setting aqueous preparations of thermally-processed milk and soya concentrates. Mechanical characterisation of the individual samples has shown similar properties in terms of rates of thermally-induced gelation, concentration-dependence of pseudo-equilibrium modulus above 20% polymer, terminal relaxation times in simple compression, and pattern of failure under large-deformation analysis (Chronakis and Kasapis, 1993). A possible interpretation of this behaviour might be that the thermal processing causes extensive denaturation to the protein molecules which, in the hydrated state, form roughly spherical colloidal clusters of comparable functionality. The positive relationship between "performance" characteristics of the protein network and polymer density in the system was used to manipulate the rheological behaviour and water immobilisation of one gelling agent at the expense of the other. Figure 11.6 demonstrates the application of de-swelling theory to milk protein continuous networks (10%) with soya inclusions as the discontinuous phase (at concentrations ranging from 6% to 16%, which reflect the minimum gelling requirement and solubility limit, respectively). When the experimental data were plotted on the calculated bounds, physically realistic results were obtained for soya concentrations in the range 6% to 10%, as expected from the concentration advantage of the milk protein component according to the analysis of the individual components. Beyond that, the deswelling argument of the composite systems becomes unrealistic, with the fitted values for solvent fraction in the milk phase increasing with increasing

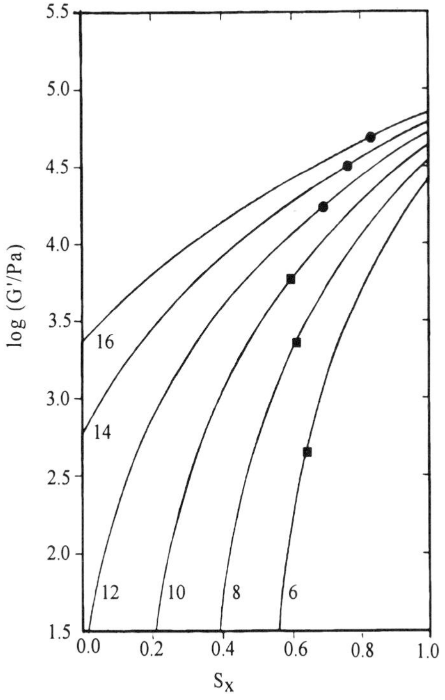

Figure 11.6 Calculated milk protein continuous bounds for 10% Promilk with soya protein at concentrations (% w/w) shown. Experimental values are shown on the bounds for physically realistic (■) and physically unrealistic (●) situations (from Chronakis and Kasapis, 1994)

concentration of soya protein. Molecular interactions in the upper half of the composition range (12% to 16% soya protein), however, could be modelled accurately by the assumption of immediate phase separation in solution, yielding in the gel-state a soya continuous network with a water to polymer ratio 1.25 times higher than in the milk protein phase (p = 0.8; milk protein is the component X) and a phase inversion point at about 11% soya in the mixture.

Discussion

The emerging picture of the effect of polymer conformation on the state of phase separation in binary mixtures with relevance to the food industry is as follows: Conformationally similar species in solution like the disordered coils of maltodextrin and gelatin or the globular structures of milk and soya macromolecules tolerate each other at low concentrations in a monophasic solution. During subsequent cold-setting, the faster-gelling polymer in each mixture develops its continuous network prior to structuring of the second component. This idea was successful in modelling the lower range of concentrations in the gelatin—maltodextrin and milk—soya protein series (Figures 11.3 and 11.6 respectively), where gelatin and milk protein were shown to employ faster rates of gelation that their partners in the blend. Maltodextrin and soya protein upon structuring claim extra solvent and as a result deswell to a certain extent the gelatin and milk protein networks but the systems remain under kinetic control as judged by repeatedly weighing the mixtures which showed a continuous, slow diffusion of water from the faster to the slower setting component with time. In both systems however, at combinations above the phase inversion point, it seems that the kinetic effect is swamped by the enthalpic disadvantage in polymer segments being surrounded by others of a different type and phase separation at equilibrium occurs in solution. Cooling of the solutions produces composite gels with a single pattern of water partition throughout the concentration range (p is equal to 1.8 and 1.25 in favour of the gelatin and soya protein respectively), but it is difficult to say if the mixed systems have gelled at thermodynamic equilibrium.

Dealing with the question of kinetic influences versus thermodynamic equilibrium in the maltodextrin–milk protein system, it was demonstrated that thermodynamic incompatibility between the conformationally dissimilar species of thermally-unfolded globular protein molecules and disordered chains of maltodextrin promotes an early phase separation in solution and then in the gel state, i.e. at both sides of the phase inversion point. As a result there is an immediate reinforcing effect of the maltodextrin filler on the milk protein gel which was not observed in the case of gelatin and maltodextrin, where composite values below the phase inversion point remain close to that of the continuous gelatin network at its nominal concentration. This is shown in Figure 11.5b with the two distinct families of water distribution, surrounding the transition from isostress to isostrain conditions in the composite gel. Furthermore, within each type of filled network, there is no systematic variation of p with polymer concentration, and the proportion of water associated with the milk phase is reduced as it ceases to be the supporting phase and becomes the discontinuous filler. A simple explanation of this behaviour is that the water binding capacity is not only a reflection of the individual properties of a polymer but it also depends on the geometrical organisation of the composite's macrostructure. Obviously, by accepting that water tries to diffuse to osmotic equilibrium, the filler would expose relatively less surface for a given volume, thus reducing its power of attraction for solvent, if it formed approximately spherical inclusions.

Figure 11.7 illustrates an attempt to predict the shape of filler particles (maltodextrin at concentrations between 0% and 12%) embedded in a milk protein gel matrix. Extraction of phase volumes from Figure 11.5b allowed the derivation of moduli of the two components at their effective concentrations and the plotting of reinforcement (the ratio of $G_{composite}$ to G_{matrix}) as a function of the maltodextrin volume fraction in the blend. Fitting of the data, using the modified Kerner equation

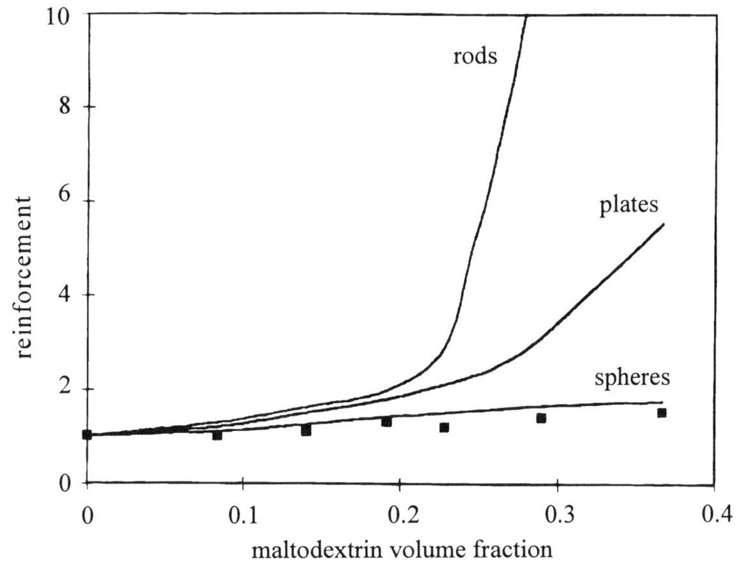

Figure 11.7 Reinforcement of the composite with a milk protein continuous phase, discussed in Figure 11.5. Extraction of phase volumes for maltodextrin concentrations between 0% and 12% in the mixed system allowed comparison of the experimental results (■) with the predicted behaviour obtained from the Kerner equation for spherical, plate and rod-like inclusions of maltodextrin in the composite gel (Chronakis, Kasapis and Richardson, in press)

(Lewis and Nielsen, 1970), suggests that the maltodextrin phase exists as discontinuous spherical particles rather than plate- or rod-like assemblies and thus it supports the idea of changing p values with phase inversion phenomena in a composite gel. It follows that the difference in p values is the result of phase separated gels trapped away from equilibrium conditions, since the equilibrium value of "relative affinity" of the two polymers for water should not be affected by the geometrical rearrangements of their phases in a binary mixture. Therefore, the p parameter, at least for the gels of this investigation, is not an equilibrium concept but a single point measurement in the experimentally accessible time—temperature continuum. Since no time—temperature superposition principle is capable of accommodating conformational transformations which lead from disordered coils at thermodynamic equilibrium to ordered structures under kinetic control, it is the opinion of this author that the water partition values derived for gelled biopolymer composites at a particular thermal regime should not be related to the Flory—Huggins equilibrium parameter (χ) for the energies of interaction between these two biopolymers and solvent in solution. Instead, the mixed gel phase continuity, phase inversion and solvent distribution in non-equilibrium arrangements should be heavily governed by the thermal history to which the blend is subjected.

Having appreciated the importance of a kinetically-controlled sequence in mixed-gel formation, we verified that the gelatin—maltodextrin system achieves a constant state within 7 hours on a weight basis, and then monitored the small deformation properties of de-swelled gelatin gels for the same time (results and modelling are shown in Figure 11.4). The slight overestimation of moduli at the intermediate maltodextrin concentrations (from 7.5% to 12.5%) might indicate that weight measurements, reaching a stable state at the end of a 7 hour period, do not necessarily reflect the establishment of a complete thermodynamic equilibrium across the system. Controlled, slower cooling of the gelatin—maltodextrin solution is needed to monitor possible changes from the viscoelastic functions reported for the rapid quenching of this system, which then could be explained on the basis of a different partition of water within the biphasic gel.

Finally, comparison of Figures 11.3b and 11.5b, in which the maltodextrin forms the continuous phase at concentrations beyond the phase inversion point, demonstrates the sensitivity of blending laws to changes in rigidity of the supporting phase. According to eqs. (1) and (2) the mechanical strength of a composite material is mainly derived from the rigidity of the continuous network, altered in a consistent manner by the presence of filler (isostrain or isostress conditions). Therefore, in the case of Paselli maltodextrin—gelatin mixed systems, where the moduli of the individual components achieve comparable values at concentrations well above the c_0 (Kasapis, Morris, Norton and Clark, 1993a), a weaker maltodextrin continuous phase surrounding the stronger beads of gelatin is well accepted from the standpoint of the Takayanagi theory (Figure 11.3b). Conversely, the overall rigidity of a composite network might be relatively insensitive to a weak filler when the moduli of the two components differ significantly. This is illustrated in Figure 11.5b and it is attributed to the network of Cerestar maltodextrin being more than one and a half orders of magnitude stronger than that of milk protein at equal concentrations (Chronakis, Kasapis and Richardson, in press).

Experimental evidence of phase separation in mixed gels

To test the validity of the theoretical analysis in terms of phase inversion phenomena and to identify the phase separated microstructure in a binary system, utilisation of several experimental techniques is needed. In the gelatin—maltodextrin study, exclusion phenomena become apparent upon mixing appropriate amounts of individual stock solutions at 45°C, a temperature at which both components have been shown to remain as disordered coils in solution over the timescale of the experiment. Cloudiness in the mixed preparation indicated the formation of phase separated droplets large enough to scatter visible light. On centrifugation, opaque solutions containing more than 2% LO-2 and 20% SA-6 resolved into two co-extant liquid layers at thermodynamic equilibrium. At intermediate concentrations, precipitation resulted from aggregation of maltodextrin particles, and at low concentrations (see the lower binodal in Figure 11.8) the components co-existed in a single phase solution (Kasapis, Morris, Norton and Gidley, 1993b). The different

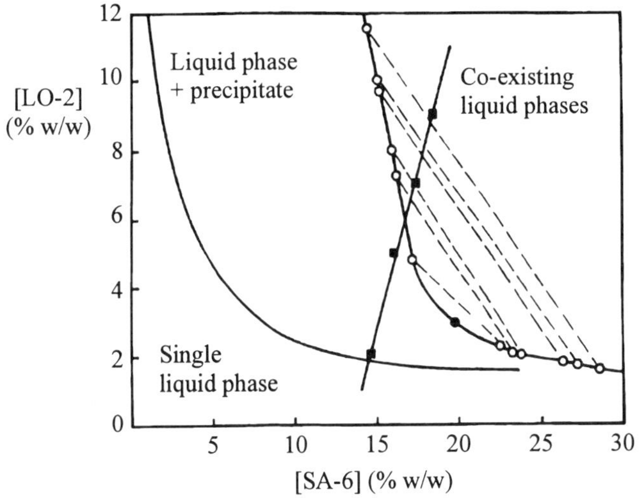

Figure 11.8 Phase diagram of the gelatin (LO-2)—maltodextrin (SA-6) system at 45°C. The binodal is defined by the composition (○) of co-existing liquid layers, joined by tie-lines (- - -), and converging to a single critical point (●). The straight line defines the compositions of phase inversion points (■) for a series of LO-2—SA-6 mixed gels (from Kasapis *et al.*, 1993b)

rates of gelation of single gelatin and maltodextrin samples also allowed visual observations to monitor polymer exclusion in the gel state. In doing so, solutions were quenched from 80°C to the experimental temperature and the time (t_g) required to form a gel strong enough to support itself was recorded. Figure 11.9 illustrates a steady decrease in gel time of composite gels at the concentration ranges associated either with a continuous gelatin network or a supporting maltodextrin matrix in Figure 11.3. The sharp discontinuity in the gelling profile, producing a dramatic decrease in the rate of gelation, was coincident with the phase inversion point suggested in Figure 11.3, since a continuous gelatin network throughout would follow the trend observed in the first part of the concentration range of maltodextrin rather than showing the abrupt change observed experimentally. The increase in gelation rate as a function of maltodextrin concentration, in both distinct sections of the graph, was attributed to thermodynamic incompatibility between the two polymers driving gelatin to the more compact ordered form, or at combinations beyond the phase inversion point, concentrating up the maltodextrin phase.

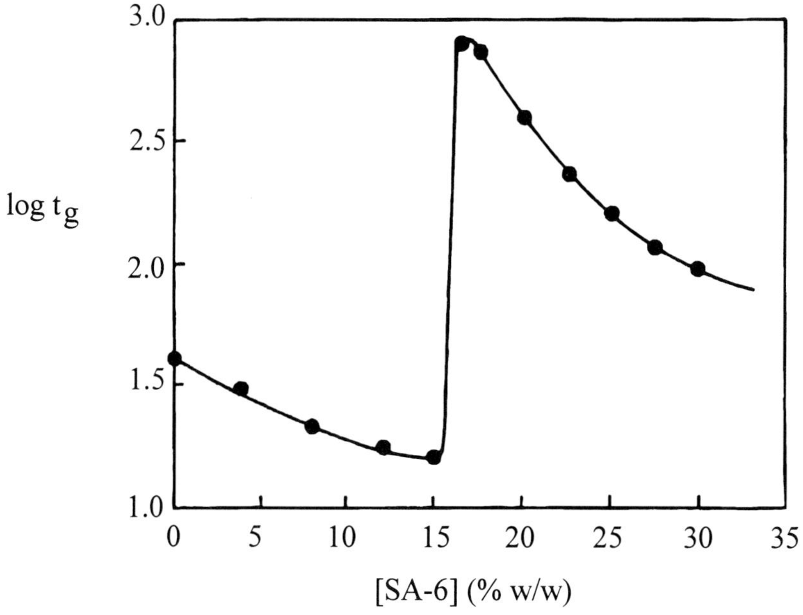

Figure 11.9 Time (t_g, in min) at 25°C required for the formation of a self-supporting network from mixed solutions of 5% (w/w) LO-2 with varying concentrations of SA-6 (from Kasapis *et al.*, 1993c)

Additional experimental characterisation of the phase separation in gelatin–maltodextrin gels included differential scanning calorimetry (DSC), light microscopy and the temperature course of gel melting under dynamic oscillation (Kasapis, Morris, Norton and Brown, 1993c). Changes accompanying the endothermic profiles of representative single and mixed preparations are shown in Figure 11.10. The thermograms observed on deconvolution of individual maltodextrin and gelatin helices, with transition mid-point temperatures (T_m) of 59°C and 23°C respectively, maintained their position and general band form in the composite gels at concentrations well below and above the phase transition from deswelling to isostress conditions, reported in the theoretical treatment of Figure 11.3. Therefore, it was concluded that there were no specific interactions between the two components, i.e. absence of co-operative binding which has been shown to create an additional, new peak in synergistic gelation of mixed systems (Williams, Day, Langdon, Phillips and Nishinari, 1991). As in the agar–gelatin work (Clark, *et al.*, 1983), light microscopy in the gelatin–maltodextrin study has taken advantage of the different transmission patterns of visible light between the transparent gels of gelatin and the three-dimensional assemblies of maltodextrin that scatter light. Gross changes in molecular organisation with increasing maltodextrin concentration in the mixture agreed with the visual observations of Figure 11.9. As shown in Figure 11.11 there is a transformation from a gelatin continuous matrix (black background) surrounding the maltodextrin spheroids (white agglomerates), to an optically dense maltodextrin network penetrated by the dark, flat image of gelatin inclusions.

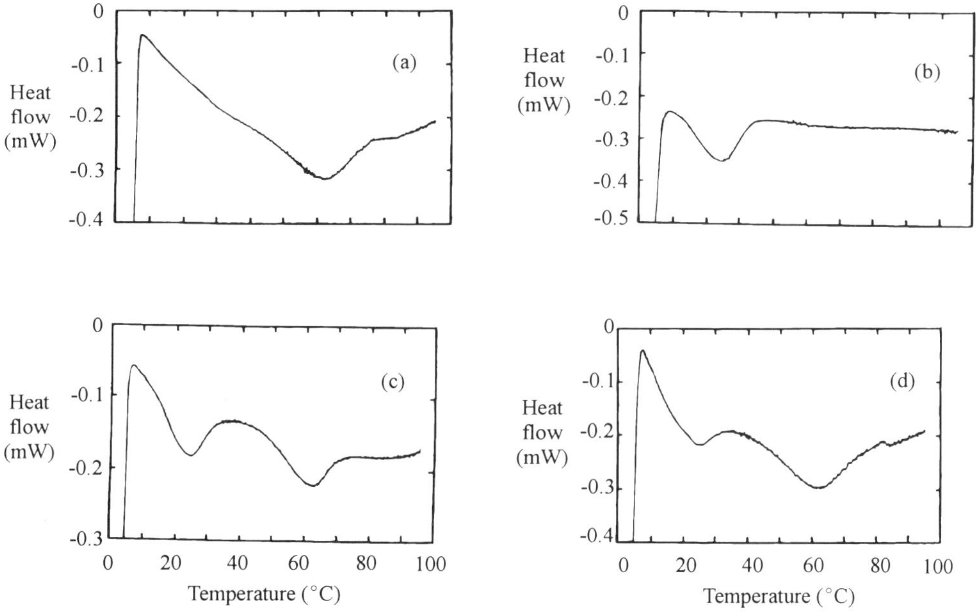

Figure 11.10 Melting endotherm of DSC heating scans from 5°C to 95°C at 0.1°C/min for: (a) 20% SA-6; (b) 2% (w/w) LO-2; (c) 2% LO-2 plus 10% (w/w) SA-6; and (d) 2% (w/w) LO-2 plus 20% (w/w) SA-6 (from Kasapis *et al.*, 1993c)

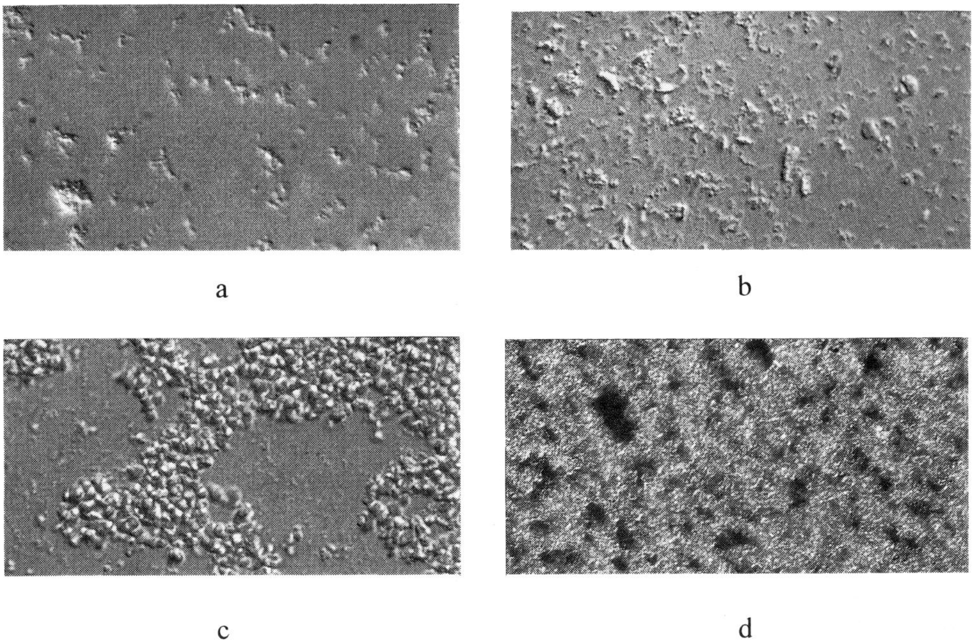

Figure 11.11 Micrographs for mixed gels of 2% (w/w) LO-2 gels with SA-6 at concentrations (w/w) of (a) 5%; (b) 10%; (c) 15%; and (d) 17.5%, obtained at magnifications of 250× (a, b, c) and 320× (d) (from Kasapis *et al.*, 1993c)

Production of new, appealing food products that imitate successfully the perceived mouthfeel of more traditional recipes relies on using suitable mixtures of biopolymers with distinct viscoelastic and textural properties. Quite often the different network characteristics of the components include a wide range of melting profiles which are then utilised to refine the physical properties of new foodstuffs in an attempt to match a desirable sensory attribute. For example, the mechanical temperature-course of gel melting in gelatin and maltodextrin samples has provided ample experimental evidence of the ability to tailor composites' properties in this way, as predicted by the theoretical studies of the previous section.

Dynamic oscillatory evidence in Figure 11.12 demonstrates that the thermal gap of about 35°C, documented for the endothermic transitions of single gelatin and maltodextrin networks (Figure 11.10), is maintained between the melting temperatures of polymeric combinations in which either a gelatin or a maltodextrin continuous network is predicted from ideas of network deswelling or initial phase separation in solution, respectively. At an SA-6 concentration of 15% (phase inversion point according to Figure 11.3), melting of gelatin weakens the network substantially but the gel remains intact until the higher temperature range associated with the mechanical collapse of the maltodextrin component. Alternatively, possible

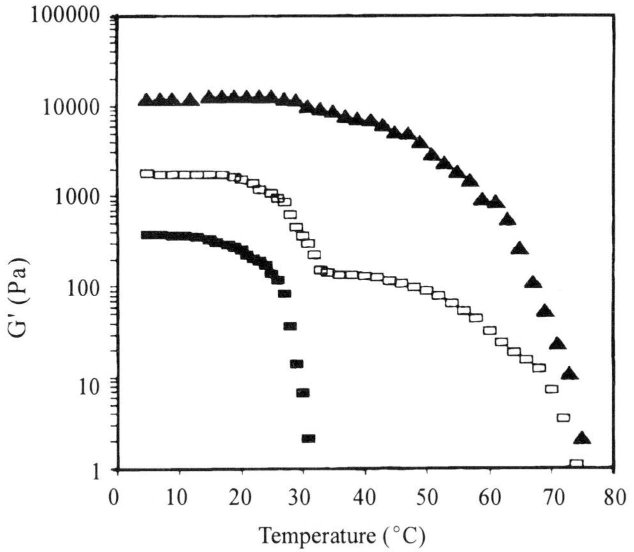

Figure 11.12 Gel melting at 1°C/min, characterised by a reduction in storage modulus (G′), for a 2% gelatin composite gel at 10% (■), 15% (□) and 27.5% (▲) SA-6 (from Kasapis *et al.*, 1993c)

changes in structural organisation can be monitored by plotting modulus values of a composite gel, at a temperature close to the melting transition of the component whose concentration is kept constant, versus the concentration of the component whose content is allowed to vary. Figure 11.13 reproduces the pattern of modulus development in the milk–soya protein system at 50°C, a temperature at which cohesion of the 10% milk–6% soya co-gel is lost (Chronakis and Kasapis, 1993). Whereas a smooth variation in the melting behaviour would argue for a continuous milk protein phase in all six preparations, the sharp discontinuity in the experimental trace supports the analysis of Figure 11.6 and locates the phase inversion point at a soya content of about 11% in the blend.

In conclusion, the combination of visual, calorimetric, microscopic and mechanical techniques seems capable of distinguishing between the structural characteristics of individual components, identifying exclusion phenomena in binary mixtures and providing an understanding of phase changes in composite gels. Eventually, quantitative interpretation of the experimental evidence disentangles the relations between polymer concentrations and moduli of the components and the composite gel, thus allowing prediction of solvent distribution between the constituent phases. Obviously direct experimental measurements of the compositions of two co-existing gelled phases is a major target for future research. Judging from the pattern of phase separation in mixed solutions of gelatin and maltodextrin, where cross-contamination occurs (phase diagram of Figure 11.8), one could argue

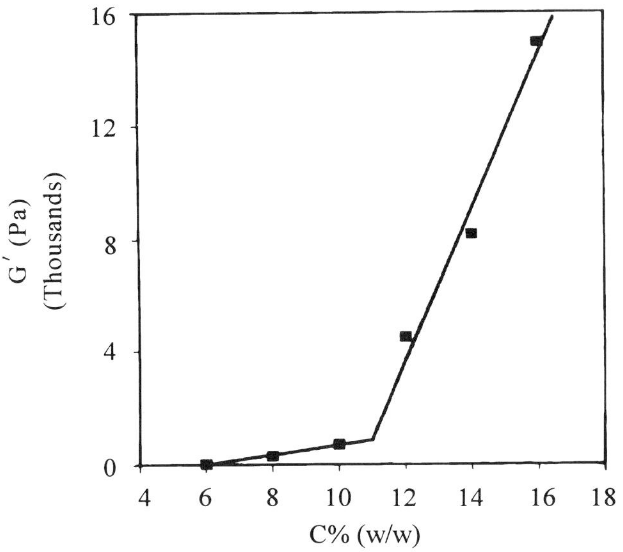

Figure 11.13 Variation in the course of gel melting, monitored from the values of G′ at 50°C, as a function of soya protein concentration during heating of the milk-soya protein gels at 1°C/min (from Chronakis and Kasapis, 1993)

that the assumption of total incompatibility between the two components in the gel state might pose a problem. On the other hand, the absence of gelatin from the precipitated maltodextrin gel at 45°C which is formed by the association of 1,4-linked α-D-glucan chains into co-axial double helices, as seen for the case of true gels at lower temperatures, suggests that at least in this system total steric exclusion might be the driving force behind the formation of phase separated gels. Furthermore, the linear increase in maltodextrin concentration at which phase inversion occurs in the mixed gel series of 2%, 5%, 7%, and 9% gelatin (Figure 11.8), passes smoothly through the binodal obtained from studies of mixed solutions in the disordered state (45°C), indicating that (at least under conditions of quench cooling) the state of dispersion of the constituent phases within the composite gel is not pre-determined by phase separation in the sol state. Obviously, additional work is needed to elucidate the state of phase separation at different time scales of gelation in mixtures of biopolymers with a wide variety of cross-linking (from temporary and physical to permanent and chemical), so that the deswelling ideas can also be developed.

A recent development in high solute mixtures with biopolymers

Over the past few years several microbial polysaccharides of commercial interest have emerged as a result of the relentless surge for versatile and effective structuring agents. One of the major members of the group is the anionic polysaccharide gellan, in the deacylated form, which is capable of forming firm, transparent gels at unusually low concentrations (down to 0.05% w/v; Papageorgiou, Kasapis and Richardson, 1994). An attractive feature of gellan gum is that in combination with gelatin it produces thermally stable water dessert and confectionery products with the required long, rubbery texture (Wolf, Lavelle and Clark, 1989). Recently, Papageorgiou, Kasapis and Richardson, (1995) have looked at single and mixed gellan–gelatin gels in the presence of sugar and corn syrup at levels up to 85% in the mixture. Whereas at sucrose concentrations between 0% and 50% in the system the usual sugar-stabilised network properties occur, as reported for κ-carrageenan (Nishinari, Watase, Williams and Phillips, 1990) and agarose (Nishinari, Watase, Kohyama, Nishinari, Oakenful, Koide, Ogino, Williams and Phillips, 1992) gels, at 60% co-solute there is a substantial reduction in network strength which however, could form at the highest experimentally accessible temperature of 90°C. Subsequent cooling of the high solids gel produces an unusual temperature dependence of the viscoelastic functions with the storage (G') and loss (G") moduli converging at low temperatures. Increasing amounts of co-solute enhance further the viscous component of the network which, for example, intercepts the G' trace at 37°C for the 50% sucrose plus 35% corn syrup gellan sample (frequency of 10 rad/sec). Mechanical spectra, taken at regular temperature intervals during a cooling scan, could be superimposed according to the time—temperature superposition principle which is expressed by the following equation:

$$\log\left(\frac{G'(T)}{G'(T_0)}\right) = \frac{-C_1\,(T - T_0)}{C_2 + T - T_0} \qquad (14)$$

where C_1 and C_2 are functions of the fractional free volume at a reference temperature T_0 and of the expansion coefficient of free volume. Ferry (1980) has used this approach to document the transition, from a rubber-like consistency to a glassy state, of high molecular weight synthetic polymers. The quality of superposition achieved for 0.5% gellan with 85% co-solute at the reference temperature of 90°C (Figure 11.14), suggests a similar transformation from a rubbery to a glassy structure. It was found that the frequency equivalent of the onset of vitrification at 90°C was about 17 Hz. In both temperature- and frequency-induced glass transitions, a gradual immobilisation (freezing out) of configurational motions, progressing from the largest segments in the gellan network to the shorter oligomeric solute molecules, has been proposed as the driving force for the transformation. Heating of the high solids gellan samples produces an initial mechanical response that overlaps the glassy part of the cooling trace (perfect thermoreversibility of behaviour in the glass state). In the second part of the heating trace however, the

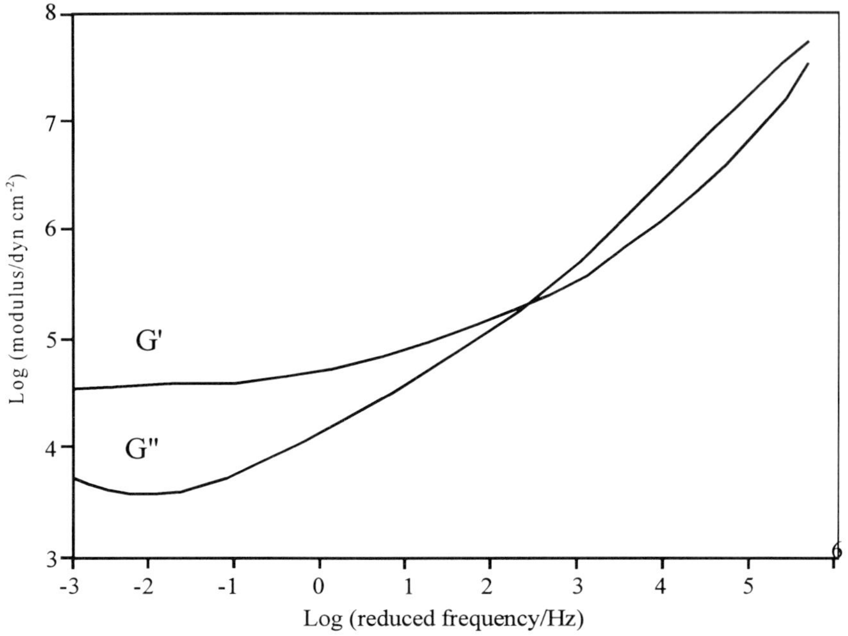

Figure 11.14 Time—temperature superposition for the mechanical spectra of 0.5% gellan sample with 85% co-solute, taken at 90°C, 70°C, 50°C, 30°C and 5°C (from Papageorgiou *et al.*, 1995)

elastic modulus increases monotonically in an atypical fashion for a biopolymer gel (Clark and Ross-Murphy, 1987) and in accordance with the thermoelasticity of a rubber-like body (Flory, 1953b). Figure 11.15 emphasises the difference in the characteristic temperature dependence of shear modulus in gellan networks with and without sucrose. Similarly for every single cold-setting biopolymer examined so far, reduction in network strength with increasing temperature produces a negative gradient of modulus change as a function of absolute temperature for aqueous gels. By contrast the positive relationship between elastic modulus and temperature (above 30°C) in the high solids gellan structures yields a family of constant gradients in the G'/T vs T graph, typical of the temperature dependence of rubber.

Surprisingly, gelatin gels at amounts of co-solute up to 70% (maximum solubility) neither show signs of rubbery behaviour nor undertake transition to a glass-like consistency (Papageorgiou and Kasapis, 1995), although the protein has been regarded as the biological prototype of a soft rubbery material (Pines and Prins, 1973). Instead, the gelation (G' > G") and melting (G" > G') temperatures or the isothermal modulus development of a 70% solids gelatin gel correspond to the upper end of a systematic variation in the viscoelastic performance of the protein with increasing concentrations of sugar. Again, the contrasting structural

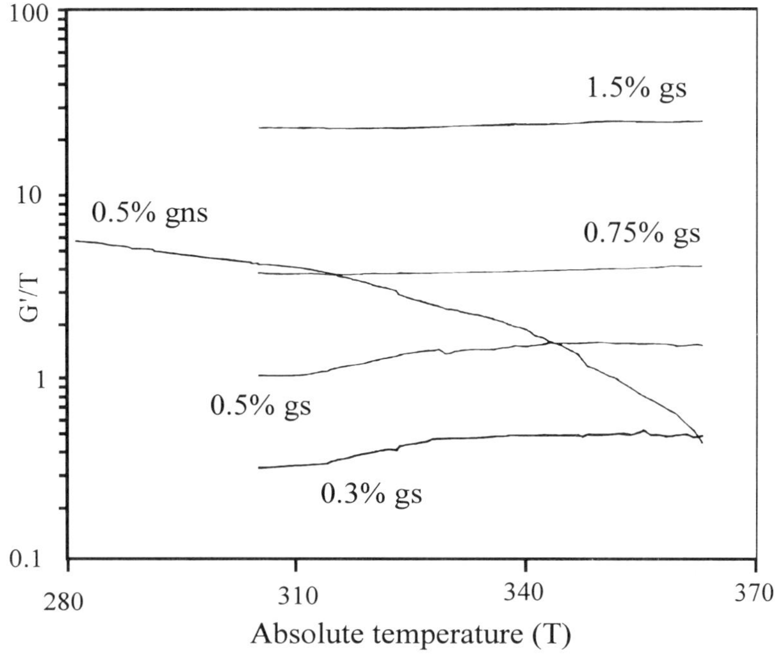

Figure 11.15 The thermal gradient of modulus development as a function of absolute temperature for gellan networks in the presence (50% sucrose plus 20% corn syrup) and absence of solute; cited as *gs* and *gns* alongside the individual traces (from Papageorgiou and Kasapis, 1995)

characteristics of gellan and gelatin networks allowed an investigation of the binary mixture, prepared at acidic pH (about 4) and moderate ionic strength (well below the level secreted by the salivary glands so that products do not taste unduly salty). Results documented that polymeric combinations, containing gelatin concentrations more than 50 times the amount of gellan in the system, form gelatin continuous networks, as judged from the temperatures of gel formation and collapse during thermal sweeps (typically, values of 30°C and 40°C respectively). However, weight ratios of gelatin to gellan below 50 in the mixture produce a gel response at the upper temperature (80°C) which during cooling shows the distinct convergence of G' and G" traces in the manner reported for the high solids gellan gels (Figure 11.16a). This is followed by a second step at a temperature regime associated with the gelling transition of the gelatin component (onset of gel formation at about 30°C). Heating of the mixed gel (Figure 11.16b), completes a "wave" of gelatin melting at about 40°C but the thermally stable gellan network prevents the system from collapsing at higher temperatures. The abrupt change in viscoelastic properties of the high solute gellan–gelatin blend, as a function of polymer composition, in association with the ability of the binary system to form gels at concentrations well below the minimal critical gelling concentration of the individual components

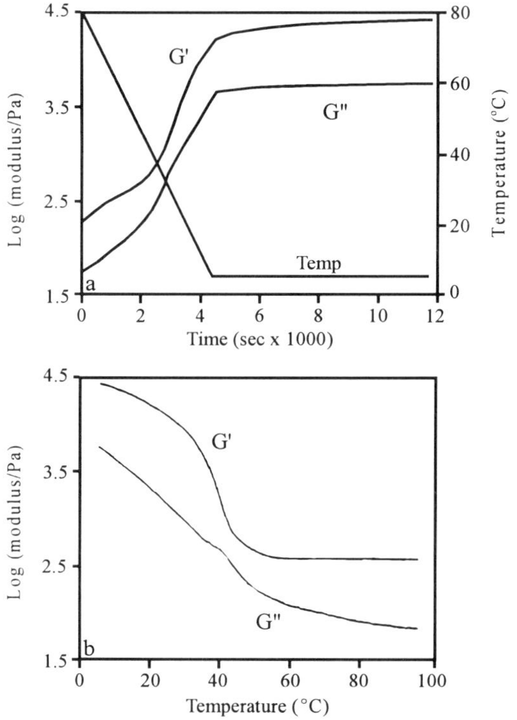

Figure 11.16 Cooling-isothermal (a) and heating (b) profiles at 1°C/min for a 5% gelatin with 0.4% gellan and 70% co-solute mixture. Between temperature sweeps samples were left for 2 hours at 5°C (from Papageorgiou and Kasapis, 1995)

suggests that a composite gel is formed which, by intelligent manipulation, will phase invert from a gelatin continuous matrix to a system where gellan constitutes the supporting phase.

The work on gelation and phase separation of gellan–gelatin blends in the presence of a wide range of co-solute in the system (from 0% to 85%) might encourage future research in the area of low-solvent mixtures of biopolymers, which remains largely unexplored. Furthermore, one should take an interest in the applicability of the time–temperature superposition principle to other polysaccharides and in its apparent unsuitability for gelatin gels. A molecular understanding that can predict or control the physical state of biopolymers in high sugar systems would be very popular with the food industry.

Application of the understanding on phase separation to product development

Gellan—gelatin mixtures at acidic pH and in the absence of salt form electrostatic coacervates (Chilvers and Morris, 1987). This proposal finds general "acceptance" for water dessert and confectionery sweets, made with batch gellan and gelatin in tap water (Shim, 1985). The investigation in this laboratory, however, suggests that the real story is not quite so simple, since under conditions of minimum positive interactions between the two components (i.e. at moderate levels of salt) the binary system is dominated by phase exclusion phenomena with the two polymers gelling independently in their respective phases. As a result, blends retain the clarity of individual components instead of the coacervate related opalescence of electrostatic interactions. Figure 11.16 demonstrates that a gellan continuous phase in the composite gel can support the liquid gelatin inclusions above 40°C, thus offering a remedy for the problem of gelatin jellies sticking together when stored at high temperature. Additionally, gellan-continuous preparations break at about the same deformation as gelatin (true yield strain of about 1), rather than having the brittleness of gellan, indicating that samples are good elastic dispersions suitable for use as table jellies (Figure 11.17; Papageorgiou *et al.*, 1994).

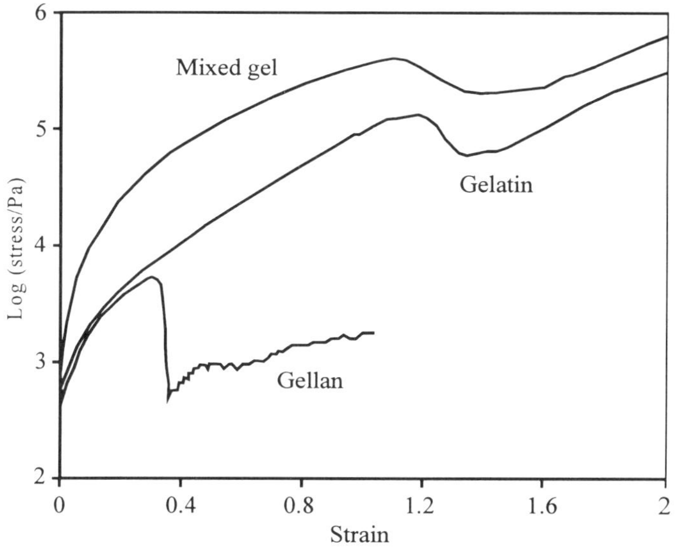

Figure 11.17 Stress—strain profiles at 5°C of 5% gelatin, 0.5% gellan at 47.5 mM NaCl and the mixed gel (5% gelatin—0.5% gellan at 47.5 mM NaCl). Samples were refrigerated for 24 h and compressed to 90% of the initial height at a rate of 0.8 mm/sec (from Papageorgiou *et al.*, 1994)

Another major area of application of phase-separated biopolymer systems is in formulation of water-continuous low-fat spreads with rheological properties that mimic those of traditional fat-continuous products such as butter or margarine. The characteristic property of spreads is that they have "plastic" rheology (Prentice, 1992), i.e. the structure is capable of maintaining its shape under gravity, but can be broken down by moderate stress, allowing the material to flow. In high-fat systems this behaviour comes from weak interactions between fat crystals. Fat-continuous spreads with similar physical properties but lower calorific value can be prepared by incorporating structured water (i.e. small particles of gel) as an inert filler. However, to achieve further reduction in fat content (or, in the limit, complete elimination of fat), water-continuous systems must be used. Structuring of water in a homogeneous biopolymer network would, of course, give gel properties rather than a spreadable plastic texture. Fat-like rheology can, however, be achieved using small microgel particles or hydrated aggregates of one biopolymer (with a size-distribution comparable to those of fat crystals) dispersed in a weak, continuous matrix formed by another biopolymer.

Spreading behaviour may be characterised rheologically by compression testing (Szczesniak, 1983), using instruments such as the Instron materials tester or the more sensitive TA-TX2 texture analyser from Stable Microsystems. To break the initial structure and follow the subsequent flow properties, compression is normally continued until the height of the sample, L, is ~15% of the original height, L_0. At such large deformations it is no longer valid to use the simple definition of strain (ϵ) as $(L_0 - L) / L_0$, because the fractional change in height relative to the absolute change increases rapidly as $L \rightarrow 0$. Instead, the "true" strain is given by (Ross-Murphy, 1984):

$$\epsilon = \ln \frac{L_0}{L} \tag{15}$$

Figure 11.18 shows an idealised stress—strain curve for a "plastic" spread, in comparison with corresponding curves for a typical hydrocolloid gel and an extremely viscous solution. Initially, stress (σ) increases with increasing strain (ϵ), until the network breaks, behaviour similar to that of a gel. At higher strains, however, the response becomes more like that of a liquid, with the resistance remaining roughly constant.

The breakdown and flow properties may be characterised by the stress and strain at the point of failure (σ_{max} and ϵ_{max}) and by the corresponding values (σ_p and ϵ_p) at the point of inflection in the "plateau" region of the stress—strain curve (i.e. where resistance reaches its minimum value after failure, before increasing again as the gap between the moving and stationary plates closes). Best fat-mimetic properties are achieved when the "overshoot" is small (σ_p/σ_{max} in the range 0.95–1.0), with a highly aggregated dispersed phase (e.g. gelling maltodextrins or milk proteins). Figure 11.19 shows experimental stress—strain curves for butter and for a biphasic dispersion of very low fat content (Gupta and Kasapis, 1995). In both

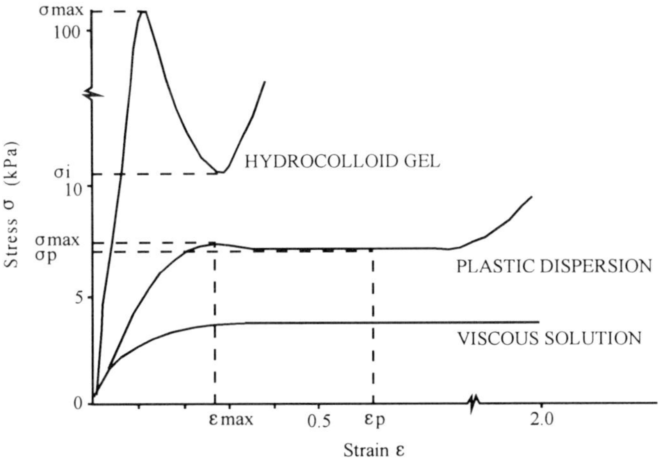

Figure 11.18 Idealised compression curves for a typical hydrocolloid gel, a spreadable "plastic dispersion" and a viscous solution

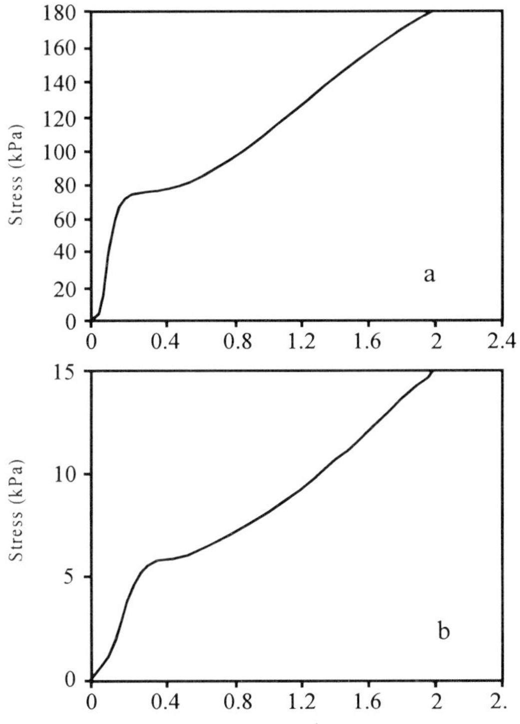

Figure 11.19 Compression curves (5°C; 0.8 mm/sec) for (a) butter and (b) a phase-separated, water-continuous, low-fat spread (Gupta and Kasapis, 1995).

cases the plateau region of plastic flow is less pronounced than in the idealised representation shown in Figure 11.18, but the success that has been achieved in matching the rheology of traditional fat-continuous spreads (80% fat) with water-continuous systems (4% fat) structured by phase-separated biopolymers is clear. Both samples were measured at 5°C, and although the stress–strain profiles are closely similar in form, the absolute stress required to deform the low-fat product is substantially smaller, demonstrating the value of biopolymers in formulation of products with good spreading characteristics at refrigerator temperature.

Acknowledgements

The author is indebted to all of his friends and colleagues from whom he has drawn both guidance and inspiration. I would like to thank in particular Prof. Edwin Morris for long, stimulating discussions lasting even until ten minutes before the college bar last orders and to Dr. Robert Richardson for showing me the wealth of information to be reaped from mechanical measurements.

References

Chilvers, G.R. and Morris, V.J. (1987) Coacervation of gelatin/gellan gum mixtures and their use in microencapsulation. *Carbohydrate Polymers*, **7**, 111–120

Chronakis, I.S. and Kasapis, S. (1993) Structural properties of single and mixed milk/soya protein systems. *Food Hydrocolloids*, **8**, 459–478

Chronakis, I.S., Kasapis, S. and Richardson, R.K. (1996) Small deformation rheological properties of maltodextrin-milk protein systems. *Carbohydrate Polymers,* in press

Clark, A.H. (1987). In *Food Structure and Behaviour*, edited by Blanshard, J.M.V. and Lillford, P.J., London: Academic Press, pp 13–34

Clark, A.H. and Ross-Murphy, S.B. (1985) The concentration dependence of biopolymer gel modulus. *British Polymer Journal*, **17**, 164–168

Clark, A.H. and Ross-Murphy, S.B. (1987) Structural and mechanical properties of biopolymer gels. *Advances in Polymer Science*, **83**, 57–192

Clark, A.H., Richardson, R.K, Ross-Murphy S.B. and Stubbs, J.M. (1983) Structural and mechanical properties of agar gelatin co-gels – small deformation studies. *Macromolecules*, **16**, 1367–1374

Dobson, G.R. and Gordon, M. (1965) Theory of branching processes and statistics of rubber elasticity. *Journal of Chemical Physics*, **43**, 705–713

Ferry, J.D. (1980). In *Viscoelastic Properties of Polymers*, New York: John Wiley and Sons, pp 264–320

Flory, P.J. (1953). *Principles of Polymer Chemistry*, New York: Cornell University Press, pp 347–398

Flory, P.J. (1953a). *Principles of Polymer Chemistry*, New York: Cornell University Press, pp 576–589

Flory, P.J. (1953b). *Principles of Polymer Chemistry*, New York: Cornell University Press, pp 432–494

Gordon, M. (1962) Good's theory of cascade processes applied to the statistics of polymer solutions. *Proceedings of the Royal Society (London) Series A*, **268**, 240–259

Gordon, M. and Ross-Murphy, S.B. (1975) The structure and properties of molecular trees and networks. *Pure and Applied Chemistry*, **43**, 1–26

Gupta, B.B. and Kasapis, S. (1995) Water-continuous spread. European Patent, Number: 0 672 350A2

Hermans, J.R. (1965) Investigation of the elastic properties of the particle network in gelled solutions of hydrocolloids. I. Carboxymethyl cellulose. *Journal of Polymer Science Part A*, **3**, 1859–1868

Kasapis, S., Morris, E.R., Norton, I.T. and Clark, A.H. (1993) Phase equilibria and gelation in gelatin/maltodextrin systems. 4. Composition dependence of mixed gel moduli. *Carbohydrate Polymers*, **21**, 269–276

Kasapis, S., Morris, E.R., Norton, I.T. and Clark, A.H. (1993a) Phase equilibria and gelation in gelatin/maltodextrin systems. 1. Gelation of individual components. *Carbohydrate Polymers*, **21**, 243–248

Kasapis, S., Morris, E.R., Norton, I.T. and Gidley, M.J. (1993b) Phase equilibria and gelation in gelatin/maltodextrin systems. 2. Polymer incompatability in solution. *Carbohydrate Polymers*, **21**, 249–259

Kasapis, S., Morris, E.R., Norton, I.T. and Brown, C.R.T. (1993c) Phase equilibria and gelation in gelatin/maltodextrin systems. 3. Phase separation in mixed gels. *Carbohydrate Polymers*, **21**, 261–268

Lewis, T.B. and Nielsen, L.E. (1970). Dynamic mechanical properties of particulate-filled composites. *Journal of Applied Polymer Science*, **14**, 1449–1471

Mageean, P. and Jones, S. (1989) Low-fat spread products. *Food Science and Technology Today*, **3**, 162–164

Manson, J.A. and Sperling, L.H. (1976). *Polymer Blends and Composites*, London: Heyden Press, pp 1–49

McEvoy, H., Ross-Murphy, S.B. and Clark, A.H. (1985) Large deformation and ultimate properties of biopolymer gels. 2. Mixed gel systems. *Polymer*, **26**, 1493–1500

McEvoy, H., Ross-Murphy, S.B. and Clark, A.H. (1985a) Large deformation and ultimate properties of biopolymer gels. 1. Single biopolymer component systems. *Polymer*, **26**, 1483–1492

Morris, E.R. (1992) The effect of solvent partition on the mechanical properties of biphasic biopolymer gels – An approximate theoretical treatment. *Carbohydrate Polymers*, **17**, 65–70

Nishinari, K., Watase, M., Williams, P.A. and Phillips, G.O. (1990) κ-carrageenan gels – Effect of sucrose, glucose, urea, and guanidine-hydrochloride on the rheological and thermal properties. *Journal of Agriculture and Food Chemistry*, **38**, 1188–1193

Nishinari, K., Watase, M., Kohyama, K., Nishinari, N., Oakenful, D., Koide, S., Ogino, K., Williams, P.A. and Phillips, G.O. (1992) The effect of sucrose on the thermoreversible gel–sol transition in agarose and gelatin. *Polymer Journal*, **24**, 871–877

Papageorgiou, M. and Kasapis, S. (1995). The effect of added sucrose and corn syrup on the physical properties of gellan-gelatin mixed gels. *Food Hydrocolloids*, **9**, 211–220

Papageorgiou, M., Kasapis, S. and Richardson, R.K. (1994) Steric exclusion phenomena in gellan/gelatin gels. 1. Physical properties of single and binary gels. *Food Hydrocolloids*, **8**, 97–112

Papageorgiou, M., Kasapis, S. and Richardson, R.K. (1995) Glassy state phenomena in gellan/sucrose corn syrup mixtures. *Carbohydrate Polymers*, **25**, 101–109

Papageorgiou, M., Gothard, M.G., Willoughby, L.E., Kasapis, S., Richardson, R.K. and Morris, E.R. (1994) Rheology and structure of gellan–alginate co-gels. In *Gums and Stabilisers for the Food Industry 7*, edited by Phillips, G.O., Williams, P.A. and Wedlock, D.J., Oxford: IRL Press, pp 345–356

Pines, E. and Prins, W. (1973). Structure-property relations of thermoreversible macromolecular hydrogels. *Macromolecules*, **6**, **888–895**

Prentice, J.H. (1992). In *Dairy Rheology – A Concise Guide*, edited by Hui, Y.H., Cambridge: VCH Publishers, pp 67–84

Richardson, R.K. and Ross-Murphy, S.B. (1981) Mechanical properties of globular protein gels. 1. Incipient gelation behaviour. *International Journal of Biological Macromolecules*, **3**, 315–322

Ross-Murphy, S.B. (1984). In *Biophysical Methods in Food Research*, edited by Chan, H.W,-S., Critical Reports on Applied Chemistry, London: Society of Chemical Industry, pp 138–199

Ross-Murphy, S.B. (1992) Physical gelation of biopolymers. In *Viscoelasticity of Biomaterials*, edited by Glasser, W. and Hatakeyma, H., American Chemical Society Symposium Series 489, Washington, D.C.: American Chemical Society, pp 204–216

Shim, J.L. (1985). Gellan gum/gelatin blends. U.S. patent 4,517,216

Stockmayer, W.H. (1943) Theory of molecular size distribution and gel formation in branched-chain polymers. *Journal of Chemical Physics*, **11**, 45–55

Szczesniak, A.S. (1983). In *Physical Properties of Foods*, edited by Peleg, M. and Bagley, E.B., Westport (Connecticut): AVI Publishing Company, pp 1–41

Takayanagi, M., Harima, H. and Iwata, Y. (1963). Viscoelastic behaviour of polymer blends and its comparison with model experiments. *Memoirs of the Faculty of Engineering, Kyushu (Imperial) University*, **23**, 1–13

Williams, P.A., Day, D.H, Langdon, M.J., Phillips, G.O. and Nishinari, K. (1991) Synergistic interaction of xanthan gum with glucomannans and galactomannans. *Food Hydrocolloids*, **4**, 489–493

Wolf, C.L., Lavelle, W.M. and Clark, R.C. (1989). Gellan gum/gelatin blends. U.S. patent 4,876,105

12

STARCH — A PHASE-SEPARATED BIOPOLYMER SYSTEM

A.-M. HERMANSSON and S. KIDMAN
SIK, The Swedish Food Research Institute, P.O.Box 5401, S-402 29 Gothenburg, Sweden

K. SVEGMARK
Lyckeby Stärkelsen, S-291 91 Kristianstad, Sweden

Introduction

Starch is one of the most abundant food components. Starch granules are semi-crystalline particles composed primarily of amylose and amylopectin. The amylose is mainly linear and composed of glucopyranosyl monomers linked by α 1–4 bindings. Amylopectin has the same composition but is also branched by α 1–6 links. Amylopectin contributes to the crystalline order of the granules, whereas amylose appears to be present in a non-ordered state within the granule (French, 1984). In aqueous solutions amylose forms ordered crystalline structures more rapidly than amylopectin (Miles, Morris, Orford and Ring, 1985). The most common sources of starch are cereals such as wheat, maize, oats, barley and rice and roots and tubers such as potatoes and tapioca.

The molecular properties of amylose and amylopectin differ in the various starches. The absolute values of molecular weights found in the literature vary according to the preparation method used, especially with regard to amylopectin (Banks, Geddes, Greenwood and Jones, 1972). Table 12.1 shows one example of differences in molecular properties between amylose and amylopectin from various sources (Swinkels, 1985). The variation between the starch sources can be considerable, for example, root starches generally contain less amylose than do cereal starches (Shannon and Garwood, 1984) and the amylose molecules from potato starch are larger than those of cereal starches. As can be seen from Table 12.1, the degree of polymerization is considerably higher for potato starch amylose. Amylose molecules can be slightly branched, and potato starch amylose has been reported to have up to 10 branches per chain, more than three times as many as for wheat starch (Hizukuri, 1991). Apart from amylose and amylopectin, the starch granule contains minor amounts of lipids, proteins, phosphorus and traces of inorganic compounds. Cereal starches contain more lipids than root and tuber starches, whereas phosphate groups bound to amylopectin are a characteristic of potato starch (Swinkels, 1985).

Table 12.1 SOME AMYLOSE AND AMYLOPECTIN PROPERTIES

Properties	*Wheat starch*	*Maize starch*	*Potato starch*
Amylose content %	28	28	21
Amylopectin content %	72	72	79
Amylose DP	800	800	3000
Amylopectin DP $\times 10^6$	2	2	2
Number of amylose molecules/g $\times 10^{20}$	130	130	30
Number of amylopectin molecules/g $\times 10^{17}$	130	130	150
Ratio: number of amylose/amylopectin	1000	1000	200

From Swinkels, 1985

The differences in molecular properties and the location of amylose and amylopectin determine many of the differences in functionality between starches. The aim of this paper is to focus on a number of differences in the behaviour of potato, wheat and, to some extent, maize starch that can be attributed to differences in phase separation between amylose and amylopectin. New information will also be provided on supermolecular structures of amylose and amylopectin.

Structural levels

In order to fully understand the structure and functional properties of heated starch dispersions, we need to study events taking place at several structural levels, as depicted in Figure 12.1.

First we can look at the native granules at the macroscopic level. Granules from different starches differ in size and shape, and some starches such as wheat starch have a bimodal distribution with bigger lenticular and smaller spherical granules.

When granules are heated in water their swelling behaviour differs, and both the swelling power and the geometry of the swollen granules have an impact on the rheological properties of the starch dispersions. During heating the crystalline structure melts and part of the starch solubilizes. If the swelling is limited, solubilized starch, mainly amylose, will leach out of the granules and the swollen starch granules

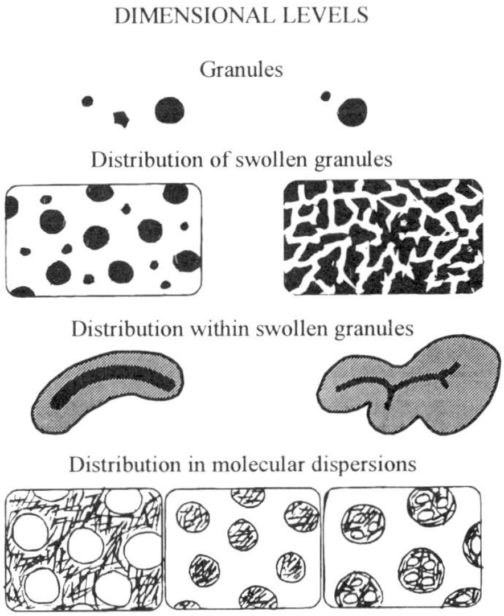

Figure 12.1 Examples of distribution of phases at different dimensional levels

will be dispersed in a continuous starch phase. On the other hand, if the swelling is faster than the leakage of solubilized starch, the swollen granules come into contact with each other and a close-packed situation is obtained. The situation that will arise depends on the swelling power of the granules as well as on the concentration of granules. Phase separation can take place due to diffusion of amylose not only out of the granules but also to the inner aqueous centre of the swollen granules. When water penetrates into the granules, the structure around the hilum in the centre is amorphous, and enrichment of solubilized amylose can often be found in the centre of swollen granules.

Interesting information can be obtained if the distribution of amylose and amylopectin in the swollen granules is characterized. Extensive research has been conducted on structural changes during early stages of gelatinization, such as glass transition and loss of crystalline and molecular order. Far less is known about the behaviour of amylose and amylopectin when the granules swell. Such studies can give information about both the organization and incompatibility of the two components within the granules and the paths of diffusion out of the granules. The structures of the outer layers, as well as weak structures such as the equatorial groove, also have a bearing on diffusion properties of swollen granules. Some examples of amylose and amylopectin distributions that will be discussed in more detail later are depicted in Figure 12.1.

The degree of separation between amylose and amylopectin differs between starches, and this has an effect both on the changes taking place within swollen granules and on the molecular dispersions of amylose and amylopectin. The distribution of amylose and amylopectin has a major effect on the rheological properties of starch dispersions. If amylose is the continuous phase, which is the case for many cereal starches, amylose will form a gel directly on cooling. On the other hand, if amylopectin is the continuous phase, as can be the case for potato starch, gel formation will not take place directly on cooling, but rheological changes will occur slowly with time.

In phase diagrams of aqueous mixtures of pure preparations of amylose and amylopectin, phase separation has been reported to take place above concentrations of 20%–30% of both components (Kalichevsky and Ring, 1987). In a recent study, Doublier and Llamas (1993) suggested that at 15% amylose, a phase inversion started, and an amylose-continuous system was achieved when the amylose concentration reached 22%. The differences in the relative amylose–amylopectin content between potato and cereal starches, shown in Table 12.1, may be one factor contributing to their different rheological properties.

Figure 12.2(a) shows the difference in rheological behaviour between wheat and maize starch dispersions and two concentrations of potato starch dispersions (Svegmark and Hermansson, 1991). The sudden rise in shear modulus for wheat and maize starch is due to gel formation of amylose outside the granules. No such increase is observed for the amylopectin-continuous potato starch dispersions. The initial increase in the shear modulus seen for the higher potato starch dispersion is due to close packing of swollen granules prior to the onset of solubilization and disintegration of the swollen granules. The behaviour of potato starch becomes similar to that of cereal starches if amylose is added. As shown in Figure 12.2(b), gel formation occurs on cooling, resulting in an increase in the shear modulus.

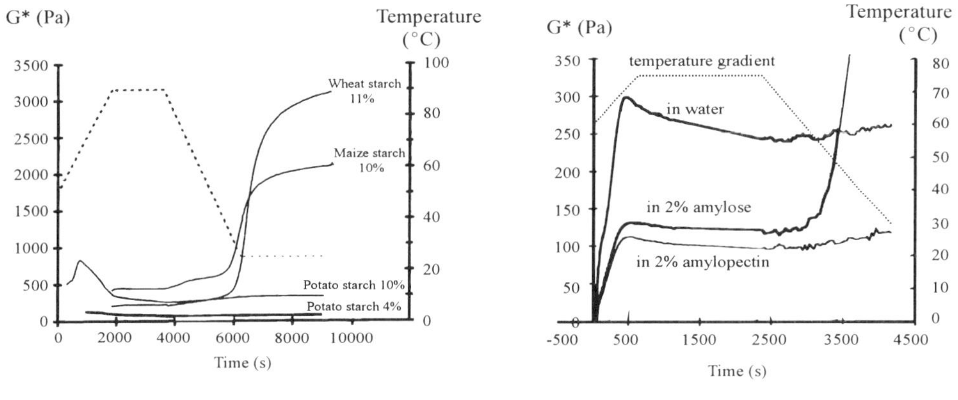

Figure 12.2 (a) Influence of temperature on the complex modulus of low-sheared wheat, maize and potato starch dispersions. The temperature is shown by the broken line (Svegmark and Hermansson, 1991a). (b) The effect on the complex modulus of adding 2% amylose and amylopectin to a 6% potato starch dispersion (Svegmark and Hermansson, 1993)

Details about phase separation will now be discussed for granular dispersions of wheat and potato starch, as well as for molecular gels and dispersions of amylose, wheat, potato and, to some extent, maize starch.

Granular dispersions

WHEAT STARCH

Leaching of amylose out of the granule during heating has been studied by a number of authors (Doublier, 1981; Ghiasi, Hoseney and Varriano-Marston, 1982; Eliasson, 1986). The rate of leaching is relatively slow during the first stage of swelling and increases drastically at temperatures above 95°C. The molecular weight and the number of branched chains also increase with extraction temperature (Ghiasi *et al.*, 1982; Hizukuri, 1991). Interesting information has been obtained by light and electron microscopy (EM) about the distribution of amylose inside the granules during swelling (Langton and Hermansson, 1989). As previously mentioned, solubilization of amylose starts in the amorphous centre of the granule. Solubilization of amylose and swelling from the centre of the granules take place in the first stage of swelling, when the degree of leakage of amylose out of the granule is low. It seems as if water first diffuses into the centre and a swelling pressure is built up. The most important diffusion path of water is through the equatorial groove and the solution inside the granule can easily be squeezed out by shear or external pressure through that path as well (Langton and Hermansson, 1989). Figure 12.3 shows a light micrograph of granules heated to 75°C, where the solubilized phase inside the granule has a lighter colour than the rest of the granules. The inner part of the granule and the path through the equatorial groove are illustrated by the transmission electron microscopy (TEM) micrograph shown in Figure 12.4.

With increasing heating temperature, more amylose leaks out of the granules and a gel is formed upon cooling, as shown in Figure 12.2(a). Iodine-stained sections reveal a blue-stained amylose phase surrounding the swollen granules. The centres of the granules are still filled with a domain of blue-stained amylose (Langton and Hermansson, 1989). Wheat starch amylopectin is less soluble than amylose, and phase separation takes place due to solubilization and diffusion of amylose out of the swollen granules. The integrity of swollen granules can be seen even after heat treatment to 120°C. However, the amylopectin-rich granular structure becomes highly hydrated, and outer layers of the granules are easily sheared off on mechanical treatment, as illustrated in Figure 12.5. This results in a relative decrease in the storage modulus after cooling, as shown in Figure 12.6. The nature of the amylose-rich outer phase will be discussed in more detail in the next section.

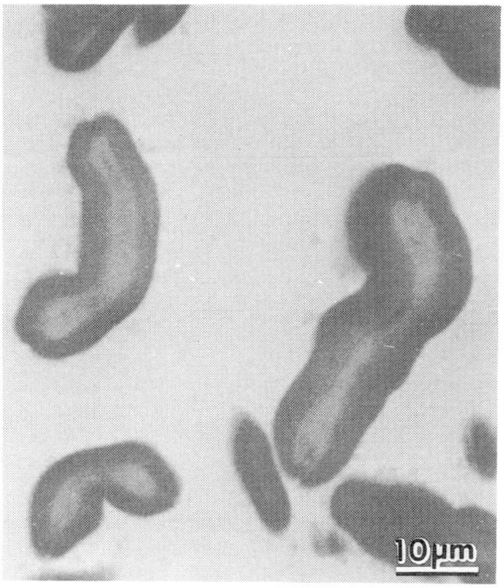

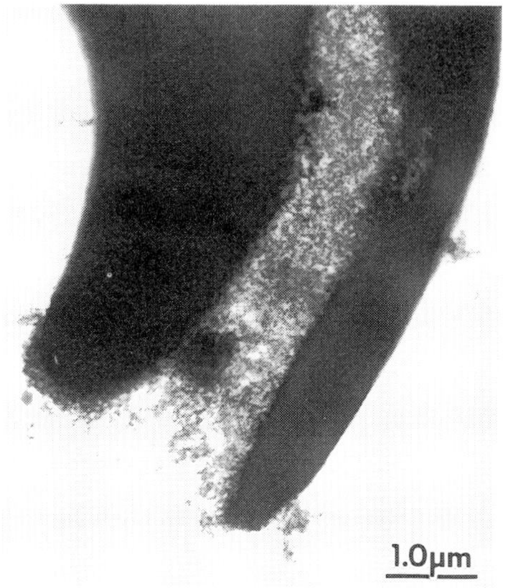

Figure 12.3 (Left) Light micrograph of a 1 μm plastic section of wheat starch granules in a 8% dispersion heated at 75°C for 30 min.

Figure 12.4 (Right) TEM micrograph of a wheat starch granule heated at 75°C for 30 min.

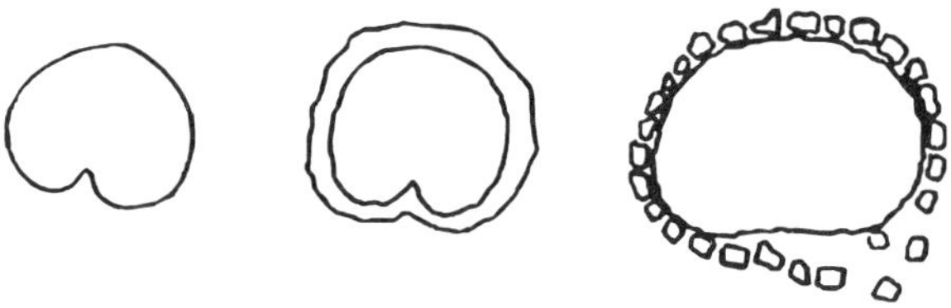

Figure 12.5 Schematic illustration of fragmentation of the outer layers of swollen granules (Langton and Hermansson, 1989)

POTATO STARCH

Swelling of potato starch dispersions has been studied in the concentration range 4%–11% (Svegmark and Hermansson, 1990; Svegmark and Hermansson, 1991a; Svegmark and Hermansson, 1991b; Svegmark and Hermansson, 1993; Svegmark, 1992). The potato starch granules seem to swell practically without restriction until a close-packed situation occurs. The swelling of potato starch granules also starts from the amorphous hilum in the centre of the granules, and pools of released

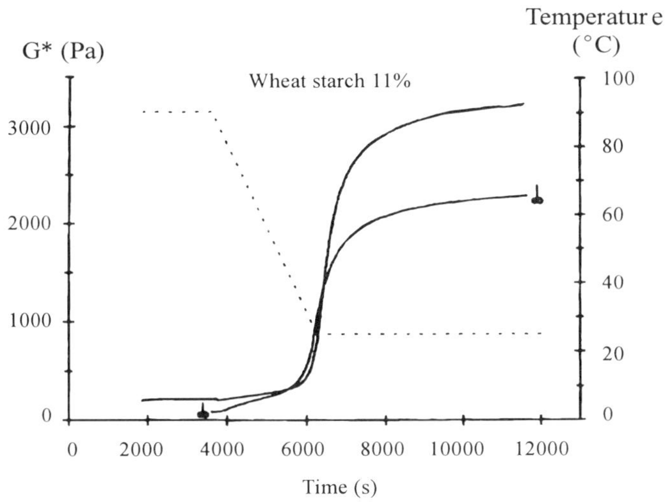

Figure 12.6 The effect of shear on the shear modulus of wheat starch (Svegmark and Hermansson, 1991a)

amylose can be seen inside the swollen granules, as depicted in Figure 12.7. The internal swelling pressure seems to cause cracks in the outer layer and folding of the walls.

The swollen granules of potato starch are more readily disintegrated or dissolved after gelatinisation than swollen wheat starch granules. It is not known whether the amylopectin of the swollen granules dissolves into its molecular state or disperses in the form of microgels or aggregates. The outer layer of potato starch granules has a denser structure than the rest of the granule and disintegrates at a higher temperature. This layer may act as a restraint on diffusion of amylose out of the granules.

Major structural changes take place on shearing and heating at 90°C. A transition in viscoelastic behaviour to a more viscous state takes place at the point where fragments of granules are dispersed in a continuous phase of solubilized starch. Increased phase separation takes place with prolonged heating and storage, where amylose is dispersed in a amylopectin-rich phase. The phase separation is marked after heat treatment to 120°C. Even at this temperature, the outer layers of the granules remain. They can be seen as long threads in Figure 12.8. The dark domains are phase-separated amylose-rich regions.

The behaviour of amylose and amylopectin differs between potato and wheat starch. Potato starch amylose does not form a continuous network structure, either in a paste of close-packed granules or in a dispersion of solubilized starch in the temperature range 75°C to 120°C. This means that there is no increase in shear modulus on cooling as observed for wheat and maize starch (Figure 12.2(a)). The higher solubility of granular amylopectin, combined with the higher swelling power

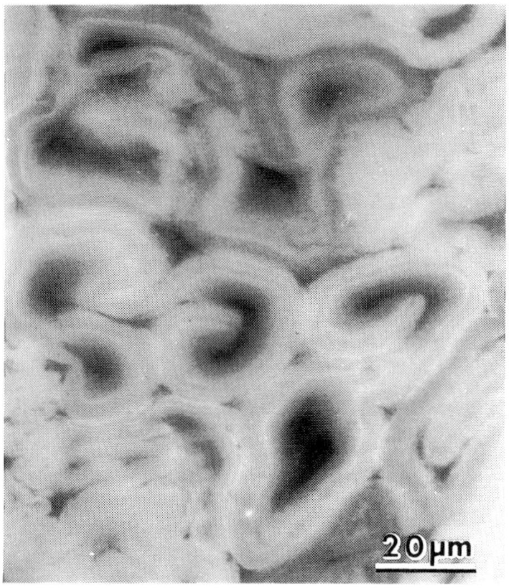

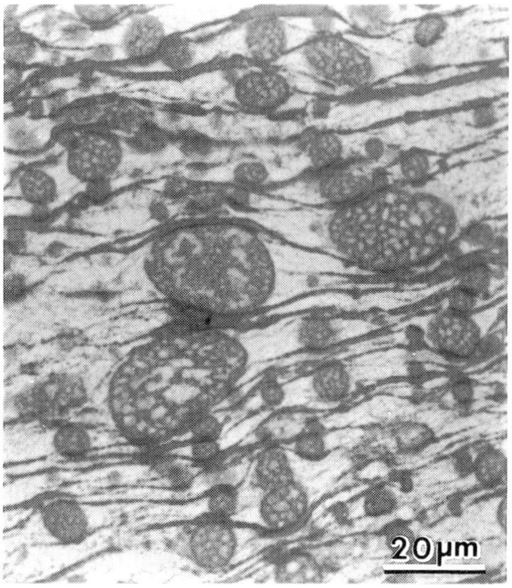

Figure 12.7 (Left) Light micrograph of a 8 μm cryo-section of a 12% dispersion of potato starch heated at 75°C for 30 min

Figure 12.8 (Right) Light micrograph of a 1 μm plastic section of a 11% dispersion of potato starch heated at 120°C for 30 min

of the granules, means that heat-treated potato starch granules are very fragile and sensitive to shearing. It seems as if there is a higher degree of compatibility between amylose and amylopectin in potato starch than in wheat starch, both in the native granule, as suggested by Zobel (1988), and in gelatinised pastes and gels.

Molecular dispersions and supermolecular structures

AMYLOSE GELS

Gel formation of pure amylose has been studied by several authors (Miles *et al.*, 1985; Clark, Gidley, Richardson and Ross-Murphy, 1989; Gidley, 1989; Leloup, Colonna, Ring, Roberts and Wells, 1992). However, it is not known whether the gel structure of pure amylose is the same as, for example, in wheat gels, or if amylose from different origins or of different molecular weights form the same type of structure. Leloup *et al.* (1992) studied the gel microstructure of amylose extracted from pea starch by leaching at 70°C and purified by n-butanol precipitation. Freeze-etching and TEM revealed a very coarse and open network structure with rigid interconnected strands of about 20 nm in diameter. Based on electron microscopy in combination with other techniques, they proposed a model for the organization of the amylose in the strand, as shown in Figure 12.9.

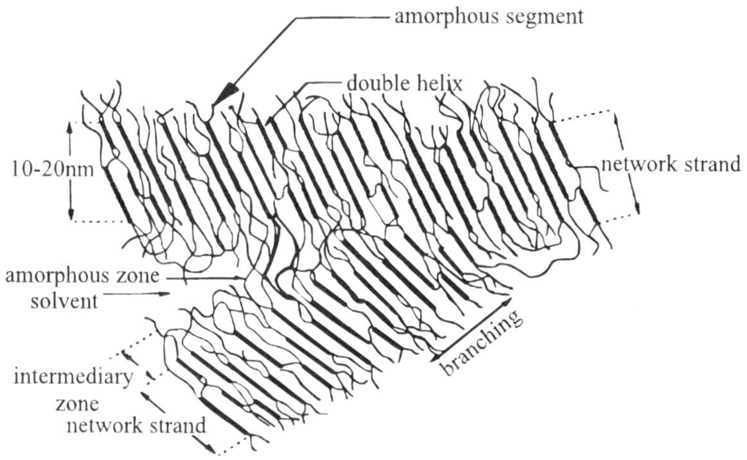

Figure 12.9 Continuous model for amylose gels proposed by Leloup *et al.* (1992)

The gel structure of amylose isolated from potato starch, purchased from Sigma with a DP_w of 2170 (P. Roger, personal communication), was studied in our laboratory. The amylose was dissolved by heating in an oven at 155°C for 130 min. Similar to the results obtained by Leloup and co-workers a very coarse and open structure was obtained with rigid bush-like strands that were 10–40 nm in diameter. The gels were prepared by freeze-etching as well as by thin-sectioning for TEM and staining with periodic acid, thiosemicarbazide and silver proteinate (Thiery, 1967; Gallant, 1974). Thin sections of 4% and 10% amylose gels are shown in Figures 12.10 and 12.11. The strands were not influenced by concentrations in the range 3% to 10%. Other factors that might have an impact on the degree of random aggregation were also investigated. The main features of the gel structure were not influenced by the rate of cooling or by the presence of salts such as 0.1 M NaCl, 0.035 M $CaCl_2$ or 0.02 M $AlCl_3$. The only salt-induced effect that was noticed was a slightly higher degree of aggregation resulting in wider strands with a more bush-like appearance. However, these structural differences did not affect the viscoelastic properties of the gels. The results show that amylose gels are very stable and not easily influenced by environmental factors, indicating that there is a strong molecular orientation responsible for the characteristic superstrands of amylose. This is in agreement with the model proposed by Leloup and co-workers.

When amylose solutions were added to a molecular dispersion of potato starch formed by shearing and heating, it was found that the presence of solubilized starch did not interfere with the formation of the typical rod-like amylose network (Svegmark, Kidman and Hermansson, 1993). The only interaction found was that solubilized amylopectin seemed to partly adsorb onto the amylose superstrands,

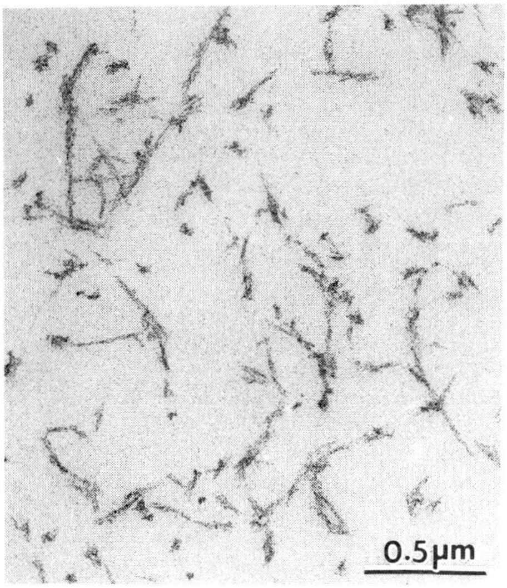

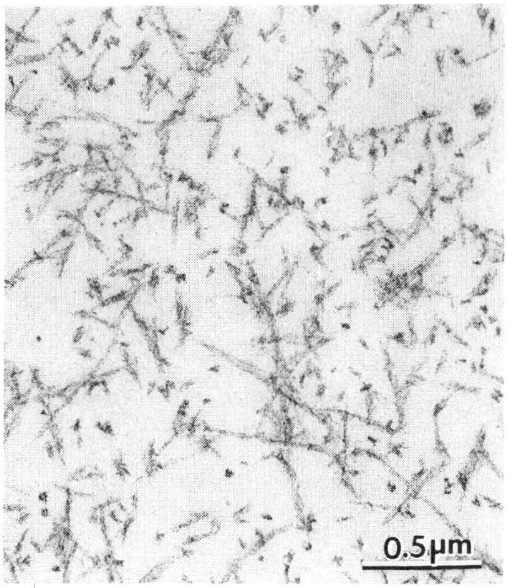

Figure 12.10 (Left) Thin section of a 4% amylose gel

Figure 12.11 (Right) Thin section of a 10% amylose gel

giving rise to a smeared appearance, as shown in Figure 12.12. Solubilized amylopectin appears as droplets and some of these have merged with the amylose strands. This implies that the presence of amylopectin does not interfere with the aggregation of amylose into superstrands.

Rheological measurements indicated that the inherent amylose of the potato starch paste did not participate in the network formation of the added amylose. The rheological data were evaluated using a model treating the added amylose and starch as a phase-separated system (Takayanagi, 1963; Clark, 1987). The question then arises as to why amylose in potato starch has different properties from amylose fractions prepared from potato starch. One possibility is that the molecular properties of amylose are influenced by the procedure of preparing pure amylose fractions. A change in molecular weight distribution, degree of branching and/or a decrease in chain length are factors that may contribute to the gel-forming properties of amylose. Clark *et al.* (1989) studied the effect of chain length on gel-forming properties of monodisperse amylose. They found that short chains lead to a more rapid gelation and a higher gel strength than longer chains in the range 250–2800 DP. Another difference is that the pure amylose was completely solubilised by heating to 150°C, whereas the potato starch was only heated to 90°C. The phase separation of amylose has been shown to be enhanced by heating potato starch to 120°C. There might be interactions between amylose and amylopectin in potato starch that hinder the phase separation.

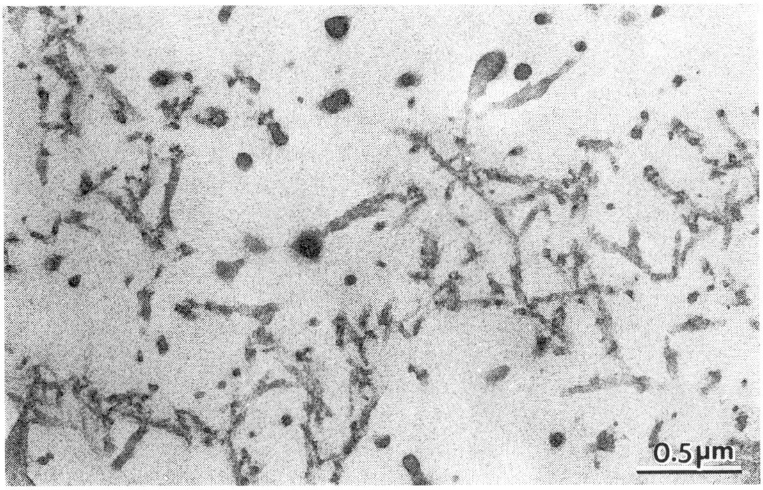

Figure 12.12 Details of a mixed system with 5% molecular dispersion of potato starch and 3% amylose

In order to obtain more information about heat-induced changes of amylose and amylopectin, a detailed structural study has been made of potato, wheat and maize starch after heat treatment at 75°C to 150°C.

SUPERMOLECULAR STRUCTURES OF POTATO STARCH DISPERSIONS

Both the temperature and the availability of water determine structural changes of potato starch. In this study 8%–11% dispersions were heated from 75°C to 150°C and cooled prior to preparation for microscopy. Figure 12.13 shows part of a granule in a 11% dispersion heated at 90°C for 30 min, cooled to room temperature and stored overnight at 4°C prior to preparation for EM. The arrow points to the compact outer layer previously discussed. The structure inside that layer has been loosened up, but there is still some ordered matrix structure. This could very well be amylopectin associated with amylose.

With more access to water part of the amylopectin starts to solubilize. This process can already be observed at 75°C. Figure 12.14 shows a 8% dispersion heated to 75°C, and Figure 12.15 shows a 10% dispersion heated to 90°C. In both cases we can see solubilized amylopectin as droplets of various sizes from 20–100 nm in diameter outside the granules. In Figure 12.15 spherical particles are attached to each other in strings. A very close look at the compact amylopectin layer reveals very small particles of about 5 nm attached to each other in some order.

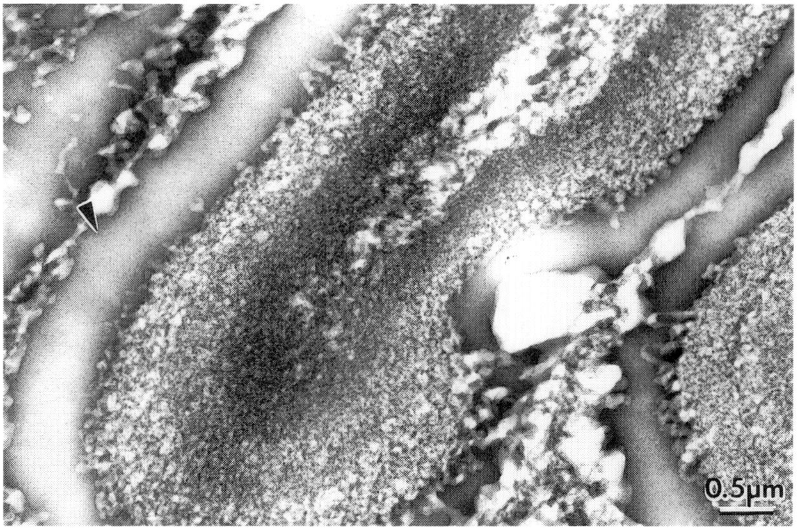

Figure 12.13 Granules in a 11% potato starch dispersion heated at 90°C for 30 min. The arrow points to the compact outer layer

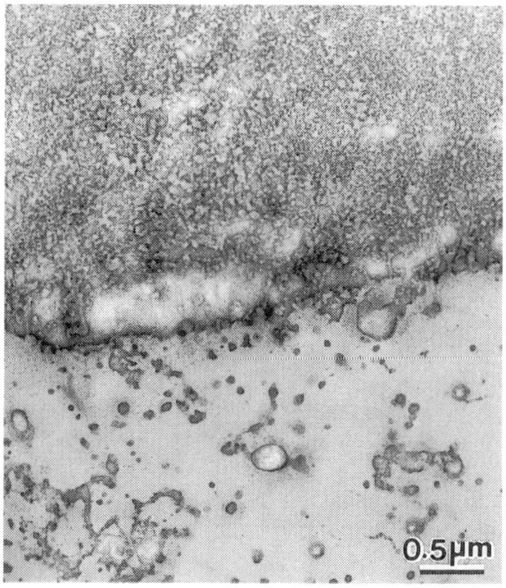

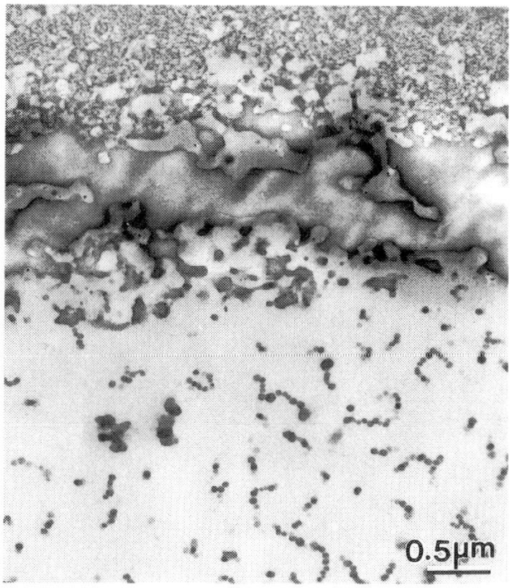

Figure 12.14 (Left) Part of a granule and the outer phase in a 8% potato starch dispersion heated at 75°C for 30 min

Figure 12.15 (Right) Part of a granule and the outer phase in a 10% potato starch dispersion heated at 90°C for 30 min

An increased degree of phase separation can be observed with increasing heating temperature. In iodine-stained sections prepared for light microscopy, blue spots appear that increase in size with prolonged heating (Svegmark and Hermansson, 1991). Figure 12.16 shows an EM micrograph of an area in a paste heated at 95°C.

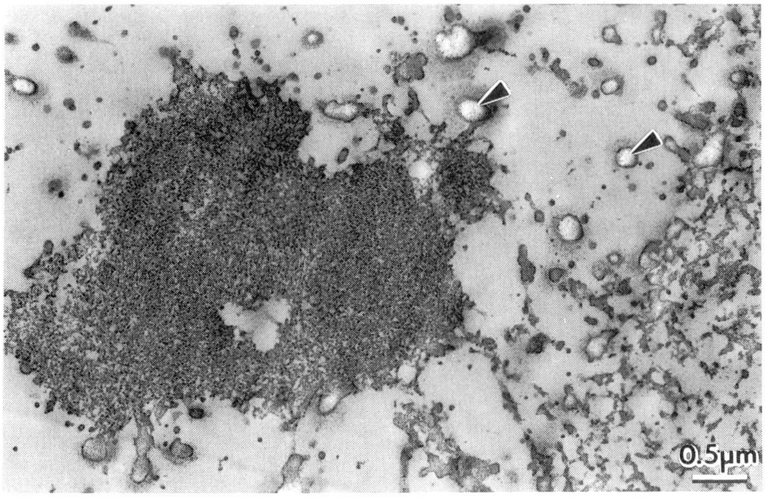

Figure 12.16 An amylose-rich aggregate surrounded by amylopectin in 8% potato starch heated at 95°C for 30 min

A densely packed amylose-rich region can be seen, as well as insoluble amylopectin particles (arrows) surrounded by solubilized amylopectin. The latter makes up the continuous phase. The phase separation increases with temperature and is most pronounced at 120°C. Figure 12.8 shows a light micrograph with darkly stained amylose-rich domains and long threads of the outer amylopectin layers that have not solubilized even at this temperature. Figures 12.17 and 12.18 show EM micrographs of oriented insoluble amylopectin, as well as densely packed amylose-rich domains. It is not possible to say whether these domains are composed of amylose alone or a mixture of amylose and amylopectin.

The typical rod-like strands observed for amylose gels could not be detected in potato pastes heated from 75°C to 120°C even if an orientation could be seen within the amylose-rich domains after prolonged cold storage. A drastic change took place on heating 11% potato starch to 140°C. A gel formed on cooling overnight. As can be seen in Figure 12.19, a rod-like network similar to that observed for pure amylose gels had formed. No phase separation could be observed by light microscopy. Thus,

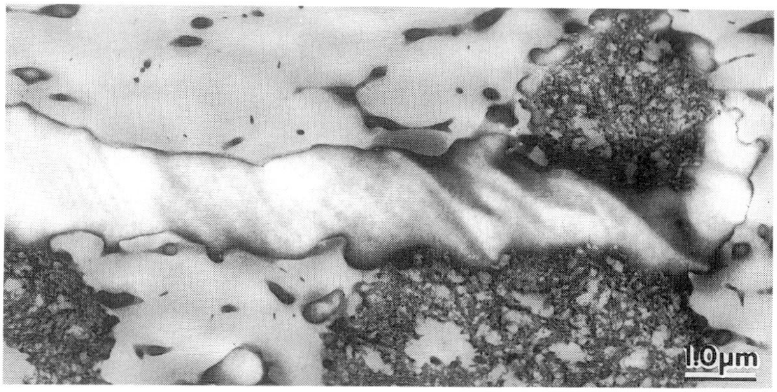

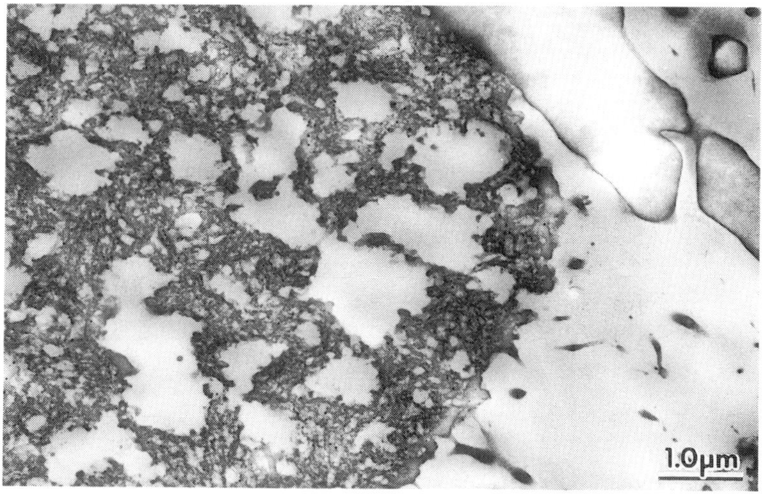

Figures 12.17 (Top) and 12.18 (Bottom) Part of the insoluble outer layer and amylose-rich aggregates in 10% potato starch heated at 120°C for 30 min

the system has changed from a phase-separated to a continuous network structure, where it is not possible to distinguish between amylose and amylopectin. The molecular changes underlying this transition need to be studied further, but the result shows that a coarse open network structure can also be formed from potato starch pastes.

Complexing with butanol is common in fractionation of amylose. Addition of butanol as well as monoglycerides is known to have a thickening effect on starch pastes. A suggested mechanism is enhanced helix formation and aggregation of helices into a network by complexation (Conde-Petit, 1992). A study was made in order to see whether addition of complexing agents had any influence on the structural state of amylose in potato starch pastes. The storage modulus of a low-sheared 8% potato starch heated to 95°C and cooled to 25°C is 60 Pa. Addition of

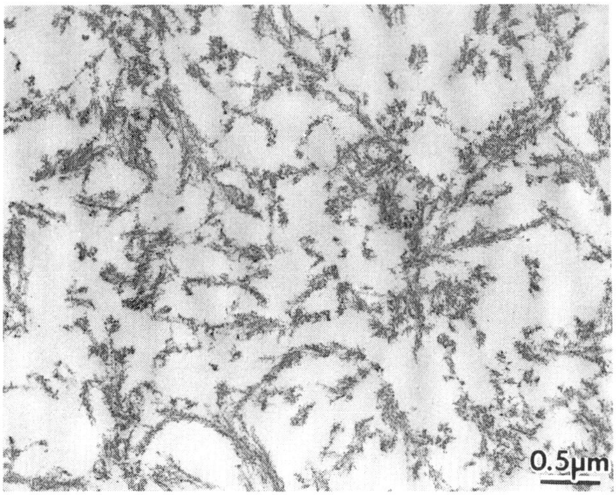

Figure 12.19 Gel structure in 11% potato starch heated at 140°C for 30 min

2% butanol results in an increase to 220 Pa, and addition of 2% glycerol monostearate (GMS) an increase to 750 Pa. The microstructure analysis revealed that the thickening effect of GMS is not due to network formation of amylose helices. Instead, amylose complexation encourages amylose out of the granules and this random aggregation of the polymer often occurs around droplets of emulsifiers. Figure 12.20 shows aggregates of amylose outside a swollen granule, and Figure 12.21 shows amylose concentrated around a droplet of GMS. Association of aggregates is probably the reason for the thickening effect.

SUPERMOLECULAR STRUCTURES OF WHEAT STARCH GELS

Wheat starch forms gels on cooling after heat treatment from 95°C to 140°C in the concentration range 8–11%. A rod-like network structure was found in the interior of the granules, as well as in the continuous phase outside the granules. Figure 12.22 shows the network inside a swollen granule surrounded by a compact amylopectin layer. The strands have a somewhat smeared appearance similar to those shown in Figure 12.12, where solubilized amylopectin adhered to strands of added amylose. Electron microscopy revealed interesting information about the continuous structure outside the swollen wheat starch granules. Figures 12.23 and 12.24 show the continuous phase of 11% wheat starch pastes heated to 95°C and 120°C respectively. The gel phase consists of a mixed structure with a rod-like

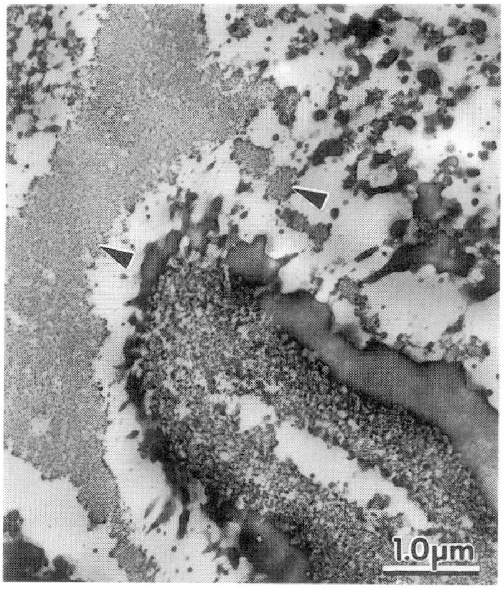

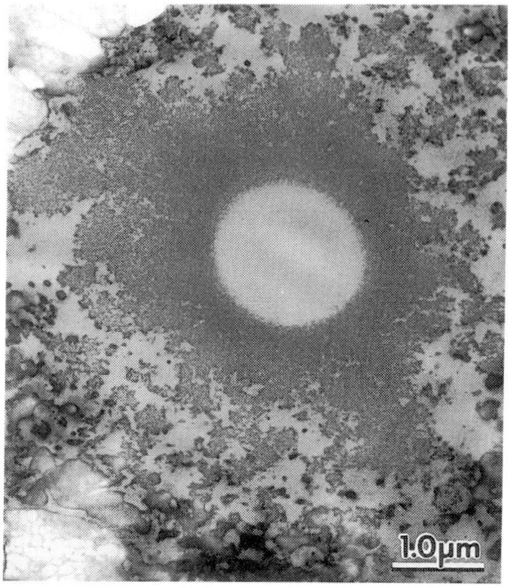

Figure 12.20 (Left) Micrograph of 8% potato starch and 2% GMS heated at 95°C for 30 min showing amylose outside the granule (arrows)

Figure 12.21 (Right) Micrograph of 8% potato starch and 2% GMS showing amylose surrounding a droplet of emulsifier

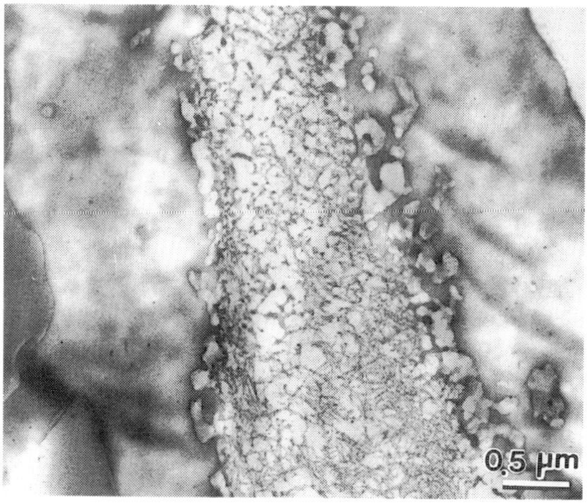

Figure 12.22 Interior of a swollen wheat starch granule heated at 120°C for 30 min

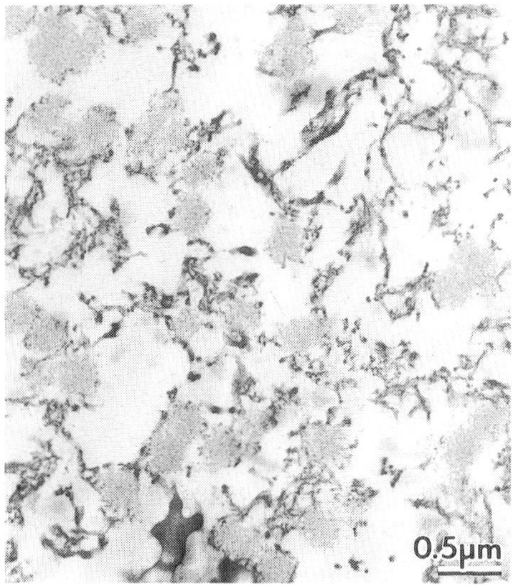

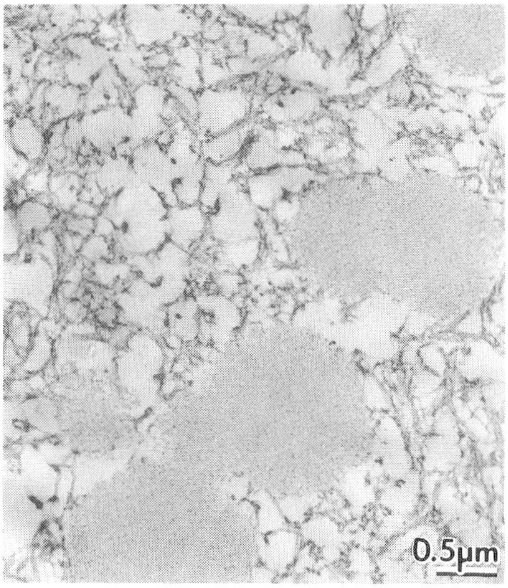

Figure 12.23 (Left) 11% wheat starch gel heated at 95°C for 30 min

Figure 12.24 (Right) 11% wheat starch gel heated at 120°C for 30 min

network and small aggregates. The network is similar to that found in the centre of the granules and has features similar to the pure amylose network, even if the strands look more smeared, as stated above. The origin of the aggregates is believed to be hydrated amylopectin fragments, but the possibility of an amylose origin cannot be completely ruled out. Figure 12.25 shows a hydrated fragment at the surface of the compact amylopectin structure. Thus, it seems as if both amylose and amylopectin contribute to the continuous gel structure of wheat starch.

SUPERMOLECULAR STRUCTURES OF MAIZE STARCH GELS

Like wheat starch, dispersions of maize starch formed gels directly on cooling, as shown in Figure 12.2. A rod-like network structure was found in the interior of swollen granules, as well as in the continuous gel phase outside the granules. Figure 12.26 shows part of a swollen granule with the typical amylose network structure both inside and outside the granule. The continuous phase did not contain the aggregates found in wheat starch gels and interpreted as hydrated amylopectin fragments. Instead, maize starch amylopectin seemed to solubilize in a manner similar to that of potato starch amylopectin. No gel formed directly on cooling when 8% maize starch was heated to 120°C. Figure 12.27 shows droplets of

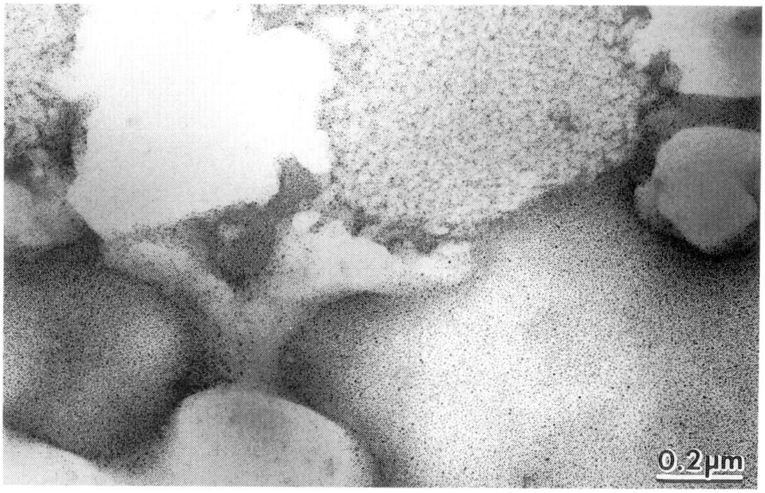

Figure 12.25 Detail showing a hydrated "amylopectin" fragment in a 11% gel heated at 120°C for 30 min

solubilized amylopectin and aggregates of amylose. All starches formed compact aggregates of retrograded starch after heating at high temperatures and/or prolonged storage after cooling.

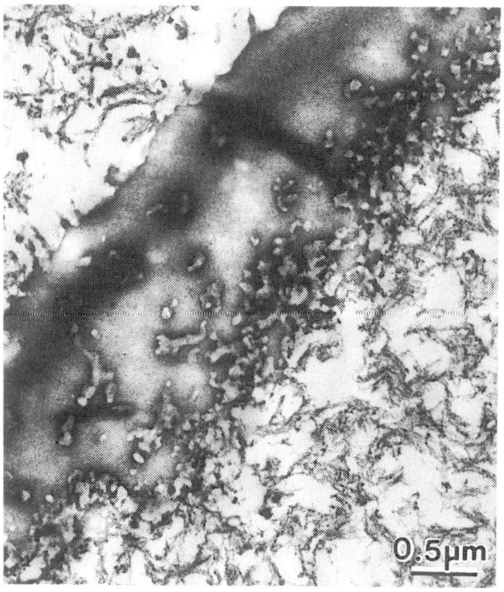

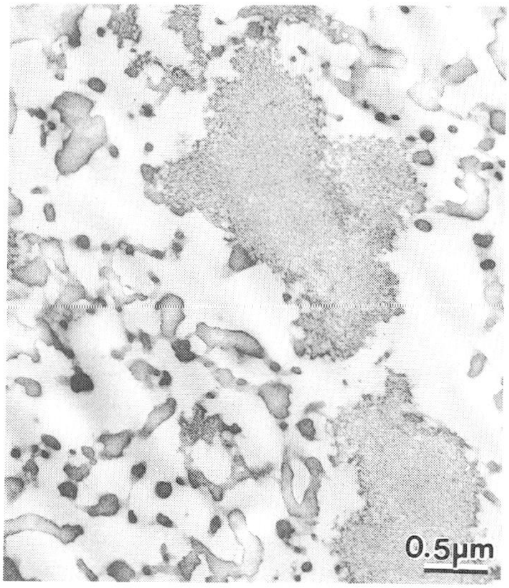

Figure 12.26 (Left) Amylose outside and inside the outer layer of a granule in 8% maize starch after heat treatment at 95°C for 30 min

Figure 12.27 (Right) 8% maize starch heated at 120°C for 30 min

Conclusions

The results have illustrated that starch is a very complex system, where phase separation plays an important role in accounting for differences in properties of starch from various sources. Phase separation can take place both in swollen granular systems and in molecular dispersions. Amylose is responsible for rapid gelation directly on cooling. Results presented in this chapter show for the first time that the open rod-like network characteristic of pure amylose gels is also found in gels of wheat, maize and even potato starch under certain conditions.

However, potato starch pastes do not form gels directly on cooling after heat treatment to 75–120°C. The behaviour of amylopectin on heating differed between potato, wheat and maize starch. Part of the potato starch amylopectin already started to solubilize at 75°C, whereas the outer layer of the granule remained insoluble even at 120°C. The swollen amylopectin-rich granules of wheat starch are less soluble. Light microscopy showed that macroscopic fragments of the outer hydrated layer can be sheared off after heat treatment to 95°C. Electron microscopy revealed that very small hydrated particles of about 0.5 μm in diameter diffused into the continuous phase and formed a mixed gel with the open amylose network. Maize starch amylopectin had properties in between those of potato and wheat starch. No hydrated fragments were found and the amylopectin seemed to solubilize in a manner similar to that of potato starch, except for the behaviour of the outer layer of the potato starch granule.

It is difficult to draw a sharp distinction between amylose and amylopectin in complex starch systems. The results have shown that solubilized amylopectin can adsorb onto amylose gel strands. On heating from 120°C to 140°C, the structure of 11% potato starch changed from a distinct phase-separated system to a continuous network structure, similar to that of pure amylose. It is likely that both amylose and amylopectin participate in the potato starch gel structure, but the underlying molecular changes necessary for the compatibility between the two components are not known. Furthermore, we do not know how much amylopectin is present in the amylose-rich phase-separated domains and the nature of interactions between amylose and amylopectin in amylopectin-rich granular structures. On severe heat treatment to 150°C and during cold storage all starches form compact aggregates of retrograded starch. Microscopy provides valuable information about phase-separated starch systems, but other techniques are needed as well, in order to elucidate the nature of interactions as well as molecular changes taking place on processing of starch.

Acknowledgements

Ina Storm and Susanne Ekstedt are gratefully thanked for skilful assistance with the preparation of the figures.

References

Banks, W., Geddes, R., Greenwood, C.T. and Jones, I.G. (1972) Physiocochemical studies on starches. Part 63. The molecular size and shape of amylopectin. *Starch/Stärke*, **24**, 245–251

Clark, A.H. (1987) The application of network theory to food systems. In *Food Structure and Behaviour*, edited by Blanshard, J.M.V. and Lillford, P.J., London: Academic Press, pp 13–34

Clark, A.H., Gidley, M.J., Richardson, R.K. and Ross-Murphy, S.B. (1989) Rheological studies of aqueous amylose gels: The effect of chain length and concentration on gel modulus. *Macromolecules*, **22**, 346–351

Conde-Petit, B. (1992). *Starch emulsifier interactions in water-containing food models*, Ph.D. thesis, ETH nr 9785, Zürich, Switzerland

Doublier, J.L. (1981) Rheological studies on starch – flow behaviour of wheat starch pastes. *Starch/Stärke*, **33**, 415–420

Doublier, J.L. and Llamas, G. (1993) A rheological description of amylose–amylopectin mixtures. In *Food Colloids and Polymers: Stability and Mechanical Properties*, edited by Dickinson, E. and Walstra, P., Cambridge: Royal Society of Chemistry, pp 138–146

Eliasson, A.C. (1986) Viscoelastic behaviour during the gelatinization of starch. 1. Comparison of wheat, maize, potato and waxy-barley starches. *Journal of Texture Studies*, **17**, 253–265

French, D. (1984) Organization of starch granules. In *Starch: Chemistry and Technology*, edited by Whistler, R.L, BeMiller, J.N. and Paschall, E.F., London: Academic Press, pp 184–247

Gallant, D.J. (1974). *Contribution a l'étude de la Structure et de l'ultrastructure du grain d'amidon*, doctoral thesis, CNRS no AO 10823 Université VI, Paris, France

Ghiasi, K., Hoseney, R.C. and Varriano-Marston, E. (1982) Gelatinization of wheat-starch. 1. Excess-water systems. *Cereal Chemistry*, **59**, 81–85

Gidley, M.J. (1989) Molecular mechanisms underlying amylose aggregation and gelation. *Macromolecules*, **22**, 351–358

Hizukuri, S. (1991) Properties of hot-water-extractable amylose. *Carbohydrate Research*, **217**, 251–253

Kalichevsky, M.T. and Ring, S.G. (1987) Incompatibility of amylose and amylopectin in aqueous solutions. *Carbohydrate Research*, **162**, 323–328

Langton, M. and Hermansson, A.M. (1989) Microstructural changes in wheat starch dispersions during heating and cooling. *Food Microstructure*, **8**, 28–39

Leloup, V.M., Colonna, P., Ring, S.G., Roberts, K. and Wells, B. (1992) Microstructure of amylose gels. *Carbohydrate Polymers*, **18**, 189–197

Miles, M.J., Morris, V.J., Orford, P.D. and Ring, S.G. (1985) The roles of amylose and amylopectin in the gelation and retrogradation of starch. *Carbohydrate Research*, **135**, 271–281

Shannon, J.C. and Garwood, D.L. (1984) Genetics and physiology of starch development. In *Starch: Chemistry and Technology*, edited by Whistler R.L., BeMiller J.N. and Paschall, E.F., London: Academic Press, pp 26–86

Svegmark, K. (1992). *Potato starch dispersions: Structure and Rheology*, Ph.D. thesis, Chalmers University of Technology and SIK

Svegmark, K. and Hermansson, A.M. (1990) Shear-induced changes in the viscoelastic behaviour of heat-treated potato starch dispersions. *Carbohydrate Polymers*, **13**, 29–45

Svegmark, K. and Hermansson, A.M. (1991a) Changes induced by shear and gel formation in the viscoelastic behaviour of potato, wheat and maize starch dispersions. *Carbohydrate Polymers*, **15**, 151–169

Svegmark, K. and Hermansson, A.M. (1991b) Distribution of amylose and amylopectin in potato starch pastes: Effects of heating and shearing. *Food Structure*, **10**, 117–129

Svegmark, K. and Hermansson, A.M. (1993) Microstructure and rheological properties of composites of potato starch granules and amylose: A comparison between observed and predicted structures. *Food Structure*, **12**, 181–193

Svegmark, K., Kidman, S. and Hermansson, A.M. (1993) Molecular structures obtained from mixed amylose and potato starch dispersions and their rheological behaviour, *Carbohydrate Polymers*, **22**, 19–29

Swinkels, J.J.M. (1985) Composition and properties of commercial native starches. *Starch/Stärke*, **37**, 1–5

Thiery, J.P. (1967) Mise en évidence des polysaccharides sur coupes fines en microscopie électronique. *Journal de Microscopie*, **6**, 987–1018

Zobel, H.F. (1988) Molecules to granules: A comprehensive starch review. *Starch/Stärke*, **40**, 44–50

PART II: INTERACTING SYSTEMS

13

POLYSACCHARIDE SYNERGISM — MORE QUESTIONS THAN ANSWERS?

E.R. MORRIS
Department of Food Research and Technology, Cranfield University, Silsoe College, Silsoe, Bedford MK45 4DT U.K.

Introduction

I must begin by thanking the Conference organisers for the title of this article, which I have interpreted as an invitation to give a personal viewpoint of the present state of play in the synergism saga, and also as an opportunity for some frank admissions of ignorance or uncertainty. To avoid undue overlap with other contributions, I have concentrated on presenting some recent results from my own laboratory (most of which are unpublished) and on critical appraisal of key experiments from other groups. In the spirit of the title, the section headings are framed as questions, and the subsequent text attempts to provide the answers, or at least to suggest what the next round of questions might be.

What is synergism?

Any definition of synergism would exclude simple additivity of the properties of individual materials, such as the surface activity of gum arabic acting together with the viscosity of guar gum to give a stable, thickened emulsion. The central requirement is for the combined effect to be greater than the sum of the individual effects.

As discussed in detail in previous contributions, the physical properties of the individual constituents in mixed biopolymer systems can be changed enormously by thermodynamic incompatibility, particularly when it results in phase separation. Although interactions of this type would satisfy the dictionary definition of "synergism", there seems to be a general consensus in the polysaccharide area that "true" synergism should involve unlike molecules sticking together rather than pushing apart.

For the purposes of the present article, I intend to restrict the term still further by excluding covalent bonding, enzyme–substrate and lectin–hapten interactions where carbohydrates are "trapped" by proteins, and non-specific "complex coacervation" between polyanions and polycations. The surviving model is confined to unlike biopolymers forming ordered intermolecular associations analogous to the co-operative junction zones in single-component polysaccharide gels.

I of course realise that this is a much narrower definition of "synergism" than the practical concept of any useful enhancement of functional properties that can be achieved by combined use of different materials. Nonetheless, it is probably a realistic reflection of the way in which the term has come to be used within the research community and it certainly encapsulates the area of greatest scientific interest and controversy.

How can we recognise it?

Most people working on biopolymers will be familiar with the much plagiarised description of gelation as "a phenomenon that is easier to recognise than to define" (Lloyd, 1926). The converse is true of synergism, at least in the narrow sense outlined above.

The first hurdle in investigation of mixed biopolymer systems is to determine whether or not there is any interaction between the individual constituents. The second is to distinguish direct binding from thermodynamic incompatibility. The third problem, which has not yet been fully solved for any synergistic polysaccharide system, is to characterise the structure of the heterotypic junctions.

In some systems interaction is obvious as a gross change in physical properties on mixing, such as the gelation that occurs between xanthan and locust bean gum, neither of which form true gels in isolation. In many cases, however, it is hard to demonstrate that any interaction has occurred, and even harder to distinguish between binding and exclusion. One criterion that certainly cannot be applied is departure from simple additivity of solution viscosities or gel moduli (although, unfortunately, claims for "synergistic" interaction are sometimes made on this basis, particularly in the patent literature). For example, finding that the viscosity of a solution containing two different polymers, each at a concentration of 1%, is higher than the sum of the viscosities for 1% solutions of the individual polymers does not imply interaction, since the viscosity of a 2% solution of either component would also be greater than twice the viscosity of a 1% solution (typically about 5 times greater). We must obviously have criteria that do not imply synergism of individual polymers with themselves!

An approach that I have advocated (Morris, 1984), and whose use is illustrated in the following section, is to prepare stock solutions of the two polymers under investigation, with viscosities matched as closely as possible, mix them in varying proportions, and look for departures from the common starting viscosity. A complication that must be borne in mind when applying this procedure to charged polysaccharides is that the counterion to one polymer may affect the behaviour of the other. For example, mixing magnesium alginate with gellan gum might well give a massive enhancement in viscosity, not because of interaction between the two polymers, but by Mg^{2+}-induced association of gellan helices. The problem can be overcome by dialysing the two stock solutions together against an appropriate salt solution, so that there is no change in ionic environment on mixing.

In mixed systems where one (or both) of the individual polymers can undergo a disorder—order transition, interactions may sometimes be detected by formation or survival of the ordered structure under conditions where it would be unstable in isolation. Such effects may be monitored by any of the techniques (Gidley and Robinson, 1990) that can be used to follow conformational changes in single-polysaccharide systems (e.g. optical rotation, circular dichroism, rheometry, DSC, loss of visible high-resolution NMR signal) and may be apparent as, for example, an increase in transition temperature or in the rate of formation of the ordered structure.

Although viscosity enhancements and displacement of conformational transitions may be diagnostic of polymer–polymer interactions, they do not, in themselves, distinguish between binding and exclusion. However, their response to changes in concentration may give an indication of the mechanism of interaction. First, exclusion effects between unlike biopolymers normally become apparent only when the total polymer concentration is comparatively high, whereas binding may persist to very low concentrations. More directly, binding is usually a stoichiometric process, and should therefore involve fixed proportions of the participating species, irrespective of their total concentration, whereas exclusion will normally become more evident as the concentration of either component is increased. Thus interactions that can be diluted out are likely to arise from thermodynamic incompatibility, whereas those that reach a maximum at a specific mixing ratio almost certainly involve direct binding.

Certain polysaccharides, notably carrageenan, alginate and pectin, can be readily cleaved into segments that are long enough to form the ordered intermolecular assemblies that act as junction zones in gelation of the parent polysaccharide, but which are themselves non-gelling because of the absence of the interconnecting sequences required to form a continuous network. In systems where the junctions involve a fixed number of strands (e.g. double helices or "egg-box" dimers), sequences of this type can inhibit gelation of the intact polysaccharide by occupying binding sites on the polymer chains without contributing to network formation (Morris, Rees, Robinson and Young, 1980). A similar "competitive inhibition" approach could obviously be used to detect and characterise heterotypic binding between unlike biopolymers, although its application to such systems has so far been extremely limited.

As discussed in greater detail later, other types of evidence for direct binding have been obtained for a few specific biopolymer mixtures. These include co-elution in gel permeation chromatography (GPC), and conformational transitions that have no obvious parentage in the behaviour of either constituent alone.

Finally, a particularly attractive route to obtaining a detailed molecular description of heterotypic junctions, as well as establishing their existence, is X-ray fibre diffraction of oriented mixed specimens. Although this is fraught with experimental difficulties, because of the background diffraction from the individual polymers, promising results have been obtained for xanthan in the presence of galactomannans or konjac glucomannan (Cairns, Miles and Morris, 1986; Cairns, Miles, Morris and Brownsey, 1987; Brownsey, Cairns, Miles and Morris, 1988). Two qualifications must, however, be made. The first is the general caveat, applicable to any fibre diffraction study, that the process of drying and stretching could, in principle, induce formation of structures that were not present in the hydrated state. The second, more important, qualification is that failure to obtain diffraction patterns diagnostic of heterotypic associations does not necessarily imply that such structures are absent, although it does demonstrate that, if present, they are difficult to orient and pack.

How common is it?

Although many mixed biopolymer systems show obvious evidence of thermodynamic incompatibility and phase separation, there are very few that are known, or even suspected, to involve heterotypic binding. One of the first to be detected was the interaction of carrageenan with κ-casein (Hansen, 1968; Grinrod and Nickerson, 1968; Lin and Hansen, 1970; Payens, 1972; Snoeren, Payens, Jeunink and Both, 1975). Despite the commercial importance of the "milk reactivity" of carrageenan, the precise mechanism of binding remains uncertain. The most likely interpretation (Snoeren *et al.*, 1975) is that the carrageenan polyanion attaches to a positively charged sequence within the κ-casein molecule (which, overall, is negatively charged at neutral pH). However, other anionic polysaccharides with similar, or higher, negative charge density do not interact with κ-casein in the same way (Lin and Hansen, 1970), suggesting that the binding process does involve specific association of sterically-compatible structures.

In an early series of investigations of the interactions of cationic homopolypeptides with glycosaminoglycans (Gelman, Glaser and Blackwell, 1973; Gelman, Rippon and Blackwell, 1973b; Gelman and Blackwell, 1973a; Gelman and Blackwell, 1973b; Gelman and Blackwell, 1973c) it was found, by circular dichroism, that these anionic polysaccharides could induce formation of the polypeptide α-helix, and that the complexes showed the sharp melting behaviour and stoichiometric relationships expected for sterically-regular, co-operative intermolecular binding. As would be anticipated, the strength of interaction was found to increase with increasing charge density on the polysaccharide. It was also, however, found to be sensitive to structural features such as the position of sulphate substituents and the relative proportions of glucuronate and iduronate residues, and to increase systematically with increasing freedom of movement of the cationic groups on the polypeptide (i.e. increasing length of the sequences linking them to the polymer backbone), through the series poly-L-arginine > poly-L-lysine > poly-L-ornithine.

A similar requirement for conformational freedom in formation of polysaccharide–polypeptide complexes has been observed more recently in investigation of the interactions of poly-L-lysine with alginates and pectins (Bystrický, Maloviková and Sticzay, 1990). The poly-L-guluronate sequences of alginate and poly-D-galacturonate sequences of pectin are almost mirror images, differing only in the configuration of the hydroxyl group on C(3), which is axial in guluronate and equatorial in galacturonate. This subtle change in geometry, however, appears to cause a significant change in chain flexibility. Under hydrated conditions, both sequences interact with Ca^{2+} ions to form dimeric "egg-box" junctions in which the chains have two-fold symmetry (Morris, Rees, Thom and Boyd, 1978; Morris, Powell, Gidley and Rees, 1982). On drying to the solid state, polyguluronate retains the two-fold conformation (Atkins, Nieduszynski, Mackie, Parker and Smolko, 1973), but polygalacturonate converts to a three-fold arrangement (Walkinshaw and Arnott, 1982), indicating greater conformational mobility.

In mixtures with poly-L-lysine, pectin binds stoichiometrically and induces complete conversion to the α-helix conformation (Bystrický *et al.*, 1990). Alginates with a low content of poly-L-guluronate are also effective in inducing α-helix formation, although less so than pectin, but the extent of ordering drops to zero as the polyguluronate content is increased, indicating that, because of their comparative inflexibility, these sequences cannot adopt the geometry required for conformationally regular heterotypic binding to the peptide helix. The periodic spacing of cationic substituents found in synthetic polypeptides is, of course, seldom encountered in the primary sequence of natural proteins, so that sterically-regular electrostatic complexes with anionic polysaccharides are likely to be correspondingly rare.

Perhaps the least controversial example of heterotypic association between different polysaccharides is the co-gelation of alginate and pectin under acidic conditions (Toft, 1982). The extent of interaction increases as the poly-L-guluronate content of the alginate is increased, indicating binding of these sequences to the near mirror-image poly-D-galacturonate sequences in pectin. Formation of the mixed gels is severely inhibited by the presence of calcium ions, suggesting that it occurs in competition with formation of homotypic "egg-box" junctions. The associations are stabilised by methyl esterification of the pectin component and by reduction in pH, both of which would be expected to promote interaction by suppressing electrostatic repulsion between the participating chains.

In mixtures where the methyl-D-galacturonate content of the pectin and L-guluronate content of the alginate are equal, maximum gel strength is attained (Figure 13.1) when the two polymers are present in equal amounts (Toft, Grasdalen and Smidsrød, 1986), consistent with the stoichiometric relationship that would be expected for direct binding. Also, the circular dichroism (CD) changes observed on acidification of mixed solutions of alginate and pectin are appreciably different (Figure 13.2) from those anticipated from the behaviour of the individual polymers in isolation (Thom, Dea, Morris and Powell, 1982), again indicative of heterotypic association. Since poly-L-guluronate and (esterified) poly-D-galacturonate have the same 1,4-diaxial linkage geometry, their association into sterically-regular mixed junctions seems entirely feasible, although the detailed geometry of the heterotypic structures has yet to be established.

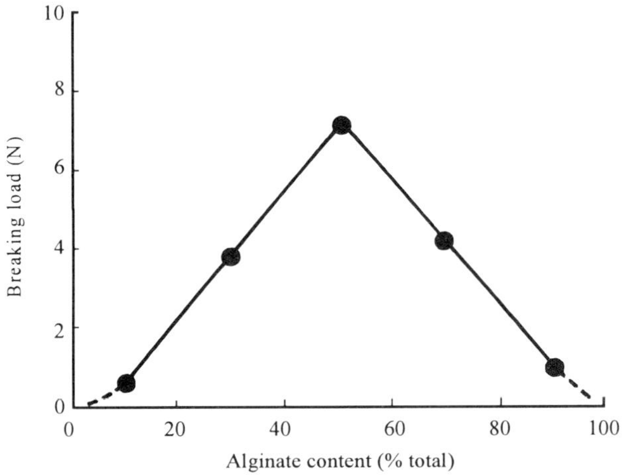

Figure 13.1 Stoichiometry of interaction in mixed gels of alginate (70% guluronate) and pectin (70% degree of esterification) under acidic conditions (pH 3.1). Yield stress (breaking load) was measured at a total polysaccharide concentration of 1.2% w/v. (From Toft, Grasdalen and Smidsrød (1986), with permission.)

The best known, and most controversial, examples of apparent synergism between different polysaccharides, however, involve interaction of galactomannans, konjac glucomannan and related plant polysaccharides (Dea and Morrison, 1975) with xanthan or with algal polysaccharides in the agar/carrageenan series. Two common features of these systems are that the plant polysaccharides all share the

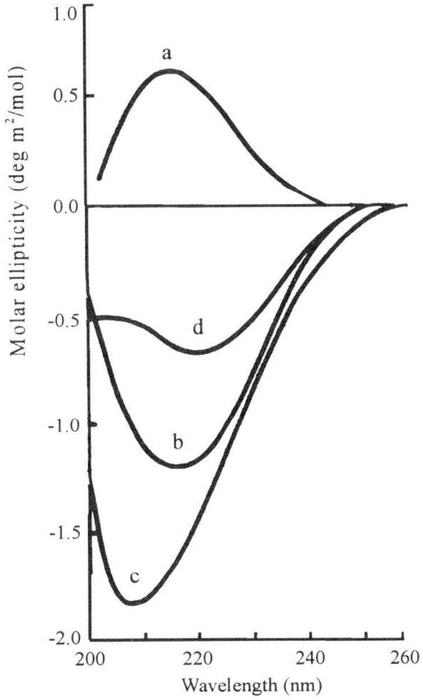

Figure 13.2 Changes in circular dichroism on acidification from pH 7 to pH 3. (a) High-methoxy pectin alone. (b) High-guluronate alginate alone. (c) A mixture containing equal concentrations of alginate and pectin. (d) Change anticipated for the alginate—pectin mixed system in the absence of interaction, obtained by addition of (a) and (b). (From Thom, Dea, Morris and Powell (1982), with permission.)

same 1,4-diequatorial linkage geometry (so that their ordered structures in the condensed phase are flat, extended ribbons), and that the other components all undergo thermoreversible disorder–order transitions under hydrated conditions. It is also well established that, within the galactomannan series, the extent of synergistic interaction increases with decreasing content of galactose sidechains, prompting an early proposal (Dea, McKinnon and Rees, 1972) that interaction occurs by attachment of unsubstituted ("smooth") regions of the mannan backbone to the ordered structure of the co-synergist (Figure 13.3), with more heavily substituted ("hairy") regions acting as interconnecting sequences in the hydrated network.

One line of reasoning behind this proposal was that conformational rigidity in one of the interacting species would lower the entropic disadvantage of intermolecular association, and thus facilitate formation of heterotypic junctions (Flory, 1956). The possibility that plant polysaccharides known to interact with xanthan, agars and carrageenans might also bind to other conformationally rigid polysaccharide structures was therefore explored as part of a Ph.D. research programme in my laboratory (Goycoolea, 1994). The plant polysaccharides chosen

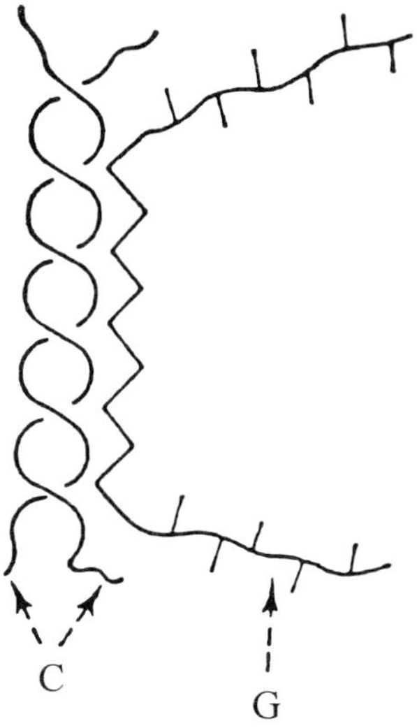

Figure 13.3 Schematic representation of a postulated model for binding of galactomannan (G) to κ-carrageenan (C) in its ordered (double helical) conformation. Analogous models have also been proposed for the synergistic interaction of galactomannans and konjac glucomannan with agarose and xanthan. (From Dea, McKinnon and Rees (1972), with permission.)

for investigation were locust bean gum (LBG), a galactomannan of low galactose content, and konjac glucomannan, commonly known as konjac mannan (KM), both of which interact strongly with xanthan and with algal polysaccharides in the agar/carrageenan family. The procedure used was the viscometric method described in the previous section, which involves preparing stock solutions to the same viscosity and ionic environment, mixing in various proportions and looking for departures from the initial common viscosity.

Figure 13.4 shows results obtained by this procedure for LBG in combination with λ-carrageenan (which cannot adopt the double-helix geometry of gelling carrageenans), and with xanthan in its five-fold ordered form. The measured viscosities for mixtures of LBG with varying proportions of disordered carrageenan lie, to within experimental error, on the straight line joining the viscosities of the individual stock solutions, indicating that, as anticipated, there is no interaction between the two constituent polysaccharides. Mixtures of LBG with ordered xanthan, by contrast, show obvious evidence of interaction, with about a 4-fold enhancement in specific viscosity in the centre of the composition range. It should be noted that the viscosities of the individual stock solutions are very low, around twice that of water (i.e. $\eta_{sp} \approx 1$), so that the polymers are well below their overlap concentration and therefore unlikely to show any significant exclusion effects.

Similar experiments were carried out using LBG and KM in combination with the new bacterial exopolysaccharides welan and rhamsan which, like xanthan, have a rigid, ordered structure in solution (Robinson, Manning and Morris, 1991), and

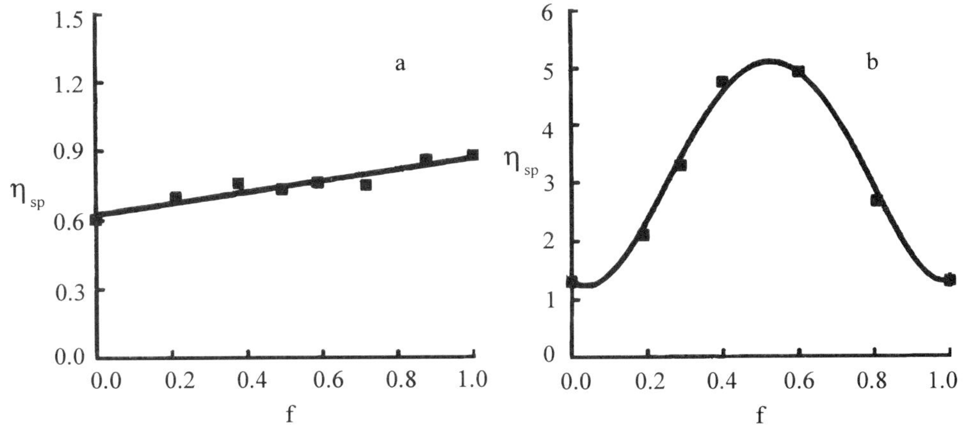

Figure 13.4 Specific viscosity (20°C; 50 s^{-1}) for dilute solutions of LBG in combination with (a) λ-carrageenan in 100 mM KCl and (b) xanthan in 10 mM KCl; f denotes the LBG content expressed as a fraction of the total polymer concentration; the concentrations (% w/w) of the individual polymer solutions were: (a) 0.063 LBG; 0.304 carrageenan; (b) 0.080 LBG; 0.008 xanthan. (From Goycoolea (1994).)

with alginate and pectin in the presence of sufficient Ca^{2+} to induce almost total conversion of polyguluronate or polygalacturonate sequences to the dimeric "egg box" form (Morris *et al.*, 1978; Morris *et al.*, 1982), but not to promote further aggregation of the dimers. The results were entirely negative. In all cases the measured viscosities remained close to the common starting viscosity, as found for mixtures of LBG with disordered carrageenan (Figure 13.4(a)). It would therefore appear that conformational rigidity is not, in itself, sufficient to induce binding of other polysaccharides to 1,4-diequatorially linked glycan chains.

In summary, "true" polysaccharide synergism seems a rare phenomenon, with documented examples confined to:

(1) Electrostatic complexes formed by certain anionic polysaccharides with cationic homopolypeptides and, perhaps, κ-casein;
(2) Association of poly-L-guluronate with esterified poly-D-galacturonate at acid pH, where electrostatic repulsions are suppressed;
(3) Interaction of xanthan and ordered agars and carrageenans with LBG, KM and related ß-1,4-glycans.

What is special about ß-1,4-glycans?

The interactions that have generated greatest scientific interest, and dispute, are those involving diequatorially-linked ß-1,4-glycans, and this is the category that I will now address exclusively.

Although an increasing amount of research effort has been devoted to KM since it became available commercially as a food hydrocolloid, most early studies of synergistic interactions concentrated on the galactomannans. Unsubstituted mannan is totally insoluble in water, although it can be dissolved in strong alkali (presumably due, predominantly, to ionisation of hydroxyl groups at high pH). The solubility of galactomannans increases progressively with increasing content of galactose substituents (Dea and Morrison, 1975). For example, guar gum, which has a mannose:galactose (M:G) ratio of 1.55, dissolves completely in water at around ambient temperature (25°C–40°C), whereas locust bean gum (M:G ≈ 3.5) is only partially soluble in cold water and can be separated into fractions of progressively lower galactose content by dissolving at progressively higher temperatures.

The ability of galactose sidechains to solubilise the mannan backbone is likely to arise in two ways; first by presenting a steric barrier to close-packing of the polymer chains, and secondly by increasing the conformational entropy of the disordered state. Solutions of galactomannans, particularly those of low galactose content, however, show obvious signs of instability. When dilute solutions are frozen and thawed, there is often substantial precipitation of material with a very high M:G ratio, indicating association of unsubstituted regions of the polymer backbone. Studies using rigorously purified ß-mannanase in combination with computer simulation of the pattern of enzymic attack indicate that the extent of precipitation is directly related to the content of unsubstituted sequences of six or more mannosyl residues (Dea, Clark and McCleary, 1986). At higher concentrations, freeze–thaw cycling of sparingly substituted galactomannans can give partially hydrated gel-like structures, but with a very pronounced tendency to contraction and syneresis (Dea, Morris, Rees, Welsh, Barnes and Price, 1977).

There is also evidence of some limited intermolecular association in galactomannan solutions at ambient temperature. For most disordered polysaccharides, double logarithmic plots of specific viscosity (measured in the Newtonian plateau region at low shear rate) versus the product of concentration and intrinsic viscosity (i.e. the "space occupancy" parameter, $c[\eta]$) show a change of slope from ~1.4 to ~3.3 at $c[\eta] \approx 4$ (Morris, Cutler, Ross-Murphy, Rees and Price, 1981), marking the transition from a solution of independent coils to an entangled network. The corresponding transition for LBG and guar gum occurs at a lower degree of space occupancy ($c[\eta] \approx 2.5$) and the slope at higher concentration is greater (~4.0). Similar behaviour has been reported (Jacon, Rao, Cooley and Walter, 1993) for KM, which contains both mannose and glucose residues in the approximate ratio of 2:1 and is solubilised by limited branching and by acetylation of about 1 residue in 17 (Nishinari, Williams and Phillips, 1992), and for tamarind seed galactoxyloglucan (Gidley, Lillford, Rowlands, Lang, Dentini, Crescenzi, Edwards, Fanutti and Reid, 1991), which has a cellulosic backbone solubilised by 1,6-linked α-D-xylose residues, approximately half of which have a ß-D-galactose residue at O(2).

When the unusual concentration-dependence of viscosity for LBG and guar gum was first observed (Morris *et al.*, 1981), it was tentatively ascribed to "hyperentanglement" (i.e. to enthalpic associations between unsubstituted mannan sequences, augmenting physical interpenetration of the polymer coils). If this interpretation is correct, strong alkali, which is capable of dissolving unsubstituted mannan, would be expected to destabilise any such "hyperentanglements", with a consequent reduction in solution viscosity.

As illustrated in Figure 13.5 for guar gum, galactomannans do indeed have substantially lower viscosities in alkali than in the same molar concentration of salt. To avoid possible complications from depolymerisation at high pH, the results shown were obtained (Goycoolea, 1994) by preparing solutions of galactomannans in alkali (1 M NaOH) and comparing their viscosities before and after neutralisation with HCl (with appropriate adjustments of volume to obtain identical polymer concentrations at high and neutral pH). The degree of viscosity enhancement on neutralisation increases with polymer concentration, as would be expected from simple mass-action considerations for formation of transient "hyperentanglements" between chains. Very similar results were obtained for LBG. It therefore seems reasonable to conclude that the galactomannans, and related polysaccharides with the same 1,4-diequatorial linkage geometry, have an unusually strong tendency to self-association at neutral pH.

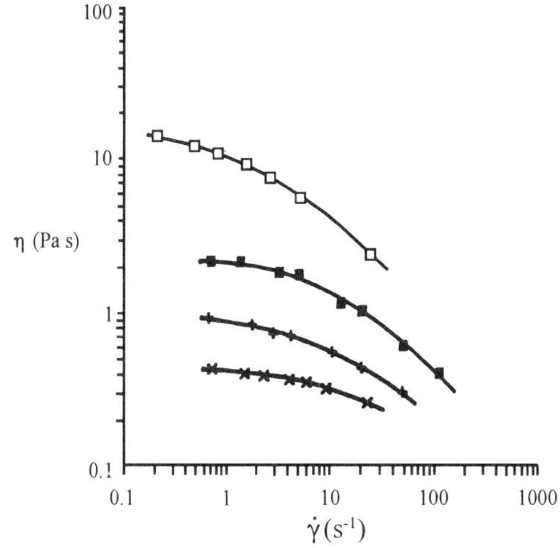

Figure 13.5 Shear-rate dependence of viscosity (20°C) for guar gum prepared in 1 M NaOH and neutralised to pH 7.0; 0.37% w/w, before (×) and after (+) neutralisation; 1.3% w/w, before (■) and after (□) neutralisation. (From Goycoolea (1994).)

Another unusual feature of the behaviour of galactomannans is that guar gum and LBG have been reported to lower surface tension in oil–water emulsions by forming an organised layer at the interface (Garti and Reichman, 1994). Surface activity is, of course, well documented for polysaccharides with a significant content of protein (gum arabic; mesquite gum) or hydrophobic substituents (methylcellulose; sugar beet pectin; emulsan) but would not be expected for such hydrophilic materials as galactomannans.

There are two factors that may facilitate association of galactomannans and related ß-1,4-glycans into larger assemblies in solution, on surfaces, or in synergistic interactions. The first is the absence of charge, with therefore no electrostatic barrier to close packing. The second is the geometry of the inter-residue linkage. Polysaccharides with 1,4-diequatorial linkages have high characteristic ratios in solution (i.e. giving more expanded coil dimensions than other polysaccharides of equivalent molecular weight). This is not because they are unusually stiff. Amylose, for example, has less freedom of rotation than cellulosic chains, but gives much more compact coils (Rees, Morris, Thom and Madden, 1982). The distinguishing feature of the 1,4-diequatorial linkage is that the bonds to and from each residue are almost co-linear, so that internal rotations (variation of ϕ and ψ angles in Figure 13.6) cause only limited changes in the overall chain contour, which remains relatively flat and extended over a length scale of many residues. Thus ordered packing of ß-1,4-glycans will involve much less conformational restriction (i.e. less reduction in entropy) than for chains with other types of inter-residue linkage.

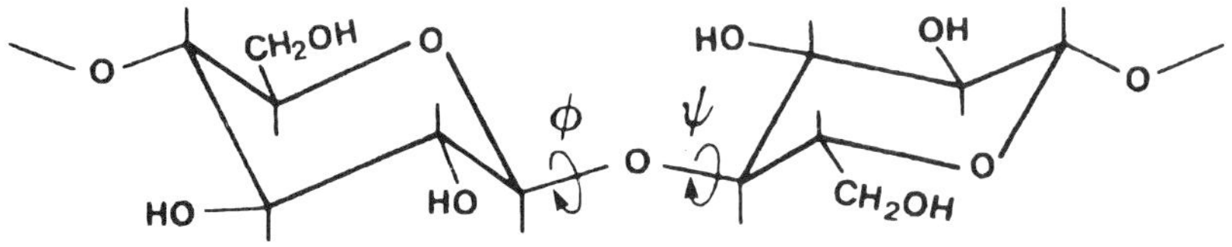

Figure 13.6 Conformational variables in carbohydrate chains

How do they interact with xanthan?

The xanthan molecule has a linear cellulosic backbone solubilised by charged trisaccharide sidechains attached at O(3) of alternate glucose residues, to give a pentasaccharide repeat unit (Jansson, Kenne and Lindberg, 1975; Melton, Mindt, Rees and Sanderson, 1976). The sidechains have the structure: ß-D-Man*p*-(1→4)-ß-D-GlcA*p*-(1→2)-α-D-Man*p*-(1→ with variable, non-stoichiometric substitution by *O*-acetate at C(6) of the inner mannose and 4,6-linked pyruvate ketal groups on terminal mannose residues.

At high temperature and low ionic strength xanthan exists in solution as a disordered coil. On cooling and/or addition of salt, however, it undergoes a cooperative conformational transition to a rigid, ordered structure (Morris, Rees,

Young, Walkinshaw and Darke, 1977). The transition is fully reversible with no detectable thermal hysteresis. As found for other polyelectrolytes, adoption of the ordered structure is facilitated by increasing ionic strength, with consequent increase in the midpoint temperature (T_m) of the conformational transition. Removal of pyruvate substituents also raises T_m by reducing electrostatic repulsion between chain segments. Acetate substituents, however, appear to contribute to the stability of the ordered structure (Smith, Symes, Lawson and Morris, 1981), and their removal lowers T_m.

Despite numerous investigations since the conformational transition of xanthan was first reported, the nature of the ordered structure remains controversial. X-ray fibre diffraction patterns from oriented specimens in the condensed phase are equally consistent with two contending models: a single-stranded five-fold helix stabilised by ordered packing of sidechains along the polymer backbone (Moorhouse, Walkinshaw and Arnott, 1977) and a co-axial double helix (Okuyama, Arnott, Moorhouse, Walkinshaw, Atkins and Wolf-Ullish, 1980) with the participating strands having essentially the same geometry as in the single-helix model. Arguments in favour of both models have been put forward from solution studies.

Irrespective of the detailed geometry of the ordered structure, there is general agreement that its presence can generate unusual rheological properties. At concentrations above ~0.3% w/v, solutions of commercial xanthan, although free-flowing, have obvious gel-like character. In small-deformation oscillatory measurements (Ross-Murphy, 1984) elastic response (storage modulus, G') substantially exceeds viscous flow (loss modulus, G'') and both moduli show only limited variation with frequency, as in conventional gels. At higher deformations the network breaks, giving rise to large "stress overshoot" effects in start-shear experiments (Richardson and Ross-Murphy, 1987) and allowing the solutions to flow. These "weak gel" properties are attributed to tenuous association of ordered chains, and are promoted by counterions (Ross-Murphy, Morris and Morris, 1983) with progressive increase in effectiveness through the series: $Na^+ < K^+ < Ca^{2+}$.

Although self-association of xanthan is limited to generation of "weak gel" character in solution, it will form true self-supporting gels with KM and certain galactomannans. One non-controversial observation is that the ability of galactomannans to interact with xanthan increases with decreasing galactose content (Dea and Morrison, 1975). Gel formation is confined to materials with a comparatively low degree of substitution, such as LBG, tara gum (M:G ≈ 2.8) and cassia gum (M:G ≈ 3.0), although guar gum does produce a substantial enhancement in solution viscosity (Lopes, Andrade, Milas and Rinaudo, 1992).

An early report of formation of mixed gels between xanthan and LBG (Kovacs, 1973) described the interaction as a "useful incompatibility". Most subsequent interpretations, however, have envisaged mechanisms involving direct association between the participating polymers rather than mutual exclusion. The first such proposal (Morris *et al.*, 1977), from Unilever's Colworth laboratory, was that the gel network is formed by attachment of unsubstituted regions of the mannan backbone to the surface of the xanthan five-fold helix. In later publications this proposal was extended to sequences devoid of galactose along one side of the mannan chain in the two-fold conformation (McCleary, 1979; Dea *et al.*, 1986), and to similar attachment of KM to the xanthan helix (Dea *et al.*, 1977).

More recently, however, it has been suggested by workers in the Institute of Food Research, Norwich, that attachment of mannan or glucomannan occurs only when the xanthan chain is in the disordered state, and that the resulting heterotypic junctions involve direct binding of the plant polysaccharide to the cellulosic backbone of the xanthan molecule in an extended conformation, rather than to the five-fold helix (Cairns *et al.*, 1986; Cairns *et al.*, 1987; Brownsey *et al.*, 1988; Morris, 1992). This proposal is based on two main lines of evidence.

(1) Solutions mixed under conditions where the xanthan helix is already present (low temperature and/or high concentration of counterions, particularly Ca^{2+}) remain fluid, but give a cohesive gel after heating and re-cooling from a temperature at which xanthan is converted to the disordered form.

(2) X-ray diffraction from oriented specimens of xanthan with galactomannans or with KM gives intensities that do not appear in the diffraction patterns of either component in isolation.

For brevity, these two proposals will be referred to as, respectively, the "Unilever" model and the "Norwich" model. A third proposal (Tako and Nakamura, 1985; Tako, 1991) invokes non-covalent bonding between specific groups on galactomannan molecules and on the xanthan sidechains. This model, however, appears to have little basis in experimental evidence and incorporates some implausible features (such as the methyl groups of acetyl substituents participating in hydrogen-bonding).

The initial "Norwich" proposal (Cairns *et al.*, 1986) envisaged ordered assemblies with xanthan and galactomannan chains both present in the 2_1 conformation, and it was argued (Cairns *et al.*, 1987) that this would provide a complementary fit absent in the "Unilever" model. However, it should be noted that, in principle, there is no reason why galactomannan or glucomannan chains should not adopt a conformation with a pitch to match that of the cellulose backbone in the xanthan five-fold helix. The ability of xanthan chains to adopt the two-fold conformation (in which the trisaccharide sidechains are clustered together along one side of the backbone) is less obvious. Computer modelling by Perez and Vergelati (1987) indicated that arrangements of this type are blocked by steric clashes between the sidechains. In a subsequent re-examination, however, Millane and Wang (1990) concluded that a 2_1 structure could be formed if the pitch was extended to give a projected residue length of ~0.526 nm along the axis of symmetry, in comparison with 0.520 nm for cellulose (Sundararajan and Marchessault, 1972) or mannan (Winter, Song and Bouckris, 1987) and 0.470 nm for the xanthan five-fold helix (Moorhouse *et al.*, 1977).

The meridional reflections attributed to heterotypic junctions in mixed specimens of xanthan with LBG or tara gum were reported (Cairns *et al.*, 1987) to correspond to a residue-rise of 0.52 nm, as was also observed for the galactomannans alone. The difference from the more extended conformation modelled by Millane and Wang (1990) may, however, be within the error limits of the measurements. Oriented fibres prepared from mixed gels of xanthan with KM, by contrast, gave an entirely different diffraction pattern (Brownsey *et al.*, 1988) for which the simplest interpretation is a six-fold structure with a pitch of 5.6 nm (Morris, 1992). This corresponds to a rise-per-residue of 0.467 nm, virtually identical to that of the five-fold helix structure of xanthan alone.

In the remainder of this section I will address the following specific questions.

(1) Is there convincing evidence that the "synergistic" interactions of xanthan with ß-1,4-glycans (particularly LBG and KM) do indeed involve direct binding rather than thermodynamic incompatibility?
(2) If so, do the mixed junctions involve association of the glycan chains with the xanthan backbone (as in the "Norwich" model), or with the surface of the five-fold helix (as in the "Unilever" model)?
(3) Does interaction require the xanthan chain to be ordered or disordered?
(4) Are there any differences in the behaviour of KM and galactomannans with xanthan in hydrated systems that might reflect the differences in molecular organisation in the condensed phase indicated by the Norwich X-ray diffraction studies?

Under salt conditions that position the xanthan disorder—order transition at high temperature, the sequence of events on cooling in the presence of ß-1,4-glycans seems reasonably straightforward. Figure 13.7 shows DSC cooling scans (Goycoolea, 1994) for xanthan (0.32% w/v in 30 mM KCl) alone, and in the presence of KM (0.24% w/v), in direct comparison with the changes in rheology of the mixed system detected by low-amplitude oscillatory measurements of G′, G″ and tan δ (the ratio of G″/G′). The DSC exotherm corresponding to the disorder—order transition of xanthan alone ($T_m \approx 73$°C; $\Delta H \approx 4.4$ J/g) also occurs in the presence of KM, and is accompanied by the small increase in G′ anticipated for formation of a xanthan "weak gel" network. The onset of synergistic gelation occurs at ~60°C, and the gelation process is accompanied by a second endotherm in DSC. The rheological changes are fully reversible on heating, with no detectable thermal

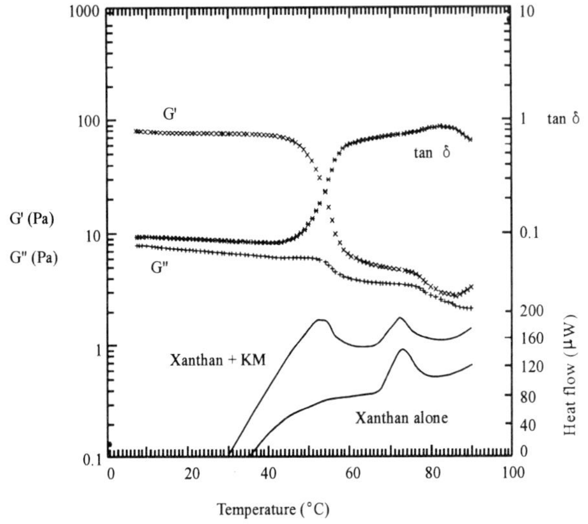

Figure 13.7 Lower traces: DSC cooling scans (0.2 deg/min) for xanthan (0.32% w/v in 30 mM KCl), alone, and in the presence of KM (0.24% w/v). Upper traces: variation of G′, G″ and tan δ (10 rad/s; 2% strain) on cooling (1 deg/min) for the xanthan—KM mixed system; virtually identical curves were obtained on re-heating. (From Goycoolea (1994).)

hysteresis. A similar sequence of events is observed (Goycoolea, Foster, Richardson, Morris and Gidley, 1994) when KM is replaced by LBG, with synergistic gelation again beginning at ~60°C. The sol–gel transition is, however, substantially broader and the accompanying thermal changes are much less evident (Williams, Day, Langdon, Phillips and Nishinari, 1991a; Foster, 1992), giving DSC exotherms that are too small to be measured reliably.

In both cases, it seems reasonable to conclude that the xanthan component first adopts the five-fold ordered conformation and then interacts with KM or LBG at lower temperatures. The simplest interpretation of such behaviour is, of course, that the glycan chains bind to the xanthan helix, as in the "Unilever" model. However, as a fair test of the "Norwich" model it is obviously essential to examine whether or not interaction can occur when the xanthan component is disordered, by displacing the disorder–order transition to a much lower temperature.

One way in which this can be done is by working at very low ionic strength. Williams, Clegg, Day and Phillips (1991b) have reported the temperature-dependence of G′ for mixtures of xanthan and KM in water, where ionic strength will come entirely from counterions to the xanthan carboxyl groups, using a total polymer concentration of 1.2% w/v and mixing ratios of 1:3, 2:2 and 3:1. In all cases the onset of gelation occurred at ~60°C (as in Figure 13.7). The transition mid-point temperature observed by DSC for 1.2% w/v xanthan alone was 51°C, and the corresponding values at the xanthan concentrations present in the mixed systems, although not measured directly, would of course be lower, indicating that interaction can also occur under conditions where the xanthan chains are predominantly disordered. We have found similar behaviour for mixtures of xanthan (0.3% w/v in water) and LBG (0.1% w/v), with gel formation again beginning at ~60°C (Foster, 1992).

However, working with a polyelectrolyte in water is not entirely satisfactory because of the variation of T_m with polymer concentration. A further complication is that, in contrast to the very fast reaction kinetics observed (Norton, Goodall, Frangou, Morris and Rees, 1984) at high ionic strengths, the disorder–order transition of xanthan under low-salt conditions appears to occur extremely slowly (Cheetham and Mashimba, 1991; Foster, 1992), making the conformational state of the chains time-dependent and ill-defined.

An alternative approach adopted in our own recent work (Foster, 1992; Goycoolea, 1994; Goycoolea *et al.*, 1994) is to remove the acetate substituents of the xanthan molecule, so that low values of T_m can be obtained at moderate ionic strength. Figure 13.8 shows the rheological and thermal changes observed using polymer concentrations identical to those in Figure 13.7, but with replacement of xanthan by deacetylated xanthan (DX) and reduction in KCl concentration from 30 mM to 10 mM. The T_m value observed for DX alone under these ionic conditions is ~39°C, but the onset of gel formation with KM again occurs at ~60°C and, as also found for unmodified xanthan at higher ionic strength, the rheological changes are fully reversible on heating and the gelation process is accompanied by an intense exotherm in DSC.

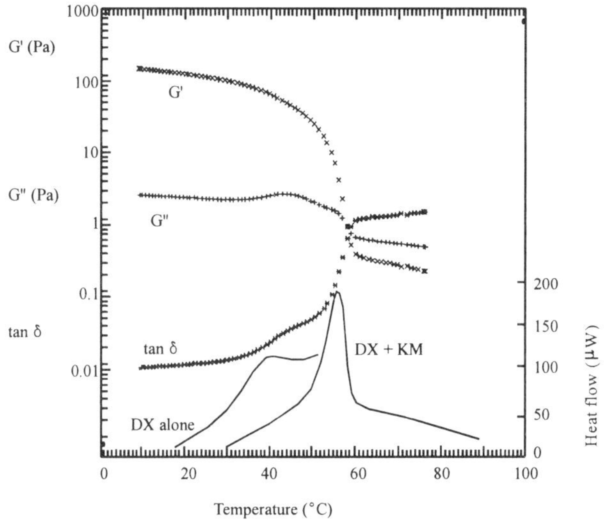

Figure 13.8 Lower traces: DSC cooling scans (0.2 deg/min) for 0.5% w/v deacetylated xanthan (DX) in 10 mM KCl, and for a mixture of DX (0.32% w/v) and KM (0.24% w/v) under the same salt conditions. Upper traces: variation of G', G'' and tan δ (10 rad/s; 2% strain) for the mixed system on cooling at 1 deg/min. (From Goycoolea (1994).)

The magnitude of this endotherm was determined (Goycoolea, 1994) for a wide range of polymer concentrations and mixing ratios, holding ionic conditions constant at 10 mM KCl. Three sets of measurements were made: first, the concentration of KM was fixed at 0.24% w/v (as in Figure 13.8) and DX was varied from 0.04% to 0.32% w/v; next, the DX concentration was held constant at 0.10% w/v and the KM content was varied from 0.01% to 0.60% w/v; finally, the DX concentration was fixed at 0.40% w/v and KM was varied from 0.04 to 1.0% w/v. A similar series of experiments was carried out using unmodified xanthan in place of DX. As shown in Figure 13.9, the enthalpy change (ΔH) per gram of DX increases linearly with increasing ratio of KM:DX until the two polymers are present in roughly equal amounts, and then remains constant (at ~29 J/g) when the KM component is present in excess. Mixtures of KM with unmodified xanthan show a similar pattern of behaviour, but the maximum value of ΔH is lower (~21 J/g) and the proportion of KM required to achieve this maximum is higher. In both cases the values of ΔH are essentially independent of total polymer concentration at the same mixing ratio, arguing strongly for a stoichiometric binding process rather than any form of exclusion mechanism.

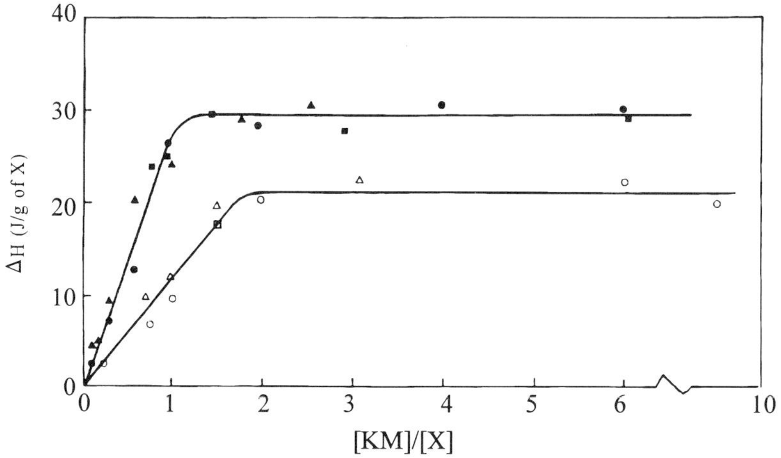

Figure 13.9 Composition-dependence of thermal changes accompanying the sol—gel transition for mixtures of KM with xanthan (open symbols) or DX (filled symbols) in 10 mM KCl. [X] denotes the concentration (% w/v) of the xanthan or DX component and [KM] the concentration of the glucomannan. Heat changes (ΔH) are expressed relative to [X] alone (i.e. as Joules per gram of xanthan or DX). Circles: [X] = 0.1% w/v, with varying [KM]; triangles [X] = 0.4% w/v, with varying [KM]; squares: [KM] = 0.24% w/v, with varying [X]. (From Goycoolea (1994).)

The rheological properties of the mixed gels are also indicative of stoichiometric association. When KM concentration is increased at fixed concentration of DX, G' rises steeply to an essentially constant value (Figure 13.10), and tan δ passes through a sharp minimum (Figure 13.11(a)), as expected for a binding process in which surplus KM forms a disordered sol fraction within the interstices of the heterotypic network (i.e. raising G″, but making little contribution to G′).

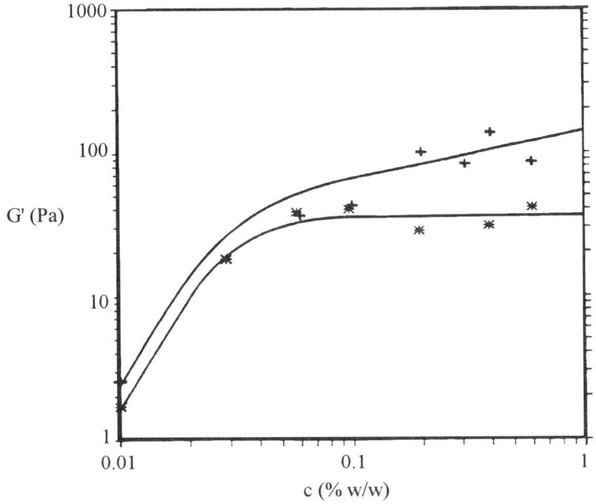

Figure 13.10 Variation of G′ (10 rad/s; 2% strain; 10°C) with concentration (c) of KM (∗) or LBG (+) in mixed gels with DX (0.1% w/v in 10 mM KCl). (From Goycoolea (1994).)

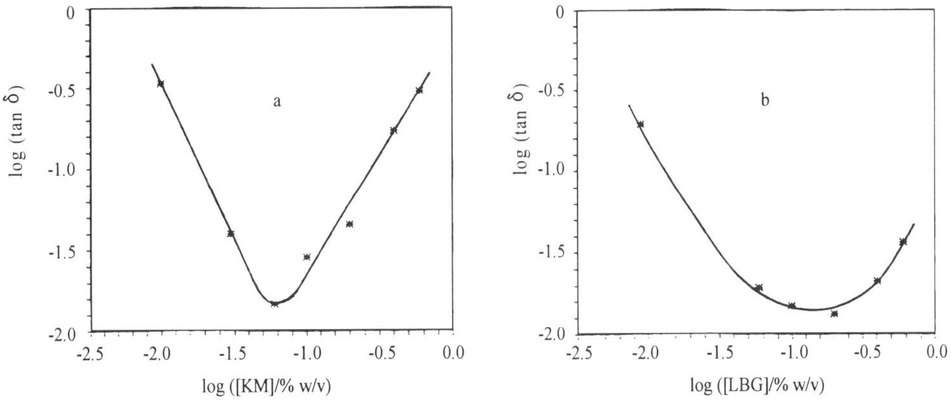

Figure 13.11 Variation of tan δ (1 rad/s; 2% strain; 10°C) with concentration of (a) KM or (b) LBG in mixed gels with DX (0.1% w/v in 10 mM KCl). (From Goycoolea (1994).)

As discussed previously, the DSC approach (Figure 13.9) cannot be applied when KM is replaced by LBG, because the enthalpy changes accompanying gel formation are much smaller and cannot be measured with the required precision.

However, the initial rise and subsequent levelling-out of G′ with increasing content of glycan at fixed concentration of DX (Figure 13.10) and the accompanying minimum in tan δ (Figure 13.11(b)) are again evident, although both effects are less sharp than in systems incorporating KM and are displaced to higher ratios of glycan:DX. It therefore seems reasonable to conclude that both KM and LBG form stoichiometric heterotypic junctions with xanthan chains. There is also further strong evidence of direct binding from GPC experiments (Cheetham, McCleary, Teng, Lum and Maryanto, 1986) in which (partially depolymerised) LBG was found to co-elute with xanthan, and from large viscosity increases (Figure 13.4) on mixing solutions at very low concentrations, where exclusion effects should be negligible.

The nature of the heterotypic junctions and the way in which they form are less straightforward. The behaviour shown in Figures 13.7 and 13.8 is difficult to reconcile with either the "Unilever" or "Norwich" models in their original form. Gel formation occurs over essentially the same temperature range for both systems, but in one case this is well below T_m for the disorder–order transition of the xanthan component (Figure 13.7), and in the other case it is well above (Figure 13.8). It could of course be argued, in an extension of the "Unilever" model, that even at temperatures well above T_m there will be a very small, but finite, proportion of five-fold order, and that this will bind to mannan or glucomannan chains, continually pulling over the disorder—order transition until full ordering and association is attained. Conversely, it can be argued that, in terms of the "Norwich" model, a small fraction of disordered sequences could act as a reaction intermediate at temperatures well below T_m.

My own view is that neither argument is necessary or valid. As one of the authors of the paper in which the "Unilever" model was first proposed (Morris *et al.*, 1977), I now believe that the "Norwich" proposal of binding of glycan chains to the cellulosic backbone of xanthan, rather than to the surface of the five-fold helix, offers the simplest, unified, interpretation of the experimental evidence currently available. I do not, however, accept that interaction can occur only when the xanthan component is disordered.

As mentioned previously, the suggestion of the necessary involvement of disordered sequences is based on the response of mixed-gel properties to ionic environment and thermal history. Figure 13.12 shows some recent measurements (Zhan, Ridout, Brownsey and Morris, 1993) of G′ at 20°C for mixed gels of xanthan (0.25% w/v) and LBG (0.25% w/v) as a function of mixing temperature and salt (NaCl) concentration. As would be anticipated from the "Norwich" model, changes that will increase the proportion of disordered xanthan present when the solutions are mixed (decreasing ionic strength and increasing mixing temperature) increase the strength of the resulting gels. However, in terms of this interpretation, it would seem reasonable to expect the major changes in gel properties to occur at mixing temperatures spanning T_m of the xanthan disorder–order transition, and therefore to move to higher temperatures with increasing salt. In practice the changes occur over approximately the same temperature range for all three salt concentrations studied (0, 20 and 400 mM NaCl) and follow closely (Figure 13.12) the temperature course of gelation and melting from our own recent work on very similar mixtures (Foster, 1992; Morris and Foster, 1994).

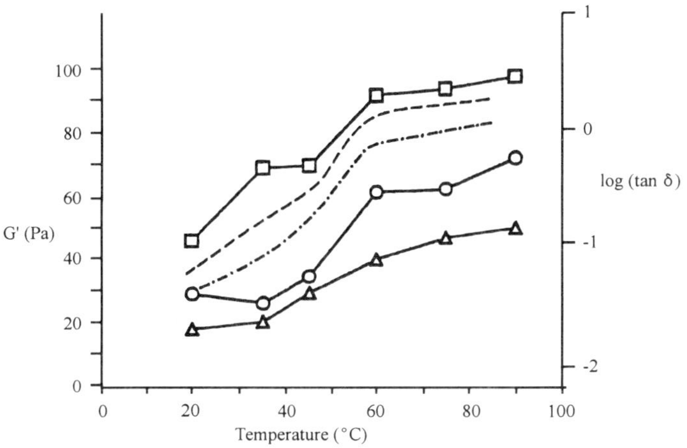

Figure 13.12 Effect of mixing temperature on final values of G' at 20°C (left-hand axis) for mixtures of xanthan (0.25% w/v) and LBG (0.25% w/v) in water (□) and in 20 mM (O) and 400 mM (Δ) NaCl (from Zhan, Ridout, Brownsey and Morris (1993), with permission), in comparison with the temperature course of gelation on cooling and gel melting on heating, as monitored by changes in tan δ (10 rad/s; 2% strain; right-hand axis) for mixtures of xanthan (0.3% w/v) and LBG (0.1% w/v) in water (- - -) and in 30 mM NaCl (- · - · -). (From Morris and Foster (1994), with permission.)

It should also be noted that the effects of mixing temperature and salt concentration on small-deformation properties are comparatively small, changing G′ by only a factor of ~2—3. By contrast, the increases in moduli that occur on initial mixing are massive (typically around two orders of magnitude), even at temperatures far below the xanthan T_m, where the proportion of disordered sequences will be vanishingly small, although the result is a viscoelastic "paste" which converts to a "true" gel only after heating and cooling (Cairns *et al.*, 1986). In systems where the xanthan T_m lies well above the sol–gel transition (e.g. Figure 13.7), heating to a temperature above the gel melting point but below the onset of the order–disorder transition has virtually the same effect on the rheological properties of the final gel as heating to a temperature where the xanthan is fully disordered (Robinson, 1989; Goycoolea *et al.*, 1994).

I would propose the following "minimum interpretation" of the above evidence.

(1) Heterotypic junctions become more stable than the xanthan five-fold helix when temperature drops to below ~60°C.

(2) Under conditions where the xanthan component is already ordered at this temperature (Figure 13.7), the glycan co-synergist interacts with the helix, and displaces the conformation to the geometry required for heterotypic association.

(3) If the xanthan chains are disordered (as in Figure 13.8), the same heterotypic junctions are formed directly by interaction of coils.

(4) Mixing at temperatures below the onset of gelation (i.e. below ~60°C) breaks the gel network as it forms, causing a gross reduction in overall cohesion (i.e. resistance to large deformation), with a smaller decrease in resistance to low-amplitude oscillation.

(5) Heating to above ~60°C melts the disrupted network formed initially and allows a continuous gel to develop on re-cooling.

It must be emphasised that this interpretation, while questioning the validity of one line of argument used to support the "Norwich" model and proposing a different reaction pathway under high-salt conditions, does not challenge the central concept of association of glycan chains with the xanthan backbone. In the absence of any compelling evidence to the contrary, it seems reasonable to assume that the diffraction patterns obtained in Norwich for oriented mixed specimens correspond to the intermolecular junctions responsible for synergistic gelation. Binding of glycans to the xanthan backbone, rather than to the surface of the helix, is also indicated by our own recent investigations, one of which was carried out in collaboration with Dr. V.J. Morris and Dr. G.J. Brownsey from the Norwich laboratory.

In these studies we examined the conformational transitions and synergistic interactions of two xanthan variants, acetan and xanthan polytetramer, both of which give X-ray fibre diffraction patterns similar to those obtained for the xanthan five-fold helix (Millane, 1992). Acetan has longer sidechains than xanthan (pentasaccharide rather than trisaccharide) and shows no sign of synergistic interaction with LBG or KM. After deacetylation, however, it forms strong mixed gels with both (Ojinnaka, Morris, Morris and Brownsey, 1994). The heptasaccharide repeat unit of acetan has two sites of acetylation. One is at the same position as in xanthan (which, as described above, interacts with LBG and KM in both its acetylated and deacetylated form), and is therefore unlikely to be responsible for preventing synergistic interaction of native acetan. The second, however, is on the polymer backbone where, in terms of the "Norwich" model, it would be expected to block association with glycan chains. The "Unilever" model, by contrast, offers no such ready explanation of the change in properties on deacetylation, since the native and deacetylated polymers both show a conformational transition indicative of conversion to the five-fold helix structure on cooling.

One unusual feature of the behaviour of deacetylated acetan is that, in contrast to xanthan and deacetylated xanthan, the synergistic gels which it forms with LBG or KM show pronounced thermal hysteresis. Completion of melting again occurs at ~60°C, but the onset of gelation on cooling is displaced to a substantially lower temperature. A possible reason for this difference is that ordered acetan has an unusually strong tendency to self-association, so that formation of heterotypic junctions with LBG or KM will involve disaggregation as well as displacement of individual chains from the five-fold conformation.

The polytetramer variant of xanthan, in which the terminal mannose residue of the sidechains is absent, does not interact with either LBG or KM, irrespective of whether or not the inner mannose residue is acetylated (Foster and Morris, 1994). This is hard to reconcile with the "Unilever" model, since the acetylated and non-acetylated polymers both undergo a disorder—order transition in solution. It could, of course, be argued that the terminal mannose residues of the sidechains in the xanthan molecule have a crucial role in attachment of ß-1,4-glycans to the five-fold helix. Such specificity, however, seems unlikely, in view of the strong interaction of LBG and KM with deacetylated acetan, where the terminal mannose is replaced by a trisaccharide sequence which is attached to the glucuronate residue in a different way (α-linkage in place of ß-).

In terms of the "Norwich" model, the absence of synergism cannot reasonably be attributed to the lack of disordered xanthan sequences available for interaction. Under the salt conditions used (10 mM NaCl), the observed value of T_m for the disorder—order transition of non-acetylated polytetramer was ~62°C, which is approximately 10°C below the T_m of xanthan in the gelling system whose behaviour is shown in Figure 13.7. The corresponding value for acetylated polytetramer was ~87°C, again well below the T_m of 115°C reported by Zhan *et al.* (1993) for xanthan in 400 mM NaCl, which also shows obvious synergism (Figure 13.12).

The main difference between xanthan and its polytetramer variant is not in the temperature of conformational ordering (and hence the proportion of residual disorder) but in the enthalpic stability of the five-fold helix. ΔH for the disorder–order transition of xanthan is ~4.0 J/g; the corresponding value for polytetramer (acetylated or non-acetylated) is ~8.5 J/g. If the mechanism that I have proposed above is correct, this difference would offer a simple explanation of why the polytetramers do not interact with LBG or KM: the stability of the five-fold helix prevents the conformational rearrangement required to allow the polymer backbone to form heterotypic junctions with other chains.

It is possible that the decrease in G' with increasing salt concentration in synergistic gelation of xanthan (Figure 13.12) can be explained in a similar way, by cation-induced aggregation of five-fold helices restricting the extent of homotypic association. For mixtures of xanthan with KM, addition of salt can also cause a substantial reduction in gelation temperature (Dalbe, 1992). In a recent systematic investigation, Annable, Williams and Nishinari (1994) have found that, as might be expected, divalent cations cause greater suppression than monovalents. Perhaps surprisingly, however, Na^+ ions were found to have a greater effect than K^+, a difference observed independently in our own recent work (Foster, 1992; Goycoolea, 1994). The influence of ionic environment on gelation temperature is reduced, or eliminated, when KM is replaced by LBG or when the xanthan component is deacetylated. Gaining a better understanding of these differences would be an obvious target for future research.

Finally, the rheological changes (Goycoolea, 1994) shown in Figure 13.8 for gelation of deacetylated xanthan (DX) with KM under low-salt conditions (10 mM KCl) indicate the involvement of a second process after the initial onset of synergistic gelation. There is a distinct shoulder in the temperature course of tan δ, accompanied by a slight, but detectable, maximum in G''. It would be tempting to attribute these effects to conformational ordering of surplus DX. However, the changes are centred at a higher temperature than the disorder—order transition of DX alone under the same ionic conditions and were found to become most clearly evident when the two polymers were present in roughly equal amounts, rather than when DX was in excess. Indeed, at KM:DX ratios of 1:1, or above, DSC cooling scans for the same samples showed a slight shoulder on the low temperature side of the main exotherm, roughly coincident with the shoulder in tan δ. No such effects were observed when KM was replaced by LBG. In both systems, however, the onset of gelation occurred at essentially the same temperature (~60°C).

A possible explanation, suggested by the Norwich X-ray work, is that similar heterotypic junctions with two-fold chain geometry are formed initially in both cases, but that the junctions involving KM then reorganise to the more compact six-fold arrangement characterised in the condensed phase (Morris, 1992). The driving force for the rearrangement may come from glucose residues in the xanthan backbone interacting more strongly with glucose residues in KM than with mannose residues in LBG (as indicated by the much higher values of ΔH in DSC cooling scans). This interpretation is, however, highly speculative, and would again merit further investigation.

Do ß-1,4-glycans bind to carrageenans and agars?

The carrageenans are linear D-galactans with a regular alternating sequence of 1,3 and 1,4 linkages. The 3-linked residues occur in the ß-configuration. The 4-linked residues are in the α-configuration and, in gelling fractions, occur mainly (but not exclusively) as the 3,6-anhydride. Commercial gelling carrageenans are hybrids (Parker, Brigand, Miniou, Trespoey and Vallée, 1993) of (predominantly) two idealised repeating structures: κ-carrageenan, which has one sulphate group per disaccharide, and ι-carrageenan, which has a sulphate group on every residue (i.e.

two charges per disaccharide). Agar polysaccharides have a similar primary structure, but with the 4-linked residues in the L-form (i.e. the mirror-image of the corresponding D-residues in carrageenan). Commercial agars have a small content of sulphate (and other) substituents but, in comparison with carrageenan, are essentially uncharged. The unsubstituted, neutral form is known as agarose.

The central process in gelation of both agars and carrageenans is formation of intermolecular double helices on cooling (Rees *et al.*, 1982). 4-linked residues in which the anhydride bridge is missing are incompatible with helix geometry, and therefore act as "kinks", promoting branching of the gel network. Segments prepared by chain cleavage at "kinking" residues form double helices, but are non-gelling. The second major determinant of gel properties is aggregation of helices into larger assemblies, with consequent thermal hysteresis between formation and melting of the gel network (Morris and Norton, 1983). Aggregation is suppressed by electrostatic repulsion between helices. Hysteresis is therefore greatest for agarose and undetectable for ι-carrageenan. Aggregation and hysteresis of κ-carrageenan are promoted specifically by K^+ and larger Group I cations, which bind to the double helix (Rochas and Rinaudo, 1980; Piculell, Nilsson, Viebke and Zhang, 1993).

The effect of ß-1,4-glycans on gelation of polysaccharides in the carrageenan/agar series is illustrated in Figure 13.13 for K^+ κ-carrageenan (0.085% w/v) and KM (0.036% w/v) in 100 mM KCl (Goycoolea, 1994). The sol—gel and gel—sol transitions are displaced to a higher temperature, with corresponding displacement of the thermal changes associated with the carrageenan helix—coil transition, and the gel strength is increased substantially. Very similar results were obtained when KM was replaced by LBG. The same general features have been observed in numerous previous investigations, using various combinations of ß-1,4-glycan and carrageenan/agar polysaccharide and different experimental approaches (e.g. characterisation of gel properties by compression testing; determination of melting and setting points by observation of the movement of ball bearings; monitoring conformational change by optical rotation or loss of high-resolution NMR signal). It has also been found that the presence of ß-1,4-glycans can cause substantial reductions in the concentration (Dea and Morrison, 1975; Turquois, Taravel and Rochas, 1993) or molecular weight (Turquois, Rochas and Taravel, 1992) of algal polysaccharide required for formation of a gel network.

In marked contrast to the synergistic gels of xanthan, where gelation temperature is unaffected by large changes in T_m for the xanthan coil—helix transition (Figures 13.7 and 13.8), the sol—gel and gel—sol transitions for carrageenan/agar polysaccharides in synergistic mixtures with ß-1,4-glycans invariably occur a few degrees above the corresponding transitions for the algal polysaccharide alone. This seems strong evidence that the normal mechanism of agar or carrageenan gelation is augmented, rather then replaced, by interaction with ß-1,4-glycans.

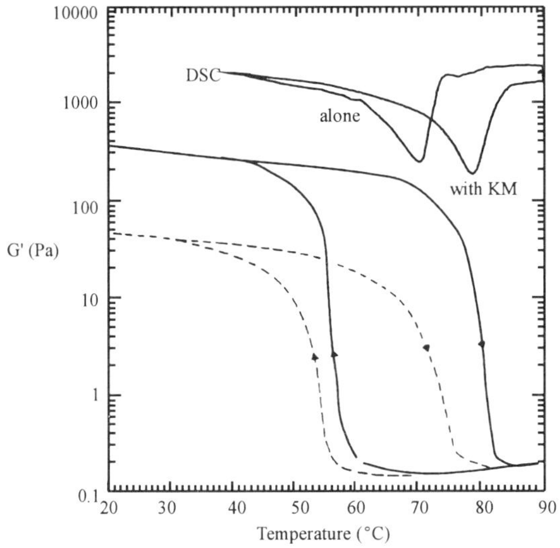

Figure 13.13 Temperature-dependence of G′ (10 rad/s; 2% strain) on cooling and heating (1 deg/min) for K^+ κ-carrageenan (0.085% w/w in 100 mM KCl) in the presence (—) and absence (---) of 0.036% w/w KM. The upper curves are a direct reproduction of DSC heating scans (0.3 deg/min) for K^+ κ-carrageenan (1.15% w/w in 100 mM KCl) in the presence and absence of 0.49% w/w KM (i.e. with the same ionic conditions and mixing ratio as in the rheological experiments, but at higher overall polymer concentration). (From Goycoolea (1994).)

Three types of interaction have been proposed:

(1) Binding of mannan or glucomannan chains to the algal polysaccharide helix (as illustrated schematically in Figure 13.3), or to aggregated assemblies of helices (Dea *et al.*, 1972; Dea and Rees, 1987);
(2) Enhancement of agar or carrageenan gelation by incompatibility with disordered glycan chains present within the interstices of the gel network (Cairns *et al.*, 1987); and
(3) A converse exclusion mechanism, in which the presence of the algal polysaccharide promotes formation of a ß-1,4-glycan network (Fernandes, Gonçales and Doublier, 1991).

The second of these proposals came from X-ray fibre diffraction analysis of an extensive series of oriented mixed specimens prepared from binary mixtures of KM, LBG or tara gum with κ-carrageenan or furcellaran (a less sulphated carrageenan). In all cases the diffraction patterns were similar to those seen for the algal polysaccharide alone, with no evidence of heterotypic association. However, while reinforcing the conclusion that self-association of carrageenan polysaccharides follows the same general mechanism in the presence or absence of ß-1,4-glycans, these findings do not disprove the involvement of other types of intermolecular association, for the reasons discussed in a previous section.

There certainly appears to be abundant evidence that ß-1,4-glycans have a more active and specific role in synergistic gelation than simply promoting self-association of the algal polysaccharide component by their presence as disordered coils.

(1) The effectiveness of galactomannans in promoting or enhancing gelation of agars and carrageenans increases systematically with increasing content of unsubstituted mannan sequences (Dea and Morrison, 1975; Dea *et al.*, 1986).
(2) When mixed gels of agarose and various galactomannans were frozen and thawed, the concentration of galactomannan in the synerised fluid was found to be substantially lower than the overall concentration, and the proportion retained within the gel network increased with decreasing galactose content (Dea *et al.*, 1972).
(3) Electron spin resonance studies of mixtures of κ-carrageenan with KM have shown that the sol—gel transition is accompanied by partial immobilisation of the KM component (Williams, Clegg, Langdon, Nishinari and Piculell, 1993).
(4) Similar immobilisation has been observed by ^{13}C NMR for partially depolymerised LBG (DP 50—60) in the presence of K^+ κ-carrageenan (Rochas, Taravel and Turquois, 1990) or agarose (Turquois *et al.*, 1993). At polymer concentrations of 3% w/v LBG and 4% w/v algal polysaccharide, the intensity of the galactomannan resonancies decreased by ~25% in the presence of agarose and by ~40% with carrageenan (in 100 mM KCl). In both cases the residual spectrum showed preferential immobilisation of unsubstituted mannan (i.e. the M:G ratio of the visible, mobile, component was lower than in the original sample). Doubling the concentration of agarose (to 8% w/v) increased the loss of visible signal by just over a factor of 2 (from ~25% to ~55%). When guar gum was substituted for LBG in the carrageenan system, the reduction in galactomannan signal on gelation was much smaller (~8%, in comparison with ~40% for LBG).

However, while these observations seem to demonstrate convincingly that ß-1,4-glycan chains are bound within the mixed gel networks and that unsubstituted regions in galactomannans associate preferentially, they do not distinguish between self-association and heterotypic binding to the algal polysaccharide.

There are similar difficulties in interpreting the results from an ingenious experiment by Turquois, Rochas and Taravel (1992). Gels of K^+ κ-carrageenan (1% w/v in 100 mM KCl) were found to dissociate completely within 5 hours of immersion in a 25-fold excess of water (because of reduction in K^+ concentration within the gel network). Complete conversion to the disordered coil state was verified by optical rotation. With LBG (0.6% w/v) present, the gels were still intact 90 hours after immersion, and ^{13}C NMR showed that the carrageenan had remained ordered and that most (~75%) of the galactomannan was retained within the gel network. It is clear, therefore, that LBG can stabilise the carrageenan helix under conditions of temperature and ionic environment where it would otherwise dissociate, and that most of the galactomannan chains are not free to diffuse out of the mixed gel. In this case the problem is in distinguishing between formation of heterotypic junctions and mutual exclusion effects promoting and stabilising self-association of both components.

There are, however, some experimental observations that are entirely consistent with heterotypic binding but hard to explain by an exclusion mechanism.

(1) On progressive addition of LBG to a fixed concentration of agarose (0.4% w/v), the elastic modulus of the resulting gels was found to increase linearly at first and then level out (Turquois *et al.*, 1993), as observed (Figure 13.10) for deacetylated xanthan in the presence of increasing concentrations of LBG or KM.

(2) In the same investigation, elastic modulus was measured as a function of the concentration of κ-carrageenan in mixed gels with 0.2% or 0.6% LBG. At carrageenan concentrations below ~0.2%, the 3-fold change in galactomannan concentration had virtually no effect on modulus.

(3) LBG (1% w/v) and tara gum (3% w/v) gave mixed gels with segmented κ-carrageenan (2% w/v), which does not gel alone at any concentration (Dea *et al.*, 1972). Since the carrageenan segments are of low molecular weight, they are unlikely to cause any significant exclusion effects at the concentration used.

(4) Galactomannan of low galactose content (M:G ≈ 4) and low molecular weight, produced by enzymic modification of guar gum, caused a decrease in gel strength when mixed with 0.2% w/v agarose (in contrast to the significant increases observed with intact galactomannans of similar, or higher, galactose content). At a lower concentration of agarose (0.1% w/v) the short galactomannan chains caused rapid precipitation (Dea and Rees, 1987). This is again difficult to explain in terms of polymer incompatibility, but is consistent with the short, sparingly substituted, galactomannan chains forming mixed junctions with agarose helices, and thus promoting aggregation without contributing significantly to network formation.

One particularly interesting new line of evidence is the behaviour seen in DSC when solutions of K^+ κ-carrageenan are cooled in the presence of KM (Williams, Clegg, Langdon, Nishinari and Phillips, 1992; Williams *et al.*, 1993). Figure 13.14 shows cooling scans recorded at a total polysaccharide concentration of 0.6% w/v

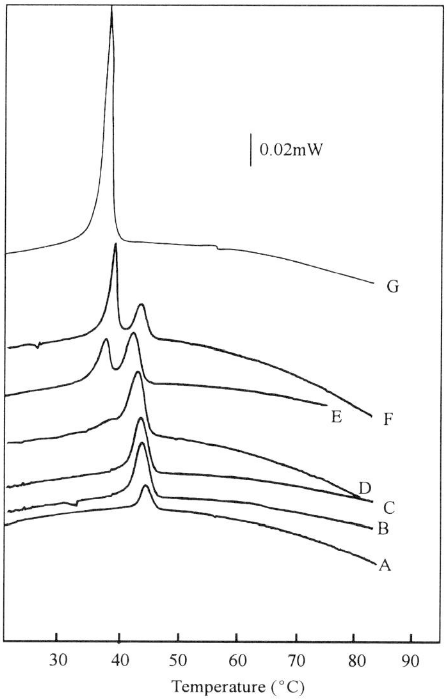

Figure 13.14 DSC cooling scans (0.1 deg/min) for mixtures of κ-carrageenan and KM in 50 mM KCl at a total polysaccharide concentration of 0.6% w/v, recorded for carrageenan:KM ratios of (A) 1:5; (B) 2:4; (C) 3:3; (D) 4:2; (E) 4.5:1.5; (F) 5:1 and (G) 6:0. (From Williams, Clegg, Langdon, Nishinari and Piculell (1993), with permission.)

in 50 mM KCl for various ratios of the two components (Williams *et al.*, 1993). Mixtures in which KM is present in excess show a single exotherm centred about 5°C above the exotherm observed for carrageenan alone. As the proportion of carrageenan is increased, a second exotherm develops at approximately the same temperature as the carrageenan transition. As shown in Figure 13.15, the intensity of the first peak passes through a maximum value when the two polymers are present in roughly equal amounts. The second transition is undetectable at this point, but increases in magnitude on further increase in the ratio of carrageenan:KM. This is, of course, exactly the pattern of behaviour that would be expected for a binding process in which the first exotherm arises from formation of heterotypic junctions and the second corresponds to the normal coil—helix transition of surplus carrageenan. Variations in KCl concentration were found to cause virtually identical displacements of both peaks, which would imply that the heterotypic junctions involve attachment of glycan chains to the carrageenan helix, rather than formation of some radically different intermolecular structure.

In terms of possible exclusion models, it seems highly unlikely that the first exotherm can arise from conformational ordering of carrageenan above its normal

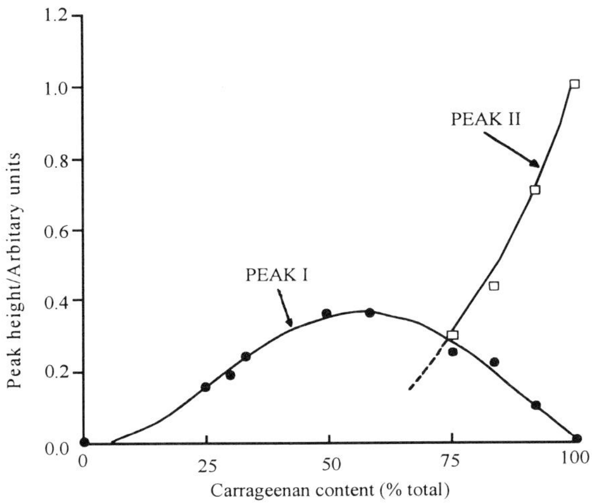

Figure 13.15 Variation in height of the first (●) and second (□) DSC exotherms (Figure 13.14) with varying proportions of κ-carrageenan and KM at a total polysaccharide concentration of 0.6% w/v in 50 mM KCl. (From Williams *et al.* (1993), with permission.)

transition temperature due to incompatibility with KM chains, because the transition, rather than increasing in size and/or moving to a higher temperature at high concentrations of KM, as would be anticipated, remains in the same position and decreases in intensity. This reduction in intensity with increasing KM content, and the response of transition temperatures to changes in K^+ concentration, both indicate strongly that the first exotherm cannot be attributed to self-association of KM in response to exclusion by carrageenan.

An alternative possibility is that thermodynamic incompatibility between the two polymers causes phase separation. Splitting of the thermal transition (Figure 13.14) could then be attributed to unequal partition of salt between the phases, with the carrageenan-rich phase having a higher K^+ concentration (because of the negative charge on the polymer), and therefore giving a disorder–order transition at a higher temperature. However, in our own recent work (Goycoolea *et al.*, 1994) we have studied mixtures of κ-carrageenan and KM at a total polysaccharide content of 0.12% w/v, in comparison with the value of 0.6% used by Williams *et al.* (1992; 1993), and have observed very similar splitting in DSC (Figure 13.16), although exclusion effects should be greatly diminished by a 5-fold reduction in concentration. We have also found evidence of two exothermic processes in DSC cooling scans for mixtures of κ-carrageenan with LBG, although in this case the additional thermal transition appears (Figure 13.16) as a shoulder on the high-temperature side of the carrageenan exotherm, rather than as a separate peak. It seems reasonable to conclude, however, that the mechanism of gelation is broadly similar in both systems.

Although the ultrastructure of the mixed gels formed by κ-carrageenan with KM has not yet been investigated, there is strong evidence that the corresponding

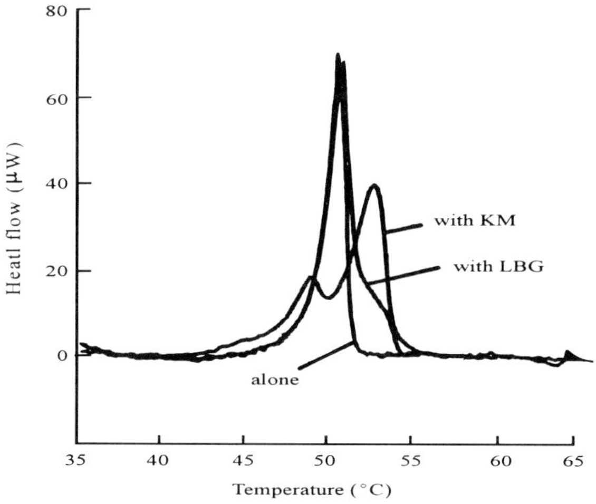

Figure 13.16 DSC cooling scans (0.3 deg/min; baselines subtracted) for K^+ κ-carrageenan (0.085% w/w in 100 mM KCl), alone, and in the presence of 0.036% w/w LBG or KM. (From Goycoolea *et al.* (1994), with permission.)

interaction with galactomannans does not involve phase separation. Studies by scanning electron microscopy (EM), using atomic mapping of sulphur and potassium to locate carrageenan and fluorescein-labelling to detect galactomannan, gave no indication of phase separation at 1 μm resolution (Cairns *et al.*, 1987), and a recent investigation (Brigham, Gidley, Hoffman and Smith, 1994) using impact-freezing ("slamming"), freeze-fracture and rotary shadowing to obtain replicas for scanning transmission EM, showed homogeneous dispersion to a resolution of 100 nm (which is about one tenth of the end-to-end length of a fully extended polysaccharide chain of typical molecular weight). It would therefore appear that direct binding of ß-1,4-glycan chains to ordered carrageenan offers the only explanation consistent with both EM and DSC.

Although the DSC traces shown in Figures 13.14 and 13.16 seem to indicate that the onset of the first exothermic process occurs at a temperature where κ-carrageenan alone would be fully disordered, van't Hoff analysis of the accompanying changes in optical rotation yields a helix fraction of ~0.25% at this point (Goycoolea *et al.*, 1994), corresponding to approximately one full turn of double helix structure per chain for carrageenan with a typical molecular weight of $\sim 5\times10^5$. The disorder—order transition may, of course, also be shifted to a slightly higher temperature by exclusion effects in mixed systems, with consequent further increase in helix fraction. A likely interpretation, therefore, is that glycan chains bind to helical sequences as they form, pulling over the coil–helix equilibrium until full stoichiometric attachment is achieved.

Adoption of the double helix structure does not, however, seem to be sufficient to allow carrageenans to form synergistic gels. In particular, the presence of ß-1,4-

glycans has little, if any, effect on the gelation of ι-carrageenan. One possible explanation (Dea and Morrison, 1975) is that the additional sulphate groups create a steric barrier to heterotypic association. However, it has also been reported (Stading and Hermansson, 1993) that LBG shows no evidence of interaction with κ-carrageenan when the cation content is confined to ions that do not bind to the double helix (Na^+; Ca^{2+}), suggesting that interaction with ß-1,4-glycans may require aggregation of carrageenan helices.

We have explored this possibility in our own recent work, using κ-carrageenan in the non-gelling tetramethylammonium salt form (Goycoolea, 1994). As shown in Figure 13.17, addition of LBG (0.43% w/v) to Me_4N^+ κ-carrageenan (1% w/v in 150 mM Me_4NI) had no discernable effect on the thermal changes accompanying the coil—helix transition, and the sample remained fluid at low temperatures where the carrageenan is fully ordered. Mechanical spectroscopy (Figure 13.18), however, showed a substantial increase in moduli, and development of some gel-like characteristics ($G' > G''$; linear decrease in log η^* with increasing log ω).

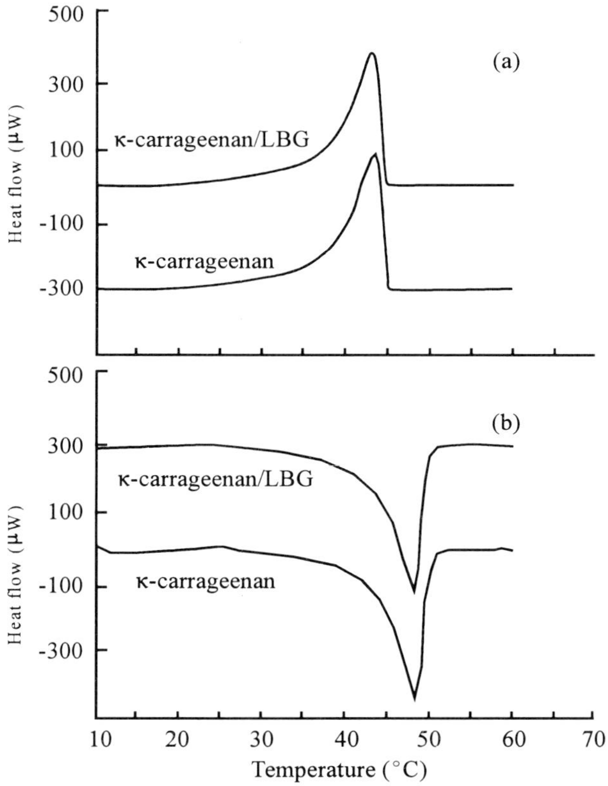

Figure 13.17 DSC scans for Me_4N^+ κ-carrageenan (1% w/w in 150 mM Me_4NI), alone, and in the presence of 0.43% LBG on (a) cooling and (b) heating at 0.5 deg/min. (From Goycoolea (1994).)

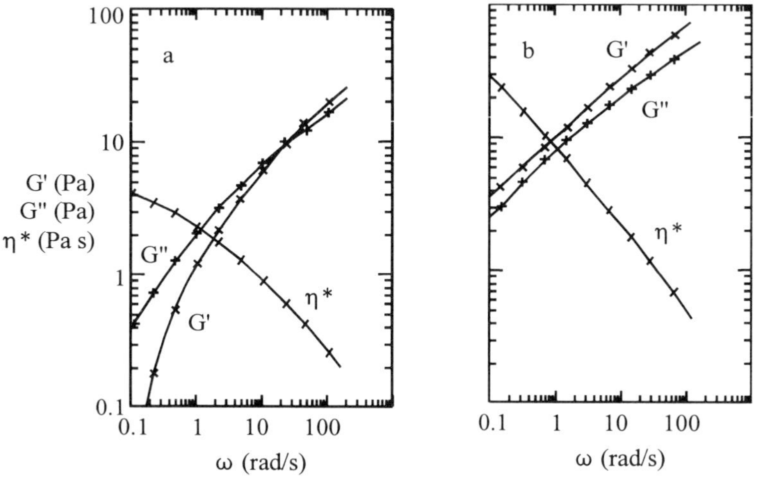

Figure 13.18 Mechanical spectra (8°C; 2% strain) for Me_4N^+ κ-carrageenan (1% w/w in 150 mM Me_4NI), (a) alone and (b) in the presence of 0.43% LBG. (From Goycoolea (1994).)

On mixing dilute solutions ($\eta_{sp} \approx 1$) of Me_4N^+ κ-carrageenan and LBG, (both equilibrated against 150 mM Me_4NI), a significant (~15%) reduction in viscosity was observed (in contrast to the large increases shown in Figure 13.4(b) for mixtures of xanthan and LBG). A similar reduction was observed for mixtures of LBG with ι-carrageenan in the presence of a high concentration of divalent cations (90 mM $CaCl_2$), but when only monovalent ions were present (100 mM KCl) the viscosity of the mixed solutions remained essentially constant at the common value for the individual stock solutions (Figure 13.19).

As found previously (Norton, Morris and Rees, 1984), the thermal hysteresis indicative of helix—helix aggregation was absent (or at least undetectable) for solutions of Me_4N^+ κ-carrageenan in Me_4NI, suggesting that the enhancements in rheology shown in Figure 13.18 might arise from binding of LBG to individual double helices. The reductions in dilute solution viscosity might similarly be explained by galactomannan chains binding to more than one helical sequence within microgel particles, and thus causing a decrease in hydrodynamic volume.

The observation of similar reductions for mixtures of LBG with ι-carrageenan in the presence of high concentrations of Ca^{2+} (Figure 13.19(b)) again indicates some form of binding interaction between the two polymers, and argues against the idea that sulphate groups block heterotypic association. The viscosities observed for mixed solutions containing a high concentration of K^+ (Figure 13.9(a)), however, show no evidence of interaction. The ι-carrageenan double helix, unlike that of the κ-form, does not appear to have any specific binding interactions with either monovalent or divalent cations (Piculell *et al.*, 1993). However, because of the

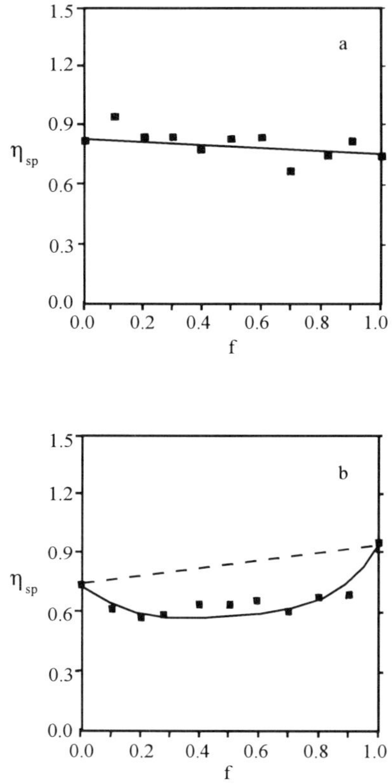

Figure 13.19 Specific viscosity (20°C; 50 s^{-1}) for dilute solutions of ι-carrageenan and LBG in (a) 100 mM KCl and (b) 90 mM $CaCl_2$; f denotes the LBG content expressed as a fraction of the total polymer concentration; the concentrations (% w/w) of the individual polymer solutions were: (a) 0.063 LBG; 0.063 carrageenan; (b) 0.069 LBG; 0.047 carrageenan. (From Goycoolea (1994).)

high charge density of the polymer, divalent cations are far more effective than monovalents in inducing helix formation by screening electrostatic repulsions between the chains. A possible explanation of the apparent interaction of LBG with ι-carrageenan in $CaCl_2$, but not in KCl, is that the enhanced screening by divalent ions is sufficient to allow some slight aggregation to occur, and that the galactomannan binds only to the aggregates, not to individual helices. It could similarly be argued that high concentrations of Me_4N^+ ions cause sufficient charge screening to induce limited aggregation of the less highly-charged κ-carrageenan helices, with binding of galactomannan again being confined to the aggregated assemblies.

These interpretations are at present entirely speculative, although they could be tested by, for example, using light scattering to screen for aggregation of carrageenan helices under different ionic conditions. There are, however, other indications that aggregation may be an important factor in the synergistic interactions of carrageenan/agar polysaccharides with ß-1,4-glycans. It has been found that galactomannans of progressively higher galactose content become effective in inducing gelation of otherwise non-gelling concentrations of carrageenans and agars as the tendency to self-association of the algal polysaccharide increases (Dea and Rees, 1987). For example guar gum, which does not form synergistic gels with κ-carrageenan, causes gelation of furcellaran, and neutral agarose can be gelled by galactomannans with an even lower content of unsubstituted mannan. However, in a recent investigation of the mixed gels formed by LBG or KM with κ-carrageen, furcellaran or an *Eucheuma gelatinae* carrageenan of even lower sulphate content (about half that of κ-carrageenan), at a total polysaccharide concentration of 1% w/v in 40 mM KCl, gel strength (yield stress) was found to decrease systematically with decreasing sulphate content (Sewall, 1992).

A possible rationalisation of these opposing trends is that the efficiency of binding increases as the size of the helix–helix aggregates increases, allowing progressively shorter sequences of unsubstituted mannan to interact with the algal polysaccharide. Such behaviour is indeed indicated by theoretical calculations of the extent of attachment of flexible chains to cylindrical rods, as a function of binding energy and radius (Piculell, Iliopoulos, Linse, Nilsson, Turquois, Viebke and Zhang, 1994). Conversely, however, the proportion of "wasted" polysaccharide in the interior of the aggregates will increase in direct proportion to aggregate size, and therefore reduce the maximum number of heterotypic associations that can be formed for a given concentration of algal polysaccharide. This may explain the observed reduction in synergistic gel strength for carrageenans of decreasing sulphate content, and therefore increasing ease of self-association.

How far have we advanced?

In the mid-1970s, synergism seemed straightforward. One type of polysaccharide formed rigid, ordered structures and another type stuck to them. Since then our understanding of the phenomenon has passed through a state of maximum entropy, and now appears to be converging towards another stable (or metastable?) position.

The picture for xanthan has changed radically. New evidence has established with virtual certainty that gelation of xanthan and related polysaccharides with ß-1,4-glycans does indeed occur by stoichiometric binding between the constituent polymers. However, despite some residual disagreements on reaction pathway, it now seems likely that the heterotypic junctions involve association of the glycan chain with the cellulosic backbone of the xanthan molecule, rather than with the surface of the five-fold helix as first suggested (Morris *et al.*, 1977).

For mixtures of ß-1,4-glycans with polysaccharides in the carrageenan/agar series, the model proposed by Dea, McKinnon and Rees (1972) still appears to offer the most simple and direct explanation of all available experimental evidence. Since the schematic illustration reproduced in Figure 13.3 has often been interpreted literally, it is perhaps worth quoting verbatim the qualifications with which it was accompanied in the original text.

'The conclusions are firmly established that (i) association exists between carrageenan double helices and sequences of unsubstituted mannose residues in the galactomannan and (ii) there is a broadly similar interaction involving the agarose tertiary structure. These associations are non-covalent and ordered but we know nothing about their geometry other than that which is implicit in the geometry of the carrageenan double helix itself and in the steric restrictions which constrain the mannan backbone to a ribbon-like shape. Neither do we know yet the "stoichiometry" of the association (the number of galactomannan-binding sites per helix, or helix-binding sites per galactomannan segment) or the length of helix involved in each association, or indeed the polymerisation state of the associated form (whether each helix can exist independently with its complement of galactomannan(s), or whether binding only occurs within some multiple of this unit). However, the conclusion is inescapable that some association must exist of the type shown schematically in Figure 13.3, even if it is only as part of more complex assemblies'.

This is probably a fair summary of our present state of understanding more than twenty years later.

References

Annable, P., Williams, P.A. and Nishinari, K. (1994) Interaction in xanthan–glucomannan mixtures and the influence of electrolyte. *Macromolecules*, **27**, 4202–4211

Atkins, E.D.T., Nieduszynski, I.A., Mackie, W., Parker, K.D. and Smolko, E.E. (1973) Structural components of alginic acid. II. The crystalline structure of poly-α-L-guluronic acid. Results of X-ray diffraction and polarized infrared studies. *Biopolymers*, **12**, 1879–1887

Brigham, J.E., Gidley, M.J., Hoffman, R.A. and Smith, C.G. (1994) Microscopic imaging of network strands in agar, carrageenan, locust bean gum and κ-carrageenan–locust bean gum gels. *Food Hydrocolloids*, **8**, 331–334

Brownsey, G.J., Cairns, P., Miles, M.J. and Morris, V.J. (1988) Evidence for intermolecular binding between xanthan and the glucomannan konjac mannan. *Carbohydrate Research*, **176**, 329–334

Bystrický, S., Maloviková, A. and Sticzay, T. (1990) Interaction of alginates and pectins with cationic polypeptides. *Carbohydrate Polymers*, **13**, 283–294

Cairns, P., Miles, M.J. and Morris, V.J. (1986) Intermolecular binding of xanthan gum and carob gum. *Nature*, **322**, 89–90

Cairns, P., Miles, M.J., Morris, V.J. and Brownsey, G.J. (1987) X-ray fibre diffraction studies of synergistic, binary polysaccharide gels. *Carbohydrate Research*, **160**, 411–423

Cheetham, N.W.H., McCleary, B.V., Teng, G., Lum, F. and Maryanto (1986) Gel permeation studies on xanthan–galactomannan interactions. *Carbohydrate Polymers*, **6**, 257–268

Cheetham, N.W.H. and Mashimba, E.N.M. (1991) Conformational aspects of xanthan–galactomannan gelation – Further evidence from optical rotation studies. *Carbohydrate Polymers*, **14**, 17–27

Dalbe, B. (1992) Interactions between xanthan gum and konjac mannan. In *Gums and Stabilisers for the Food Industry 6*, edited by Phillips, G.O., Williams, P.A. and Wedlock, D.J., Oxford: IRL Press, pp 201–208

Dea, I.C.M., McKinnon, A.A. and Rees, D.A. (1972) Tertiary and quaternary structure in aqueous polysaccharide systems which model cell wall adhesion: Reversible changes in conformation and association of agarose, carrageenan and galactomannans. *Journal of Molecular Biology*, **68**, 153–172

Dea, I.C.M. and Morrison, A. (1975) Chemistry and interactions of seed galactomannans. *Advances in Carbohydrate Chemistry and Biochemistry*, **31**, 241–312

Dea, I.C.M., Morris, E.R., Rees, D.A., Welsh, E.J., Barnes, H.A. and Price, J. (1977) Associations of like and unlike polysaccharides: Mechanism and specificity in galactomannans, interacting bacterial polysaccharides and related systems. *Carbohydrate Research*, **57**, 249–272

Dea, I.C.M., Clark, A.H. and McCleary, B.V. (1986) Effect of the molecular fine structure of galactomannans on their interaction properties – The role of unsubstituted sides. *Food Hydrocolloids*, **1**, 129–140

Dea, I.C.M. and Rees, D.A. (1987) Affinity interactions between agarose and ß-1,4-glycans – A model for polysaccharide associations in algal cell walls. *Carbohydrate Polymers*, **7**, 183–224

Fernandes, P.B., Gonçales, M.P. and Doublier, J.L. (1991) A rheological characterisation of κ-carrageenan–galactomannan mixed gels – A comparison of locust bean gum samples. *Carbohydrate Polymers*, **16**, 253–274

Flory, P.J. (1956) Phase equilibria in solutions of rod-like particles. *Proceedings of the Royal Society, London, Series A*, **234**, 73–89

Foster, T.J. (1992). *Conformation and Properties of Xanthan Variants*, Ph.D. thesis, Cranfield Institute of Technology, Silsoe College, Silsoe, Bedford, U.K.

Foster, T.J. and Morris, E.R. (1994) Xanthan polytetramer: conformational stability as a barrier to synergistic interaction. In *Gums and Stabilisers for the Food Industry 7*, edited by Phillips, G.O., Williams, P.A. and Wedlock, D.J., Oxford: IRL Press, pp 281–289

Garti, N. and Reichman, D. (1994) Surface properties and emulsification activity of galactomannans. *Food Hydrocolloids*, **8**, 155–173

Gelman, R.A. and Blackwell, J. (1973a) Interactions between mucopolysaccharides and cationic polypeptides in aqueous solution: Chondroitin 4-sulfate and dermatan sulfate. *Biopolymers*, **12**, 1959–1974

Gelman, R.A. and Blackwell, J. (1973b) Mucopolysaccharide–polypeptide interactions – Effect of the position of the sulfate group. *Biochimica et Biophysica Acta*, **297**, 452–455

Gelman, R.A. and Blackwell, J. (1973c) Heparin–polypeptide interactions in aqueous solution. *Archives of Biochemistry and Biophysics*, **159**, 427–433

Gelman, R.A., Glaser, D.N. and Blackwell, J. (1973a) Interaction between chondroitin-6-sulfate and poly-L-arginine in aqueous solution. *Biopolymers*, **12**, 1223–1232

Gelman, R.A., Rippon, W.B. and Blackwell, J. (1973b) Interactions between chondroitin-6-sulfate and poly-L-lysine in aqueous solution: Circular dichroism studies. *Biopolymers*, **12**, 541–558

Gidley, M.J. and Robinson, G. (1990) Techniques for studying interactions between polysaccharides. In *Methods in Plant Biochemistry*, Vol. 2, edited by Dey, P.M., London: Academic Press, pp 607–642

Gidley, M.J., Lillford, P.J., Rowlands, D.W., Lang, P., Dentini, M., Crescenzi, V., Edwards, M., Fanutti, C. and Reid, G. (1991) Structure and solution properties of tamarind-seed polysaccharide. *Carbohydrate Research*, **214**, 299–314

Goycoolea, F.M. (1994). *Rheology and Interactions of Plant Polysaccharides*, Ph.D. thesis, Cranfield University, Silsoe College, Silsoe, Bedford, U.K.

Goycoolea, F.M., Foster, T.J., Richardson, R.K., Morris, E.R. and Gidley, M.J. (1994) Synergistic gelation of galactomannans or konjac glucomannan: binding or exclusion? In *Gums and Stabilisers for the Food Industry 7*, edited by Phillips, G.O., Williams, P.A. and Wedlock, D.J., Oxford: IRL Press, pp 333–344

Grinrod, J. and Nickerson, T.A. (1968) Effect of various gums on skimmilk and purified milk proteins. *Journal of Dairy Science*, **51**, 834–841

Hansen, P.M.T. (1968) Stabilization of α_s-casein by carrageenan. *Journal of Dairy Science*, **51**, 192–195

Jacon, S.A., Rao, M.A., Cooley, H.J. and Walter, R.H. (1993) The isolation and characterisation of a water extract of konjac flour gum. *Carbohydrate Polymers*, **20**, 35–41

Jansson, P.-E., Kenne, L. and Lindberg, B. (1975) Structure of the extracellular polysaccharide from *Xanthomonas campestris*. *Carbohydrate Research*, **45**, 275–282

Kovacs, P. (1973) Useful incompatibility of xanthan gum with galactomannans. *Food Technology*, **27**, March 1973, 26–30

Lin, C.F. and Hansen, P.M.T. (1970). *Macromolecules*, **3**, 269–274

Lloyd, D.J. (1926). In *Colloid Chemistry*, Vol. 1, edited by Alexander, J., New York: Chemical Catalogue Company, pp 767–782

Lopes, L., Andrade, C.T., Milas, M. and Rinaudo, M. (1992) Role of conformation and acetylation of xanthan on xanthan–guar interaction. *Carbohydrate Polymers*, **17**, 121–126

McCleary, B.V. (1979) Enzymic hydrolysis, fine structure, and gelling interaction of legume-seed D-galacto-D-mannans. *Carbohydrate Research*, **71**, 205–230

Melton, L.D., Mindt, L., Rees, D.A. and Sanderson, G.R. (1976) Covalent structure of the extracellular polysaccharide from *Xanthomonas campestris*: Evidence from partial hydrolysis studies. *Carbohydrate Research*, **46**, 245–257

Millane, R.P. (1992). In *Frontiers in Carbohydrate Research 2*, edited by Chandrasekaran, R., New York: Elsevier, pp 168–190

Millane, R.P. and Wang, B. (1990) A cellulose-like conformation accessible to the xanthan backbone and implications for xanthan synergism. *Carbohydrate Polymers*, **13**, 57–68

Moorhouse, R., Walkinshaw, M.D. and Arnott, S. (1977). *American Chemical Society Symposium Series*, **45**, 90–100

Morris, E.R., Rees, D.A., Young, G., Walkinshaw, M.D. and Darke, A. (1977) Order–disorder transition for a bacterial polysaccharide in solution. A role for polysaccharide conformation in recognition between *Xanthomonas* pathogen and its plant host. *Journal of Molecular Biology*, **110**, 1–16

Morris, E.R., Rees, D.A., Thom, D. and Boyd, J. (1978) Chiroptical and stoichiometric evidence of a specific, primary dimerisation process in alginate gelation. *Carbohydrate Research*, **66**, 145–154

Morris, E.R., Rees, D.A., Robinson, G. and Young, G.A. (1980) Competitive inhibition of interchain interaction in polysaccharide systems. *Journal of Molecular Biology*, **138**, 363–374

Morris, E.R., Cutler, A.N., Ross-Murphy, S.B., Rees, D.A. and Price, J. (1981) Concentration and shear rate dependence of viscosity in random coil polysaccharide solutions. *Carbohydrate Polymers*, **1**, 5–21

Morris, E.R., Powell, D.A., Gidley, M.J. and Rees, D.A. (1982) Conformations and interactions of pectins. 1. Polymorphism between gel and solid states of calcium polygalacturonate. *Journal of Molecular Biology*, **155**, 507–516

Morris, E.R. and Norton, I.T. (1983). In *Aggregation Processes in Solution*, edited by Wyn-Jones, E. and Gormally, J., Amsterdam: Elsevier, pp 549–593

Morris, E.R. (1984) Rheology of hydrocolloids. In *Gums and Stabilisers for the Food Industry 2*, edited by Phillips, G.O., Wedlock, D.J. and Williams, P.A., Oxford: Pergamon Press, pp 57–78

Morris, E.R. and Foster, T.J. (1994) Role of conformation in synergistic interactions of xanthan. *Carbohydrate Polymers*, **23**, 133–135

Morris, V.J. (1992) Designing polysaccharides for synergistic interactions. In *Gums and Stabilisers for the Food Industry 6*, edited by Phillips, G.O., Williams, P.A. and Wedlock, D.J., Oxford: IRL Press, pp 161–171

Nishinari, K., Williams, P.A. and Phillips, G.O. (1992) Review of the physicochemical characteristics and properties of konjac mannan. *Food Hydrocolloids*, **6**, 199–222

Norton, I.T., Morris, E.R. and Rees, D.A. (1984) Lyotropic effects of simple anions on the conformation and interactions of κ-carrageenan. *Carbohydrate Research*, **134**, 89–101

Norton, I.T., Goodall, D.M., Frangou, S.A., Morris, E.R. and Rees, D.A. (1984) Mechanism and dynamics of conformational ordering in xanthan polysaccharide. *Journal of Molecular Biology*, **175**, 371–394

Ojinnaka, C., Morris, E.R., Morris, V.J. and Brownsey, G.J. (1994) Effect of acetyl substituents on the conformational stability and functional interactions of acetan polysaccharide. In *Gums and Stabilisers for the Food Industry 7*, edited by Phillips, G.O., Williams, P.A. and Wedlock, D.J., Oxford: IRL Press, pp 15–26

Okuyama, K., Arnott, S., Moorhouse, R., Walkinshaw, M.D., Atkins, E.D.T. and Wolf-Ullish, Ch. (1980). *American Chemical Society Symposium Series*, **141**, 411–427

Parker, A., Brigand, G., Miniou, C., Trespoey, A. and Vallée, P. (1993) Rheology and fracture of mixed ι-carrageenan and κ-carrageenan gels – two-step gelation. *Carbohydrate Polymers*, **20**, 253–262

Payens, T.A.J. (1972) Light scattering of protein reactivity of polysaccharides especially carrageenans. *Journal of Dairy Science*, **55**, 141–150

Perez, S. and Vergelati, C. (1987) Molecular modelling of the xanthan chain conformations. *International Journal of Biological Macromolecules*, **9**, 211–218

Piculell, L., Nilsson, S., Viebke, C. and Zhang, W. (1993) Gelation of (some) seaweed polysaccharides. In *Food Hydrocolloids*, edited by Nishinari, K. and Doi, E., New York: Plenum Press, pp 35–44

Piculell, L., Iliopoulos, I., Linse, P., Nilsson, S., Turqois, T., Viebke, C. and Zhang, W. (1994) Association and segregation in ternary polymer solutions and gels. In *Gums and Stabilisers for the Food Industry 7*, edited by Phillips, G.O., Williams, P.A. and Wedlock, D.J., Oxford: IRL Press, pp 309–322

Rees, D.A., Morris, E.R., Thom, D. and Madden, J.K. (1982). In *The Polysaccharides*, Vol. 1, edited by Aspinall, G.O., New York: Academic Press, pp 195–290

Richardson, R.K. and Ross-Murphy, S.B. (1987) Non-linear viscoelasticity of polysaccharide solutions. 2. Xanthan polysaccharide solutions. *International Journal of Biological Macromolecules*, **9**, 257–264

Robinson, G. (1989). *Functional Interactions of Industrial Polysaccharides*, Ph.D. thesis, Cranfield Institute of Technology, Silsoe College, Silsoe, Bedford, U.K.

Robinson, G., Manning, C.E. and Morris, E.R. (1991) Conformation and physical properties of the bacterial polysaccharides gellan, welan, and rhamsan. In *Food Polymers, Gels and Colloids*, edited by Dickinson, E., Cambridge: Royal Society of Chemistry, pp 22–33

Rochas, C. and Rinaudo, M. (1980) Activity coefficients of counterions and conformation in κ-carrageenan systems. *Biopolymers*, **19**, 1675–1687

Rochas, C., Taravel, F.R. and Turquois, T. (1990) NMR studies of synergistic κ-carrageenan–carob galactomannan gels. *International Journal of Biological Macromolecules*, **12**, 353–358

Ross-Murphy, S.B., Morris, E.R. and Morris, V.J. (1983) Molecular viscoelasticity of xanthan polysaccharide. *Faraday Symposia of the Chemical Society*, **18**, 115–129

Ross-Murphy, S.B. (1984) Rheological methods. In *Biophysical Methods in Food Research*, Society of Chemical Industry Critical Reports on Applied Chemistry, Vol. 5, edited by Chan, H.W.-S., Oxford: Blackwell, pp 138–199

Sewall, C.J. (1992) Gelling interactions of phycocolloids extracted from red algae with a galactomannan from locust bean gum and a glucomannan from konjac tuber. *Journal of Applied Phycology*, **4**, 347–351

Smith, I.H., Symes, K.C., Lawson, C.J. and Morris, E.R. (1981) Influence of the pyruvate content of xanthan on macromolecular association in solution. *International Journal of Biological Macromolecules*, **3**, 129–134

Snoeren, Th. H.M., Payens, T.A.J., Jeunink, J. and Both, P (1975). *Milchwissenschaft*, **30**, 393–396

Stading, M. and Hermansson, A.-M. (1993) Rheological behaviour of mixed gels of κ-carrageenan–locust bean gum. *Carbohydrate Polymers*, **22**, 49–56

Sundararajan, P.R. and Marchessault, R.H. (1972) Conformational terminology for crystalline cellulose. *Canadian Journal of Chemistry*, **50**, 792–794

Tako, M. and Nakamura, S. (1985) Synergistic interaction between xanthan and guar gum. *Carbohydrate Research*, **138**, 207–213

Tako, M. (1991) Synergistic interaction between xanthan and tara-bean gum. *Carbohydrate Polymers* **16**, 239–252

Thom, D., Dea, I.C.M., Morris, E.R. and Powell, D.A. (1982) Interchain associations of alginate and pectins. *Progress in Food and Nutritional Science*, **6**, 97–108

Toft, K. (1982) Interactions between pectins and alginates. *Progress in Food and Nutrition Science*, **6**, 89–96

Toft, K., Grasdalen, H. and Smidsrød, O. (1986) Synergistic gelation of alginates and pectins. *American Chemical Society Symposium Series*, **310**, 117–132

Turquois, T., Rochas, C. and Taravel, F.R. (1992) Rheological studies of synergistic κ-carrageenan–carob galactomannan gels. *Carbohydrate Polymers*, **17**, 263–268

Turquois, T., Taravel, F.R. and Rochas, C. (1993) Synergy of the agarose–carob galactomannan blend inferred from NMR and rheological studies. *Carbohydrate Research*, **238**, 27–38

Walkinshaw, M.D. and Arnott, S. (1982) Conformations and interactions of pectins. 2. Models for junction zones in pectinic acid and calcium pectate gels. *Journal of Molecular Biology*, **153**, 1075–1085

Williams, P.A., Day, D.H., Langdon, M.J., Phillips, G.O. and Nishinari, K. (1991a) Synergistic interaction of xanthan gum with glucomannans and galactomannans. *Food Hydrocolloids*, **4**, 489–493

Williams, P.A., Clegg, S.M., Day, D.H. and Phillips, G.O. (1991b) Mixed gels formed with konjac mannan and xanthan gum. In *Food Polymers, Gels and Colloids*, edited by Dickinson, E., Cambridge: Royal Society of Chemistry, pp 339–348

Williams, P.A., Clegg, S.M., Langdon, M.J., Nishinari, K. and Phillips, G.O. (1992) Studies on the synergistic interaction of konjac mannan and locust bean gum with κ-carrageenan. In *Gums and Stabilisers for the Food Industry 6*, edited by Phillips, G.O., Williams, P.A. and Wedlock, D.J., Oxford: IRL Press, pp 209–216

Williams, P.A., Clegg, S.M., Langdon, M.J., Nishinari, K. and Piculell, L. (1993) Investigation of the gelation mechanism in κ-carrageenan–konjac mannan mixtures using differential scanning calorimetry and electron spin resonance spectroscopy. *Macromolecules*, **26**, 5441–5446

Winter, W.T., Song, B.K. and Bouckris, H. (1987) Structural studies on galactomannans and their complexes. *Food Hydrocolloids*, **1**, 581–582

Zhan, D.F., Ridout, M.J., Brownsey, G.J. and Morris, V.J. (1993) Xanthan–locust bean gum interactions and gelation. *Carbohydrate Polymers*, **21**, 53–58

14

SYNERGISTIC INTERACTIONS WITH GALACTOMANNANS AND GLUCOMANNANS

V.J. MORRIS
Institute of Food Research, Norwich Laboratory, Norwich Research Park, Colney, Norwich. NR4 7UA U.K.

Introduction

Certain binary mixtures of polysaccharides exhibit synergistic effects. Such effects may be recognised as an enhanced viscosity in dilute solution, or as enhanced (or even unexpected) gelation at higher polysaccharide concentrations. Many of these mixtures, including those used as industrial gelling agents, consist of a mixture of a galactomannan or glucomannan with a second polysaccharide (Morris, 1986; Brownsey and Morris, 1988; Morris, E.R., 1990). Such mixtures can be divided into two classes of mixed gels depending upon the nature of the second polysaccharide component. For Class I gels the second polysaccharide is a gelling polysaccharide such as agarose, furcellaran or κ-carrageenan (Table 14.1). For Class II gels (Table 14.1) both polysaccharide components are non-gelling under the conditions for which the mixed systems do gel.

Table 14.1 BINARY SYNERGISTIC GELLING SYSTEMS

Class I	*Class II*
κ-carrageenan–carob	
κ-carrageenan–tara	alginate–pectin
furcellaran–carob	
furcellaran–tara	xanthan–carob
agarose–carob	xanthan–tara
agarose–tara	xanthan–konjac mannan
κ-carrageenan–konjac mannan	
furcellaran–konjac mannan	
agarose–konjac mannan	

If two polysaccharides are mixed together and the binary mixture gels then it is possible to define idealised structures for the mixed gels (Cairns, Miles, Morris and Brownsey, 1987). Although idealised, these simple pictures do divide mixed gels into discrete types for which the interrelationship between the properties of the mixed gel and those of its components can be defined. These simple pictures can also suggest experimental methods for investigating the types of gels formed by mixtures of polysaccharides. At least four types of gels can be described. These have been named swollen networks, interpenetrating networks, phase-separated networks and coupled networks (Cairns *et al.*, 1987) (Figure 14.1). In the case of swollen networks only one of the polysaccharide components forms a gel network. The second component residues within and swells the gel network. Interpenetrating networks are formed when each polysaccharide component gels independently, resulting in the formation of two independent networks spanning the entire sample and interpenetrating through each other. Phase-separated networks occur when the two components are incompatible. Gelation within each or one of the phases arrests phase separation at an early stage, giving rise to a composite or filled gel. The gel consists of "droplets" of one gelled phase acting as filler particles within the second continuous gelled phase. The most intriguing type of mixed gel is the coupled system. In this type of gel the two polysaccharides are considered to bind to each other forming a new type of heterotypic junction zone.

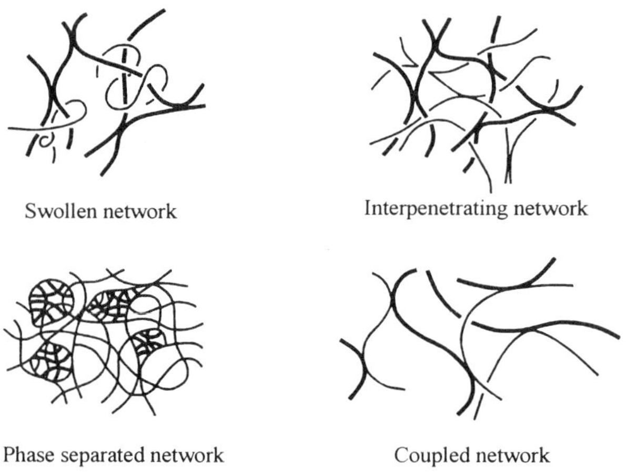

Figure 14.1 Idealised models for the types of gels formed by binary mixtures of polysaccharides: (a) swollen gel; (b) interpenetrating networks; (c) phase-separated gels; and (d) coupled networks

The early established models for synergistic gelation involving galactomannans or glucomannans assumed that synergism implied intermolecular binding and the formation of coupled gels. In these early models (Dea and Morrison, 1975; Dea and Morris, 1977) it was considered that the intermolecular binding occurred between unsubstituted, or partially substituted regions of the galactomannan backbone and the helix of the second polysaccharide (Figure 14.2). Likewise, gelation with glucomannans was considered to involve a similar type of interaction. Thus the model suggested a coupled network with heterotypic junction zones (Dea, 1987). Novel junction zones of this type should be observable by X-ray diffraction studies of oriented fibres prepared from these mixed gels. Such X-ray fibre diffraction studies (e.g. Cairns *et al.*, 1987) questioned the validity of these early models for intermolecular binding. This chapter reviews the results of those X-ray fibre diffraction studies and discusses subsequent experimental studies which have led to the current views on the gelation of these mixed systems. It is convenient to discuss the gelation of Class I and Class II gels separately.

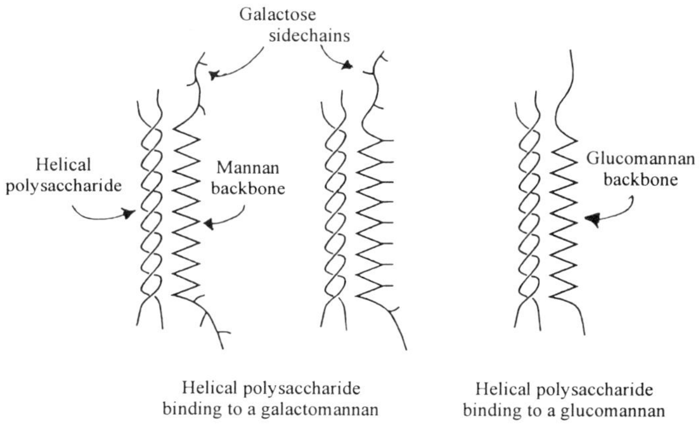

Figure 14.2 Early model proposed for the interaction between helix forming polysaccharides and either galactomannans or glucomannans

Class I gels

Class I gels consist of mixtures of gel-forming polysaccharides such as agarose, furcellaran or κ-carrageenan with other plant polysaccharides such as certain types of galactomannan or glucomannan (Table 14.1). The early model for the gelation of these mixed gels is shown in Figure 14.2. This model was based on observations of the self-association of galactomannans. It was observed that the level of self-association increased as the level of galactose substitution decreased. In addition it was noticed that galactomannans with similar levels of galactose substitution could show markedly different self-association behaviour and synergistic interactions with other polysaccharides (Dea and Morrison, 1975; Dea and Morris, 1977; Dea, Clark and McCleary, 1986a; Dea, Clark and McCleary, 1986b). Thus it was argued that unsubstituted faces of the mannan backbone bound to the helix of the other polysaccharide. Other polysaccharides with backbones exhibiting similar stereochemistry, such as glucomannans, were considered to bind in a similar manner.

X-ray diffraction studies of oriented fibres prepared from mixed gels of κ-carrageenan–tara, κ-carrageenan–carob, κ-carrageenan–konjac mannan, furcellaran–carob, furcellaran–konjac mannan (Carroll, Miles and Morris, 1984; Miles, Morris and Carroll, 1984; Cairns, Miles and Morris, 1986a; Cairns, Morris, Miles and Brownsey, 1986c; Cairns *et al.*, 1987; Cairns, Atkins, Miles and Morris, 1991) consistently gave patterns equivalent to those of the pure gelling component. In similar unpublished studies on furcellaran–tara, agarose–carob and agarose–tara gels it also proved to be impossible to detect the proposed intermolecular binding shown in Figure 14.2. These experiments were carried out at different ratios of the two polysaccharides, different cation forms for furcellaran and κ-carrageenan, and X-ray diffraction data was collected for fibres at a range of controlled relative humidity values. The failure to detect any new X-ray patterns suggested that the simple intermolecular model (Figure 14.2) could not be confirmed. It was noticed that the X-ray patterns were crystalline except at very low levels of added "gelling polysaccharides" (Cairns *et al.*, 1991). Thus it was suggested that a modified version of the model shown in Figure 14.2 could be consistent with the X-ray data; thus the gels could be considered to contain aggregates or crystallites of the helix-forming gelling polysaccharide to which surface attachment of galactomannans or glucomannans occurred. Such surface binding would need to be random with no preferential alignment of the adsorbed molecules and the attachment regions would need to be small or dynamic in character (Cairns *et al.*, 1987).

In the absence of any evidence for intermolecular binding and coupled network formation, additional experiments were carried out in order to try to investigate the spatial location and any possible orientation of each polysaccharide within these mixed gels (Cairns *et al.*, 1987). Based on the observation that oriented fibres of carob samples, which are initially poorly crystalline, will crystallise slowly when stored (Cairns, Miles and Morris, 1986b; Cairns *et al.*, 1987), attempts were made to follow carob crystallisation in aligned fibres prepared from κ-carrageenan–carob mixed gels (Cairns *et al.*, 1987). Crystallisation of the carob resulted in diffraction rings superimposed on the aligned fibre pattern. These results argued against aligned co-crystallisation of carob onto carrageenan junction zones and suggested random orientation of carob even within aligned fibres. In order to locate the individual polysaccharides within the mixed gel it is necessary to label the polysaccharides. Galactomannans can be labelled with fluorescein but such labelling of carrageenans interferes with gelation. The sulphur present in the sulphate residues of carrageenan can also be used as a label. Evidence for site-binding of certain cations within the junction zones of κ-carrageenan (Belton, Morris and Tanner, 1985, 1986) means that potassium can be used as a label for κ-carrageenan. Potassium and sulphur maps of mixed κ-carrageenan–tara gels were obtained by rapidly quenching gels to the temperature of liquid nitrogen, transferring the gels to a scanning electron microscope and recording the characteristic X-ray emission of each element resulting from electron bombardment upon scanning the gels. At a resolution of ~1 μm, potassium and sulphur mapping suggested uniform distribution of carrageenan within the gel. Fluorescence microscopy was used to examine κ-carrageenan mixed gels containing fluorescein-labelled tara. At a resolution of 1 μm the galactomannan appeared to be uniformly distributed within the gel. Recent studies cited by Goycoolea and co-workers (Goycoolea, Foster, Richardson, Morris and Gidley, 1994) using impact freezing, freeze fracture and rotary shadowing to produce replicas for scanning transmission electron microscopy suggest that the gels are homogeneous to a resolution of ~100 nm. In view of the lack of evidence for specific intermolecular binding and the lack of evidence for phase separation, the simplest model for these mixed gels was considered to be a swollen network (Cairns *et al.*, 1987). For such gels it should be possible to relate the rheological properties of the mixed gels to those of the gelling component. Limited rheological studies (Cairns, Morris, Miles and Brownsey, 1986d) suggested that data on fracture stress and modulus for the mixed gels could be normalised relative to the properties of a gel formed from the gelling polysaccharide alone at the same polymer concentration. The normalised curves differed for different galactomannans and could be used to predict behaviour of the mixtures when the composition of the gel was known.

Since these studies described above, there have been a number of articles published which seem to support the idea of binding between the two polysaccharides (Rochas, Taravel and Turquois, 1990; Turquois, Rochas and Taravel, 1992; Williams, Clegg, Langdon, Nishinari and Phillips, 1992; Williams, Clegg, Langdon, Nishinari and Piculell, 1993; Goycoolea *et al.*, 1994; Piculell, Iliopoulos, Linse, Nilsson, Turquois, Viebke and Zhang, 1994). The earliest reported experimental observations in favour of intermolecular binding were described by Dea and co-workers (Dea, McKinnon and Rees, 1972). Carrageenans contain irregularities in their chemical structure. The molecules can be chemically cleaved at these points, forming low molecular weight helix-forming, but non-gelling carrageenan segments. Dea and co-workers (Dea *et al.*, 1972) reported that carrageenan segments could be gelled by the addition of certain galactomannans, although there was no detailed experimental characterisation of such "interactions". If such gelation does occur then it is difficult to explain without invoking intermolecular binding. Studies of κ-carrageenan–carob gels (Rochas *et al.*, 1990; Turquois *et al.*, 1992) and agarose–carob gels (Turquois, Rochas and Taravel, 1993) by NMR and rheological studies find evidence for immobilisation of a portion of the galactomannan during gelation. This is taken to favour intermolecular binding although the possibility of galactomannan–galactomannan interactions cannot be excluded. Further evidence favouring intermolecular binding has come from DSC studies (Williams *et al.*, 1992; Williams *et al.*, 1993; Goycoolea *et al.*, 1994), DSC and ESR studies (Williams *et al.*, 1992; Williams *et al.*, 1993) and further NMR studies (Piculell *et al.*, 1994). In particular, factors which inhibit κ-carrageenan helix–helix association and gelation are found to inhibit synergistic interactions (Williams *et al.*, 1993; Piculell *et al.*, 1994). Thus, whereas potassium κ-carrageenan forms mixed gels with galactomannans or glucomannans, the sodium salt form does not form mixed gels (Williams *et al.*, 1993; Piculell *et al.*, 1994). These types of experiments may explain why κ-carrageenan forms mixed gels whereas ι-carrageenan, for which little if no helix—helix aggregation appears to occur, does not. Indeed, the ability to form mixed gels does seem to follow the degree of insolubility or tendency to aggregate: agarose > furcellaran > κ-carrageenan >> ι-carrageenan (Morris, E.R., 1990).

The model which appears to be becoming accepted is that the galactomannan or glucomannan adsorbs to the surface of aggregates or crystallite junction zones within the gelling polysaccharide network thus consolidating this network (Williams *et al.*, 1993; Goycoolea *et al.*, 1994; Piculell *et al.*, 1994). Such extra linkages would permit gelation of segmented carrageenan, permit gelation of high molecular weight material at lower polymer concentrations and increase the modulus of the gel. It is not clear why such interactions would modify the fracture stress of the gels. This model is a refinement of the earliest models proposed by Dea and co-workers (Dea *et al.*, 1972; Dea and Morrison, 1975; Dea and Morris, 1977) which was suggested (Cairns *et al.*, 1987) to reconcile the possibility of intermolecular binding with the failure to detect new heterotypic junction zones by X-ray diffraction studies. A pleasing feature of the present studies is that factors which enhance gelation of the gelling component, such as specific cation effects, or high ionic strength, should also enhance synergisms with galactomannan or glucomannans. Thus in real applications where κ-carrageenan is an acceptable gelling agent, galactomannans or glucomannans can be added in order to alter texture or reduce total polysaccharide concentrations needed for gelation. One difficulty with such a model is its reliance on a non-specific binding to crystallites or aggregates. If such a model is correct why are there not similar synergisms between galactomannans or glucomannans and polysaccharides such as gellan which are also believed to form junction zones composed of aggregated helices (Morris, 1991; Robinson, Manning and Morris, 1991)? In addition one might expect to be able to fractionate galactomannans on the basis of their adsorption to resin beads (e.g. sepharose) into fractions with different mannose:galactose ratios and/or substitution patterns. There may still be interesting aspects of the gelation of these mixed gels to be discovered!

Class II gels

Class II gels are formed from mixtures of two polysaccharides, each of which alone will not gel under similar experimental conditions (Table 14.1). Pectin–alginate gels are a good example of such gels and are discussed in detail elsewhere (Morris, E.R., 1990). Mixed gels of this type are also formed between the bacterial polysaccharide xanthan and certain galactomannans or certain glucomannans (Table 14.1). These sorts of gels involving galactomannans or glucomannans will be considered in more detail in this article.

It is useful to consider the chemical and stereochemical structures of xanthan, galactomannans and glucomannans. The primary structure of xanthan consists of a cellulosic backbone of ß 1–4 linked D-glucose solubilised by 03 substitution on alternate glucose residues with a charged trisaccharide sidechain (Figure 14.3(a)) (Jansson, Kenne and Lindberg, 1975; Melton, Mindt, Rees and Sanderson, 1976). Galactomannans consist of a mannan backbone of ß 1–4 linked D-mannose solubilised by irregular 06 substitution with α D-galactose (Figure 14.3(b)). The primary structure of glucomannans such as konjac mannan consist of a backbone containing an irregular distribution of both ß 1–4 D-glucose and ß 1–4 D-mannose (Figure 14.3(c)). The backbone is solubilized by irregular acetylation (Takahashi, Kusakabe, Kusama, Sakurai, Murakami, Maekawa and Suzuki, 1984). The only difference between mannose and glucose is the orientation of the hydroxyl group at the C2 position. Thus the backbones of the galactomannans, glucomannans and

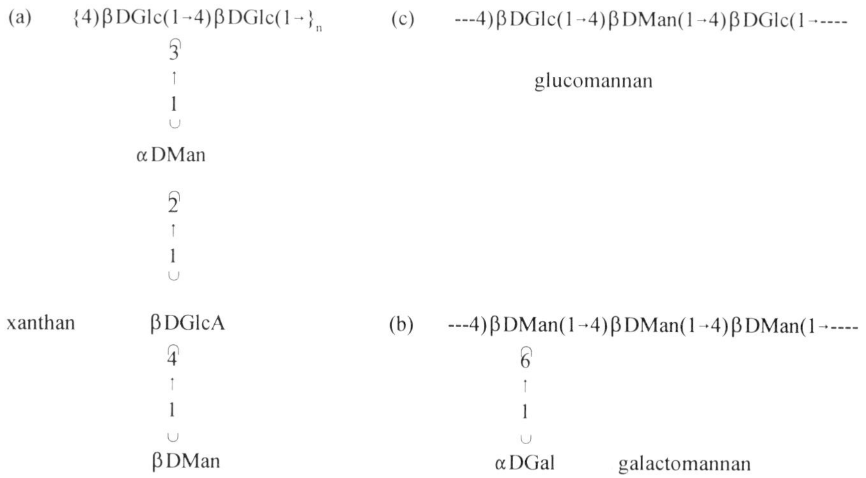

Figure 14.3 Chemical structures of (a) xanthan, (b) galactomannans and (c) glucomannans. The mannose-galactose ratio (MG) can be used to measure the level of galactose substitution

xanthan would alone form two-fold ribbon structures with similar shape and dimensions. Whereas acetylation in glucomannans, or galactose substitution in galactomannans does not alter the shape of the backbone, the substitution in xanthan alters the helical conformation. Addition of the sidechain causes the xanthan molecule to adopt a five-fold helical structure (Moorhouse, Walkinshaw and Arnott, 1977; Okuyama, Arnott, Moorhouse, Walkinshaw, Atkins and Wolf-Ullish, 1980).

Xanthan–galactomannan and xanthan–glucomannan mixtures form thermoreversible gels over a wide compositional range (Dea and Morrison, 1975; Dea, Morris, Rees, Welsh, Barnes and Price, 1977; Morris, Rees, Young, Walkinshaw and Darke, 1977). Dea and co-workers (Dea *et al.*, 1977; Morris, *et al.*, 1977; Dea, 1987) proposed a model for intermolecular binding in these mixed gels in which the glucomannan backbone, or an unsubstituted backbone or an unsubstituted face of the galactomannan backbone, bound to the xanthan helix. This is essentially the model illustrated in Figure 14.2, which was also proposed to account for binding in Class I gels. This model became well accepted and is widely quoted in review articles and text books. Once again X-ray diffraction studies of oriented fibres prepared from mixed gels provided a means of testing these models. It is convenient to discuss xanthan–glucomannan gels first and then to discuss xanthan–galactomannan gels.

XANTHAN–GLUCOMANNAN GELS

The discussion of xanthan–glucomannan gels will be concerned with only xanthan–konjac mannan gels. Mixtures of xanthan and konjac mannan form thermoreversible gels (Dea and Morrison, 1975). Native konjac mannan is acetylated and does not gel although gelation can be induced upon de-esterification under alkaline conditions (Maekaji, 1974). Xanthan does not form true gels (Ross-Murphy, Morris and Morris, 1983). The rheology of xanthan samples is complex and sensitive to preparation conditions, but is indicative of a weakly structured medium which breaks down under shear (weak gel) (Ross-Murphy *et al.*, 1983). Thus neither xanthan nor konjac mannan alone will gel, and the gelation of xanthan–konjac mannan mixtures is in itself suggestive of intermolecular binding.

On the basis of the type of model shown in Figure 14.2 factors that disrupt the xanthan helix should inhibit gelation. This is in fact what is observed for galactomannan or glucomannan mixed gels with algal polysaccharides (Class I gels discussed earlier). It is possible to manipulate the xanthan conformation by varying ionic strength and temperature (Lambert, Milas and Rinaudo, 1985) and monitor such conformational changes by optical rotation (Figure 14.4). Three regimes can be used for preparing xanthan–konjac mannan gels. Aqueous samples can be mixed at room temperature (25°C) with xanthan in the helical form. Aqueous samples of each polysaccharide can be mixed at 95°C with xanthan in the disordered form or, in the presence of 0.5 M $CaCl_2$, with xanthan in the helical form, and then these mixtures can be cooled to 25°C and allowed to set. Samples prepared under such conditions were stored at 25°C for about 12 hours and then compared. Comparisons were made using tests designed to emphasise the differences between the samples. Figure 14.5 shows the effect of inverting tubes containing the three samples (Brownsey, Cairns, Miles and Morris, 1988). Both samples for which xanthan was in the helical form when mixed with konjac mannan are pourable fluids, whereas the mixture prepared with xanthan in the disordered form formed a true gel upon cooling. This effect can also be demonstrated by creep studies (Figure 14.6). Once again the mixtures prepared with xanthan in the ordered state showed evidence of flow whereas the mixture prepared with xanthan in the disordered state formed a true gel on cooling. The observation that factors which destabilise the xanthan helix promote gelation is contrary to the model shown in Figure 14.2 and favours an interaction between the glucomannan and the stereochemically similar cellulosic backbone exposed upon denaturation of the xanthan helix.

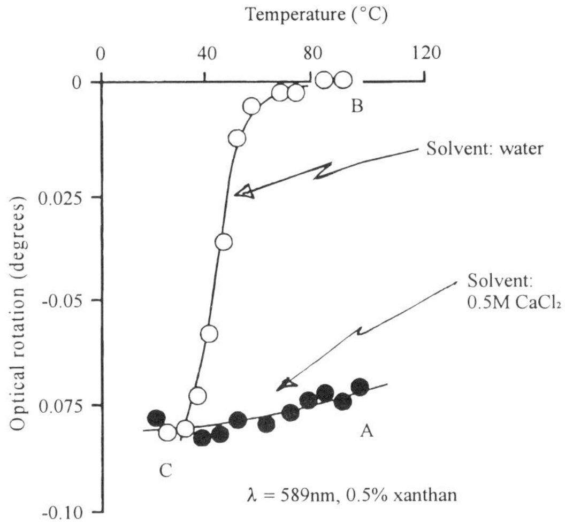

Figure 14.4 Optical rotation data illustrating the helix–coil transition of aqueous xanthan (0.5% w/v) solutions (o). The data was collected at a wavelength of 589 nm. The additional data (•) is for a xanthan solution (0.5% w/v) in 0.5 M $CaCl_2$

X-ray diffraction was used to probe the structure present in xanthan–konjac mannan gels. Oriented fibres can be prepared for konjac mannan samples, xanthan samples and xanthan–konjac mannan mixed gels. X-ray diffraction patterns were obtained for the pure components and the mixed gels (Brownsey *et al.*, 1988). The patterns obtained for the mixed gels correspond to a new structure and are not

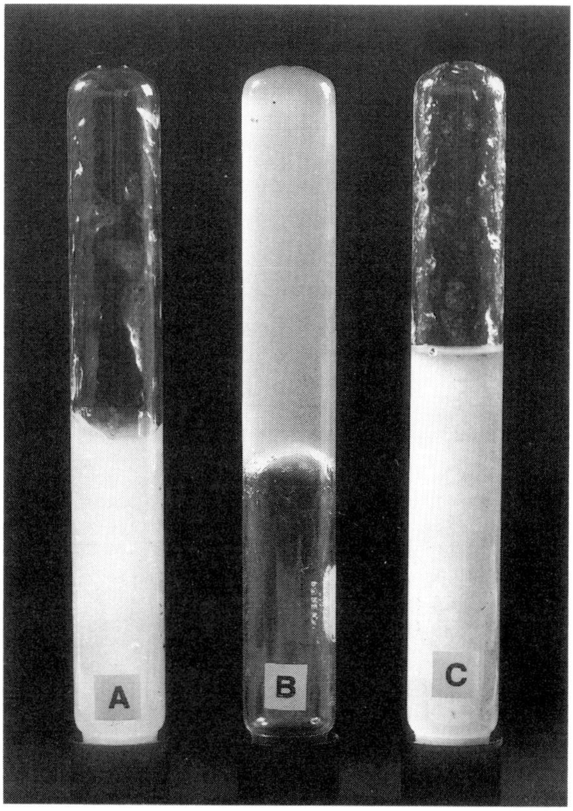

Figure 14.5 Effects of inverting tubes containing xanthan–konjac mannan mixtures prepared under different conditions. C: mixed at 25°C; and B: mixed at 95°C for aqueous xanthan and konjac mannan samples. A: mixture prepared in the presence of 0.5 M $CaCl_2$ at 95°C. 1:1 mixtures at a 1% total polymer concentration

simply superpositions of the patterns for the two components. The mixed gel patterns show a meridional spacing of 0.94 nm which is essentially the samc as for xanthan. However, the layer line spacing corresponds to a helical pitch of 5.6 nm with the first meridional reflection on the sixth layer line suggestive of a six-fold helix. It is tempting to suggest that the patterns result from a combination of a glucomannan backbone with the free face of the denatured xanthan backbone and that the duplex forms a novel heterotypic mixed helix. However, in view of the uncertainty concerning whether xanthan forms a single or double helix, and the difficulty with modelling heterotypic structures, particularly those involving complex and irregular structures such as konjac mannan, it is difficult to see how such a model could be tested in the near future.

The X-ray diffraction studies do suggest intermolecular binding and the mixing experiments seem to favour binding between the stereochemically similar structures

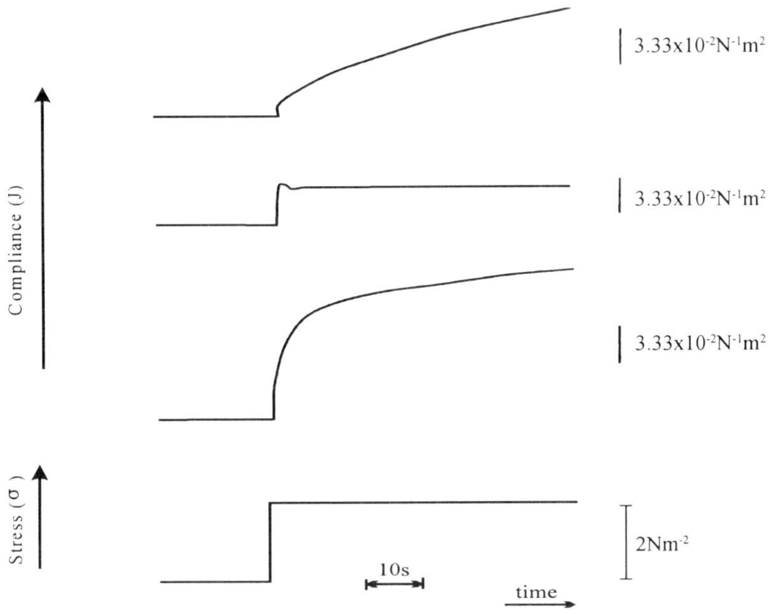

Figure 14.6 Creep compliance (J) studies of xanthan–konjac mixtures prepared under conditions defined for Figure 14.5. Applied stress (σ) was 2 N/m^2. Measurements made using an Instron 3250 mechanical spectrometer

of the glucomannan backbone and the denatured cellulosic face of the xanthan backbone. Molecular modelling studies (Millane and Wang, 1990) have shown that it is sterically feasible for xanthan to adopt a two-fold helical structure with a pitch close to that of the mannan and cellulose structures. Such interactions could provide a starting point for formation of the final heterotypic junction zones. There is certainly experimental evidence for an affinity between glucomannan structures and cellulose. There are reports of the growth of glucomannan crystals upon cellulose fibrils (Chanzy, Grosrenaud, Joseleau, Dabe and Marchessault, 1982) and of reprecipitation of glucomannans on wood pulp during the pulping of softwoods (Rydholm, 1965). There is also now considerable evidence favouring intermolecular binding between xanthan and konjac mannan. Foster and Morris (1994) have studied dilute mixtures of xanthan and konjac mannan. They observed substantial increases in viscosity attributed to intermolecular association. The mechanical spectra of such dilute mixtures show an obvious gel-like character (Goycoolea *et al.*, 1994). Studies using DSC and rheology show very large enthalpy changes on the melting and setting of the mixed gels, a direct dependence of the measured enthalpy change on mixing ratio at different overall concentrations, and a saturation of the enthalpy change and the storage modulus above ~1:1 stoichiometry. Using DSC and ESR experiments together with rheological studies, Williams and co-workers (Williams, Day, Langdon, Phillips and Nishinari, 1991a; Williams, Clegg, Day and Phillips, 1991b; Annable, Williams and Nishinari, 1994) have demonstrated xanthan–konjac mannan interaction and binding. A number of researchers have reported that deacetylation of xanthan enhances the formation of mixed gels with konjac mannan (Shatwell, Sutherland, Ross-Murphy and Dea, 1991a; Foster and Morris, 1994).

There does therefore appear to be strong evidence for intermolecular binding in xanthan–konjac mannan gels. There is still confusion and controversy regarding the nature and formation of the heterotypic junction zones. As a result of the mixing experiments carried out by Brownsey and co-workers (Brownsey *et al.*, 1988) these authors concluded that denaturation of the xanthan helix was a crucial step for gelation to occur. It has been proposed that the differences in interaction between cold mixed and hot mixed systems are due to disruption of the "cold-setting" gel on mixing, resulting in a weaker smashed gel network (Goycoolea *et al.*, 1994). Such an explanation could not account for the differences in the behaviour of hot mixed systems in the absence and presence of electrolyte (Figures 14.5 and 14.6). Several authors have reported that gelation can occur for mixtures prepared at temperatures below the helix–coil transition temperature for xanthan (Williams *et al.*, 1991a; Williams *et al.*, 1991b; Annable *et al.*, 1994; Goycoolea *et al.*, 1994). This has led to the suggestion that the interaction arises as a result of binding to the helical form or, depending upon the mixing conditions, binding to both the ordered and disordered forms of xanthan (Williams *et al.*, 1991a; Williams *et al.*, 1991b; Annable *et al.*, 1994; Goycoolea *et al.*, 1994). The inhibition of gelation upon addition of electrolyte is attributed to xanthan–xanthan aggregation. Improvements in gelation upon heating are then attributed to disruption of such aggregates rather than denaturation. Goycoolea and co-workers (Goycoolea *et al.*, 1994) have calculated the degree of denaturation of xanthan from analysis of studies on xanthan in salt solutions, and argue that in the presence of electrolyte the degree of denaturation is insignificant. Thus they argue that interaction occurs with the helix and the heterotypic structure then reverts to a new heterotypic junction zone.

It seems likely that the controversy and "array of models" is semantic rather than real. If one accepts that a new structure arises due to intermolecular binding then it is clear that factors weakening the xanthan helix will promote intermolecular binding. It would seem to be the stability of the helix which is important. Furthermore, it is the stability of the helix in the presence of the glucomannan which is important. A possible analogy is the action of protein denaturants on α-helices. In this case it is suggested that the glucomannan acts as a denaturant, displacing xanthan from its helical structure and leading to the formation of a more stable heterotypic structure.

XANTHAN–GALACTOMANNAN GELS

Gelation of xanthan–galactomannan mixtures is dependent upon the chemical structure of the galactomannan (Dea and Morrison, 1975). Galactomannans from different sources can be characterised by their mannose:galactose ratio (MG). Certain galactomannans such as guar are quite highly substituted with galactose (MG ≈ 1.6) whereas others have a lower levels of substitution (e.g. tara (MG ≈ 3.0), carob (MG ≈ 3.6)). The samples are generally mixtures of molecules with different MG ratios: thus carob can be fractionated into hot water soluble and cold water soluble fractions with the hot water fractions having lower galactose content. Mixtures of xanthan and guar do not gel. However, some form of interaction does occur because the mixtures are synergistic, showing an enhanced viscosity. Gelation must be sensitive to galactose distribution along the backbone as well as galactose content. Thus the galactomannan from *Leucaena leucocephala*, which has a similar MG ratio to guar, forms weak gels with xanthan at concentrations as low as 0.1% w/v, whereas xanthan–guar mixtures only show enhanced viscosity (McCleary, 1979; Dea *et al.*, 1986a). Similarly, *Sophora japonica* galactomannan has a similar MG ratio to hot water soluble carob but shows a noticeably weaker degree of self-association and of interaction and gelation with xanthan (Dea *et al.*, 1986a; Dea *et al.*, 1986b). Enzymic debranching of galactomannans can be used to enhance their gelling ability and such studies seem to suggest a range of MG ratios, subject to the detailed galactose distribution, within which gelation can be optimised (McCleary and Neukom, 1982; McCleary, Dea, Windust and Cooke, 1984). These types of observations resulted in the model for intermolecular binding and gelation shown in Figure 14.2.

As has been discussed previously for xanthan–glucomannan mixtures, it is possible to test whether helix denaturation alters gelation of xanthan–galactomannan mixtures (Cairns *et al.*, 1986b; Cairns *et al.*, 1987). The conformation of xanthan can be altered by varying temperature or ionic strength (Lambert *et al.*, 1985) and the changes monitored by optical rotation (Figure 14.4). Mixtures were prepared by cold mixing with xanthan in the ordered state (C), or by mixing hot (95°C) with xanthan in the disordered state (B) or the ordered conformation (A). Hot mixed samples were then cooled to room temperature. Figure 14.7 shows the effect of inverting tubes containing such samples of xanthan—carob mixtures. The sample (A) prepared with xanthan in the ordered state flows. The sample (B) prepared with xanthan in the disordered state is clearly an elastic gel. Experiments such as these were taken to suggest that experimental conditions favouring denaturation of the xanthan helix promoted gelation (Cairns *et al.*, 1986b; Cairns *et al.*, 1987; Brownsey and Morris, 1988).

Oriented fibres were prepared from mixed gels and studied by X-ray diffraction (Morris and Miles, 1986; Cairns *et al.*, 1986b; Cairns *et al.*, 1987). Such studies

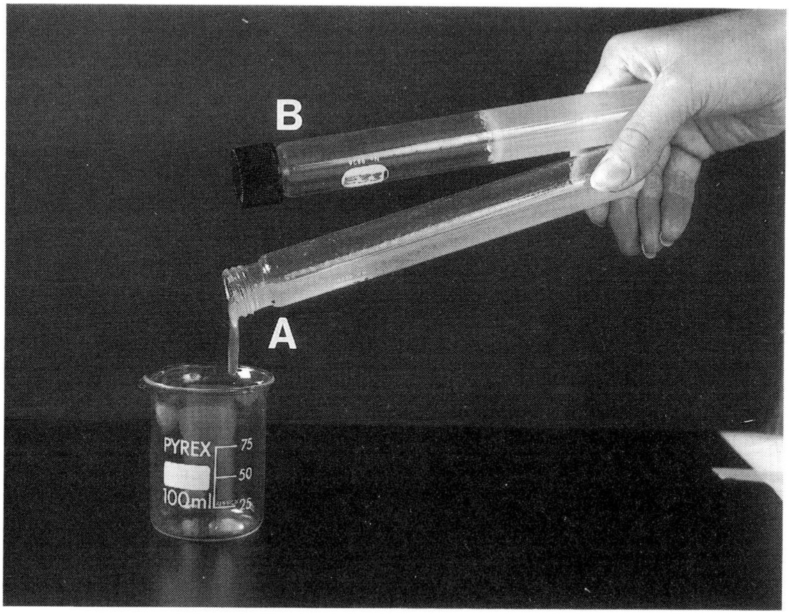

Figure 14.7 Effects of inverting tubes containing xanthan–carob mixtures prepared under different conditions. 1:1 xanthan–carob mixtures prepared at 95°C, total polymer concentration 1% w/v. The sample prepared by mixing with xanthan in the coil conformation (B in Figure 14.4) behaves as a gel. The sample prepared with xanthan in the helical conformation (A in Figure 14.4) flows

resulted in new X-ray patterns for mixed gels indicative of intermolecular binding. Examples of such data are shown in Figure 14.8. These new patterns provide a basis for a discussion of the nature of the junction zones in xanthan—galactomannan mixed gels. Analysis of the patterns has been discussed in detail elsewhere (Cairns *et al.*, 1986b; Cairns *et al.*, 1987). The X-ray data, together with the mixing experiments discussed above, were taken to suggest that the denatured xanthan backbone bound to the mannan backbone of the galactomannan. The meridional spacing of the X-ray patterns corresponds to an axial advance per repeat unit of 0.52 nm which is characteristic of cellulose (Blackwell, Gardener, Kolpack, Minke and Claffey, 1980) and mannan (Frei and Preston, 1968). Simple intermolecular binding schemes of the types shown in Figures 14.9(a), (b) and (c) are not allowable because they would give an axial advance per repeat unit of 1.04 nm. Schemes would need to be devised, such as the sandwich structure of Figure 14.9(d), in which, through staggering the sidechains of the xanthan chain, the observed repeat distance can be reduced to 0.52 nm. Models of this kind have been explored by Cheetham and co-workers (Cheetham and Mashimba, 1988; Cheetham and Mashimba, 1991). A variety of such models can be produced. However, it should be remembered that the X-ray patterns are crystalline; simple models of the type

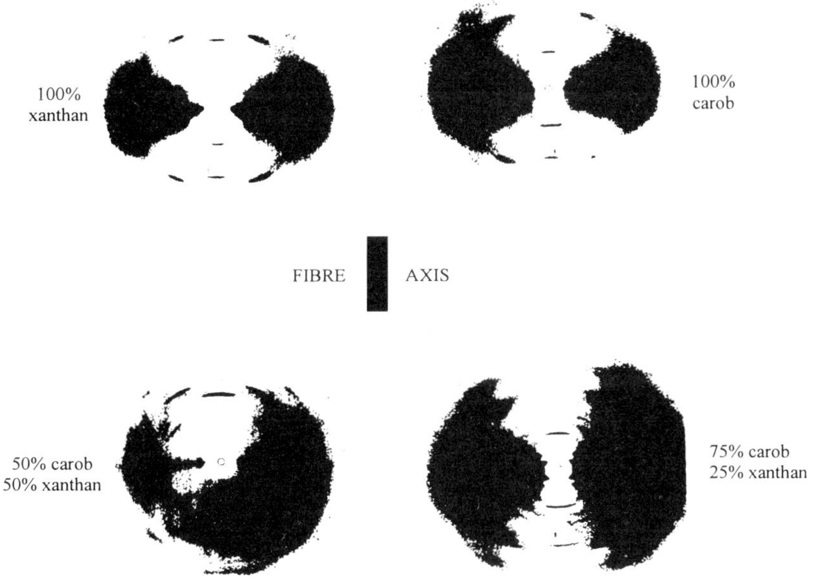

Figure 14.8 X-ray fibre diffraction patterns for xanthan, carob and xanthan—carob mixed gels

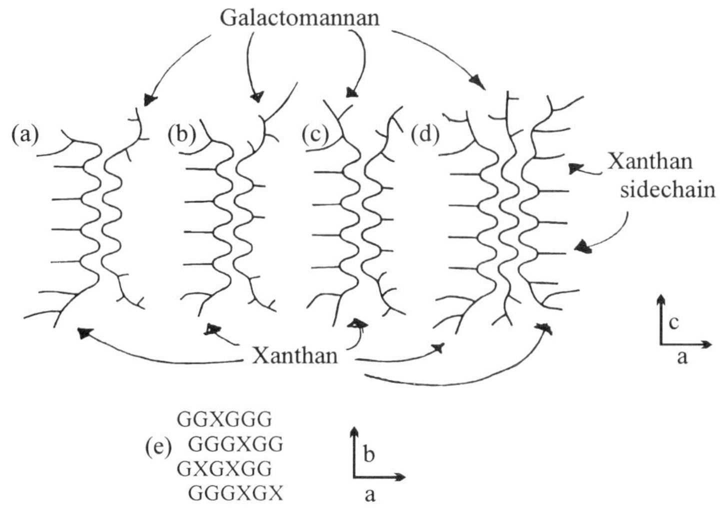

Figure 14.9 Possible types of xanthan–galactomannan binding. Binding of a xanthan backbone with (a) bare mannan segment of galactomannan, (b) irregularly galactose-substituted mannan backbone, and (c) alternately galactose-substituted mannan backbone; (d) Sandwich structure-bare mannan backbone sandwiched between two xanthan backbones; (e) random co-crystallisation of xanthan and galactomannan. Schematic diagram of a–b plane. The diagrams are highly schematic illustrating principles of suggested binding

shown in Figure 14.9(d) which are molecular structures would produce molecular transforms, not crystalline patterns. In fact the mixed gel patterns are equivalent to pure galactomannan patterns for which only Okl reflections are observed. These missing reflections are unlikely to be systematic absences related to highly specific molecular arrangements within the unit cell. They are more likely to be due to selective and extensive broadening of particular reflections which then appear to be absent. Broadening normally results from a decrease in crystal size or an increase in crystal disorder. Galactomannans crystallise in a modified mannan crystal lattice (Winter, Chien and Bouckris, 1984). The mannan backbones lie in the b—c plane with the side chains poking into the interplanar spacing. The interplanar spacing (a) depends on the level of galactose substitution and the relative humidity. It is reflections corresponding to this interplanar spacing that are absent in the mixed gel patterns. A simple explanation would be to consider that the galactomannans crystallise but that growth only occurs normally in the "b" and "c" directions. Surface attachment of xanthan molecules via the unsubstituted cellulosic face of the xanthan backbone could poison growth in the "a" direction (Cairns *et al.*, 1987). Such a micellar model with the hydrophilic xanthan side chains immersed in the aqueous environment, with cellulosic and mannan segments buried in the interior of the sandwich, is attractive. However, the laminate structure should lead to a streaking of layer lines. This was not observed experimentally. An alternative is to consider that the "a" dimension is aperiodic (Cairns *et al.*, 1987). The simplest model (Cairns *et al.*, 1987; Morris, V.J., 1990) is to consider a co-crystallisation of the xanthan and galactomannan in which denatured segments of the xanthan chain displace galactomannan chains within the crystal lattice (Figure 14.9(c)). The trisaccharide sidechains of the xanthan are larger than the galactose sidechains of the galactomannans and may be considered to act as defects in the "a" dimension, leading to aperiodicity. Such a model could also account for the MG range within which gelation occurs. If the galactose substitution is too low then large insoluble blocks of the galactomannan chains will crystallise into a mannan structure. If the galactose substitution is too high then there will be insufficient space, due to the presence of galactose sidechains within the lattice, to accommodate the more bulky xanthan sidechains. Thus co-crystallisation would account for a dependence on MG ratio and the galactose substitution pattern. In the case of highly substituted galactomannans such as guar, intermolecular binding may occur, resulting in aggregation and enhanced viscosity, but the structures may not be able to grow large enough to become permanent cross-links. Thus mixed gel formation would be envisaged as a conversion of the weak gel network of xanthan into a permanent structure through formation of permanent junction zones due to co-crystallisation. One test of this model is to determine whether the denatured xanthan backbone can adopt a two-fold helical structure which is compatible with co-crystallisation. Molecular modelling studies (Millane and Wang, 1990) have shown that such a structure is feasible. It is clearly not possible at the present time to build molecular models of such complex gels in order to test this mixed gel model. However, it may be possible to model such systems by introducing disorder into a simpler mannan lattice, or by considering the effects of restricting growth to a laminate structure, and investigating whether the qualitative changes in the diffraction patterns (i.e. selective broadening of certain reflections) would be consistent with the above models.

There is now a considerable body of evidence, obtained by a variety of techniques, which supports intermolecular binding. In addition to the evidence discussed above this includes enzyme inhibition (McCleary and Neukom, 1982), suggested binding of galactomannans to cellulose (Marchessault, Buléon, Deslandes and Goto, 1979), optical rotation and gel permeation chromatography (Cheetham, McCleary, Teng, Lum and Maryanto, 1986; Cheetham, 1987; Cheetham and Mashimba, 1988; Cheetham and Mashimba, 1991; Cheetham and Punruckrong, 1989), rheology (Tako, Asato and Nakamura, 1984; Shatwell, Sutherland, Ross-Murphy and Dea, 1991b; Foster and Morris, 1994; Goycoolea *et al.*, 1994) and DSC studies (Foster and Morris, 1994; Goycoolea *et al.*, 1994). In particular there is strong evidence to suggest stoichiometric binding between xanthan and galactomannans. Where experiments have been carried out which can discriminate between the effects of mixing galactomannans with xanthan in the disordered or ordered forms they have tended to favour enhanced binding under conditions that destabilise the xanthan helix (McCleary and Neukom, 1982; Cheetham *et al.*, 1986; Cairns *et al.*, 1986b; Cairns *et al.*, 1987; Cheetham, 1987; Cheetham and Mashimba, 1988; Cheeham and Mashimba, 1991; Cheetham and Punruckrong, 1989; Zhan, Ridout, Brownsey and Morris, 1993). Recently, a number of papers have shown that gelation can occur if the two polysaccharides are mixed at temperatures below the xanthan helix–coil transition (Williams *et al.*, 1991a; Mannion, Melia, Launay, Cuvelier, Hill, Harding and Mitchell, 1992; Foster and Morris, 1994; Goycoolea *et al.*, 1994). These studies have been taken to suggest that an interaction (binding) can occur to either the xanthan helix or the disordered xanthan backbone, depending upon the experimental conditions of mixing (Williams *et al.*, 1991a; Mannion *et al.*, 1992; Goycoolea *et al.*, 1994). Recently Zhan and co-workers (Zhan *et al.*, 1993) have looked at the effect of varying the mixing temperature upon gelation of the mixtures. It was shown that the melting temperatures (T_m) of the mixed gels are independent of ionic strength. The helix—coil transition of xanthan varies with ionic strength. Hence mixing could be carried out at temperatures above both the xanthan helix–coil transition temperature (T_c) and T_m, between T_c and T_m, or below both T_c and T_m. The results of such studies were taken to show that the modulus of the mixed gels increased with increasing disordering of the xanthan helix. Such experiments have been taken to imply that formation of the new heterotypic structure results from the galactomannan acting rather like a denaturant and perturbing the helix–coil equilibrium, favouring disruption of the helix (Zhan *et al.*, 1993; Morris,

Brownsey and Ridout, 1994). Conditions which enhance denaturation of the xanthan thus favour intermolecular binding.

At the present time, therefore, there appears to be a spectrum of models describing mixed gel formation. Tako and co-workers (Tako *et al.*, 1984) suggest that binding occurs to xanthan in the helical form and involves binding to the xanthan sidechains. The stoichiometry of the proposed interaction is unclear and the model does not seem to be compatible with the X-ray data. At the other extreme (Doublier, Castelain and Lefebvre, 1993), it has been suggested that intermolecular binding is not necessary for gelation. It is, however, this author's view that the evidence for intermolecular binding is very strong and that heterotypic junction zones are formed in these systems.

There are two alternative explanations for the mixing experiments. One view is that the differences observed by Cairns and co-workers (Cairns *et al.*, 1986b; Cairns *et al.*, 1987) between hot and cold mixed samples are due to disruption of the gel network upon mixing for the cold mixed samples (Morris, E.R., 1990; Morris and Foster, 1994; Goycoolea *et al.*, 1994). The effects of increased ionic strength are attributed to helix–helix association hindering binding of galactomannan to the helix (Morris, E.R., 1990; Morris and Foster, 1994). The alternative argument is that it is the degree of denaturation of the xanthan helix which is important (Morris *et al.*, 1994), in particular the degree of denaturation of the helix in the presence of the galactomannan. In practice what seems to be important is that the heterotypic interaction is favoured and leads to disruption of the xanthan helix. Factors stabilising the helix thus inhibit intermolecular binding and gelation (Goycoolea *et al.*, 1994; Morris *et al.*, 1994).

The effects of helix stability upon mixed gel formation can be assessed by studying the effects of variations of the xanthan chemical structure upon mixed gel formation.

XANTHAN-LIKE POLYSACCHARIDES

It is possible to produce a range of bacterial polysaccharides with chemical structures similar to xanthan (Figure 14.10). The biosynthetic pathway for the production of xanthan is known and the possibility for genetically controlling the carbohydrate structure and the substitution pattern of xanthan has been demonstrated (Betlach, Capage, Doherty, Hassler, Henderson, Vanderslice, Marrelli and Ward, 1987). Of particular interest are polymers produced by removal of sidechain sugars or removal of acetate substituents. Another family of xanthan-like polysaccharides can be produced based on the polysaccharide acetan. Acetan, which is secreted by the bacterium *Acetobacter xylinum*, is an anionic heteropolysaccharide (Figure 14.10) composed of a cellulosic backbone solubilised by attachment of a charged pentasaccharide sidechain to alternate glucose residues (Jansson, Lindberg,

{4)βDGlc(1→4)βDGlc(1→}n   backbone

∩
3
↑
R
∪

(a) R ≡ βDMan(1→4)βDGlcA(1→2)αDMan(1→
(b) R ≡ βDGlcA(1→2)αDMan(1→
(c) R ≡ αDMan(1→
(d) R ≡ αLRha(1→6)βDGlc(1→6)αDGlc(1→4)βDGLcA(1→2)αDMan(1→
(e) R ≡ βDGLcA(1→2)αDMan(1→

Figure 14.10 Chemical structures of: (a) xanthan; (b) xanthan tetramer; and (c) xanthan trimer. The xanthan molecules are partially acetylated at 06 of the (1, 2) Man residue. Chemical structures of: (d) acetan; and (e) CR1/4. Acetan and CR1/4 are partially acetylated at 06 of the (1, 34) Glc and the (1,2) Man residues

Wilmalasiri and Dankert, 1993). Chemical mutagenesis of cellulose minus acetan positive *A. xylinum* strains has produced the strain CR1/4, which produces a polysaccharide (CR1/4) with a sidechain terminating in glucuronic acid (Figure 14.10) (MacCormick, Harris, Gunning and Morris, 1993). Both acetan and CR1/4 have been shown to adopt the five-fold helical structure characteristic of xanthan, and to show evidence for a helix–coil transition in solution (Morris, Brownsey, Cairns, Chilvers and Miles, 1989; Ojinnaka, Morris, Morris and Brownsey, 1994; Ridout, Brownsey, Morris and Cairns, 1994).

Deacetylation of xanthan destabilises the xanthan helix (Foster and Morris, 1994). Studies of the interaction of xanthan with either carob or konjac mannan, made by measuring the viscosity of dilute solutions, suggest that deacetylation favours intermolecular binding and gelation (Foster and Morris, 1994). The helical structure formed by the xanthan tetramer is more stable than the xanthan helix, and mixtures of xanthan tetramer with either carob or konjac mannan show no evidence for intermolecular binding as detected by dilute solution viscosity measurements (Foster and Morris, 1994). These studies appear to confirm the belief that the stability of the xanthan helix is an important factor determining intermolecular association.

Preliminary studies on mixtures of acetan with carob or konjac mannan can also be used to test this view (Ojinnaka *et al.*, 1994). As judged from dilute solution viscosity measurements, there is no significant detectable interaction between acetan and either carob or konjac mannan. However, after deacetylation an interaction can be detected with both carob and konjac mannan (Ojinnaka *et al.*, 1994). In this case deacetylation stabilises the helix (Ojinnaka *et al.*, 1994) and might not be expected to promote gelation. This discrepancy can still be explained away because acetan has additional acetylation on the branched backbone residue (Colquhoun, Defernez and Morris, 1995). It is known that acetylation of konjac mannan inhibits intermolecular binding and crystallisation on steric grounds (Millane, Hendrixson, Morris and Cairns, 1992). Steric factors could therefore account for both destabilisation of the helix and inhibition of gelation. In this case deacetylation may then favour gelation. Deacetylated acetan has been shown to form gels with both carob and konjac mannan at higher polysaccharide concentrations (Ojinnaka *et al.*, 1994). Thus the unusual behaviour of acetan can be attributed to the acetylation of the backbone. A troubling observation is the fact that, as judged by dilute solution viscosity measurements, the acetylated acetan tetramer (CR1/4) interacts with carob but not with konjac mannan (Ridout *et al.*, 1994), whereas the xanthan tetramer (differing only in the absence of one acetylation site on the backbone) does not show any interaction with carob or konjac mannan (Foster and Morris, 1994). Perhaps there is still much to learn about these systems! However, it is necessary to be wary of overinterpreting dilute solution viscosity data. Although such systems should go to equilibrium it is possible that the mode of association may differ in dilute solution, or be such that the aggregates do not differ significantly enough in shape or size from the individual molecules to cause large changes in viscosity. Positive changes in dilute solution probably mean gelation at higher concentration, but negative results may not preclude gelation.

Conclusions

For Class I gels current opinion favours an interaction involving adsorption of the non-gelling polysaccharide to aggregates or crystallites of the gelling component. This non-specific interaction enhances gelation or aids gelation at lower concentrations. Factors which enhance helix formation and gelation of the gelling component enhance mixed gel formation. For Class II gels there is now very strong evidence for intermolecular binding in the xanthan–galactomannan and xanthan–glucomannan systems. There is still controversy regarding the mechanisms of gelation. However, the emerging view is that a new heterotypic interaction is involved and factors stabilising the xanthan helix inhibit intermolecular binding and gelation.

References

Annable, P., Williams, P.A. and Nishinari, K. (1994) Interaction in xanthan–glucomannan mixtures and the influence of electrolyte. *Macromolecules*, **27**, 4202–4211

Belton, P.S., Morris, V.J. and Tanner, S.F. (1985) Interaction of group I cations with ι, κ and λ-carrageenans studied by multinuclear NMR. *International Journal of Biological Macromolecules*, **7**, 53–56

Belton, P.S., Morris, V.J. and Tanner, S.F. (1986) ^{39}K and ^{87}Rb NMR study of chemical shift effects in ι-carrageenan and κ-carrageenan gels. *Macromolecules*, **19**, 1618–1621

Betlach, M.R., Capage, M.A., Doherty, D.H., Hassler, R.A., Henderson, N.M., Vanderslice, R.W., Marrelli, J.D. and Ward, M.B. (1987) Genetically engineered polymers – Manipulation of xanthan biosynthesis. In *Industrial Polysaccharides, Genetic Engineering, Structure/Property Relations and Applications*, edited by Yalpini, M., London: Elsevier, pp 35–50

Blackwell, J., Gardener, K.H., Kolpack, F.-J., Minke, R. and Claffey, W.B. (1980). *American Chemical Society Symposium Series*, **141**, 315–334

Brownsey, G.J. and Morris, V.J. (1988) Mixed and filled gels – Models for foods. In *Food Structure - Its Creation and Evaluation*, edited by Blanshard, J.M.V. and Mitchell, J.R., London: Butterworths, pp 7–23

Brownsey, G.J., Cairns, P., Miles, M.J. and Morris, V.J. (1988) Evidence for intermolecular binding between xanthan and the glucomannan konjac mannan. *Carbohydrate Research*, **176**, 329–334

Cairns, P., Miles, M.J. and Morris, V.J. (1986a) X-ray diffraction studies of κ-carrageenan tara gum mixed gels. *International Journal of Biological Macromolecules*, **8**, 124–127

Cairns, P., Miles, M.J. and Morris, V.J. (1986b) Intermolecular binding of xanthan gum and carob gum. *Nature*, **322**, 89–90

Cairns, P., Morris, V.J., Miles, M.J. and Brownsey, G.J. (1986c) Synergistic behaviour in κ-carrageenan–tara gum mixed gels. In *Gums and Stabilisers for the Food Industry 3*, edited by Phillips, G.O., Wedlock, D.J. and Williams, P.A., London: Elsevier Applied Science, pp 597–604

Cairns, P., Morris, V.J., Miles, M.J. and Brownsey, G.J. (1986d) Comparative studies of the mechanical properties of mixed gels formed by κ-carrageenan and tara gum or carob gum. *Food Hydrocolloids*, **1**, 89–93

Cairns, P., Miles, M.J., Morris, V.J. and Brownsey, G.J. (1987) X-ray fibre diffraction studies of synergistic, binary polysaccharide gels. *Carbohydrate Research*, **160**, 411–423

Cairns, P., Atkins, E.D.T., Miles, M.J. and Morris, V.J. (1991) Molecular transforms of κ-carrageenan and furcellaran from mixed gel systems. *International Journal of Biological Macromolecules*, **13**, 65–68

Carroll, V., Miles, M.J. and Morris, V.J. (1984) Synergistic interactions between κ-carrageenan and locust bean gum. In *Gums and Stabilisers for the Food Industry 2*, edited by Phillips, G.O., Wedlock, D.J. and Williams, P.A., Oxford: Pergamon, pp 501–506

Chanzy, H.D., Grosrenaud, A., Joseleau, J.P., Dabe, M. and Marchessault, R.H. (1982) Crystallisation behaviour of glucomannan. *Biopolymers*, **21**, 301–309

Cheetham, N.W.H., McCleary, B.V., Teng, G., Lum, F. and Maryanto (1986) Gel permeation studies on xanthan–galactomannan interactions. *Carbohydrate Polymers*, **6**, 257–268

Cheetham, N.W.H. (1987). In *Industrial Polysaccharides*, edited by Stivala, S.S., Crescenzi, V. and Dea, I.C.M., London: Gordon and Breach, pp 325–332

Cheetham, N.W.H. and Mashimba, E.N.M. (1988) Conformational aspects of xanthan–galactomannan gelation. *Carbohydrate Polymers*, **9**, 195–212

Cheetham, N.W.H. and Punruckrong, A. (1989) Gel permeation and optical rotational studies on xanthan–galactomannan interactions. *Carbohydrate Polymers*, **10**, 129–141

Cheetham, N.W.H. and Mashimba, E.N.M. (1991) Conformational aspects of xanthan galactomannan gelation – Further evidence from optical rotation studies. *Carbohydrate Polymers*, **14**, 17–27

Colquhoun, I.J., Defernez, M. and Morris, V.J. (1995) NMR studies of acetan and the related bacterial polysaccharide, CR1/4, produced by a mutant strain of *acetobacter xylinum*. *Carbohydrate Research*, **269**, 319-331

Dea, I.C.M., McKinnon, A.A. and Rees, D.A. (1972) Tertiary and quaternary structure in aqueous polysaccharides systems which model cell wall adhesion: Reversible changes in conformation and association of agarose, carrageenan and galctomannans. *Journal of Molecular Biology*, **68**, 153–172

Dea, I.C.M. and Morrison, A. (1975) Chemistry and interactions of seed galactomannans. *Advances in Carbohydrate Chemistry and Biochemistry*, **31**, 241–312

Dea, I.C.M. and Morris, E.R. (1977). Synergistic xanthan gels. *American Chemical Society Symposium Series*, **45**, 174–182

Dea, I.C.M., Morris, E.R., Rees, D.A., Welsh, E.J., Barnes, H.A. and Price, J. (1977) Associations of like and unlike polysaccharides: Mechanism and specificity in galactomannans, interacting bacterial polysaccharides and related systems. *Carbohydrate Research*, **57**, 249–272

Dea, I.C.M., Clark, A.H. and McCleary, B.V. (1986a) Effect of galactose substitution patterns on the interaction properties of galactomannans. *Carbohydrate Research*, **147**, 275–294

Dea, I.C.M., Clark, A.H. and McCleary, B.V. (1986b) Effect of the molecular fine structure of galactomannans on their interaction properties – The role of unsubstituted sides. *Food Hydrocolloids*, **1**, 129–140

Dea, I.C.M. (1987). Mixed polysaccharide systems. In *Industrial Polysaccharides*, edited by Stivala, S.S., Crescenzi, V. and Dea, I.C.M., London: Gordon and Breach, pp 367–285

Doublier, J.L., Castelain, C. and Lefebvre, J. (1993). Viscoelastic properties of mixed polysaccharide systems. In *Plant Polymeric Carbohydrates*, edited by Meuser, F., Manners, D.J. and Seibel, W., Cambridge: Royal Society of Chemistry Special Publication Number 134, pp 76–85

Foster, T.J. and Morris, E.R. (1994) Xanthan polytetramer: conformational stability as a barrier to synergistic interaction. In *Gums and Stabilisers for the Food Industry 7*, edited by Phillips, G.O., Wedlock, D.J. and Williams, P.A., Oxford: IRL Press, pp 281–289

Frei, E. and Preston, R.D. (1968) Non-cellulosic structural polysaccharides in algal cell walls. III. Mannan in siphoneous green algae. *Proceedings of the Royal Society, London, Series B*, **169**, 127–145

Goycoolea, F.M., Foster, T.J., Richardson, R.K., Morris, E.R. and Gidley, M.J. (1994) Synergistic gelation of galactomannans or konjac glucomannan: binding or exclusion? In *Gums and Stabilisers for the Food Industry 7*, edited by Phillips, G.O., Wedlock, D.J. and Williams, P.A., Oxford: IRL Press, pp 333–344

Jansson, P.-E., Kenne, L. and Lindberg, B. (1975) Structure of the extracellular polysaccharide from *xanthomonas campestris*. *Carbohydrate Research*, **45**, 275–282

Jansson, P.-E., Lindberg, J., Wilmalasiri, K.M.S. and Dankert, M.A. (1993) Structural studies of acetan, an exopolysaccharide elaborated by *acetobacter xylinum*. *Carbohydrate Research*, **245**, 303–310

Lambert, F., Milas, M. and Rinaudo, M. (1985) Sodium and calcium counterion activity in the presence of xanthan polysaccharide. *International Journal of Biological Macromolecules*, **7**, 49–52

MacCormick, C.A., Harris, J.E., Gunning, A.P. and Morris, V.J. (1993) Characterisation of a variant of the polysaccharide acetan produced by a mutant of *acetobacter xylinum* strain CR1/4. *Journal of Applied Bacteriology*, **74**, 196–199

Maekaji, K. (1974) The mechanism of gelation of konjac mannan. *Agricultural and Biological Chemistry*, **38**, 315–321

Mannion, R.O., Melia, C.D., Launay, B., Cuvelier, G., Hill, S.E., Harding, S.E. and Mitchell, J.R. (1992) Xanthan–locust bean gum interactions at room temperature. *Carbohydrate Polymers*, **19**, 91–98

Marchessault, R.H., Buléon, A., Deslandes, Y. and Goto, T. (1979) Comparison of X-ray diffraction data of galactomannans. *Journal of Colloid and Interface Science*, **71**, 375–382

McCleary, B.V. (1979) Enzymic hydrolysis, fine structure, and gelling interaction of legume-seed D-galacto-D-mannans. *Carbohydrate Research*, **71**, 205–230

McCleary, B.V. and Neukom, H. (1982) Effect of enzymic modification on the solution and interaction properties of galactomannans. *Progress in Food and Nutrition Science*, **6**, 109–118

Melton, L.D., Mindt, L., Rees, D.A. and Sanderson, G.R. (1976) Covalent structure of the extracellular polysaccharide from *xanthomonas campestris*: Evidence from partial hydrolysis studies. *Carbohydrate Research*, **46**, 245–257

Miles, M.J., Morris, V.J. and Carroll, V. (1984) Carob gum–κ-carrageenan mixed gels – Mechanical properties and X-ray fibre diffraction studies. *Macromolecules*, **17**, 2443–2445

Millane, R.P., Hendrixson, T.L., Morris, V.J. and Cairns, P. (1992) Chain associations of konjac mannan in fibres and gels. In *Gums and Stabilisers for the Food Industry 6*, edited by Phillips, G.O., Wedlock, D.J. and Williams, P.A., Oxford: IRL Press, pp 531–534

Millane, R.P. and Wang, B. (1990) A cellulose-like conformation accessible to the xanthan backbone and implications for xanthan synergism. *Carbohydrate Polymers*, **13**, 57–68

Moorhouse, R., Walkinshaw, M.D. and Arnott, S. (1977). Xanthan gum – molecular conformation and interaction. *American Chemical Society Symposium Series*, **45**, 90–102

Morris, E.R., Rees, D.A., Young, G., Walkinshaw, M.D. and Darke, A. (1977) Order–disorder transition for a bacterial polysaccharide in solution. A role for polysaccharide conformation in recognition between *xanthomonas* pathogen and its plant host. *Journal of Molecular Biology*, **110**, 1–16

Morris, E.R. (1990) Mixed polymer gels. In *Food Gels*, edited by Harris, P., London: Elsevier, pp 291–359

Morris, E.R. and Foster, T.J. (1994) Role of conformation in synergistic interactions of xanthan. *Carbohydrate Polymers*, **23**, 133–135

Morris, V.J. (1986) Multicomponent gels. In *Gums and Stabilisers for the Food Industry 3*, edited by Phillips, G.O., Wedlock, D.J. and Williams, P.A., London: Elsevier Applied Science, pp 87–99

Morris, V.J. and Miles, M.J. (1986) Effect of natural modifications on the functional properties of extracellular bacterial polysaccharides. *International Journal of Biological Macromolecules*, **8**, 342–348

Morris, V.J., Brownsey, G.J., Cairns, P., Chilvers, G.R. and Miles, M.J. (1989) Molecular origins of acetan solution properties. *International Journal of Biological Macromolecules*, **11**, 326–328

Morris, V.J. (1990) Biotechnically produced carbohydrates with functional properties for use in food systems. *Food Biotechnology*, **4**, 45–57

Morris, V.J. (1991) Weak and strong polysaccharide gels. In *Food Polymers, Gels and Colloids*, edited by Dickinson, E., Cambridge: Royal Society of Chemistry Special Publication Number 82, pp 310–321

Morris, V.J., Brownsey, G.J. and Ridout, M.J. (1994) Role of conformation in synergistic interactions of xanthan – Reply. *Carbohydrate Polymers*, **23**, 139–140

Ojinnaka, C., Morris, E.R., Morris, V.J. and Brownsey, G.J. (1994) Effect of acetyl substituents on the conformational stability and functional interactions of acetan polysaccharide. In *Gums and Stabilisers for the Food Industry 7*, edited by Phillips, G.O., Wedlock, D.J. and Williams, P.A., Oxford: IRL Press, pp 15–26

Okuyama, K., Arnott, S., Moorhouse, R., Walkinshaw, M.D., Atkins, E.D.T. and Wolf-Ullish, C.H. (1980). Fiber diffraction studies of bacterial polysaccharides. *American Chemical Society Symposium Series*, **141**, 411–427

Piculell, L., Iliopoulos, I., Linse, P., Nilsson, S., Turquois, T., Viebke, C. and Zhang, W. (1994) Association and segregation in ternary polymer solutions and gels. In *Gums and Stabilisers for the Food Industry 7*, edited by Phillips, G.O., Wedlock, D.J. and Williams, P.A., Oxford: IRL Press, pp 309–322

Ridout, M.J., Brownsey, G.J., Morris, V.J. and Cairns, P. (1994) Physicochemical characterisation of an acetan variant secreted by *acetobacter xylinum* strain CR1/4. *International Journal of Biological Macromolecules*, **16**, 324–330

Robinson, G., Manning, C.E. and Morris, E.R. (1991) Conformation and physical properties of the bacterial polysaccharides gellan, welan and rhamsan. In *Food Polymers, Gels and Colloids*, edited by Dickinson, E., Cambridge: Royal Society of Chemistry Special Publication Number 82, pp 22–33

Rochas, C., Taravel, F.R. and Turquois, T. (1990) NMR studies of synergistic κ-carrageenan–carob galactomannan gels. *International Journal of Biological Macromolecules*, **12**, 353–358

Ross-Murphy, S.B., Morris, V.J. and Morris, E.R. (1983) Molecular viscoelasticity of xanthan polysaccharide. *Faraday Symposia of the Chemical Society*, **18**, 115–129

Rydholm, S. (1963). *Pulping Processes*, New York: Interscience

Shatwell, K.P., Sutherland, I.W., Ross-Murphy, S.B. and Dea, I.C.M. (1991a) Influence of the acetyl substituent on the interaction of xanthan with plant polysaccharides – III. Xanthan–konjac mannan systems. *Carbohydrate Polymers*, **14**, 131–147

Shatwell, K.P., Sutherland, I.W., Ross-Murphy, S.B. and Dea, I.C.M. (1991b) Influence of the acetyl substituent on the interaction of xanthan with plant polysaccharides – I. Xanthan–locust bean gum systems. *Carbohydrate Polymers*, **14**, 29–51

Takahashi, R., Kusakabe, I., Kusama, S., Sakurai, Y., Murakami, K., Maekawa, A. and Suzuki, T. (1984) Studies on the mannanase of *streptomyces*. 4. Structures of glucomanno-oligosaccharides from the hydrolytic products of konjac glucomannan produced by a ß-mannanase from *streptomyces sp. Agricultural and Biological Chemistry*, **48**, 2943–2950

Tako, M., Asato, A. and Nakamura, S. (1984) Rheological aspects of the intermolecular interaction xanthan and locust bean gum in aqueous media. *Agricultural and Biological Chemistry*, **48**, 2995–3000

Turquois, T., Rochas, C. and Taravel, F.R. (1992) Rheological studies of synergistic κ-carrageenan–carob galactomannan gels. *Carbohydrate Polymers*, **17**, 263–268

Turquois, T., Rochas, C. and Taravel, F.R. (1993) Synergy of the agarose–carob galactomannan blend inferred from NMR and rheological studies. *Carbohydrate Research*, **238**, 27–38

Williams, P.A., Day, D.H., Langdon, M.J., Phillips, G.O. and Nishinari, K. (1991a) Synergistic interaction of xanthan gum with glucomannans and galactosemannans. *Food Hydrocolloids*, **4**, 489–493

Williams, P.A., Clegg, S.M., Day, D.H. and Phillips, G.O. (1991b) Mixed gels formed with konjac mannan and xanthan gum. In *Food Polymers, Gels and Colloids*, edited by Dickinson, E., Cambridge: Royal Society of Chemistry Special Publication Number 82, pp 339–348

Williams, P.A., Clegg, S.M., Langdon, M.J., Nishinari, K. and Phillips, G.O. (1992) Studies on the synergistic interaction of konjac mannan and locust bean gum with κ-carrageenan. In *Gums and Stabilisers for the Food Industry 6*, pp 209–206, edited by Phillips, G.O., Wedlock, D.J. and Williams, P.A., Oxford: IRL Press, pp 209–206

Williams, P.A., Clegg, S.M., Langdon, M.J., Nishinari, K. and Piculell, L. (1993) Investigation of the gelation mechanism in κ-carrageenan–konjac mannan mixtures using differential scanning calorimetry and electron spin resonance spectroscopy. *Macromolecules*, **26**, 5441–5446

Winter, W.T., Chien, Y.Y. and Bouckris, H. (1984) Structural aspects of food galactomannans. In *Gums and Stabilisers for the Food Industry 2*, edited by Phillips, G.O., Wedlock, D.J. and Williams, P.A., Oxford: Pergamon, pp 535–539

Zhan, D.F., Ridout, M.J., Brownsey, G.J. and Morris, V.J. (1993) Xanthan–locust bean gum interactions and gelation. *Carbohydrate Polymers*, **21**, 53–58

15

RHEOLOGY AND PHASE SEPARATION IN PROTEIN–POLYSACCHARIDE MIXTURES

J.-L. DOUBLIER, C. CASTELAIN, G. LLAMAS and J. LEFEBVRE
Institut National de la Recherche Agronomique, Rue de la Géraudière, BP 1627, 44316 Nantes Cedex 03, France

Introduction

The phase behaviour of protein–polysaccharide–water mixtures has been the subject of a considerable number of investigations. As many as eighty protein–polysaccharide–water ternary systems have been described by Tolstoguzov and co-workers (see, for example, Tolstoguzov, 1986; Tolstoguzov 1991). Phase separation normally takes place as a result of the limited thermodynamic compatibility between the two types of macromolecule. In fact, only one system, a serum albumin–pectin mixture (Tolstoguzov, 1991), has been reported to exhibit compatibility. Of course, incompatibility is a general tendency exhibited by numerous ternary systems composed of one solvent and two unlike polymers and this has been described for many other polymer mixtures in aqueous media (Hefford, 1984; Inamura, Toki, Tamae and Araki, 1984; Kalichevsky, Orford and Ring, 1986; Medin and Janson, 1993). In the case of protein–polysaccharide systems, many parameters such as solvent conditions (pH, ionic strength, etc.), the chemical structure of the polysaccharide, its molecular weight and conformation, the type of protein, etc. are known to be involved in the separation process.

The present contribution deals with the phase behaviour of bovine serum albumin–hydroxyethylcellulose (BSA–HEC) systems in aqueous media at neutral pH. We chose HEC as a neutral polysaccharide in order to avoid the electrostatic effects that would be encountered with ionic polysaccharides. This cellulose derivative was preferred to dextran as a neutral polysaccharide since it is known that the degree of branching of the latter, and hence its behaviour in aqueous solution, varies as a function of molecular weight. The first objective was to describe the effect of the polysaccharide's molecular weight on the phase diagrams. Once these phase diagrams were obtained, we investigated the viscosity of the mixed system both well below and at the vicinity of the separation line. This addresses the second objective, which was to evaluate how the overall properties of protein—polysaccharide mixtures are modified even within the "one-phase" region of the phase diagram as compared to those of the solutions of the individual macromolecules.

Materials and methods

MATERIALS

Five HEC samples of the Natrosol series from Aqualon (France) with the same molar degree of substitution (M.S. = 2.5) were used. Their intrinsic viscosities (in water at 25°C) ranged from 1.64 to 16.1 dL/g (Table 15.1). Another HEC sample from Polysciences referred to as MV in Table 15.1, of medium viscosity, was also employed. Its degree of substitution was M.S. = 1.1.

The bovine serum albumin (BSA) used was from Sigma Chemical. Before its use in the experiments, this sample was cold defatted with pentane.

Table 15.1 PHYSICOCHEMICAL CHARACTERISTICS OF HEC SAMPLES

Sample	*[η] dL/g*	DP_v [1]
250 L	1.64	316
250 G	4.61	1030
250 M	11.5	2950
250 H	14.3	3800
250 HH	16.1	4350
MV	8.07	2000

[1] From Mark-Houwink equation (K= 1.1 x 10^{-2}; a = 0.87) (Wirick, 1968)

METHODS

Preparations of solutions and mixtures

Stock solutions of HEC were prepared by throwing the polysaccharide powder into water which was rapidly stirred with a magnetic stirrer. Stirring was continued for 12 hours and then the solution was filtered and deaerated under vacuum. HEC solutions were prepared at the highest possible concentration. The upper limit was actually determined by the viscosity of the solutions and thus depended upon the molecular weight. Therefore concentrations ranged from 3% to 20% (w/w) depending upon the HEC sample. Stock solutions of BSA (concentrations: ≈10–25%) were prepared by dispersing BSA powder in water. Magnetic stirring was performed gently in order to avoid foaming. In all experiments, 0.1% sodium azide was used as a preservative.

Phase diagrams

Stock solutions of BSA and HEC were mixed in different ratios and thoroughly stirred at 10°C for 1 hour. The mixtures were then poured into closed tubes and stored for 21 days at 10°C. At the end of this period, the tubes were examined and, in the case of high enough protein and polysaccharide concentrations, two liquid phases were found to have separated in the sample. The volumes of the two separated phases were determined, and the phases were quantitatively recovered and analysed for protein and polysaccharide content. Total protein content was determined spectrophotometrically using the decadic extinction coefficient at 280 nm ($A_{1\%,\,1\,cm} = 6.67$) (Reynolds, Herbert, Polet and Steinhard, 1967). HEC concentrations were estimated by the phenol–sulphuric method (Dubois, Gilles, Hamilton, Rebers and Smith, 1956).

Viscometry

Equilibrium flow curves of BSA–HEC mixtures in the one-phase region or close to the separation line were established at 25°C using a Contraves Low Shear 40 viscometer equipped with coaxial cylinders. Shear rates ranged between 0.01 s^{-1} and 100 s^{-1}. Measurements were performed pointwise at decreasing shear rates. For each shear rate, the shear stress was estimated from the torque value obtained at equilibrium. Additional viscosity measurements were performed with the same procedure using a Rheomat 30 over the shear rate range 0.1–700 s^{-1}.

Results and discussion

PHASE DIAGRAMS

Figure 15.1 illustrates the BSA–HEC phase diagram we obtained with HEC-L, the lowest molecular weight HEC sample. Separation could be relatively fast or extremely slow depending upon the position in the two-phase part of the diagram. For the more concentrated systems, i.e. well inside the two-phase region of the phase diagram, the mixture appeared cloudy immediately. For the systems in the vicinity of the separation line, cloudiness appeared more slowly. With time the mixtures separated into two clear liquid layers, giving rise to two coexisting phases. The upper phase was enriched in HEC and depleted in BSA and the lower one was BSA-enriched and HEC-depleted. Furthermore, the final concentration of BSA in the lower phase was much higher than the concentration of HEC in the upper phase. This separation, of course, results from thermodynamic incompatibility.

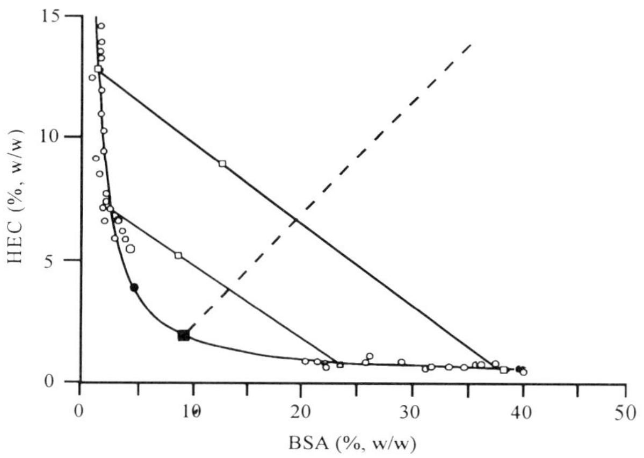

Figure 15.1 Phase diagram of BSA–HEC-L ([η] = 1.64 dL/g) systems in water (temperature 10°C). Two tie-lines are shown. Continuous line: binodal. Dashed line: mid-points of the tie-lines. Critical point (■); phase separation threshold (●)

This finding demonstrates that the compatibility which has been reported for BSA–pectin systems (Tolstoguzov, 1991) should not be generalized to other BSA–polysaccharide mixtures. It is likely that the overall phase behaviour of BSA–polysaccharide systems depends on the type of polysaccharide and resembles the general behaviour of protein–polysaccharide systems, i.e. thermodynamic incompatibility. From the initial composition and the composition of the separated phases, the tie-lines and the binodal could easily be drawn and the latter described by an empirical hyperbolic function allowing the limiting solubilities to be determined (Figure 15.1).

These limiting concentrations were estimated to be approximately 0.5% and 1.5% for HEC and BSA respectively. The critical point was obtained from the intercept of the binodal with the curve linking the mid-points of each tie-line. Its coordinates were 1.94% for HEC and 9.4% for BSA. The phase separation threshold, i.e the minimum total concentration required for phase separation to occur, was 4.2% HEC and 4.3% BSA. Because of the co-ordinates of the critical point, the distance between this point and the phase separation threshold and the slopes of the tie-lines, this diagram can be described as highly asymmetric (Tolstoguzov, 1988).

This asymmetry can originate from two features. It could result from the fact that the polysaccharide is much more hydrophilic than BSA (Tolstoguzov, 1993). It could also be due to the difference in molecular weight between the two biopolymers, the binodal being nearer the axis of the lower molecular weight molecule (Tolstoguzov, 1988).

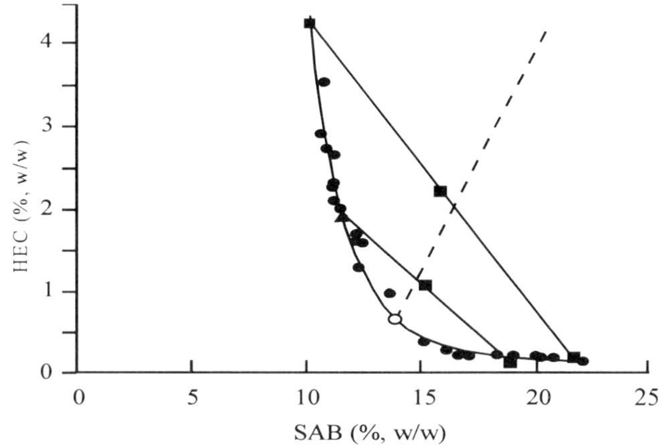

Figure 15.2 Phase diagram of BSA–HEC-HH ([η] = 16.1 dL/g) systems in water (temperature 10°C). Two tie-lines are shown. Continuous line: binodal. Dashed line: mid-points of the tie-lines. Critical point (○); phase separation threshold (▲)

Figure 15.2 shows the BSA–HEC phase diagram obtained with HEC-HH, the highest molecular weight HEC sample. The behaviour exhibited was the same as for HEC-L but with two main differences. The binodal line position was shifted towards lower HEC concentrations and the line asymptote along the ordinate shifted the opposite way. These features were also reflected in the co-ordinates of the critical point: 0.64% and 13.9% for HEC and BSA respectively. Asymmetry is thus significantly increased. This clearly illustrates the fact that miscibility tends to decrease with increasing molecular weight of one of the polymers.

The results obtained with the other HEC samples followed a similar pattern. Each diagram was characterized by co-ordinates of the critical point, and of the phase separation threshold and the limiting solubility of each component (Table 15.2). HEC concentration at the critical point decreased significantly as the molecular weight increased, whereas BSA concentration increased but varied to a lesser extent. The limiting solubility of HEC also decreased with molecular weight. In contrast, BSA limiting solubility tended to increase slightly as HEC molecular weight increased. Actually, variations of this latter parameter were mostly noted for the HEC with the two lowest molecular weights (L and G), while no clear variation was seen for the other three samples with relatively higher DP_v. The fact that the ordinate of the critical point, as well as limiting solubility, decreases when molecular weight increases, is the expected behaviour of polymer mixtures with limited compatibility in a common solvent. Similar observations have been reported for different polymers mixtures in aqueous solution (Hefford, 1984; Inamura *et al.*, 1984; Medin and Janson, 1993).

Table 15.2 COMPOSITIONS CORRESPONDING TO THE CRITICAL POINT, PHASE SEPARATION THRESHOLD AND LIMITING SOLUBILITIES

HEC Sample	*Critical point*		*Limiting solubility*		*Phase separation threshold*	
	HEC (%, w/w)	*BSA (%, w/w)*	*HEC (%, w/w)*	*BSA (%, w/w)*	*HEC (%, w/w)*	*BSA (%, w/w)*
L	1.94	9.37	0.5	1.4	4.2	4.3
G	1.21	14.7	0.32	7.5	4.0	8.2
M	0.94	12.4	0.25	9.6	2.5	9.5
H	0.84	14.2	0.23	10.6	2.3	11.6
HH	0.64	13.9	0.2	10	1.9	11.6
MV	1	12	0.23	8	1.8	10.3

In order to account for the effects of molecular weight of the two types of macromolecule, an attempt has been made to plot the phase diagrams in terms of reduced concentrations, i.e. the concentration of HEC or BSA times the intrinsic viscosity. These diagrams are not shown but the characteristic parameters are given in Table 15.3. For the critical point, the abscissa related to BSA was found at reduced concentrations of the order of 0.5. In contrast, the $c[\eta]$ ordinate related to HEC ranged from $\approx$3 to $\approx$10, showing that HEC concentration at this critical point was much higher than the concentration required for incipient coil overlap ($c[\eta] \approx 0.8$). Limiting solubility of BSA appeared to take place at an almost fixed reduced concentration ($\approx$0.3–0.35) with however the exception of the diagram with HEC-L ($\approx$0.05). These values are of the same order of magnitude as the critical reduced concentration for the transition from the dilute to the semi-dilute regime in the case of globular proteins: $c^*[\eta] \approx 0.2$–0.3 as determined from viscosity measurements (Monkos, 1994). For HEC, this reduced limiting parameter ranged from 0.8 to 3, which is still higher than the reduced concentration for coil overlap, again with the exception of the lowest molecular weight sample (HEC-L: $c[\eta] = 0.8$). Therefore, it is clear that the minimum HEC concentration required for phase separation in BSA—HEC mixed systems is significantly higher than the concentration corresponding to the transition from a dilute to a semi-dilute solution according to the classical criterion ($c[\eta] \approx 0.8$). This means that phase separation takes place well within the entangled regime of HEC. This in turn indicates that the repulsive interaction between BSA and HEC which induces separation is relatively weak (Kalichevsky *et al.*, 1986).

Table 15.3 CRITICAL POINT AND LIMITING SOLUBILITY IN REDUCED CONCENTRATIONS

HEC Sample	*Coordinates of the critical point*		*Limiting solubility*	
	$HEC^*[\eta]_{HEC}$	$BSA^*[\eta]_{BSA}$	$HEC^*[\eta]_{HEC}$	$BSA^*[\eta]_{BSA}$
L	3.14	0.35	0.81	0.052
G	5.68	0.54	1.48	0.28
M	10.9	0.46	2.89	0.35
H	11.6	0.52	3.17	0.39
HH	10.5	0.52	3.27	0.37

FLOW BEHAVIOUR OF BSA–HEC MIXTURES

In the first series of measurements, HEC concentration was kept constant and the BSA concentration was increased stepwise so that the binodal was crossed. Figure 15.3 shows a series of flow curves obtained with 2% HEC-L, this concentration corresponding to that at the critical point, and with varying BSA concentrations: 5%, well within the one-phase region; 10%, just above the critical point (9.4%); and 15%, inside the two-phase region but still relatively close to the binodal. The flow curve with 2% HEC showed a Newtonian behaviour ($\eta \approx 0.01$ Pa.s) within the shear rate range investigated (<100 s^{-1}). The flow curves of BSA–HEC systems displayed a general behaviour which differed from that of the polysaccharide alone at the same concentration in two respects: firstly, at high shear rate the apparent viscosity was slightly increased as BSA concentration increased from 0.01 Pa.s for HEC alone to 0.025 Pas with 15% BSA, and secondly, the behaviour became non-Newtonian at low shear rate, the apparent viscosity increasing as the shear rate decreased. The viscosity increase upon BSA addition at high shear rate cannot be simply ascribed to the presence of BSA – this protein in itself does not contribute significantly to the viscosity of the medium. For instance, the viscosity of a 15% BSA solution was of the order of 0.003 Pa.s, i.e. much lower than the increase experienced in the corresponding BSA–HEC mixture, 0.015 Pa.s. The second phenomenon can be related to the existence of an apparent yield stress. This suggests that the BSA–HEC mixture is a disperse system, even in the one-phase region of the diagram.

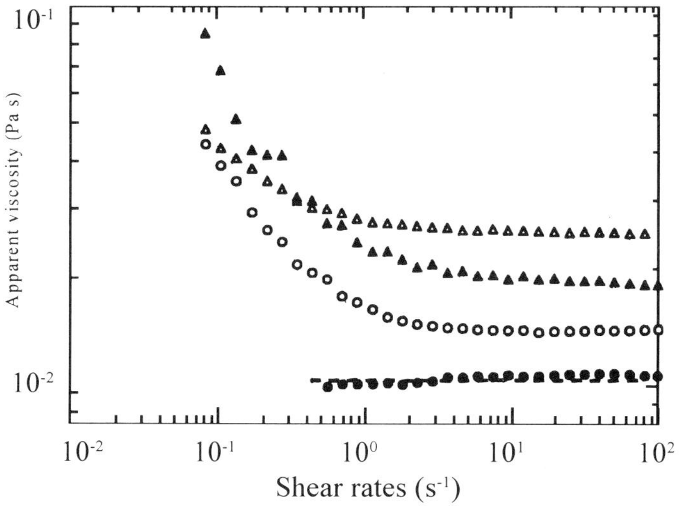

Figure 15.3 Flow curves of 2% HEC-L and BSA systems. Temperature: 25°C. HEC alone (●); BSA:HEC = 5:2 (○); BSA:HEC = 10:2 (▲); BSA:HEC = 15:2 (△)

In a second series of experiments, the protein:polysaccharide ratio was kept constant (2:1) and the total biopolymer concentration was varied from 0.1% to 4.5%. In these experiments, the HEC sample (HEC-MV) had a higher molecular weight than in the previous ones. With this HEC sample the co-ordinates of the critical point were approximately 11.6% for BSA and 0.98% for HEC. BSA and HEC concentrations ranged from 0.07% to 3.06% (w/w) and from 0.03% to 1.53% (w/w) respectively. The composition was therefore well inside the one-phase region. The results are shown in Figure 15.4. The flow curves of HEC alone were typical of polysaccharide solutions: Newtonian behaviour below the critical concentration c^* (0.07%, w/w) and, above c^*, shear-thinning with a limiting Newtonian viscosity η_0 at low shear rate. The flow curves of the mixtures were almost superimposed on those of the solutions of HEC alone at high shear rate, but for the lowest concentrations and as described above, a steep rise in apparent viscosity was observed as the shear rate decreased. Hence, no limiting viscosity could be determined. However, each flow curve exhibited a "plateau" viscosity η_p in an intermediate shear rate range that could be compared to η_0. Here again, an apparent yield stress was exhibited, the origin of which has to be discussed.

When plotting the shear stress as a function of shear rate this low shear rate behaviour is clearly highlighted as illustrated in Figure 15.5 for the 2:1 BSA (0.8%) : HEC (0.4%) of Figure 15.4. The flow curve of the corresponding 0.4% HEC solution was linear in this shear rate range (Figure 15.5). In contrast, the curve of the mixed system was much more complex and can be separated into three parts schematically as drawn in Figure 15.6. Similar flow curves have been found for all mixed systems. They display many similarities with flocculated dispersions (Patel

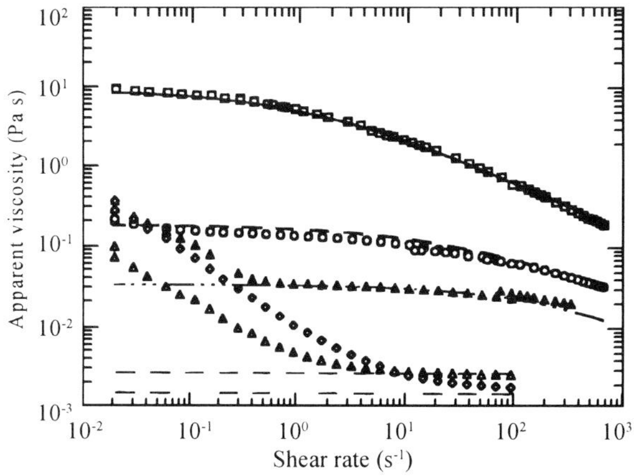

Figure 15.4 Flow curves of BSA–HEC-MV system (BSA:HEC = 2:1) in 0.1 M NaCl. Temperature: 25°C. (◇)0.1%:0.05% (△); 0.2%:0.1% (▲); 0.8%:0.4% (○); 1.2%:0.6% (□); 3.1%:1.55%. Lines: flow curves of HEC-MV alone at the same HEC concentration

and Russel, 1987) and we can attempt to interpret the present behaviour on a similar basis. At shear rates higher than $\dot{\gamma}_C$, the curve is linear and is characterized by the plastic viscosity η_{PL}, the extrapolation of this linear part yielding the Bingham yield stress σ_B. Between $\dot{\gamma}_1$ and $\dot{\gamma}_C$ the shear stress is almost constant, while it decreases rapidly for $\dot{\gamma} < \dot{\gamma}_1$; this latter part corresponds to a very high viscosity ("infinite viscosity") at zero shear rate: the BSA–HEC systems present a solid-like behaviour at rest resembling that of a very weak gel. Furthermore, since the flow curves have been obtained at equilibrium and are reversible, this means that the infinite network at rest breaks up during shear and then recovers after shear stops. The weak gel character can originate from flocculated particles. Increasing the shear rate beyond $\dot{\gamma}$ results in the rupture of the flocs until they reach their ultimate size in the third zone. This is consistent with the fact that η_{PL} was found to be very close to the viscosity of HEC alone.

The fact that a common overall behaviour was found in the two series of experiments suggests that the same mechanism underlies these processes. Furthermore, a comparable behaviour has been described for BSA–anionic polysaccharides in neutral conditions (CMC, λ-carrageenan, alginate) (Gonçalves, 1984; Castelain, Lefebvre and Doublier, 1986) and BSA–polyethylene oxide (PEO) (Castelain and Lefebvre, unpublished). Such solid-like properties suggest the existence of an apparent yield stress in BSA–polymer systems which is independent of the nature of the polymer, provided no specific interaction between protein and polymer occurs.

In order to interpret this phenomenon, we suggest that "water in water" emulsions are formed with droplets of a BSA-rich liquid phase dispersed in the polymer liquid

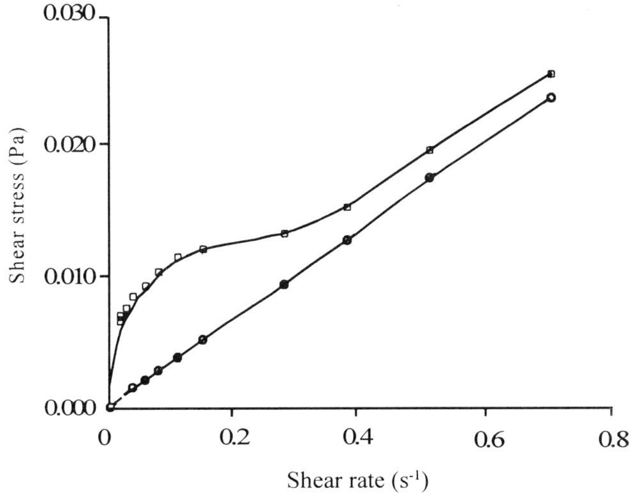

Figure 15.5 Flow curve at low shear rate of the BSA—HEC-MV (0.8%:0.4%) system (symbols and continous line). Comparison with HEC-MV (0.4%) (dashed line)

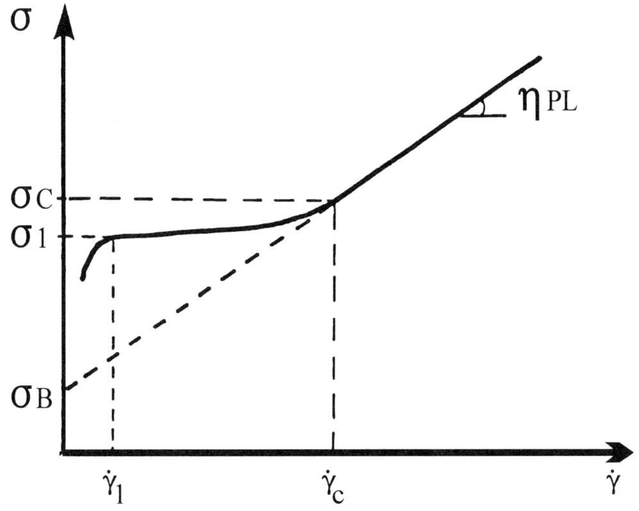

Figure 15.6 Schematic plot at low shear rate of the flow curves of the mixed systems. η_{PL}: plastic viscosity and σ_B: yield stress (Bingham model)

phase. At low shear rate ($\dot{\gamma} < \dot{\gamma}_1$ the BSA-rich droplets will tend to flocculate and to form large flocs. At rest these large flocs are responsible for the solid-like rheological behaviour. Beyond $\dot{\gamma}_c$ the applied stress reduces the size of the flocs down to single units. Deflocculation of the BSA aggregates will thus be the mechanism for the shear thinning behaviour in the intermediate shear rate range (between $\dot{\gamma}_1$ and $\dot{\gamma}_c$).

This mechanism can quite easily explain the properties of the systems within the two-phase region and close to the binodal. At 15% BSA with 2% HEC-L, for instance, immediate phase separation occurred, but the two phases separated very slowly on the macroscopic scale. A period of several days was required for the two phases to separate from each other in the tube. Therefore, the system was a quasi-stable "water in water" emulsion. Within the timescale of the viscosity measurements the resulting rheological behaviour was that of a disperse system. For 10% BSA and 2% HEC-L, a system still lying within the two-phase region of the phase diagram but too close to the critical point for the separation process to occur at the macroscopic scale during the time of observation, the system can be regarded as metastable but can still unambiguously be described as a "water in water" emulsion of the same type. The results in the single phase region were unexpected. They suggest that here again the same kind of dispersion has been formed. A logical explanation would be that in this region of the diagram the liquid—liquid phase separation would proceed on the microscopic scale, resulting in an emulsion which flocculates, but remains kinetically stable as far as sedimentation and coalescence are concerned.

Conclusions

Phase separation in BSA–HEC systems in aqueous media should arise from thermodynamic incompatibility between the two types of biopolymers as has been reported for many other protein–polysaccharide systems. The process is strongly dependent upon the molecular weight of the polymer. The rheological properties of the system which, in the absence of interactions, should reflect those of the polysaccharide solution, are strongly modified by the presence of the protein even in the single phase region and even though phase separation does not occur at the macroscopic scale. This change in the flow behaviour at very low shear rates is particularly dramatic at low polysaccharide concentrations. Another consequence of the addition of BSA to the polysaccharide solution is a drastic increase of the elasticity of the medium (Castelain *et al.*, 1986). Both phenomena obviously indicate solid-like properties. This suggests the formation of a weak and labile network that would originate from flocculated BSA-rich droplets. The fact that such peculiar behaviour is experienced even at very low polysaccharide concentrations may provide an explanation for reported results on the diffusion coefficients of BSA in PEO aqueous solutions, which were found to be much greater than expected from measurements in the absence of the polymer (Phillies, 1985).

References

Castelain, C. and Lefebvre, J. (Unpublished)

Castelain, C., Lefebvre, J. and Doublier, J.L. (1986) Rheological behaviour of BSA–cellulose derivative mixtures in neutral aqueous medium. *Food Hydrocolloids*, **1**, 141–151

Dubois, M., Gilles, K.A., Hamilton, J.K., Rebers, P.A. and Smith, F. (1956) Colorimetric method for determination of sugars and related substances. *Analytical Chemistry*, **28**, 350–356

Gonçalves, M.P. (1984). *Coprécipitation des protéines de plasma bovin avec des polyosides anioniques. Etude rhéologique des coprécipités et des intéractions entre la SAB et le λ-carraghénane ou l'alginate de sodium*, thesis, Université Clermont Ferrand II, France

Hefford, R.J. (1984) Polymer mixing in aqueous solution. *Polymer*, **25**, 979–984

Inamura, S., Toki, K., Tamae, T. and Araki, T. (1984) Effects of molecular weight on the phase equilibrium of a poly(vinyl alcohol)–poly(ethylene glycol)–water system. *Polymer Journal*, **16**, 657–660

Kalichevsky, M.T., Orford, P.D. and Ring, S.G. (1986)The incompatibility of concentrated aqueous solutions of dextran and amylose and its effect on amylose gelation. *Carbohydrate Polymers*, **6**, 145–154

Medin, A.S. and Janson, J.-C. (1993) Studies on aqueous polymer two-phase systems containing agarose. *Carbohydrate Polymers*, **22**, 127–136

Monkos, K. (1994) Viscometric study of human, bovine and ovine haemoglobin in aqueous solution. *International Journal of Biological Macromolecules*, **16**, 31–34

Patel, P.D. and Russel, W.B. (1987) The rheology of polystyrene latices phase separated by dextran. *Journal of Rheology*, **31**, 599–618

Phillies, G.D.J. (1985) Diffusion of bovine serum albumin in a neutral polymer solution. *Biopolymers*, **24**, 379–386

Reynolds, J.A., Herbert, S., Polet, H. and Steinhard, J. (1967) The effect of diverse detergent anions to bovine serum albumin. *Biochemistry*, **6**, 937–947

Tolstoguzov, V.B. (1986) Functional properties of protein–polysaccharide mixtures. In *Functional Properties of Food Macromolecules*, edited by Mitchell, J.R. and Ledward, D.A., London: Elsevier, pp 385–415

Tolstoguzov, V.B. (1988) Some physicochemical aspects of protein processing into foodstuffs. *Food Hydrocolloids*, **2**, 339–370

Tolstoguzov, V.B. (1991) Functional properties of food proteins and role of protein—polysaccharide intereaction. *Food Hydrocolloids*, **4**, 429–468

Tolstoguzov, V.B. (1993) Thermodynamic aspects of food protein functionality. In *Food Hydrocolloids: Structure, Properties and Functions*, edited by Nishinari, K. and Doi, E., London and New York: Plenum Press, pp 327–340

Wirick, M.G. (1968) A study of the enzymic degradation of CMC and other cellulose ethers. *Journal of Polymer Science, A1*, **6**, 1965–1974

16

SYNERGISM AND INTERACTIONS IN MIXED PROTEIN SYSTEMS

NAZLIN K. HOWELL
University of Surrey, School of Biological Sciences, Guildford, Surrey. GU2 5XH U.K.

Introduction

The aim of this chapter is to elucidate the mechanism of interactions of a number of food proteins which result in synergism, aggregation or phase separation in heat-set gels, precipitates and in foams.

Protein–protein interactions occur in many physical and chemical processes not only *in vivo* but also in mixtures of ingredients used in the processing, manufacture and storage of food products. There are several reasons for investigating protein–protein interactions including acquiring knowledge of structure–function relationships, optimising product constituents (thereby improving quality, nutrition and safety), cost reduction and new product applications. Previous studies have been empirical and have displayed interesting and technologically useful phenomena such as synergistic behaviour or phase separation of protein mixtures (Howell and Lawrie, 1984; Howell and Taylor, 1991). More recently, however, advances have been made in the study of protein interactions by the author with funding from the Ministry of Agriculture, Fisheries and Food / Industry LINK scheme, the Agricultural and Food Research Council (now the Biotechnology and Biological Sciences Research Council) and the E.U.

Interactions of proteins with other proteins which affect texture lead to three main phenomena:

(a) Synergistic interactions and compatible gel formation and stable foams;
(b) Precipitation through predominantly electrostatic interactions; and
(c) Phase separation.

These interactions are governed by molecular forces which involve the stronger covalent bonds such as disulphide and peptide bonds as well as the weaker non-covalent forces including electrostatic linkages, hydrogen bonds and hydrophobic interactions. The nature of these bonds has been widely reviewed (Howell, 1992).

Synergistic interactions

Synergistic interactions can be defined as those interactions that result in an effect which is greater than that produced by each protein in isolation. The effect manifests itself in greater viscosity, cohesiveness or gel strength values and the structure indicates compatible gel networks. This phenomenon has been observed in many empirical systems including, for example, bakery and meat products. One such system which has been studied in detail has been used to illustrate the interaction of plasma proteins with egg albumen (egg white) proteins in a cake-type model system.

PLASMA–EGG ALBUMEN PROTEIN INTERACTIONS

Blood plasma is a by-product of the meat industry and is currently incorporated in meat products. However, it has potential for use in bakery products as an egg replacer for economic and functional reasons. Howell and Lawrie (Howell and Lawrie, 1983; Howell and Lawrie, 1987) investigated the viscous and gelling properties of egg albumen and plasma proteins using a cake-type model system comprising 6% protein and 45% sucrose in distilled water (of pH 8.0). The interaction of egg albumen with porcine and bovine plasma was examined by heating mixtures over 20°—79°C, followed by cooling to 20°C prior to measurement on a Ferranti–Shirley Cone and Plate Viscometer. Synergistic interactions were observed for mixtures heated above 73°C. The interaction was also observed and quantified in heat-set gels made from protein mixtures comprising porcine or bovine plasma and egg albumen mixed in the ratios 4:2, 3:3 and 2:4.

Method

Protein solutions (6% w/w) in 45% sucrose solution were heated at 85°C, 90°C and 95°C for 15, 30 or 60 minutes. After cooling and ageing overnight, the gels were compressed on the Instron Universal Testing Machine according to Howell and Lawrie (1984).

Results

The gel strengths of blood plasma gels were found to be greater, increased with both time and temperature of heating, compared with egg albumen proteins (Table 16.1).

Table 16.1 EFFECT OF HEATING TIME (min) AND OF PROTEIN CONCENTRATION ON THE GEL STRENGTH (g) OF EITHER COMMERCIAL PORCINE PLASMA (P) OR EGG ALBUMEN (E) OR A MIXTURE OF BOTH (P+E), ALL IN 45% SUCROSE SOLUTION AT 85°C

Time of heating	*% protein (w/w)*				
	P 6	*P + E* 4 + 2	*P + E* 3 + 3	*P + E* 2 + 4	*E* 6
15 min (n)	8	8	8	6	8
Mean gel strength	428	437	411	318	101
SD±	18.0	8.8	21.3	31.6	24.8
CV%	4.2	2.0	5.1	9.9	24.4
Additive value	428	319	264	210	101
Interaction index		+37	+55	+51	
30 min (n)	8	9	9	8	8
Mean gel strength	582	510	433	338	137
SD±	21.2	23.9	18.0	20.3	11.4
CV%	3.6	4.7	4.1	5.9	8.2
Additive value	582	434	360	285	137
Interaction index		+17	+20	+18	
60 min (n)	8	8	7	7	8
Mean gel strength	633	534	467	367	172
SD±	34.6	52.8	46.4	17.9	29.8
CV%	5.4	9.8	9.9	4.9	17.3
Additive value	633	422	403	326	172
Interaction index		+26	+15	+12	

Note: n, number of replicates; SD, standard deviation; CV, coefficient of variation; additive value, the value derived from summing the contributions of the component proteins, as measured in isolation, in proportion to the concentration of each in the mixtures;

$$\text{Interaction index} = \frac{\text{actual value additive value}}{\text{additive value}} \times 100$$

Synergistic interactions were observed depending on the type of protein, species, the concentration of plasma in the mixture and the time and temperature of heating. This was clearly and reproducibly illustrated using an interaction index devised by the author (Howell and Lawrie, 1984) (Figure 16.1).

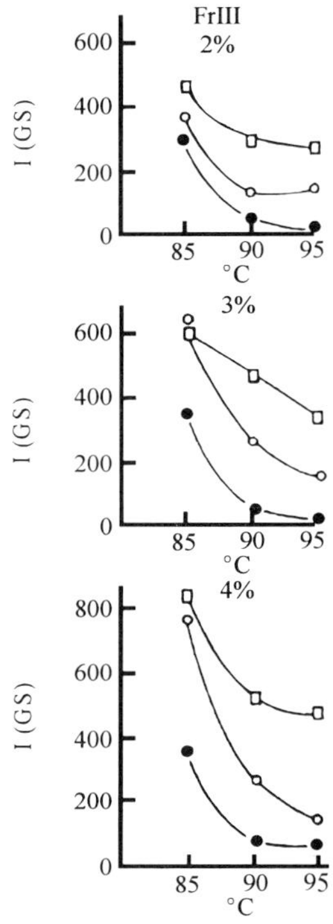

Figure 16.1 Variation of interaction index I in terms of gel strength (GS) with temperature (°C) for 2, 3 and 4% (w/w) porcine plasma Fraction III (FR III) protein levels. In all cases the plasma fractions were made up to 6% (w/w) total protein in 45% (w/w) sucrose solution with egg albumen proteins.

MECHANISM OF INTERACTION

The currently held view of protein gelation supports the theory that partially unfolded globular proteins associate linearly at low temperatures by non-covalent bonds, and randomly to form aggregates at high temperatures through covalent and hydrophobic interactions (Figure 16.2).

An explanation for the differences between egg albumen gels which are considerably weaker than the plasma protein gels (Table 16.1) is attributed to the differences in the number and position of disulphide/sulphydryl groups. A major

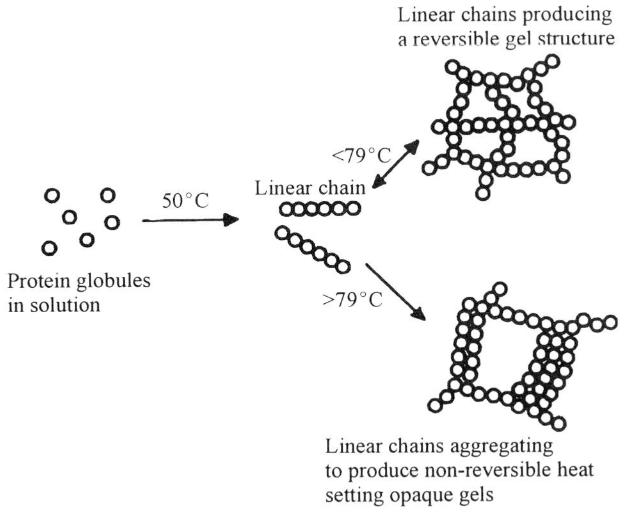

Figure 16.2 Proposed mechanism for the gelation of blood plasma proteins

difference between BSA and ovalbumin (and also bovine blood plasma and egg albumen) is their sulphydryl/disulphide content. BSA contains 17 disulphide bonds and one free sulphydryl group (Hoshi and Yamauchi, 1983) whilst ovalbumin contains only one disulphide bond and four unreactive sulphydryl groups (Fothergill and Fothergill, 1970; Fernandez-Diez, Osuga and Feeney, 1964). In addition egg albumen and ovalbumin had low surface hydrophobicity (S_0) compared to BSA and bovine blood plasma, as reported by Kato and Nakai (1980) and Kato and co-workers (Kato, Osako, Matsudomi and Kobayashi, 1983; Kato, Tsutsui, Matsudomi, Kobayashi and Nakai, 1981).

If the general theory of protein denaturation and unfolding to expose reactive groups is applied to the study of interactions, then it appears that at low temperatures and short heating times, the egg albumen proteins undergo substantial conformational changes as indicated by the viscosity increases (Howell and Lawrie, 1987). In this state they appear to interact with the partially unfolded plasma proteins. At higher temperatures of heating and at higher plasma protein concentrations, plasma–plasma interaction becomes dominant (Howell and Lawrie, 1985).

Elucidation of protein interactions can be achieved by electron microscopy. For blood plasma and egg albumen proteins, and mixtures of the two proteins heated at 85° or 95°C for 15 minutes, a similar type of gel network was shown by transmission electron microscopy, confirming the compatibility of these proteins in gel formation (Plate 16.1) (Howell and Lawrie, 1985). Electron microscopy to date does not indicate the disposition of the individual proteins in the gel. However with the aid of immunocytochemical techniques using gold labelled antibodies raised to egg albumen, preliminary studies showed that the egg albumen proteins (seen as black dots) are uniformly distributed along the gel network (Plate 16.2). This technique, at present, does not identify all of the targeted protein or very fine details of the network. It is envisaged that with advances in the development of the technique more information will be obtained.

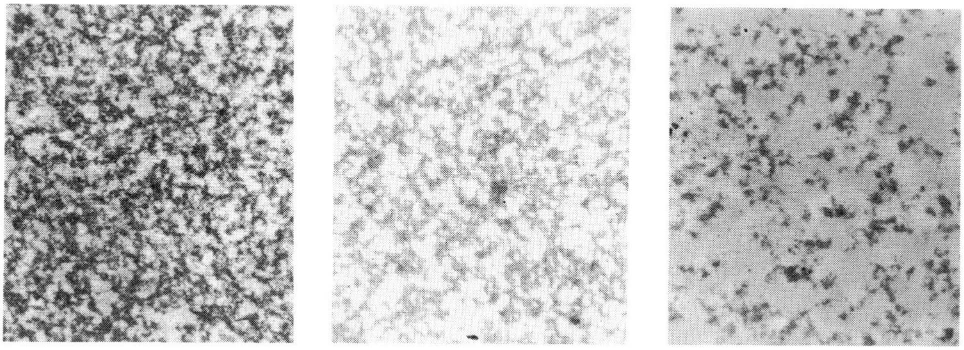

(a) (b) (c)

Plate 16.1 Transmission electron micrographs of (a) 6% plasma; (b) 6% egg albumen; and (c) a mixture of 3% plasma and 3% egg albumen protein gels heated at 95°C for 20 min

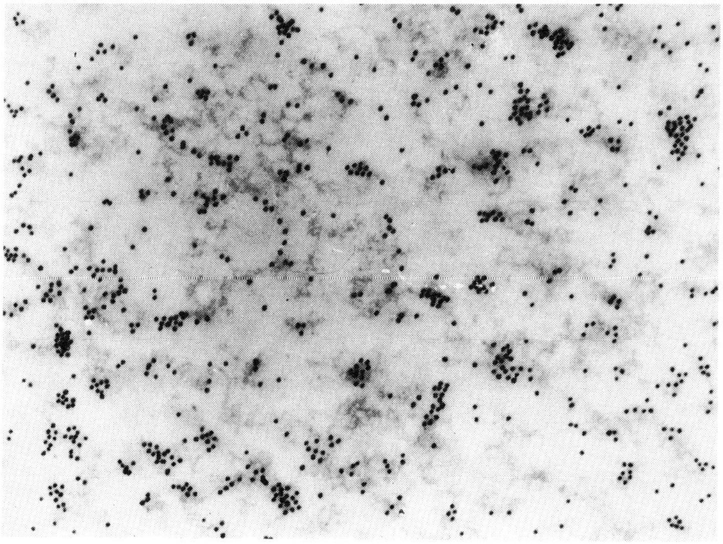

Plate 16.2 Transmission electron micrograph of a mixed protein gel (3% egg albumen : 3% plasma protein) indicating the distribution of egg albumen proteins using anti-egg albumen immunogold labelling (black gold particles 15 μm diameter)

GELATION VIA ELECTROSTATIC COMPLEXES

In contrast to the synergistic compatible single gel network illustrated above, gelation can also occur when negatively charged proteins are mixed with positively charged proteins. In this case, the reaction may be synergistic i.e. gelation may occur even if the individual proteins are unable to gel at the particular processing conditions. An example of this is the interaction between negatively charged deamidated gluten proteins and positively charged lysozyme (pI 10.3) (Friedli and Howell, 1995). On its own the deamidated gluten protein forms a gel when heated at over 100°C and about 19% deamidated gluten (15% protein) concentration. In the presence of only 0.1% lysozyme the gelation temperature of deamidated gluten was lowered to 80°C. The mechanism proposed by Friedli and Howell was that deamidated gluten when heated to at least 60°C disaggregates and fragments as observed by phase contrast microscopy. In the presence of 0.1% lysozyme and at 80°C where partial unfolding of the proteins occurs the two proteins aggregate electrostatically to form a gel. It is likely that a very small quantity of lysozyme can form junction points to cause the aggregation and network formation of the deamidated undenatured gluten (Plate 16.3). This mechanism is in contrast to the gel formed by the deamidated gluten heated at 120°C, when the proteins unfold and undergo denaturation. The physical-chemical properties and distribution of the proteins in the different types of gel networks are currently being studied.

PRECIPITATION

Interaction between α-lactalbumin, ß-lactoglobulin and lysozyme

When larger concentrations of positively charged proteins such as lysozyme (5%) were mixed with negatively charged proteins such as whey proteins (5%) in water they immediately precipitated at room temperature. The amount of precipitate varied with the concentration of proteins in the mixture, the type of protein pH and ionic strength (Howell *et al.*, 1995). The whey proteins make up 0.6% milk proteins and comprise α-lactalbumin (22%), ß-lactoglobulin (50%), bovine serum albumin (5%), immunoglobulins (12%) and proteose peptones (10%).

α-lactalbumin is homologous in sequence and backbone conformation with hen egg white lysozyme. Both proteins are globular (MW 14300) and contain four disulphide bonds. There is also a hydrophobic box region in α-lactalbumin which is homologous to the hydrophobic region associated with the monosaccharide binding subsites D and E in lysozyme (Richardson and Brew, 1980). α-lactalbumin binds calcium in a calcium binding fold and has a specific binding site for zinc.

On the other hand, lysozyme very specifically cleaves bacterial cell wall mucopolysaccharides between the C-1 of N-acetylmuramic acid (NAM) and the

C-4 of N-acetylglucoseamine (NAG). The hexasaccharide is bound to lysozyme with the involvement of Asp 52 and Glu 35.

ß-lactoglobulin is a major whey globular protein (MW 18,000). Between pH=5 to pH=7 the protein exists as a dimer of 36,700 Da because of electrostatic interactions between Asp 130 and Glu 134 of one monomer with lysyl residues Lys 135, Lys 138 and Lys 141 of another monomer, to produce a stable α-helical segment (Creamer, Parry and Malcolm, 1983). Above pH=6.5 the dimer dissociates due to excessive electrostatic repulsion (Brown and Farrell, 1978). At pH=4–5 and also above pH=8.0 the dimer self-associates to form an octamer especially when the temperature is reduced to 2°C (Pessen, Purcell and Farrell, 1985). There are two disulphide bonds (Cys 66—Cys 160 and Cys 106—Cys119) and a free thiol group Cys 121 which is occluded in the dimer but becomes more reactive above pH=6.5. The structure of ß-Lg has been determined by high resolution X-ray crystallography (2.8Å) and comprises a protein fold composed of two slabs of anti-parallel ß-sheet (Papiz, Sawyer, Eliopoulos, North, Findlay, Sivaprasadarao, Jones and Newcomer, 1986). The core of the fold is an eight stranded, cross-hatched ß-barrel with hydrophobic amino acids lining the internal cavity.

Methodology

ANALYSIS OF INTERACTION IN THE SUPERNATANT BY ION-EXCHANGE FAST PROTEIN LIQUID CHROMATOGRAPHY (IE–FPLC)

Supernatant solutions obtained from the mixtures (1:1) of α-lactalbumin–lysozyme and ß-lactoglobulin–lysozyme and the individual protein solutions were analysed by IE–FPLC using a Pharmacia–LKB MonoQ HR5/5 anion exchange column and a gradient elution of 0 M–0.7 M sodium acetate as described by Manji, Hill, Kakuda and Irvine (1985) and Hill, Manji, Kakuda, Myers and Irvine (1987).

The supernatant solutions and interaction products were analysed by the following methods.

SDS–polyacrylamide gel electrophoresis

For this, a 5%–20% (w/v) acrylamide gradient containing 4 M urea and 25 mM dithiothreitol and tris-glycine buffer of pH 8.6 was used. Samples were reduced with 1% ß-mercaptoethanol in the presence of 8 M urea prior to electrophoresis.

Proton NMR

Studies were undertaken on a Brucker AC 300 Fourier Transform spectrometer at 300 MHz. A total of 16000 data points was collected with a 30° pulse, a 15 ppm spectral width and a repetition time of 3.4 s. Between 128 and 512 scans were accumulated, as required.

Circular dichroism (CD)

Spectra were obtained using a JASCO J600 spectro-polarimeter equipped with a thermostated cell-holder. Protein supernatant solutions were measured at 185 nm–260 nm and 260 nm–340 nm.

Raman spectroscopy

Raman spectra of 15% heated (90°C, 30 min) and unheated protein solutions in D_20 were recorded on a JASCO Model NR-1100 laser Raman spectrophotometer with excitation from the 488 nm line of a Spectra-Physics Model 168B argon ion laser according to the method described by Li-Chan and Nakai (1991) and Nonaka, Li-Chan and Nakai (1993).

Molecular modelling

Molecular modelling studies were undertaken using an Evans and Sutherland ESV30 high resolution graphics work-station running the SYBYL molecular modelling package (Tripos Associates under Unix) to aid the visualizations of α-lactalbumin, ß-lactoglobulin and lysozyme. The crystal structure for lysozyme and α-lactalbumin were obtained from the Brookhaven Protein Databank. The ß-lactoglobulin crystal structure to date was kindly provided by Prof. A. North of the University of Leeds (Papiz *et al.* ,1986). In view of the occurrence of electrostatic interactions shown by ion exchange–FPLC, the acidic and basic residues in the proteins were examined. The structures were manipulated in real time and 3-D to visualize possible interaction between lysozyme and α-lactalbumin and between lysozyme and ß-lactoglobulin.

RESULTS

Studies on the interaction between α-lactalbumin, ß-lactoglobulin and lysozyme have shown that, in aqueous solution of pH 6.8, a higher level of interaction occurred between lysozyme and ß-lactoglobulin compared with lysozyme and α-lactalbumin. A similar interaction was not observed between α-lactalbumin and ß-lactoglobulin. The predominant product of the interaction was an insoluble precipitate. In addition, a small amount of soluble complex was also recovered by ion exchange–FPLC of the supernatant (Figure 16.3). The soluble complex was shown by SDS–PAGE to comprise the same components as the insoluble precipitate i.e. lysozyme and α-lactalbumin or lysozyme and ß-lactoglobulin.

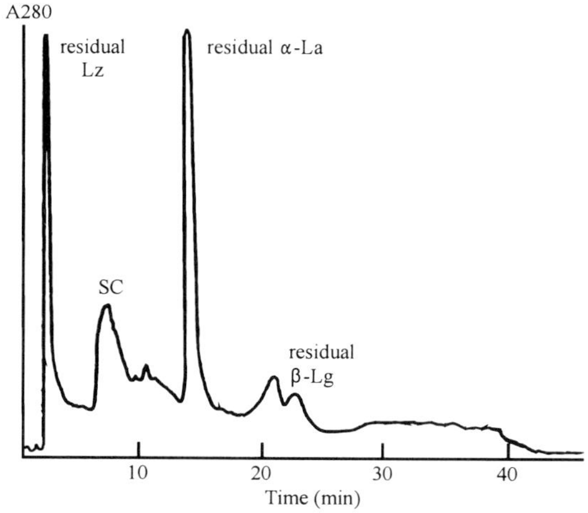

Figure 16.3 IE-FPLC chromatogram of a mixture of lysozyme (Lz), α-lactalbumin (α-LA) and β-lactoglobulin (β-Lg) showing preferential binding of Lz with β-Lg and formation of a soluble complex (SC). (Howell *et al.*, 1995)

The addition of up to 100 μg of lysozyme to a mixture of α-lactalbumin and ß-lactoglobulin led to a faster decrease in ß-lactoglobulin A and B peaks compared with the α-lactalbumin peak indicating preferential binding of ß-lactoglobulin by lysozyme probably due to its greater size, charge or hydrophobicity (Howell *et al.*, 1995).

MOLECULAR MODELLING

Visualisation of the possible interactions between lysozyme and ß-lactoglobulin indicated optimal fit where lysines 138 and 141 on ß-lactoglobulin formed ion pairs with Glutamate 35 and Aspartate 53 on lysozyme. It is interesting to note that these are the same carboxylates that are involved in the active site of lysozyme. Additionally there are about four or five acidic residues on ß-lactoglobulin which appear to match up with basic residues on lysozyme (Figure 16.4) (Howell *et al.*,

1995). The proposal that the ß-lactoglobulin is able to interact as a monomer with lysozyme is feasible because above a pH of 6.5 the dimer, normally prevalent over the pH range 3–7 dissociates due to excessive electrostatic repulsion (Brown and Farrell, 1978). In addition, it is quite likely, that non-specific interactions between the two proteins also occur. In contrast to ß-lactoglobulin, α-lactalbumin did not indicate any specific areas of interaction. A tentative visualisation for possible interactive docking based on electrostatic interaction between α-lactalbumin and lysozyme is shown in Figure 16.5 (Howell *et al.*, 1995).

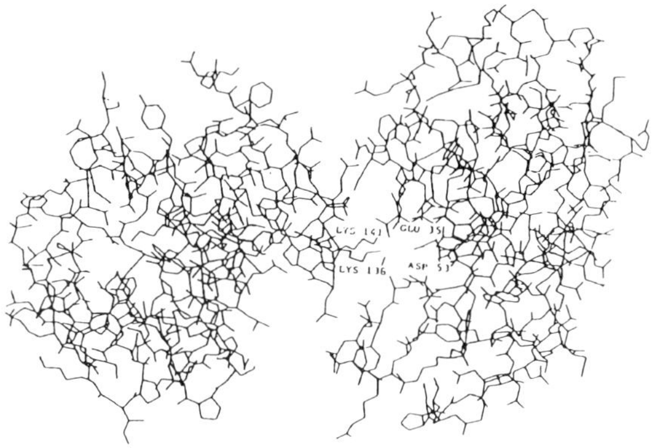

Figure 16.4 Possible interaction for ß-lactoglobulin (left) and lysozyme (right) based on electrostatic interactions of acidic and basic groups (Howell *et al.*, 1995)

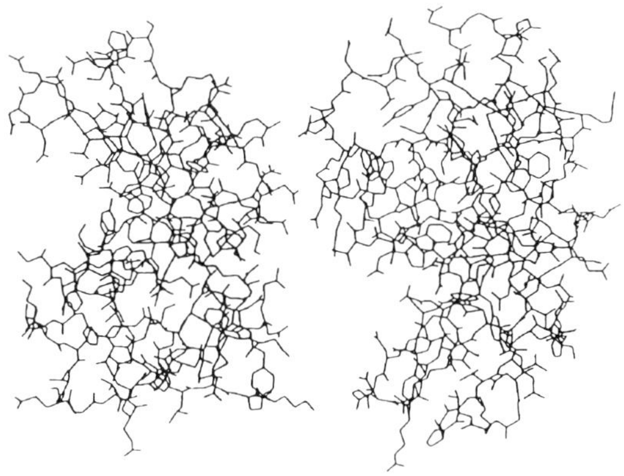

Figure 16.5 Possible interaction for α-lactalbumin (left) and lysozyme (right) based on electrostatic interactions of acidic and basic groups (Howell *et al.*, 1995)

Proton NMR studies on supernatants obtained from heated and unheated proteins and mixtures on a Brucker AC300 indicated changes in the aromatic and aliphatic regions. Like the NMR spectra, circular dichroism (CD) studies indicated that, fully folded proteins were present in the supernatant solutions of unheated proteins.

HEATING OF PRECIPITATES

Heating the lysozyme–ß-lactoglobulin or lysozyme–α-lactalbumin mixtures to 90°C resulted in an aggregated precipitate. The Raman spectra of the lysozyme–ß-lactoglobulin mixture indicated differences when compared to the simulated spectra based on 1:1 combination showing evidence of interaction between the two proteins, both before and after heating (Figure 16.6). The Raman spectra of the unheated and heated precipitates indicated not only the presence of electrostatic interactions but also hydrophobic interactions and changes in the secondary structure. On heating, the involvement of disulphide bonds was shown (Howell and Li-Chan, submitted).

Foams

Foaming properties of proteins refer mainly to foam expansion and foam stability. Foam formation requires surface active agents such as proteins to lower the surface tension and hence promote a large surface area and the formation of small air bubbles. Subsequently, the air bubbles are stabilised by the formation of a viscoelastic film at the bubble surface. The strength of the film is improved by proteins which denature and unfold at the surface and undergo electrostatic and hydrophobic interactions. Interactions of proteins in foam formation and stability have been investigated previously by the author and co-workers by chemically modifying proteins particularly bovine serum albumin (BSA).

AMIDATION

Howell and Taylor (1991) amidated BSA using a carbodiimide mediated reaction using ammonium sulphate as the nucleophile. It was found that whilst extensively amidated BSA (pI=9) on its own had poor foam expansion and foam stability, its combination with native BSA and blood plasma enhanced these properties when mixed in the ratio of 1:1 (Plate 16.3).

Partial amidation of BSA resulted in a product with a pI range of 5.3–7.5 which had excellent foam expansion and stability at pH=7.0. The interaction was due to electrostatic as well as hydrophobic interactions (Howell and Taylor, 1992).

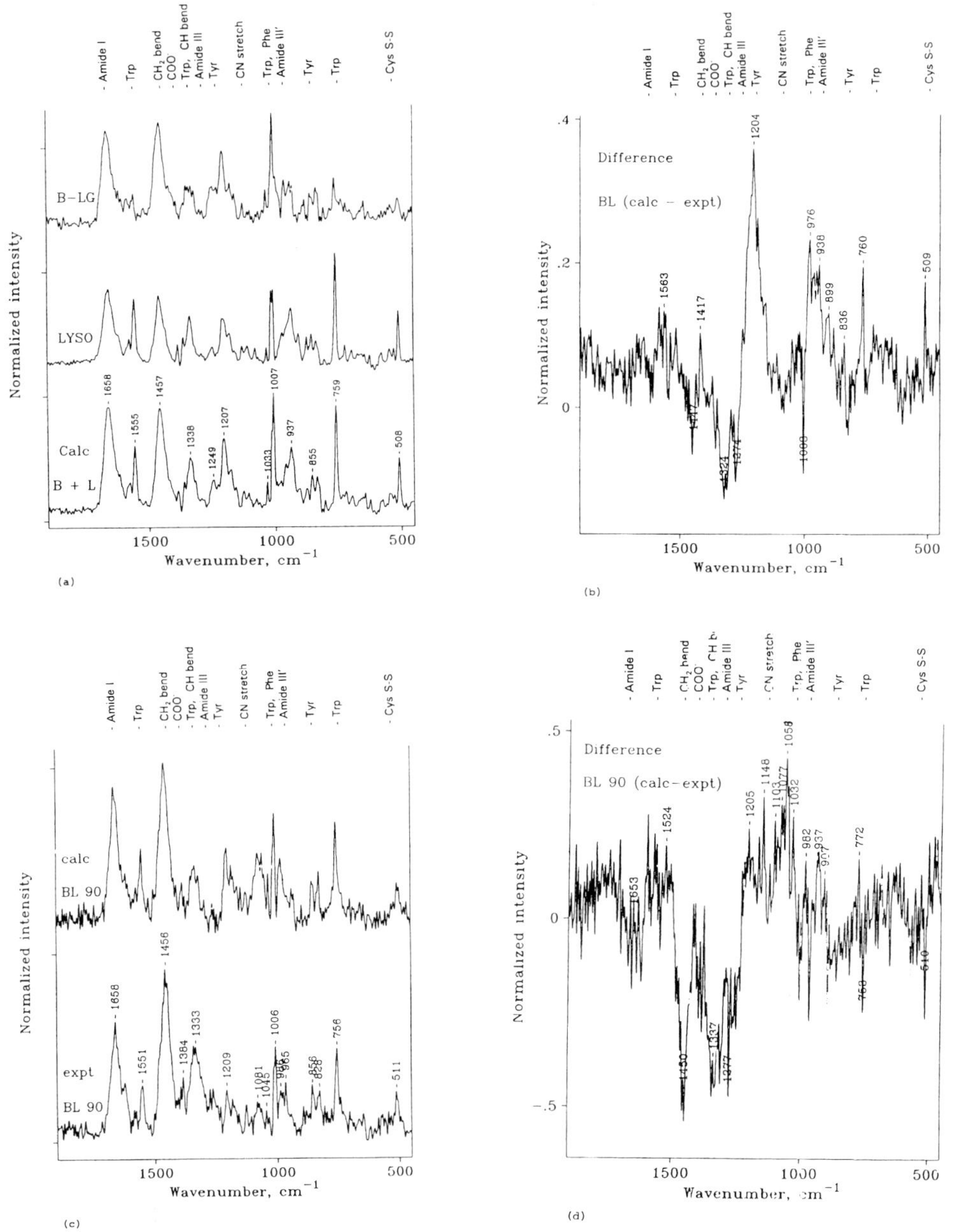

Figure 16.6 Raman spectra of aggregates from unheated and heated mixtures of ß-lactoglobulin and lysozyme (1:1 ratio by weight) (15% total protein in D_2O): (a) Spectra of ß-lactoglobulin (B-LG), lysozyme (LYSO) and calculated average (Calc B+L) of ß-lactoglobulin–lysozyme precipitate; (b) Spectra of the difference between calculated and experimental spectra (BL calc - expt) ß-lactoglobulin; (c) Calculated (calc BL90) and experimental (expt BL90) spectra of 90°C heated ß-lactoglobulin–lysozyme precipitate; (d) Difference spectra of calc BL90 - expt BL90. (Howell and Li-Chan, submitted)

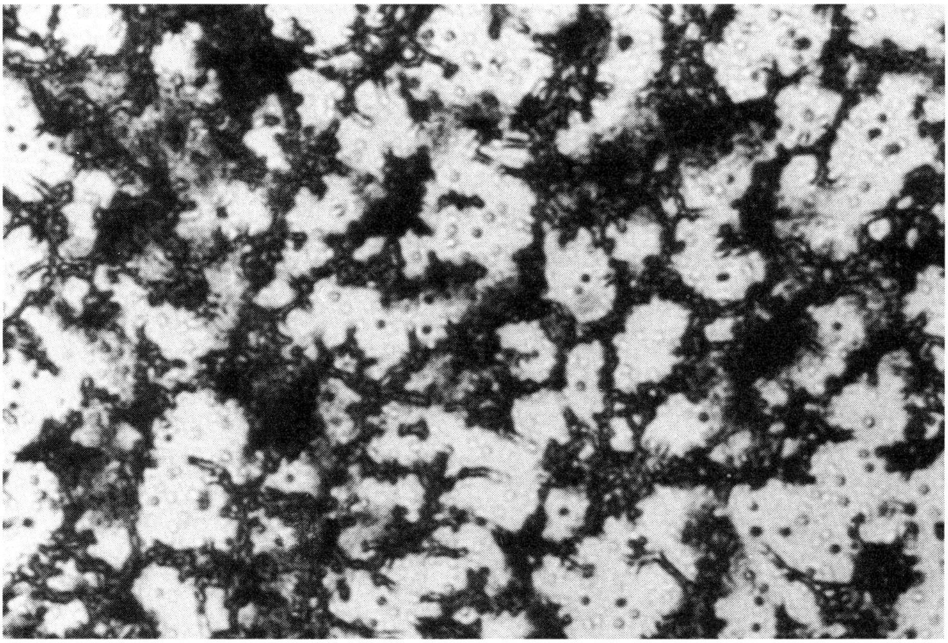

Plate 16.3 Phase contrast micrograph of the gelation of soluble wheat protein (19%) and 0.1% lysozyme heated at 80°C

The interactions between differently chemically modified BSA and their foaming properties in 1:1 mixtures by weight were investigated by Murphy and Howell (1991) in order to determine whether a combination of the most negative and most positive protein had superior functional properties. Succinylated proteins showed inferior foaming properties both individually and in mixtures. The generally held theory of increased electrostatic interactions between positive and negative proteins was not upheld for succinylated or thiolated BSA with lysyl-BSA. However an interaction was evident between EDC-modified polylysyl-BSA and native BSA.

Murphy and Howell (1991) concluded that the interactions are between positive and native negatively charged proteins and not proteins chemically modified to provide the negative charge by blocking the amino groups Thus the interaction of the available amino groups and not the overall charge o the protein provides the basis for the mechanism of the molecular association which improves foaming properties.

Phase separation and gel inhibition

In contrast to the compatible gel networks formed mainly by the globular protein mixtures, mixtures of blood plasma proteins with soya isolate or sodium caseinate either did not form gels or the gels were weak (Howell and Lawrie, 1984). This negative interaction or gel inhibition was attributed to the formation of incompatible gel structures due to interference of the plasma gel network with the large protein aggregates and micelles of soya isolate or caseinate. In this respect, the behaviour was similar to mixtures of plasma and a number of polysaccharides all of which considerably lowered the gelation properties of plasma proteins (Howell and Lawrie, 1984).

SOYA ISOLATE—WHEY PROTEIN MIXTURES

Similarly, the gelling properties of mixtures of commercial soya isolates and whey protein isolates in 2% NaCl were examined by heating at 80° or 90°C for 30 minutes (Lee and Howell, 1985). Both isolates formed gels at a concentration of 10%. Mixtures of 9% soya isolate with 1% whey isolate and 8% soya isolate with 2% whey isolate both formed gels which appeared as a single phase. However, additional increase in whey concentrations resulted in phase separation and the formation of three layers with top and bottom soya proteins layers and a middle layer which was predominantly whey.

Table 16.2 INTERACTION OF THE WATER-SOLUBLE FRACTIONS OF SOYA ISOLATE WITH WHEY PROTEIN ISOLATE

Protein concentration (%) *Soluble soya (SS) : Whey isolate (W)*	*Gel strength*
SS 10%	76
SS 8 : W 2	357
SS 6 : W 4	573
SS 2 : W 8	824
W 10%	450

In order to test the theory proposed above, i.e. that it is the large molecular aggregates of soya isolates which are involved in phase separation, the soya isolate was separated into the water soluble and the water insoluble fractions. Dispersions prepared from the freeze-dried water soluble fraction were heated at 90°C for 30 minutes in isolation (10%) and in combination with whey proteins mixed in the ratios 8:2, 6:4 and 2:8. The gels were tested by compression on the Instron Testing Machine (Table 16.2) as described previously.

Single phase gels and synergistic interactions were observed for all combinations of the soya isolate–whey protein mixtures compared to 10% gels formed by each protein in isolation. Thus phase separation in mixtures of globular proteins with high molecular weight proteins is proposed by the author to be due to the incompatible molecular size and configuration of the proteins and physical movement of the dense material to the bottom and of lighter material to the top. Further detailed studies in this area by rheological, physico-chemical and immunocytochemistry–electron microscopy techniques are being conducted by S. Comfort and N. Howell (1995).

Conclusion

Protein–protein interactions are highly complex and result in three main types of interactions:

(1) *Synergistic interdctions* occur between mainly globular proteins of similar size and charge and with two or more proteins in the mixture. This results in a single phase with greater gelling ability, foam formation and foam stability. For gels, the mechanism is dependent on the differing rates of unfolding of the different proteins in the mixtures and exposure and types of reactive groups. The gelation process is sensitive to protein concentration, time and temperature of heating and the presence of salts.

(2) *Aggregation* arises mainly from electrostatic interactions between oppositely charged proteins which may be non-specific or highly specific. Conformational changes and hydrophobic interactions are also involved. On heating the aggregates, disulphide bonds also occur.

(3) *Phase separation* and incompatibility may be caused by the repulsion of like charges on the protein molecules. The presence of large molecules or aggregates cause steric hindrance and result in gel inhibition, separation into two or more layers and the inclusion of a disperse phase in a continuous phase.

These complex interactions require systematic studies using a range of modern techniques, currently being undertaken by the author, to elucidate the structure–function relationships of proteins which can be used to enhance gelation, foaming and emulsification.

Acknowledgements

I am very grateful to all the past and present members of my research team and collaborators for their various contributions. I should also like to thank the S.E.R.C., A.F.R.C. and MAFF/Industry (LINK) Scheme for funding the projects.

References

Brown E.M. and Farrell H.M. (1978) Interaction of ß-lactoglobulin and cytochrome C: complex formation and iron reduction. *Archives of Biochemistry and Biophysics*, **185**, 156–164

Brown, J.R. (1975) Structure of bovine serum albumin. *Federation Proceedings* [Federation of American Societies for Experimental Biology], **34**, 591

Creamer L.K., Parry D.A.D. and Malcolm G.N. (1983) Secondary structure of bovine ß-lactoglobulin. *Archives of Biochemistry and Biophysics*, **227**, 98–105

Fernandez-Diez M.J., Osuga, D.T. and Feeney, R.E. (1964) The sulphydryls of avian ovalbumin, bovine ß-lactoglobulin and bovine serum albumin. *Archives of Biochemistry and Biophysics*, **107**, 449–458

Fothergill, L.A. and Fothergill, J.E. (1970) Thiol and disulphide contents of hen ovalbumin. C-terminal sequence and location of disulphide bonds. *Biochemical Journal*, **116**, 555–561

Hill A.R., Manji, B., Kakuda, Y., Myers, C. and Irvine, D.M. (1987) Quantification and characterization of whey protein fractions separated by anion exchange chromatography. *Milchwissenschaft*, **42**, 693–696

Hoshi, Y. and Yamauchi, F. (1983) Determination of sulphydryl and disulphide contents of soyabean 11 S globulin and their change by lyophilisation. *Agricultural and Biological Chemistry*, **47**, 2435–2440

Howell, N.K. (1992) Protein–protein interactions. In *Biochemistry of Food Proteins*, edited by B.J.F. Hudson, London: Elsevier Applied Science, pp 35–74

Howell, N. and Lawrie, R.A. (1983) Functional aspects of blood plasma proteins. 1. Separation and characterisation. *Journal of Food Technology*, **18**, 747–762

Howell, N.K. and Lawrie, R.A. (1984) Functional Aspects of blood plasma proteins. 2. Gelling properties. *Journal of Food Technology*, **19**, 289–295

Howell, N.K. and Lawrie, R.A. (1984) Functional Aspects of blood plasma proteins. 3. Interaction with other proteins and stabilisers. *Journal of Food Technology*, **19**, 297–313

Howell, N.K. and Lawrie, R.A. (1985) Functional aspects of blood plasma proteins. 4. Elucidation of the mechanism of gelation of plasma and egg albumen proteins. *Journal of Food Technology*, **20**, 489–504

Howell, N.K. and Lawrie, R.A. (1987) Functional aspects of blood plasma proteins. 5. Viscosity. *Journal of Food Technology*, **22**, 145–151

Howell, N. and Li-Chan, E. (1995). Elucidation of the interactions of lysozyme and whey proteins by Raman spectroscopy (Submitted)

Howell N.K. and Taylor, C. (1991) Effect of amidation on the foaming and physicochemical properties of bovine serum albumin. *International Journal of Food Science and Technology*, **26**, 385–395

Howell, N.K., Yeboah, N. and Lewis, D. (1995) Studies on the electrostatic interactions of lysozyme with α-lactalbumin and β-lactoglobulin. *International Journal of Food Science and Technology*, **30**, 321–334

Kato, A. and Nakai, S. (1980) Hydrophobicity determined by a fluorescent probe method and its correlation with surface properties of proteins. *Biochimica et Biophysica Acta*, **623**, 13–20

Kato, A., Osako,Y., Matsudomi, N. and Kobayashi, K. (1983) Changes in the emulsifying and foaming properties of proteins during heat denaturation. *Agricultural and Biological Chemistry*, **47**, 33–37

Kato, A., Tsutsui, N., Matsudomi, N., Kobayashi, K. and Nakai, S. (1981) Effects of partial denaturation on surface properties of ovalbumin and lysozyme. *Agricultural and Biological Chemistry*, **45**, 2755–2760

Li-Chan, E. and Nakai, S.(1991) Raman spectroscopic study of thermally and/or dithiothreitol induced gelation of lysozyme. *Journal of Agriculture and Food Chemistry*, **39**, 1238–1245

Manji, B., Hill, A., Kakuda Y. and Irvine, D.M. (1985) Rapid separation of milk whey proteins by anion exchange chromatography. *Journal of Dairy Science*, **68**, 3176–3179

McKenzie, H.A. and White, F.H. (1991) Lysozyme and α-lactalbumin: structure, function and interrelationships. *Advances in Protein Chemistry*, **41**, 173–315

Murphy, M.C. and Howell, N.K. (1990a) Effect of succinylation on the functional and physicochemical properties of bovine serum albumin. *Journal of the Science of Food and Agriculture*, **51**, 109–123

Murphy, M.C and Howell, N.K. (1990b) Effect of thiolation on the physicochemical and functional properties of bovine serum albumin. *Journal of the Science of Food and Agriculture*, **53**, 549–558

Murphy, M.C. and Howell, N.K. (1991) Functional properties of native and positively and negatively charged modified bovine serum albumin. *Journal of the Science of Food and Agriculture*, **55**, 489–492

Nonaka, M., Li-Chan, E. and Nakai, S. (1993) Raman spectroscopic study of thermally induced gelation of whey proteins. *Journal of Agriculture and Food Chemistry*, **41**, 1176–1181

Papiz, M.Z., Sawyer, L., Eliopoulos, E.E., North, A.C.T., Findlay, J.B.C., Sivaprasadarao, R., Jones, T.A. and Newcomer, M.E. (1986) The structure of bovine ß-lactoglobulin and its similarity to plasma retinol-binding protein. *Nature*, **324**, 383–385

Pessen, H., Purcell, J.M. and Farrell H.M. (1985) Proton relaxation rates of water in dilute solutions of ß-lactoglobulin. Determination of cross relaxation and correlation with structural changes by the use of two genetic variants of a self-associating globular protein. *Biochimica et Biophysica Acta*, **828**, 1–12

Polak, J.M. and Van Norden, S. (1984). *An introduction to immunocytochemistry: current techniques and problems*, Royal Microscopical Society

Richardson, R.H. and Brew, K. (1980) Lactose synthase: an investigation of the interaction site of -lactalbumin for galactosyltransferase by differential kinetic labelling. *Journal of Biological Chemistry*, **255**, 3377–3385

17

MIXED BIOPOLYMERS AT INTERFACES

ERIC DICKINSON
Procter Department of Food Science, University of Leeds, Leeds. LS2 9JT U.K.

Introduction

Mixtures of biopolymers are to be found in a wide range of food colloids (Dickinson, 1992a). The physico-chemical stability properties of these systems are determined in large part by the nature of the biopolymer interactions at the surface of the dispersed particles, bubbles or droplets. In a typical food emulsion of the oil-in-water type, the primary macromolecular stabilizing agent is a mixture of proteins adsorbed as a thin layer at the oil/water interface (Dickinson, 1992b). Additionally, food polysaccharides are commonly present as thickening or gelling agents, and there may be direct interaction between these polysaccharides and adsorbed proteins, possibly leading to the formation of a secondary macromolecular stabilizing layer (Dickinson and Euston, 1991).

The stability and rheology of a food colloid are sensitive to the nature and strength of the protein–protein and protein–polysaccharide interactions in the mixed biopolymer layer. Whether these interactions are net attractive or net repulsive (weak or strong) is determined by the detailed chemistry of the biopolymers and the solution conditions (pH, temperature, etc.). The overall structure and composition of the layer depends on the balance between the different kinds of biopolymer interactions, i.e. with the surface, with the aqueous medium and with other biopolymer molecules in the system.

The local macromolecular concentration in an adsorbed biopolymer layer is similar to that in a highly concentrated biopolymer solution or gel. In terms of theoretical understanding, therefore, there seems much to be gained in trying to relate the properties of mixed biopolymer layers to the thermodynamics of concentrated bulk solutions containing the same biopolymer components. From discussions elsewhere in this volume, we know that many mixed biopolymer systems are difficult to maintain in a homogeneous state at high solute concentrations. This is due to the predominance of energetic factors over entropy of mixing terms

(Dickinson and Stainsby, 1988). A binary aqueous solution of protein plus polysaccharide may exhibit either complex coacervation or thermodynamic incompatibility depending on whether the overall molecular protein–polysaccharide interaction is, respectively, net attractive or net repulsive. Complex coacervation (Bungenberg de Jong, 1949) involves spontaneous separation of the system into coexisting solvent-rich and solvent-depleted phases, the latter being predominantly composed of an insoluble co-precipitate of the two biopolymers. Thermodynamic incompatibility (Albertsson, 1971; Tolstoguzov, 1991) involves the system separating into two solvent-rich phases, each composed predominantly of just one of the biopolymer components.

Interaction with the adsorbent surface provides an additional driving force towards heterogeneity in colloids containing mixed biopolymers. It is convenient mechanistically to distinguish between competitive and cooperative phenomena. What we mean by competitive adsorption is the partial or complete displacement from the surface of one biopolymer species by another, e.g. casein displacing gelatin (Mussellwhite, 1966). Cooperative adsorption involves the transient or permanent association of different species together at the interface, e.g. alginate interacting with milk protein (Bergenståhl, Fäldt and Lassen, 1992). In practice, for many mixed systems, there is evidence for both competitive and cooperative effects in the same layer (Dickinson, Murray, Murray and Stainsby, 1987; Bergenståhl, Fäldt and Malmsten, 1995).

A combination of various direct and indirect physico-chemical techniques is providing valuable information on the organization in mixed biopolymer layers at fluid interfaces:

(i) Ellipsometry and radio-labelling studies at planar interfaces;
(ii) Surface tension and surface shear rheology at planar interfaces;
(iii) Electrophoretic mobility of polymer-coated oil droplets;
(iv) Stability studies of emulsions and foams.

This chapter describes some recent evidence from these techniques as applied to some well-defined systems containing binary mixtures of food biopolymers. An attempt is made to explain the origin of the generic types of adsorbed layer structure in terms of the biomolecular interactions involved.

Theoretical Background

According to de Gennes (1987), two different configurational arrangements can usefully be distinguished in polymer chain adsorption at a solid flat surface. One situation, called uniform adsorption, occurs when simple homopolymers are adsorbed due to a preference of the individual chain segments to interact with the surface rather than with bulk solvent. An alternative situation arises with so-called 'functional polymers' (e.g. associative thickeners or block copolymers) where one part of the chain (the anchor) becomes permanently bonded to the surface and the remainder (the buoy) is either free to stick out into the solution or to adsorb reversibly, as the case may be. Features of both these models are useful for understanding the main physico-chemical factors influencing the structure of pure and mixed biopolymer layers.

In bulk solution under Θ-solvent conditions, it is well known that linear polymers can be regarded as random walk chains. This description breaks down, however, near an impenetrable surface because half the space is inaccessible to the polymer segments. The spatial restriction leads to a lower conformational entropy. As a consequence, if the segments have no binding affinity, a depletion zone develops near the surface (Fleer, Cohen Stuart, Scheutjens, Cosgrove and Vincent, 1993). Alternatively, in the common situation where there *is* a binding affinity, the polymer can adsorb onto the surface in spite of the entropy loss because the adsorption energy u_{a0} of a solvent molecule is less negative than the adsorption energy u_{a1} of a polymer segment. Following the nomenclature of Silberberg (1968), it is conventional to express the net polymer segment binding affinity in terms of the dimensionless adsorption energy parameter

$$\chi_s = \frac{u_{a0} - u_{a1}}{kT} \tag{1}$$

where k is Boltzmann's constant and T is temperature. The parameter P_s is related to the difference between the two Flory–Huggins parameters, P_{0s} and P_{1s}, corresponding to solvent–surface and polymer–surface interactions, respectively. When P_s is negative, zero, or only slightly positive (i.e. $P_s < P_s^c$), the homopolymer does not adsorb because of the unfavourable configurational entropy change. But it only takes a relatively weak segment binding energy to overcome this entropy effect and thus allow the polymer to adsorb. For long chains on a tetrahedral lattice, the value of the 'critical' adsorption energy is $P_s^c = 0.29$ (Fleer *et al.*, 1993). The structure of the homopolymer adsorbed layer is usually described in terms of the so-called train–loop–tail model. Influences of polymer concentration, chain length and segment binding energy on proportions of homopolymer segments in trains, loops and tails are now well known from both mean field theory and computer simulation (Lipatov and Sergeeva, 1974; Dickinson and Lal, 1980; Fleer *et al.*, 1993).

In a system containing a mixture of two homopolymers of similar chain length, the component with the largest positive value of χ_s will predominate at the interface. Let us denote component 1 as the polymer with the higher surface affinity. Its adsorption energy parameter with respect to the solvent is χ_{s10}. Polymer species 2 has a lower affinity χ_{s20}, and the difference

$$\chi_{s12} = \chi_{s10} - \chi_{s20} \tag{2}$$

represents the adsorption energy (in units of kT) of segments of type 1 from a solvent of type 2. When the polymers are of high molecular weight, a positive value of χ_{s12} of just 0.1 or maybe even 0.01 is sufficient to induce essentially complete displacement from the surface of polymer 2 by polymer 1. That is, the energetic driving force for equilibrium polymer exchange between bulk solution and the adsorbed layer is generally much less than that for total polymer adsorption–desorption. The reason why the layer composition is so very sensitive to the parameter χ_{s12} is simply a reflection of the accumulative effect of a very large number of potentially adsorbing segments which can exchange with one another at the various adsorption sites on the surface. While the difference in adsorption energy may be small per segment, it is nevertheless large per chain. (The argument here is somewhat equivalent to that used elsewhere in this volume to explain the incompatibility of two polymers in a concentrated solution when the mutual solvency parameter χ_{12} is only very slightly positive.)

To obtain good steric stabilization by adsorbed polymer, three main conditions should be satisfied (Dickinson, 1992a):

(i) There should be full surface coverage (otherwise polymer bridging may occur);
(ii) The molecules should be strongly attached (but some parts of the chain should stick away from the surface in order to give the layer significant thickness);
(iii) Regions away from the surface should be in a good solvent environment (otherwise attraction between different layers may occur).

All of these conditions cannot be met simultaneously for a single homopolymer component, since the polymer either adsorbs too strongly (in poor solvents) to give all-train flat configurations and hence thin layers, or too weakly (in good solvents) leading to poor coverage and incipient polymer bridging. This means that homopolymers do not make good colloid stabilizers. Much more effective are permanently grafted polymer chains, or alternatively AB block copolymers each composed of a small fraction of strongly adsorbing type-A segments and a large fraction of non-adsorbing type-B segments.

Properties of systems containing tethered chains have attracted considerable theoretical interest recently (Halperin, Tirrell and Lodge, 1992). An ideal tethered chain has one end permanently attached to the surface. (In a real system the attachment point may not occur exactly at the end; nor is it necessarily restricted to just a single segment along the chain.) Figure 17.1 shows highly schematic representations of terminally anchored flexible chains at a flat plane. Where the separation between anchors is much greater than the radius of gyration R_g, the resulting configuration is called (a) a 'mushroom' or (b) a 'pancake', depending on the value of χ_s (de Gennes, 1976). In the mushroom state ($\chi_s < \chi_s^c$), the chain is repelled entropically from the surface (up to a distance $\sim 2R_g$) and the segment

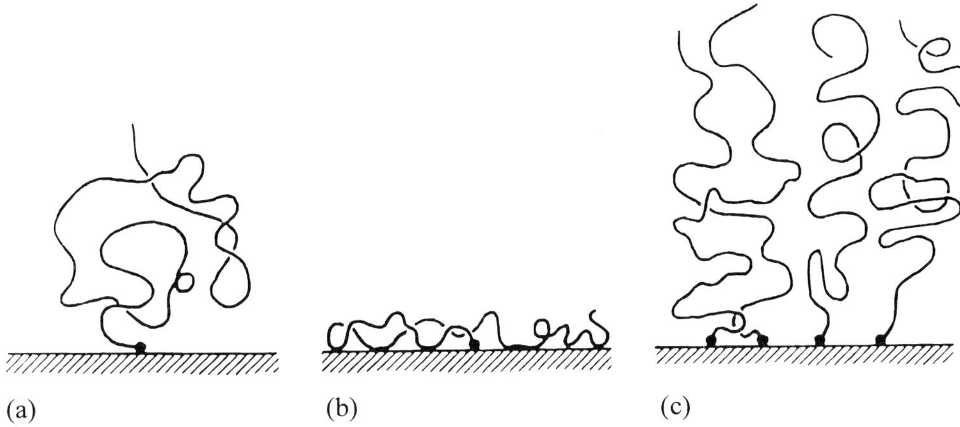

Figure 17.1 Diagrammatic representation of three alternative idealized configurations of terminally anchored chains at a hard planar surface: (a) mushroom; (b) pancake; (c) brush. (Reproduced from Fleer *et al.* (1993) with permission.)

density profile $\phi(z)$ goes through a maximum at $z \sim R_g$. In the pancake state ($\chi_s > \chi_s^c$), the configurational structure of the adsorbed chain resembles the train–loop–tail representation of an adsorbed homopolymer with an exponentially decaying segment density profile $\phi(z)$. When the surface density of anchor points is high, the structure of the layer of tethered chains is rather different, as illustrated in Figure 17.1(c). Under conditions of substantial chain overlap, there is a large excluded volume interaction between neighbouring chains which leads to a stretched configuration called a 'brush'. The transition from pancake to brush with increasing anchor surface density becomes sharper as the chain length increases (Chakrabarti, 1994).

Polymer brushes may contain chains of different sizes or different χ_s parameters. In a brush made from chains of two different lengths, the shorter chains are harder to stretch (Milner, 1991). This leads to a vertical segregation of the chain free ends, with short-chain free ends closer to the surface. In this situation the short chains act as spacers between the long chains, reducing the excluded volume interaction between the long chains and so reducing their need to stretch so far away from the surface. Another kind of segregation effect occurs in brushes made of chemically dissimilar chains of the same length (Milner, 1991). In a brush containing a mixture of randomly tethered A and B chains, phase separation in the polymer layer will take place if the A—B interaction is energetically unfavourable. For example, B-chain ends might be mainly confined to the portion of the end-density distribution closest to the surface, leaving a region composed of nearly pure A chains further away (Figure 17.2(a)). Alternatively, one could envisage some sort of micro-phase separation in the plane of the layer, but this would be limited spatially due to the lack of translational mobility of segments close to the anchors (Figure 17.2(b)). On the other hand, at a fluid interface where anchors would be more able to diffuse freely in the plane of the interface, extensive two-dimensional phase separation would be possible (see Figure 17.9(a) later).

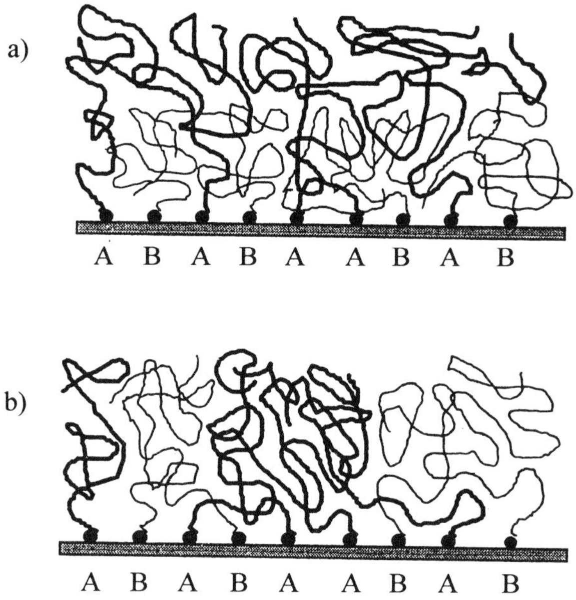

Figure 17.2 Diagrammatic representation of two types of compositional inhomogeneity in brushes containing a mixture of two different terminally anchored chains (A and B): (a) micro-phase separation perpendicular to the surface; (b) micro-phase separation parallel to the surface.

The thermodynamics of a bulk system containing solvent (0) + polymer 1 + polymer 2 can be described in terms of three Flory–Huggins parameters (χ_{10}, χ_{20} and χ_{12}) (Flory, 1953). The overall mixing free energy is

$$\Delta G_{mix} = \frac{RTV}{V_0}\left(\phi_0 \ln \phi_0 + \frac{\phi_1}{m_1} \ln \phi_1 + \frac{\phi_2}{m_2} \ln \phi_2 + \chi_{10}\,\phi_1\,\phi_0 + \chi_{20}\,\phi_2\,\phi_0 + \chi_{12}\,\phi_1\,\phi_2\right) \quad (3)$$

where ϕ_i (i = 0, 1, 2) and m_i (i = 1, 2) are respectively the volume fraction and chain length of component i, R is the gas constant, V is the volume of the solution, and V_0 is the solvent molar volume. If the free energy change is not negative for the mixing together of ternary solutions of different composition, the homogeneous mixed system is materially unstable. Two kinds of phase separation are then possible: complex coacervation or thermodynamic incompatibility. Which one actually occurs depends predominantly on the value of the polymer–polymer Flory–Huggins parameter χ_{12}. Complex coacervation takes place when there is sufficiently strong net attraction between the two polymers ($\chi_{12} < 0$), and thermodynamic incompatibility (originally called 'simple' coacervation) occurs when there is sufficient net repulsion ($\chi_{12} > 0$). The larger are the two solute polymer volume fractions (ϕ_1 and ϕ_2), the smaller is the absolute value of χ_{12} required for phase separation. For long chains ($m_i \sim 10^4$) in concentrated solutions ($\phi_0 \leq 0.95$), the theory predicts that there is phase separation for $|\chi_{12}| \leq 0.2$.

In a typical system containing a mixed biopolymer layer, the local segment concentration in the immediate vicinity of the surface is very high compared with that normally encountered in an aqueous solution. This means that the onset of thermodynamic incompatibility or complex coacervation becomes important in thin layers at small values of $|\chi_{12}|$. If both polymers are present together at the surface, then phase separation may occur within the mixed adsorbed layer. When only one of the polymeric species is actually adsorbed, this may be because of thermodynamic incompatibility between the polymer component on the surface and the less surface-active polymer component in the bulk. For any specific binary polymer system, the detailed structure and composition of the adsorbed layer depends on the balance of competitive and cooperative interactions at the interface.

Mixed Biopolymer Adsorption: Competition *versus* Cooperation

Prior to some recent systematic studies on milk proteins, a survey of the literature showed (Dickinson, 1986) that one of the most carefully investigated multicomponent biopolymer systems from the point of view of adsorption is blood plasma. The compositions of adsorbed layers on various surfaces in contact with a solution of blood plasma proteins are known to change with time due to the competitive adsorption between the different components (Vroman and Adams, 1986). The most abundant small molecules adsorb first. These are then slowly and sequentially replaced by the less abundant high-molecular-weight species. This sequence of proteins indicating the order of adsorption at a glass surface is now commonly known as the Vroman series: albumin→immunoglobulin→fibrinogen and fibronectin→high-molecular-weight kininogen.

Three points of more general significance are illustrated by the blood plasma protein experiments. First, it is not necessarily the biopolymer that predominates in the freshly formed initial layer that is preferentially adsorbed under equilibrium conditions. Secondly, if sufficient time is allowed for them to diffuse to the interface and then compete with previously adsorbed species, even components present at very low concentration may predominate at the interface at long times. Thirdly, the nature of the adsorbent surface has a strong influence on the amount adsorbed and the adsorbed layer composition.

There may be a delicate balance between competitive and cooperative (associative) interactions at the interface. Consider the situation where one biopolymer A is already adsorbed at a plane solid surface and a second biopolymer B is introduced into the bulk phase at a similar overall composition. What happens next depends on a number of factors, including most notably the nature and strength of the biopolymer–surface and biopolymer–biopolymer interactions. In Figure 17.3 are illustrated a few simple scenarios for the idealized case of biopolymers represented as equal-sized spheres. In situation (a) the original monolayer of type-A molecules is unaffected by the presence of type-B molecules. Assuming this corresponds to the equilibrium state, it means that biopolymer A has a significantly higher surface affinity than biopolymer B (i.e. $\chi_{sAB} > 0$). In situation (b) the original monolayer of type-A molecules is replaced by a monolayer of type-B molecules. This competitive displacement of A by B implies now that biopolymer B has the higher equilibrium surface activity (i.e. $\chi_{sAB} < 0$ or $\chi_{sBA} > 0$). When the two binding affinities are very closely matched ($\chi_{sAB} \sim 0$), we will get a mixed monolayer at equilibrium as illustrated in Figure 17.3(c).

Whereas situations (a), (b) and (c) are examples of simple competition, other scenarios can be envisaged involving cooperative or associative adsorption. In situation (d) the type-B molecules adsorb on top of the original monolayer of type-A molecules to form a distinct secondary layer. This implies that there is a net attraction between the two kinds of biopolymer molecules ($\chi_{AB} < 0$) which causes biopolymer B to adsorb at a surface composed of type-A molecules without competing with the latter for a place in the primary layer ($\chi_{sAB} > 0$). Situation (e) is rather like (d) but with the roles of the two biopolymers reversed. That is, type-B molecules displace type-A molecules from the primary layer ($\chi_{sAB} < 0$), as in scenario (b), but now the latter adsorb back onto the molecules of the former ($\chi_{AB} < 0$) to produce a distinct two-layer structure. Finally, in situation (f), instead of having distinct primary and secondary layers composed of individual adsorbate species, as in (d) and (e), we have mixed layers which contain both species. This corresponds to a system with closely matched binding affinities ($\chi_{sAB} \sim 0$) and a net attraction between the two kinds of biopolymer molecules ($\chi_{AB} < 0$). Here the build-up of multilayers is equivalent to incipient complex coacervation at the surface.

Examples of the situations illustrated schematically in Figure 17.3 can be found in recent experiments on model systems containing two proteins that are introduced sequentially to a hydrophobic surface. Case (a) turns out to be rather common. This is due in large part to the fact that, in practice, the equilibrium state is rarely attained. For instance, radiotracer experiments at the air–water interface have shown (Hunter, Carbonell and Kilpatrick, 1991) that ß-casein cannot displace previously adsorbed lysozyme at pH 7; nor can lysozyme displace previously adsorbed ß-casein.

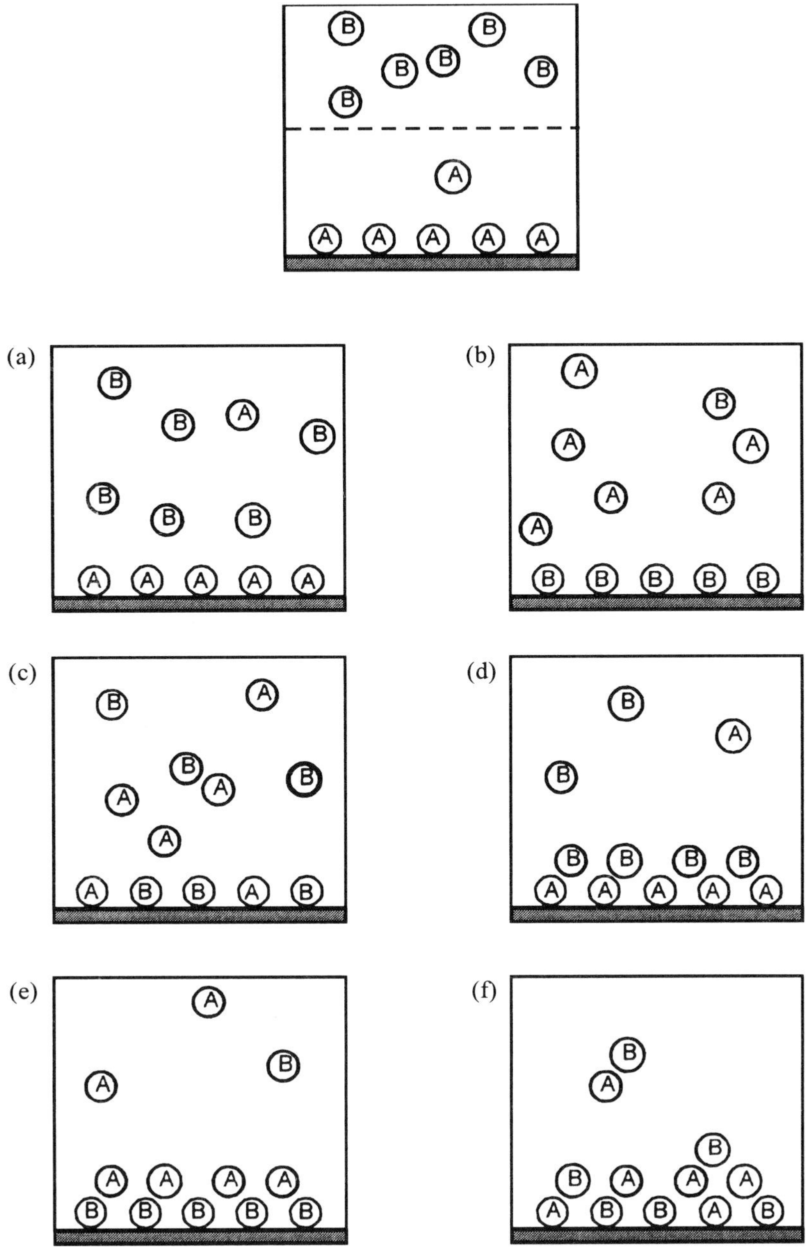

Figure 17.3 Competitive and cooperative adsorption of spherical molecules A and B at a planar surface. Component A is present at the surface as a saturated monolayer, and component B is introduced into the bulk phase in the vicinity of the surface: (a) no change; (b) monolayer B replaces monolayer A; (c) formation of mixed monolayer; (d) component B forms secondary layer on top of monolayer A; (e) component B displaces component A into secondary layer; (f) formation of mixed multilayer.

Similarly, ellipsometry experiments at the hydrophobized silica surface have shown (Nylander and Wahlgren, 1994) that a pre-adsorbed layer of ß-casein prevents sequential adsorption of ß-lactoglobulin. In the reverse experiment at the same hydrophobized silica surface, the adsorbed amount increases when a pre-adsorbed layer of ß-lactoglobulin is exposed to a solution of ß-casein at pH 7, although it never reaches the same level as for pure ß-casein under similar conditions. This agrees with light-scattering measurements on polystyrene latex particles (Dalgleish and Leaver, 1991) and measurements of adsorbed amounts of ß-lactoglobulin and ß-casein at the surface of emulsion droplets (Dalgleish, Euston, Hunt and Dickinson, 1991). At the emulsion droplet surface, there is also evidence for incorporation of ß-lactoglobulin into a pre-adsorbed layer of ß-casein without any significant displacement of the latter from the oil–water interface. In relation to the idealized situations illustrated in Figure 17.3, this seems close to type (d) behaviour, although as yet we have no direct structural evidence for any distinct two-layer arrangement.

In each of the experimental studies mentioned above, it is believed that the measured sequential adsorption behaviour is not governed by the principles of equilibrium thermodynamics. This main reason for this is that, once globular proteins have become established at the interface, they interact strongly with their adsorbed neighbours, and this makes them difficult to displace by other proteins (globular or disordered). Using radiolabelling, Hunter *et al.* (1991) have demonstrated that the flexible structure of ß-casein allows it to exchange between the bulk phase and the air–water interface, whereas the more rigid globular structure inhibits such exchange. The actual picture is then, of course, very much more complicated than implied by the simple spherical representations of adsorbate species in Figure 17.3. This is because the differing flexibilities of the disordered and globular protein allow the former to become accommodated in intermolecular spaces which the latter cannot access when both are competing for adsorption sites at the surface.

In practice, even if one protein is much more surface-active than another, complete displacement of the latter from the interface by the former is unlikely due to the influence of interfacial protein–protein interactions. As another example, let us consider the mixed system of ß-casein (pI ≈ 5) and gelatin (pI ≈ 9) at pH 7. Recent measurements at 25 °C at the air–water interface (Galazka and Dickinson, 1995) give steady-state tension values of 50.3 mN/m for ß-casein and 60.2 mN/m for gelatin at the same bulk protein concentration of 10^{-3} wt%. The tension for a 10^{-3} wt% solution of ß-casein and gelatin (1:9 by weight) was found to be the same as for the pure ß-casein solution within the estimated experimental error (±0.3 mN/m). This would seem to suggest that the film adsorbed from the mixture of ß-casein and gelatin under these solvent conditions consists entirely of ß-casein. However, surface shear viscosity measurements at the oil–water interface (Figure 17.4) show that this cannot be the case because the surface viscosity of the film adsorbed from ß-casein–gelatin (1:9) is very much higher than that for pure ß-casein at the same total bulk protein concentration (10^{-3} wt%). For example, after

30 hours, the apparent surface viscosity values are 1000 mN.s/m for gelatin alone, 200 mN.s/m for ß-casein–gelatin, and less than 10 mN.s/m for ß-casein alone.

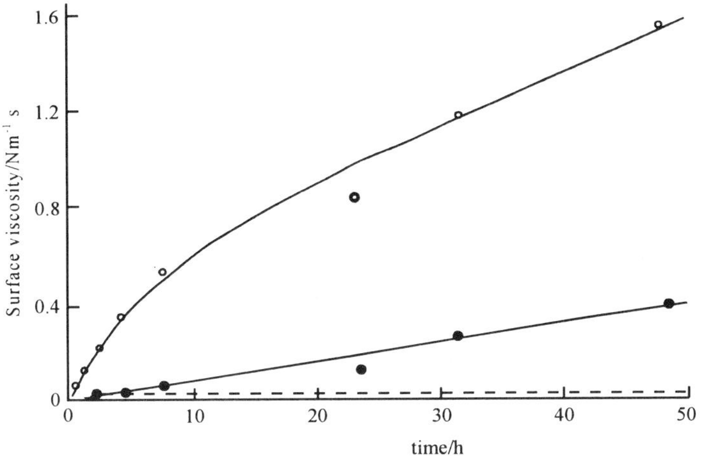

Figure 17.4 Time-dependent surface shear viscosities of adsorbed layers of disordered proteins at the n-tetradecane—water interface (pH 7, ionic strength 0.05 M, 25 °C): - - - 10^{-3} wt% ß-casein; ○ 10^{-3} wt% gelatin; ● 10^{-4} wt% ß-casein + 9×10^{-4} wt% gelatin.

The combined surface tension and interfacial viscosity data for ß-casein–gelatin (or α_{s1}-casein–gelatin) can be interpreted in terms of a primary protein layer composed of predominantly casein (which mainly determines the tension) and a secondary layer composed mainly of gelatin (which mainly determines the surface rheology). Interfacial electrostatic interaction between molecules in the primary and secondary layers is expected in this system due to the opposite net charges of the ß-casein (pH > pI) and gelatin (pH < pI). However, we also have evidence for accumulation of a secondary layer of negatively charged gelatin (pI≈5) below a primary layer of casein at neutral pH, again based on surface shear viscosity experiments (Castle, Dickinson, Murray and Stainsby, 1987), but involving sodium caseinate instead of ß-casein. These results would seem to imply that intermolecular hydrogen bonding interactions, as well as ionic forces, are responsible for holding together the structure of the composite casein–gelatin film.

Based on model systems like those described above, we now have some insight into the likely factors affecting competitive and cooperative adsorption at the surfaces in real food colloids. Though the latter contain not just two but several protein components, many such systems can be regarded as quasi-two-component protein mixtures. For instance, dairy emulsions contain a mixture of whey proteins and caseins. Model studies have shown (Hunt and Dalgleish, 1994) that caseinate is adsorbed in preference to whey protein isolate in model emulsions containing an excess of protein. On the other hand, when protein concentration is the limiting factor, the two quasi-components are adsorbed more-or-less to the same extent at the emulsion droplet surface. It has been suggested (Hunt and Dalgleish, 1994) that competition under conditions of excess protein results from the inability of the whey protein to produce multilayers due to steric hindrance by loops and tails from the adsorbed caseinate.

Interfacial Protein–Polysaccharide Interactions

Control of protein—polysaccharide interactions is considered to be potentially important for developing new food formulations (Samant, Singhal, Kulkarni and Rege, 1993; Ledward, 1994). If the mixing of protein and polysaccharide is a significantly exothermic process, there is a net attraction between the two biopolymers which leads to the formation of soluble and/or insoluble complexes. At pH values below the isoelectric point of the protein, the mixing of even dilute solutions of proteins and anionic polysaccharides leads to electrostatic complex formation between biopolymers of opposite net charge (Ledward, 1979). But complex formation is inhibited at high ionic strength. Polysaccharides containing carboxyl groups do not complex with globular proteins at pH > pI or at ionic strengths above ~0.2 M. With sulphated polysaccharides, on the other hand, complexation does occur at pH > pI.

Interaction with anionic polysaccharides is generally considered to involve the positively charged residues on the protein (ϵ-amino, guanidinium, imidazole, α-amino). The overall strength of the protein–polysaccharide interaction depends on the protein net charge, as well as on the number and distribution of the positively charged side-chains. Following denaturation, the number of available interacting sites on the protein increases as the buried basic groups are liberated (Ledward, 1994). In addition, the greater flexibility of the denatured state permits configurational adjustments to maximize favourable electrostatic interactions, resulting in more stable complexes than with the native proteins. Partial unfolding of globular proteins in the adsorbed state makes them more susceptible to complexation with polysaccharides than in aqueous solution at the same pH and ionic strength.

In certain special cases, a polysaccharide species can act as the sole emulsion stabilizer by adsorbing directly at the fresh oil–water interface during emulsification (Dickinson, 1988; Garti and Reichman, 1993). More commonly, however, the emulsion is primarily stabilized by adsorbed protein, with any association of the polysaccharide component with the oil–water interface involving some kind of protein–polysaccharide complex (Dickinson and Euston, 1991). Whether the interfacial complexation actually has a positive or negative effect on emulsion stability depends on the nature and strength of the interaction and the extent of original droplet surface coverage by the polysaccharide. Where the net protein–polysaccharide interaction is either weakly attractive or repulsive, the emulsion is susceptible to enhanced creaming instability due to flocculation of droplets by, respectively, either a bridging mechanism or a depletion mechanism (Cao, Dickinson and Wedlock, 1990). In the case of depletion flocculation, the incipient thermodynamic incompatibility between the polysaccharide in aqueous solution and the adsorbed protein around the droplets induces separation of the system into a polymer-rich aqueous serum phase and a polymer-depleted oil-rich cream phase. Similar behaviour is exhibited by systems containing non-adsorbing polysaccharide and surfactant-coated droplets (Dickinson, Goller and Wedlock, 1993) or non-adsorbing polysaccharide and polysaccharide-coated droplets (Dickinson, Elverson and Murray, 1989).

Evidence for complexation of bovine serum albumin (BSA) and dextran sulphate (DS) at the oil–water interface at neutral pH has been provided by surface shear viscometry (Dickinson and Galazka, 1992). Figure 17.5 shows data for BSA plus DS (5×10^5 Da) at the n-hexadecane–water interface; also shown are corresponding data for BSA alone and DS alone. Films adsorbed from 10^{-3} wt% solutions of pure BSA and pure DS have surface viscosities after 24 h of 280 mN.s/m and <1 mN.s/m, respectively. (Surface tension measurements indicate no significant adsorption of DS in the absence of protein.) Addition of 10^{-3} wt% DS to the aqueous phase below the 24-hour-old BSA film produces an increase in apparent surface viscosity after a further 24 hours to a value approximately twice that for the pure BSA film. A similar high value is reached on repeating the experiment with both biopolymers present together in solution at the creation of the fresh oil–water interface (Dickinson and Galazka, 1992).

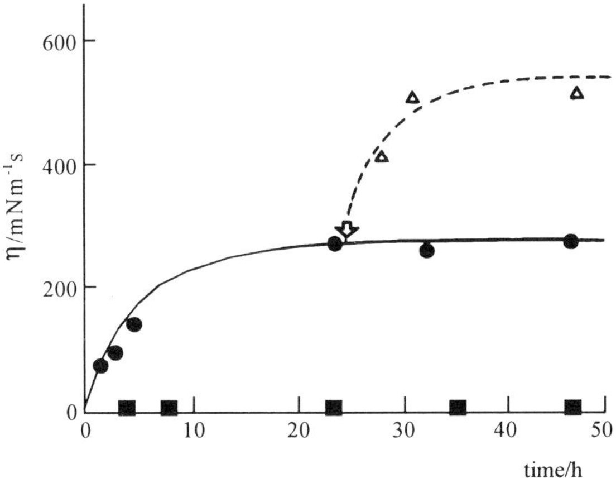

Figure 17.5 Influence of anionic polysaccharide on the time-dependent apparent surface shear viscosity of an adsorbed film of globular protein at the n-tetradecane—water interface (pH 7, ionic strength 0.05 M, 25 °C): ■ 10^{-3} wt% DS; ● 10^{-3} wt% BSA; Δ 10^{-3} wt% DS added (⇓) to the subphase below aged BSA film after 24 hours.

Interfacial complexation between BSA and DS at neutral pH can lead to substantial improvements in the stability of emulsions and foams. Oil-in-water emulsions made with 0.5 wt% BSA and 0.5 wt% DS have been found to be much more stable with respect to creaming and coalescence than those made with 0.5 wt% BSA (Galazka, 1991). This is attributed to better steric stabilization of the emulsion droplets by the mixed BSA–DS layer than by the pure BSA layer. If the polysaccharide does form a secondary layer, then one would expect to find bridging flocculation of BSA-coated droplets at DS concentrations well below that giving full polysaccharide surface coverage (Dickinson and Eriksson, 1991). This is indeed what is observed (White, 1994): addition of DS at a concentration corresponding to a surface coverage of 0.2 mg/m^2 leads to an increase in average effective droplet size from d_{32} = 1.26 μm to d_{32} = 4.3 μm, whereas addition at an excess concentration corresponding to 10 mg/m^2 gives no change in average size (±0.02 μm). We have also investigated (Izgi and Dickinson, 1995) the influence of DS on the stability of the foam produced from an aqueous solution of 0.05 wt% BSA at pH 7 by sparging with nitrogen gas, where stability is defined in terms of the time $t_{½}$ for the foam to collapse to 50% of its initial volume. Addition of 0.05 wt% DS to the aqueous protein solution leads to a doubling in the measured value of the foam half-life $t_{½}$, which is consistent with more effective stabilization of foam lamellae by the adsorbed BSA–DS complex than by the protein alone.

The strength of interaction of an anionic polysaccharide with an adsorbed protein layer is dependent on the native protein structure. For instance, replacing BSA by ß-lactoglobulin (BLG) in the above experiments leads to quite different behaviour. The addition of 10^{-3} wt% DS to the aqueous phase below a BLG adsorbed layer at pH 7 produces no discernible change in surface shear viscosity (cf. the substantial change for DS–BSA shown in Figure 17.5). The stability of a foam produced at pH 7 by sparging a solution of 0.3 wt% BLG does not change significantly when 0.3 wt% DS is included in the aqueous phase. The addition of DS to an emulsion of BLG-coated droplets at concentrations corresponding to surface coverages in the range 0.2—2.0 mg/m^2 produces no discernible change in the average effective droplet size (i.e. no bridging flocculation). These combined results suggest that, in contrast to BSA, there is no significant electrostatic complex formed at pH 7 between BLG and dextran sulphate at the oil–water or air–water interfaces.

Adsorption of charged polysaccharide onto the surface of protein-coated emulsion droplets would be expected to produce a change in electrical potential around the droplets at the surface of shear (Dickinson, 1992a). This means that particle electrophoretic mobility measurements are useful as an indicator of interfacial protein–polysaccharide interactions (Bergenståhl *et al.*, 1992; Ward-Smith, Hey and Mitchell, 1994). Figure 17.6 shows the mobility of emulsion droplets stabilized by BSA and BLG as a function of DS concentration at pH 7 and constant low ionic strength (White, 1994). The results are expressed as values of the relative mobility, $\mu_r = \mu/\mu_0$, where μ_0 is the measured droplet mobility in the absence of polysaccharide. We see that, when dispersed in 10^{-5} wt% DS, the mobility of BSA-coated droplets increases by more than 10%, whereas the mobility of BLG-coated droplets remains essentially unchanged. At DS concentrations 2 or 3 orders of magnitude higher, there is a slight increase in μ_r for BLG-coated droplets, but even at 0.1 wt% DS the value of μ_r for the BLG system is substantially lower than that for the BSA system. These results are again consistent with electrostatic complexation of dextran sulphate with adsorbed bovine serum albumin at neutral pH, but not with adsorbed ß-lactoglobulin. The difference is due presumably to subtle differences in molecular charge density distribution for these two globular proteins.

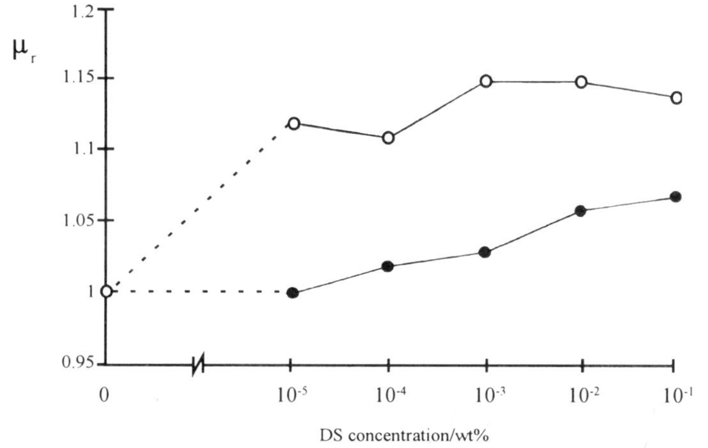

Figure 17.6 Influence of anionic polysaccharide on the electrophoretic mobility of protein coated emulsion droplets (pH 7, ionic strength 0.005 M, 25 °C). Relative electrophoretic mobility μ_r is plotted against DS concentration: ● ß-lactoglobulin; ○ BSA.

Upon lowering pH towards the isoelectric point, the number of positively charged groups on the protein increases. This is expected to lead to an enhanced strength of electrostatic interaction with anionic polysaccharides. Figure 17.7 shows recently reported electrophoretic mobility data (Ward-Smith *et al.*, 1994) for BSA-coated emulsion droplets as a function of pH in the presence or absence of sodium alginate at constant ionic strength 0.15 M. In accordance with expectation, we can see that, as the pH is lowered below neutral, the presence of alginate has a strongly increasing effect on the absolute value of the droplet electrophoretic mobility.

The strongest form of protein–polysaccharide interaction in a mixed biopolymer layer occurs when protein and polysaccharide molecules are linked together covalently to form a new hybrid amphiphilic macromolecule (Dickinson 1992a). Without using any chemical reagents, covalent protein–polysaccharide conjugates suitable for stabilizing emulsions and foams can be made by linking protein ε-

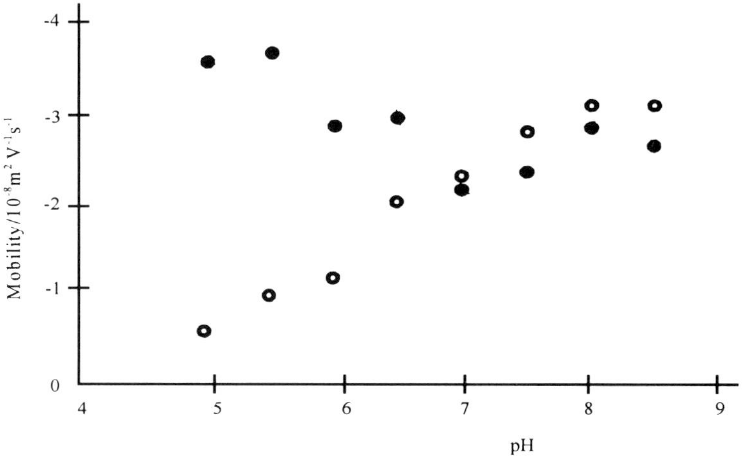

Figure 17.7 Complexation between sodium alginate and adsorbed BSA at the surface of emulsion droplets as determined by electrophoretic mobility measurement as a function of pH at constant ionic strength (0.15 M): ○ without polysaccharide; ● with polysaccharide.

amino-groups to the reducing-end carbonyl groups of polysaccharides through controlled Maillard reaction (Kato, Sasaki, Furuta and Kobayashi, 1990; Dickinson and Galazka, 1991, 1992; Dickinson and Semenova, 1992; Nakamura, Kato and Kobayashi, 1993). The mixture of protein and polysaccharide (e.g. BSA–dextran) is heated at a temperature below the normal protein denaturation temperature under conditions of low water activity. The resulting soluble conjugates can be made to have very good emulsifying properties if the heating time–temperature and relative proportions of the two biopolymers are carefully optimized. Based on computer simulations (Dickinson and Euston, 1992), it can be concluded that optimum functionality of the mixed biopolymer layer occurs when there is sufficient polysaccharide bound to protein to enhance solubility and steric stabilization, whilst there still remain enough accessible hydrophobic groups on the conjugate to allow strong and rapid binding to the oil–water interface. A key advantage of permanent covalent linkage over electrostatic protein–polysaccharide complexation is the maintenance of a strong interaction without precipitation over a wide range of solution conditions.

Gelation and Phase Separation in Two-Dimensional Mixed Layers

Adsorbed layers formed from most globular proteins (and also gelatin) may be regarded as thin films of protein gel with values of the bulk shear modulus equal to the surface shear modulus divided by the layer thickness (Graham and Phillips,

1980). The rate of protein cross-linking in the adsorbed layer is very greatly enhanced over that in bulk solution because of (i) the high local macromolecular concentration in the adsorbed layer and (ii) the exposure of reactive groups on partially unfolded globular proteins at the interface.

In emulsions made with pure ß-lactoglobulin, it has been demonstrated (Dickinson and Matsumura, 1991) that slow polymerization takes place in the adsorbed layer via sulphydryl–disulphide interchange. Interfacial polymerization does not occur, however, in emulsions made with another whey protein α-lactalbumin. This is presumably because α-lactalbumin does not contain a free –SH group. In mixed aged films containing both whey proteins, however, cross-linking does occur between α-lactalbumin (ALA) and ß-lactoglobulin (BLG) molecules, as well as between pairs of BLG molecules. As was inferred previously from surface viscosity measurements (Dickinson, Rolfe and Dalgleish, 1990), it seems that adsorbed ALA inhibits disulphide bond formation amongst adsorbed BLG molecules by itself linking to BLG through disulphide bonds. As ALA has no free –SH group, its effective functionality is lower than that of BLG, which means that, if it becomes connected to BLG, it may make further cross-linking more difficult by acting as a sort of blocking group in the polymerization process. One therefore expects the mixed whey protein gel layer to have a rather different network structure from that of the pure BLG layer.

Figure 17.8 shows pictorial representations of two simulated particle gel structures produced from a two-dimensional aggregating system consisting of a concentrated binary mixture of equal-sized particles. The network structures were generated by the computational technique of Brownian dynamics starting from a pseudo-random distribution of attractive particles in a periodic simulation cell (Dickinson, 1994). In the pictures in Figure 17.8, the two components are present at equal molar concentrations; the particles of components 1 and 2 are denoted by filled and open circles, respectively. Picture (a) refers to a system in which the two species are identical in every respect, whereas in picture (b) the type 1 particles can bond only to type 2 particles and not to other type 1 particles. If we suppose that component 1 represents ALA and component 2 represents BLG, then we can see how the presence of ALA molecules in an adsorbed layer of BLG molecules can inhibit the development of a highly cross-linked viscoelastic film at the oil–water interface (Dickinson *et al.*, 1990).

The network structures illustrated in Figure 17.8 are the result of a simulation of a dense system of mutually attracting particles having a high bonding reactivity (except for interactions between the like type 1 species in Figure 17.8(b)). Aggregation is very fast, and so the gel structure is kinetically controlled. This does not allow for large-scale compositional inhomogeneities in the mixed particle layer. The position becomes rather different, however, under simulation conditions where the packing is less dense, the bond reactivity is much lower, and there is net

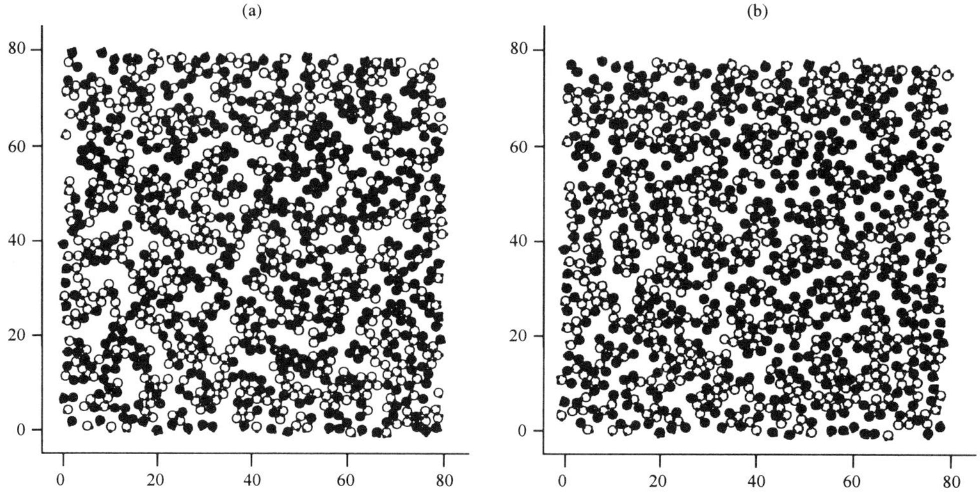

Figure 17.8 Two-dimensional simulated gel generated from a binary mixture of equal-sized particles. (a) Two kinds of particle are identical in every respect. (b) Black particles (representing α-lactalbumin) can bond only to white particles (representing β-lactoglobulin) and not to other black particles.

repulsion between the unlike species ($\chi_{12} > 0$). Three possibilities are illustrated in Figure 17.9, again for systems containing equal concentrations of the two components, but now at lower total particle concentration. Picture (a) is obtained by simulating a system with no cross-linking and a strong net repulsion between the unlike species. Being materially unstable, the mixture tends to separate into coexisting phases by a process of spinodal decomposition. The system in picture (b) is identical to that in picture (a) except that now there is a low rate of cross-linking between the like particles. (Cross-linking between the unlike particles is inhibited by the strong interparticle repulsion.) In this case, the homogeneous mixture is again thermodynamically unstable, but now there is competition between the kinetics of phase separation and the kinetics of cross-linking. Picture (b) therefore shows a phase-separating system consisting of two types of stringy aggregates, held together irreversibly by extensive cross-linking between the like particles, but with no overall macroscopic network structure (at this finite simulation time).

A different kind of heterogeneous cross-linked structure is illustrated in Figure 17.9(c). In this case the type 1 particles attract each other without bonding, whereas the type 2 particles repel each other moderately strongly but nevertheless can form bonds both with type 1 and type 2 particles. This latter type of interparticle interaction leads to flexible chain-like network structures as previously discussed in detail for one-component systems (Dickinson, 1994). To this network is added a second phase-separating component which reinforces the two-dimensional gel by forming additional unlike cross-links. The net result is a complex heterogeneous mixed particle gel system having large variations in local density and composition and

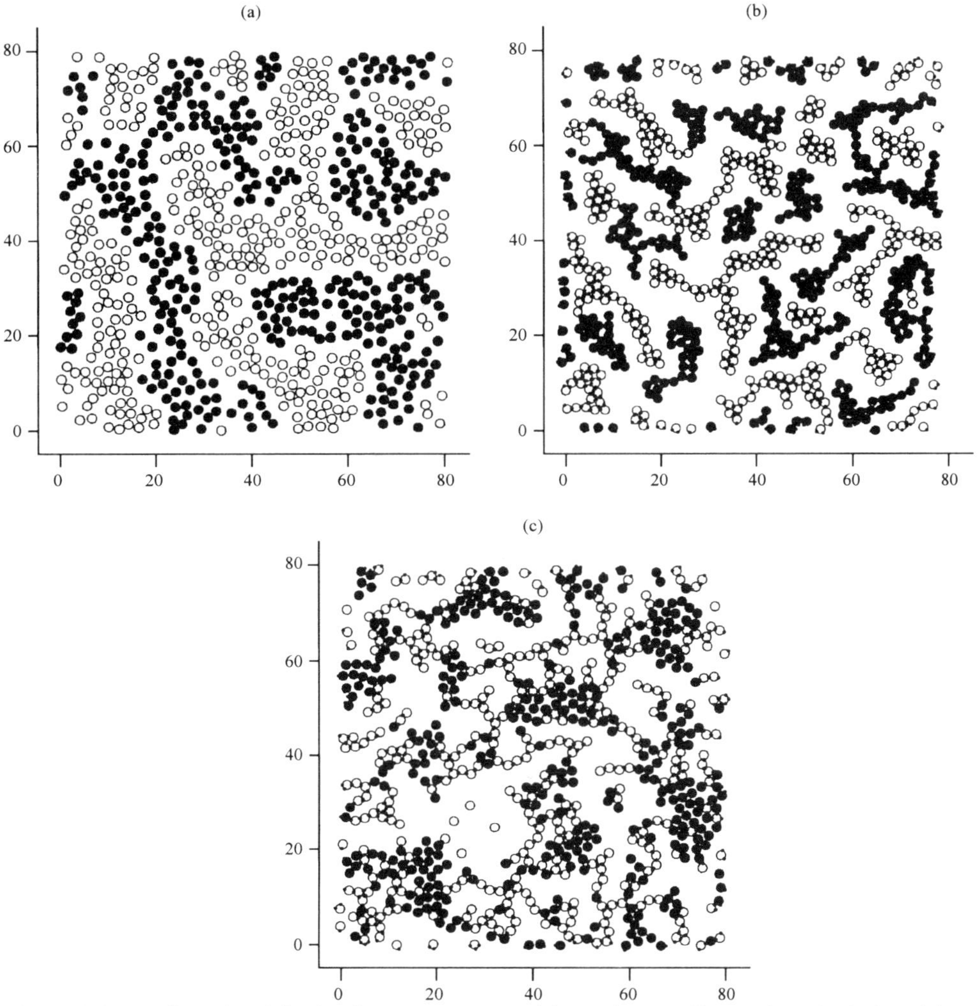

Figure 17.9 Two-dimensional simulated structures generated from mixtures of interacting spherical particles. (a) Phase separation without cross-linking. (b) Phase separation with cross-linking between like particles. (c) Complex particle gel with a mixture of strong links (between white particles) and weak reversible links (between black particles).

containing both weak (reversible) and strong (irreversible) structural units. To what extent this simulated structure, or any of the others shown in Figures 17.8 and 17.9, is close to actual structures in real mixed protein layers, or in lipid–protein membranes (Vaz, 1994), remains an open question at this stage. What is quite clear from the simulations, however, is that there do exist many possible types of structures for binary biopolymer adsorbed layers. Which ones actually occur in practice depends on the nature of the interactions between the various participating macromolecules.

Conclusions

This article has discussed some of the important considerations influencing the organization of mixed biopolymers at interfaces. Specific attention has been directed towards the role of protein–protein and protein–polysaccharide interactions in adsorbed layers at fluid interfaces because of the relevance of these to the stability properties of food emulsions and foams. The concepts described here are nevertheless also relevant to a wide range of other (non-food) applications such as fabrication of self-assembled ordered protein monolayers (Yamaki, Matsubara and Nagayama, 1993) or modification of polymer surfaces to improve biocompatibility (Hubbell, 1994).

The evolving structure and composition of a mixed biopolymer layer depends on a balance of competitive and cooperative processes. In essence, the adsorbent acts as a sort of localized external field that perturbs the structure of the bulk phase solution (Dickinson and Pelan, 1993). This produces large gradients in segment density and composition perpendicular to the surface, with the less strongly adsorbing segments (or molecules) distributed further away from the position of maximum adsorbent field intensity.

Under conditions approximating to monolayer coverage, a mixed biopolymer layer may be regarded as a two-dimensional dense liquid mixture or a two-dimensional cross-linked viscoelastic gel. Due to the high local adsorbate density, effects of phase separation and complexation are much more important in the adsorbed layer than in the equivalent aqueous solution. Small changes in solvent conditions (e.g. pH) can produce large changes in adsorbed layer structure and composition. Existing experimental data indicate useful correlations between mixed biopolymer layer properties and the thermodynamics of the corresponding biopolymer solutions. However, in many cases the adsorbed layer structure is controlled by kinetic processes such as the rate of macromolecular rearrangement at the interface or the rate of protein cross-linking in the layer. In practice, this means that, in addition to the thermodynamic considerations, mixed layer properties tend to be strongly dependent on time-related factors such as the age of the adsorbed film or the sequence of exposure of the different biopolymers to the interface.

References

Albertsson, P.-A. (1971). *Partition of Cell Particles and Macromolecules*, 2nd edition, New York: Wiley

Bergenståhl, B., Fäldt, P. and Lassen, B. (1992) Adsorption structures in food emulsions. In *Emulsions: A Fundamental and Practical Approach*, edited by Sjöblom, J., Dordrecht: Kluwer, pp 51–60

Bergenståhl, B., Fäldt, P. and Malmsten, M. (1995) Surface structures and surface-active components in food emulsions. In *Food Macromolecules and Colloids*, edited by Dickinson, E. and Lorient, D., Cambridge: Royal Society of Chemistry, pp 201–214

Bungenberg de Jong, H.G. (1949) Crystallization, coacervation and flocculation. In *Colloid Science*, edited by Kruyt, H.R., Amsterdam: Elsevier, Vol. 2, pp 232–258

Cao, Y., Dickinson, E. and Wedlock, D.J. (1990) Creaming and flocculation in emulsions containing polysaccharide. *Food Hydrocolloids*, **4**, 185–195

Castle, J., Dickinson, E., Murray, B.S. and Stainsby, G. (1987) Mixed protein films adsorbed at the oil/water interface. *American Chemical Society Symposium Series*, **343**, 118–134

Chakrabarti, A. (1994) Monte Carlo study of pancake to brush transition. *Journal of Chemical Physics*, **100**, 631–635

Dalgleish, D.G. and Leaver, J. (1991) Dimensions and possible structures of milk proteins at oil/water interfaces. In *Food Polymers, Gels and Colloids*, edited by Dickinson, E., Cambridge: Royal Society of Chemistry, pp 113–122

Dalgleish, D.G., Euston, S.E., Hunt, J.A. and Dickinson, E. (1991) Competitive adsorption of ß-lactoglobulin in mixed protein emulsions. In *Food Polymers, Gels and Colloids*, edited by Dickinson, E., Cambridge: Royal Society of Chemistry, pp 485–489

de Gennes, P.G. (1976) Scaling theory of polymer adsorption. *Journal de Physique (Paris)*, **37**, 1445–1452

de Gennes, P.G. (1987) Polymers at an interface: a simplified view. *Advances in Colloid and Interface Science*, **27**, 189–209

Dickinson, E. (1986) Mixed proteinaceous emulsifiers: review of competitive protein adsorption and the relationship to food colloid stabilization. *Food Hydrocolloids*, **1**, 3–23

Dickinson, E. (1988) The role of hydrocolloids in stabilizing particulate dispersions and emulsions. In *Gums and Stabilizers for the Food Industry*, edited by Phillips, G.O., Wedlock, D.J. and Williams, P.A., Oxford: IRL Press, Vol. 4, pp 249–263

Dickinson, E. (1992a). *An Introduction to Food Colloids*, Oxford: University Press

Dickinson, E. (1992b) Structure and composition of adsorbed protein layers and the relationship to emulsion stability. *Journal of the Chemical Society, Faraday Transactions*, **88**, 2973–2983

Dickinson, E. (1994) Computer simulation of particle gel formation. *Journal of the Chemical Society, Faraday Transactions*, **90**, 173–180

Dickinson, E. and Eriksson, L. (1991) Particle flocculation by adsorbing polymers. *Advances in Colloid and Interface Science*, **34**, 1–29

Dickinson, E. and Euston, S.R. (1991) Stability of food emulsions containing both protein and polysaccharide. In *Food Polymers, Gels and Colloids*, edited by Dickinson, E., Cambridge: Royal Society of Chemistry, pp 132–146

Dickinson, E. and Euston, S.R. (1992) Computer simulation model of the adsorption of protein/polysaccharide complexes. *Food Hydrocolloids*, **6**, 345–357

Dickinson, E. and Galazka, V.B. (1991) Emulsion stabilization by ionic and covalent complexes of ß-lactoglobulin with polysaccharides. *Food Hydrocolloids*, **5**, 281–296

Dickinson, E. and Galazka, V.B. (1992) Emulsion stabilization by protein/polysaccharide complexes. In *Gums and Stabilizers for the Food Industry*, edited by Phillips, G.O., Wedlock, D.J. and Williams, P.A., Oxford: IRL Press, Vol. 6, pp 351–362

Dickinson, E. and Lal, M. (1980) Statistical mechanics of physical adsorption in condensed phases. *Adv. Mol. Relax. Interact. Proc.*, **17**, 1–87

Dickinson, E. and Matsumura, Y. (1991) Time-dependent polymerization of ß-lactoglobulin through disulphide bonds at the oil/water interface in emulsions. *International Journal of Biological Macromolecules*, **13**, 26–30

Dickinson, E. and Pelan, E.G. (1993) Molecular dynamics simulation of competitive adsorption of spherical particles of different sizes. *Journal of the Chemical Society, Faraday Transactions*, **89**, 3435–3457

Dickinson, E. and Semenova, M.G. (1992) Emulsifying properties of covalent protein/dextran hybrids. *Colloids and Surfaces*, **64**, 299–310

Dickinson, E. and Stainsby, G. (1988) Emulsion stability. In *Advances in Food Emulsions and Foams*, edited by Dickinson, E. and Stainsby, G., London: Elsevier Applied Science, pp 1–44

Dickinson, E., Elverson, D.J. and Murray, B.S. (1989) On the film-forming and emulsion-stabilizing properties of gum arabic: dilution and flocculation aspects. *Food Hydrocolloids*, **3**, 101–114

Dickinson, E., Goller, M.I. and Wedlock, D.J. (1993) Creaming and rheology of emulsions containing polysaccharide and nonionic or anionic surfactants. *Colloids and Surfaces*, **75**, 195–201

Dickinson, E., Murray, A., Murray, B.S. and Stainsby, G. (1987) Properties of adsorbed layers in emulsions containing a mixture of caseinate and gelatin. In *Food Emulsions and Foams*, edited by Dickinson, E., London: Royal Society of Chemistry, pp 86–99

Dickinson, E., Rolfe, S.E. and Dalgleish, D.G. (1990) Surface shear viscosity as a probe of protein/protein interactions in mixed protein films adsorbed at the oil/water interface. *International Journal of Biological Macromolecules*, **12**, 189–194

Fleer, G.J., Cohen Stuart, M.A., Scheutjens, J.M.H.M., Cosgrove, T. and Vincent, B. (1993). *Polymers at Interfaces*, London: Chapman and Hall

Flory, P.J. (1953). *Principles of Polymer Chemistry*, Ithaca (New York): Cornell University Press

Galazka, V.B. (1991). *Protein/polysaccharide interactions in food emulsions*, Ph.D. thesis, University of Leeds

Galazka, V.B. and Dickinson, E. (1995) Surface properties of protein layers adsorbed from mixtures of gelatin with various caseins. *Journal of Texture Studies* (in press)

Garti, N. and Reichman, D. (1993) Hydrocolloids as food emulsifiers and stabilizers. *Food Structure*, **12**, 411–426

Graham, D.E. and Phillips, M.C. (1980) Proteins at liquid interfaces: shear properties. *Journal of Colloid and Interface Science*, **76**, 240–250

Halperin, A., Tirrell, M. and Lodge, T.P. (1992) Tethered chains in polymer microstructures. *Advances in Polymer Science*, **100**, 31–71

Hubbell, J.A. (1994) Chemical modification of polymer surfaces to improve biocompatibility. *Trends in Polymer Science*, **2**, 20–25

Hunt, J.A. and Dalgleish, D.G. (1994) Adsorption behaviour of whey protein isolate and caseinate in soya oil-in-water emulsions. *Food Hydrocolloids*, **8**, 175–187

Hunter, J.R., Carbonell, R.G. and Kilpatrick, P.K. (1991) Coadsorption and exchange of lysozyme/ß-casein mixtures at the air/water interface. *Journal of Colloid and Interface Science*, **143**, 37–53

Izgi, E. and Dickinson, E. (1995) Determination of protein foam stability in the presence of polysaccharide. In *Food Macromolecules and Colloids*, edited by Dickinson, E. and Lorient, D., Cambridge: Royal Society of Chemistry, 312–315

Kato, A., Sasaki, Y., Furuta, R. and Kobayashi, K. (1990) Functional protein/polysaccharide conjugate prepared by controlled dry-heating of ovalbumin/dextran mixtures. *Agricultural and Biological Chemistry*, **54**, 107–112

Ledward, D.A. (1979) Protein/polysaccharide interactions. In *Polysaccharides in Food*, edited by Blanshard, J.M.V. and Mitchell, J.R., London: Butterworths, pp 205–217

Ledward, D.A. (1994) Protein/polysaccharide interactions. In *Protein Functionality in Food Systems*, edited by Hettiarachchy, N.S. and Ziegler, G.R., New York: Dekker, pp 225–259

Lipatov, Yu.S. and Sergeeva, L.M. (1974). *Adsorption of Polymers*, New York: Wiley

Milner, S.T. (1991) Polymer brushes. *Science*, **251**, 905–914

Mussellwhite, P.R. (1966) The surface properties of an oil-in-water emulsion stabilized by mixtures of casein and gelatin. *Journal of Colloid and Interface Science*, **21**, 99–102

Nakamura, S., Kato, A. and Kobayashi, K. (1993) Novel macromolecular emulsifier having improved antimicrobial action elaborated by covalent attachment of polysaccharides to lysozyme. In *Food Proteins: Structure and Function-*

ality, edited by Schwenke, K.D. and Mothes, R., Weinheim: VCH Publishers, pp 29–39

Nylander, T. and Wahlgren, N.M. (1994) Competitive and sequential adsorption of ß-casein and ß-lactoglobulin on hydrophobic surfaces and the interfacial structure of ß-casein. *Journal of Colloid and Interface Science*, **162**, 151–162

Samant, S.K., Singhal, R.S., Kulkarni, P.R. and Rege, D.V. (1993) Protein/polysaccharide interactions: a new approach in food formulations. *International Journal of Food Science and Technology*, **28**, 547–562

Silberberg, A. (1968) Adsorption of flexible macromolecules: effect of solvent/solute interactions, solute concentration and molecular weight. *Journal of Chemical Physics*, **48**, 2835–2851

Tolstoguzov, V.B. (1991) Functional properties of food proteins and role of protein/polysaccharide interaction. *Food Hydrocolloids*, **4**, 429–468

Vaz, W.L.C. (1994) Diffusion and chemical reactions in phase-separated membranes. *Biophysical Chemistry*, **50**, 139–145

Vroman, L. and Adams, A.L. (1986) Adsorption of proteins out of plasma and solutions in narrow spaces. *Journal of Colloid and Interface Science*, **111**, 391–402

Ward-Smith, R.S., Hey, M.J. and Mitchell, J.R. (1994) Protein/polysaccharide interactions at the oil/water interface. *Food Hydrocolloids*, **8**, 309–315

White, M. (1994) *Effect of polysaccharide on the creaming and electrophoretic mobility of protein-stabilized emulsions*, B.Sc. project thesis, University of Leeds

Yamaki, M., Matsubara, K. and Nagayama, K. (1993) A thin liquid layer on the surface of mercury as a matrix of a flow-mediated two-dimensional assembly of proteins. *Langmuir*, **9**, 3154–3158

18

METHODS USED TO DEVELOP MUCOADHESIVE DRUG DELIVERY SYSTEMS: BIOADHESION IN THE GASTROINTESTINAL TRACT

I. FIEBRIG and S.S. DAVIS
Department of Pharmaceutical Sciences, School of Pharmacy, University of Nottingham, University Park, Nottingham. NG7 2RD U.K.

S.E. HARDING
National Centre for Macromolecular Hydrodynamics, University of Nottingham, Sutton Bonington Campus, Sutton Bonington. LE12 5RD U.K.

Introduction

The term "adhesion" is defined by the physicist as "the molecular force of attraction in the area of contact between unlike bodies that acts to hold them together" (Webster, 1989). "*Bio*adhesion" refers to adhesive phenomena where at least one of the adherents is of biological nature (Kaelbe and Moacanin, 1977). The materials are attached to each other by interfacial forces for an "extended" period of time (Gu, Robinson and Leung, 1988; Duchêne, Touchard and Peppas, 1988). The term "mucoadhesion" is often employed when the biological substrate is a mucosal surface (Robinson, 1990).

In nature, the adhesion of living beings to surfaces is often an imperative for survival, keeping the organism at a favourable site and avoiding its being transferred to a less favourable environment. The numerous forms of symbiosis generally require some sort of permanent attachment of one organism to the other. The attachment of micro-organisms to the gut is one good example of biological adhesion between two organisms, either symbiotic or infectious. These interactions are often mediated by bacterial adhesins, lectin-like, carbohydrate-binding proteins, expressed on the bacterial surface which bind specifically to sugar residues of mucins or other carbohydrates of the host cell surface (Beachey, 1980; Boedecker, 1984; Mergenhagen and Rosan, 1985; Hörstedt, Danielsson, Nyhlin, Stenling and Suhr, 1989). Although these are all examples of "biological adhesion" the term bioadhesion has been used mainly to describe adhesive phenomena related to the ability of some synthetic and biological macromolecules as well as hydrocolloids to adhere to biological tissues for therapeutic purposes in medicine (Kaelbe and Moacanin, 1977; Peppas and Buri, 1985). The adhesion mechanisms involved are usually a combination of physical and chemical phenomena.

The present review deals with bioadhesion aimed at improving drug therapy using man-made materials as adhesives, focusing on oral drug administration. Bioadhesive systems have been used for many years for medical applications other than drug delivery in the area of dentistry for denture adhesives (Wright, 1981; Hollingsbee and Timmins, 1990), stoma adhesives such as karaya gum in Stomahesive® or synthetic polypectins (Winkler, 1986) and for surgical applications such as the cyanoacrylates used as "surgical glue" (Wang, 1974, Harper and Ralson, 1983). It is only about 15 years ago that pharmaceutical researchers began to explore more intensively the possible applications of bioadhesion with the aim of improving drug treatment.

Oral administration of drugs

Local treatment of diseases can be unsatisfactory, because the drug may not stay at the site of action long enough for the desired effect (typically eye, mouth or vaginal cavity). Similarly systemic treatment via the oral route can be hampered because the drug may not stay at the site of absorption long enough. Oral administration of drugs is the most popular route of administration for both health care personnel and patients. Taking pharmaceutical dosage forms via the mouth is generally well accepted. It can easily be undertaken almost anywhere and it is generally safe. In contrast, invasive methods (e.g. injection) usually require the assistance of trained personnel and the procedure always involves certain risks.

Oral drug administration begins with ingestion of the dosage form through the mouth. From there it passes down the oesophagus and into the stomach. Once released, little drug is absorbed from the stomach because of its relatively small surface area, particularly in the case of cationic drugs which will be mainly ionised in the acidic conditions of the stomach. It has been recognized that the major site of drug absorption is the small intestine. Its large surface area (approximately 100 m^2) makes it very efficient for the uptake of solutes (Bowman and Rand, 1980). Theoretically, drug absorption can occur along the entire length of the small intestine, nevertheless the majority of drugs are actually absorbed from the proximal small intestine (Booth, 1967). However, if the drug is poorly soluble or is in the form of a controlled release dosage form, significant absorption of the drug may also occur in the large intestine (Davis, 1989), despite its limited surface area. More recent studies with a once-a-day preparation (e.g. theophylline) have shown that therapeutic drug levels can be maintained for periods of up to 24 h (Gruber, Longer and Robinson, 1987) even though these systems are expected to have emptied from the stomach, passed through the small intestine and have arrived at the ileocaecal junction after 4–12 h (Davis, 1985). Oral drug delivery ceases eventually with faecal excretion of any unabsorbed drug.

Absorption enhancement in Oral Drug Delivery Systems

Low appearance of the drug in the systemic circulation (bioavailability) can be due to (i) rapid transit of the drug-containing delivery system past the ideal absorption site, (ii) rapid degradation of the drug in the gastrointestinal tract once it has been released (peptide drugs) and, (iii) low transmucosal permeability due to size, ionisation, solubility or other characteristics of the drug molecule. There are several strategies that have been proposed in an attempt to overcome some of these problems, for example delaying the transit of a pharmaceutical dosage form within the gastrointestinal tract. This approach will be beneficial for drugs which are only poorly absorbed from selected regions of the small intestine and whose bioavailability is believed to be dependent on the residence time of the dosage form at or upstream of its small intestinal absorption window. Such a drug is hydrochlorothiazide (Beerman, Groschinsky-Grind and Rosen, 1976; Lynch, Pownall and Taylor, 1987). It is particularly important in the case of controlled release drug delivery systems (DDS), designed to release drugs over extended periods of time (e.g. 12–24 h). Once in the colon, these DDS may be delivering a proportion of drug to a non-optimal site for absorption (Davis, 1985). An ideal oral sustained release dosage form should be comparable to an intravenous infusion, which continuously supplies the amount of drug needed to maintain constant plasma levels once steady state is reached (Förster and Lippold, 1982)

Proposed strategies to delay gastrointestinal transit

Various attempts have been made to delay gastrointestinal transit. These have involved pharmacological and physiological as well as pharmaceutical approaches. Pharmacological means involve the co-administration or incorporation into the dosage form of a drug that delays gastrointestinal emptying such as antimuscarinics, e.g. propantheline, which are relaxants of the smooth muscle (Beermann and Groschinsky-Grind, 1978; Manninen, Apajalahti, Melin and Karesoja, 1973) or a drug that changes motility, e.g opiate analgesics or derivatives such as loperamide, (Minami and McCallum, 1984). However, the potential side effects that may occur from such treatments on a routine basis would not be acceptable for regulatory approval.

The use of passage-delaying excipients which activate a physiological mechanism to delay gastrointestinal transit has also been proposed. This may involve stimulation of the duodenal or jejunal receptors using a fat derivative like triethanolamine myristate (Gröning and Heun, 1984, 1989) or the use of large amounts of a polymer like polycarbophil that may induce a volume related fed-like state (Harris, Fell,

Sharma and Taylor, 1990) and delay gastric emptying due to a blocking effect. In fact, in some of the studies involving so called bioadhesive polymers it is not quite clear whether the gastric retention shown in animal models is due to this, or due to true adhesion, or a combination of phenomena (Russel and Bass, 1985; Harris, Fell, Taylor, Lynch and Sharma, 1989, 1990; Harris *et al.*, 1990).

Pharmacological and physiological approaches set out to delay gastrointestinal transit by modification of the rate of gastric emptying with passage-delaying agents. The pharmaceutical strategies instead attempt to achieve the same objective by actually retaining the dosage form at or upstream of its absorption site for as long as possible by virtue of its special physical or physicochemical characteristics.

Size-related retention of a dosage form in the stomach has been studied with various systems. If large enough, the formulation would not be expelled from the stomach even when the pyloric sphincter is in its non-contracted state. Their size has to increase after ingestion to such an extent that gastric emptying is totally inhibited (Moës, 1993). According to Moës uncommon delivery systems such as swelling balloon hydrogels (Park and Park, 1987) or unfolding stratified medicated polymer sheets (BE Patent No. 867, 692) or non-erodible or erodible tetrahedron shaped devices (Cargill, Caldwell, Engle, Fix, Porter, and Gardner, 1988; Cargill, Engle, Gardner, Porter, Sparer and Fix, 1989) have never passed beyond the experimental stage and clinical data are unavailable. In any case these gastric retention devices may not be safe. The hazard of lodging in the oesophagus (Kikendall, Friedman, Oyewole, Fleischer and Johnson, 1983; Al-Dujaili, Salole and Florence, 1983; Wilson, 1990) or permanent retention in the stomach with cumulative effects (Brahams, 1984; Vere, 1984) could lead to life-threatening problems.

Another approach uses dosage forms of moderately high density, based on the premise that high density formulations remain in the stomach longer than conventional formulations since they would be localised in the lower part of the antrum provided the density exceeds that of the normal stomach contents, i.e. > 1.4 g/cm^3 (Bechgaard and Ladefoged, 1978). The effectiveness of this approach has not been confirmed on a broad basis and the evidence remains controversial (Moës, 1993).

A further retention principle is based on the use of buoyant dosage forms which float on the gastric contents as a result of their relatively low density. Floating dosage forms have been discussed extensively by Moës (1993) in his recent review, of which we will give a summary here. The first floating dosage forms (F forms) were described by Sheth and Tossounian (1984). These F forms, called "hydrodynamically balanced systems" (HBS), maintain their low density while the polymer hydrates and builds a gelled barrier at the outer surface. Research at Hoffmann-LaRoche has led to various patents for floating drug delivery systems and *in vivo* studies on diazepam HBS capsules such as Valium® CR and Valrelease®

and the L-dopa plus benserazide containing formulation Madopar® HBS (Prolopa® HBS). Numerous other rather sophisticated buoyancy mechanisms have been invented too. In his critical review Moës clarified the conflicting views on the gastric retention capabilities of floating systems resulting from a number of *in vivo* trials by different authors (Müller-Lissner and Blum, 1981; Davis, Stockwell, Taylor, Hardy, Whalley, Wilson, Bechgaard and Christensen, 1986; Timmermans and Moës, 1990; Timmermans, 1991; Kaus, 1987; Sangekar, Vadino, Chaudry, Parr, Beihn and Digenis, 1987; Lippold and Günther, 1991). It was concluded that amongst the various strategies proposed for prolongation of gastric residence time of oral DDS, floating delivery systems offer the best protection against early and random gastric emptying to date. In order to achieve adequate floating of the dosage form, it should be administered after a meal and the patient should not lie down after dosing.

The retention of a formulation via bioadhesion has been another strategy aiming at retention of dosage forms at or upstream of their absorption site via attachment to the mucous surfaces lining the gastrointestinal tract ("indirect bioadhesion"). It has been proposed that a mucoadhesive polymer could act as drug carrier and adhesive, thus retaining the delivery system either in the stomach or in the small intestine.

The greatest challenge for the concept of bioadhesion would be the successful delivery of orally administered proteins and peptide drugs. These endogenous compounds regulate numerous endocrine and other physiological processes in the human body. They are highly active drugs with a low toxicity, if dosed correctly. Recent advances in recombinant DNA-technology and production technology have made it possible to produce oligopeptides on a large scale. Insulin, growth hormones and interferons, amongst others can be used broadly for therapeutic and diagnostic purposes. Despite the advances and achievements in gene technology, therapeutic use of these substances still remains difficult (with a few exceptions, e.g. insulin). Due to their low absorption characteristics and instability (enzymatic degradation and biotransformation) parenteral application of peptides is usually unavoidable (Junginger and Verhoef, 1992).

We can talk of "direct bioadhesion" when very intimate contact to the absorbing epithelial tissue occurs and an increased concentration gradient is achieved. A decrease in diffusion path from the oral DDS to the absorbing biological membrane is an additional advantage for improving absorption particularly in intestinal delivery of peptide drugs. At the same time this would prevent previous dilution and possible degradation in the luminal fluids (Hayton, 1980). The addition of penetration enhancers to an adhering dosage form could enable alteration of membrane permeability. Inclusion of specific enzyme inhibitors could prevent early degradation of the peptide (Wearly, 1991; Junginger and Verhoef, 1992) and consequently increase bioavailability.

The development of gastrointestinal bioadhesive drug delivery systems

Gastrointestinal retention of dosage forms through adhesion to the mucosa has been studied for over a decade now, mainly *in vitro* or *ex vivo* with few *in situ* or *in vivo* studies and even fewer trials in man (Table 18.1). Despite the fact that bioadhesion, or more specifically mucoadhesion, has led to some success in drug delivery for ocular, buccal, nasal, [transdermal[1]], vaginal and cervical applications (Chen and Cyr, 1970; Schor, Davis, Nigalaye and Bolton, 1983; Nagai, Nishimoto, Nambu, Suzuki and Sekine, 1984; Nagai, 1986; Duchêne *et al.*, 1988; Greaves and Wilson, 1993; Smart, 1993; Bouckaert, Temmerman, Dhont and Remon, 1994), gastrointestinal mucoadhesive drug delivery systems have yet to be established (Helliwell, 1993). There follows below

(i) A summary of the physiological aspects of the gastrointestinal tract that have to be considered in an attempt to evaluate the potential of bioadhesion; and
(ii) A general overview of the different methods that have been used to study mucoadhesion outlining the particular difficulties in finding a suitable method capable of giving meaningful and relevant data.

THE SUBSTRATE: ADHERENT MUCUS AND ITS MUCINS

Mucus, known for its viscoelastic properties, its stickiness and the characteristic *Spinnbarkeit*[2] contains between 95%–99.5% water. Its most important polymeric, gel-forming component is the mucus glycoprotein or mucin (0.5%–5%) (Carlstedt and Sheehan, 1988; Neutra and Forstner, 1987; Gibbons, 1972). In the gastrointestinal tract, mucus is secreted by specialized cells as a polymer of high molecular weight. Unlike other gastrointestinal secretions, it adheres to the mucosal epithelial surfaces as a water insoluble gel until degradation and erosion takes place (Allen, 1989). Chemical analysis of the mucus gives evidence of a rather heterogeneous material which also contains small amounts of a variety of proteins, lipids, bacteria, sloughed-off epithelial cells and in some cases nucleic acids (Creeth, 1978). It becomes clear that mucoadhesion is a process that involves large amounts of water, or more vividly, it could be seen as “adhesion to water in a semisolid form” where the mucins play a key role in maintaining the gel-like properties of

---

[1] The transdermal drug delivery systems such as the skin adhesives are bioadhesive systems in a wider sense and not discussed here. Currently commercially available products are Estraderm® TTS (estradiol), Scopoderm® TTS (scopolamine), Nicotinell® TTS (nicotine) and Deponit T® (glyceroltrinitrate)

[2] Spinnbarkeit: the inherent ability of the material to form a thread

the substrate for a potential drug delivery platform. Mucin displays considerable heterogeneity that has been well described (e.g. Carlstedt and Sheehan, 1984; Neutra and Forstner, 1987; Allen, 1989; Sheehan and Carlstedt, 1989; Harding 1989). The size and shape of the molecules (macrostructure) as well as their biochemical composition and assembly (microstructure) may vary considerably, however, there are some common features. Mucins are large molecules with molecular weights ranging from 0.5×10^6 to over 4×10^6 g/mol. They contain high amounts of carbohydrate (for gastrointestinal mucins 70%–80% carbohydrate, 12%–25% protein and up to ~5% ester sulphate). Undegraded mucins are made up of multiples of a basic unit (M~400,000–500,00), linked together to give linear arrays (Silberberg and Mayer, 1982). The basic unit is made from a single chain polypeptide backbone with two distinct regions: (i) a heavily glycosylated polypeptide chain rich in serine, threonine and proline, to which a large number of carbohydrate side chains are attached, followed by (ii) one or two terminal peptide segments that bear very little or no carbohydrate side chains and are therefore often referred to as "naked protein sections". The carbohydrate chains may contain up to five different monosaccharides, namely D-galactose, L-fucose, N-acetylglucosamine, N-acetylgalactosamine and sialic acid. As multi-branched oligosaccharides they are covalently attached via O-glycosidic linkages from N-acetylgalactosamine to serine and threonine residues of the protein core. The absence of uronic acid and only trace amounts of mannose (<1%) distinguish mucin glycoproteins from the proteoglycans of connective tissue and serum glycoproteins, respectively. Sialic acid residues which belong to a family of acidic sugars (Schauer, 1992) (in gastrointestinal mucins usually either N-acetyl or N-glycollyl-neuraminic acid) are usually in a terminal position on the carbohydrate chain, whereas ester sulphate residues occur in a more internal position, e.g. as N-acetylglucosamine-6-sulphate in pig gastric mucus (Allen, 1978; Slomiany and Meyer, 1972). They both contribute in giving the molecule a net negative charge, thought to be of importance in interactions with polycationic materials (Lehr, Bouwstra, Schacht and Junginger, 1992b; Fiebrig, Harding and Davis, 1994). The heavily glycosylated regions ("T-domains"), resistant to proteolysis, have often been referred to as "bottle brush regions" in analogy to the bristles (carbohydrate side chains) attached to a wire (central protein core). They are linked linearly to one another via the "naked" regions into "subunits" (M ~ 2.5×10^6) and then further by disulphide bridging into the mucin macrostructure, which is able to form a gel.

Secretion and Function of Mucus

The adherent mucus layer in the gastrointestinal tract is secreted by specialised cells. They are the surface epithelial cells found mostly in the stomach but also in other parts of the gut and the goblet cells of the small and large intestine, glands in the stomach as well as Brunner's glands in the duodenum (Neutra and Forstner, 1987; Allen, 1989; Ito, 1981). We distinguish two kinds of mucus relevant to the concept of mucoadhesion. Firstly the aforementioned water insoluble mucus gel lining the mucosa of the gastrointestinal tract which forms the target substrate and has a variable thickness, 50–450 μm, in man and about half that in the rat (Allen, 1978; Kerss, Allen and Garner, 1982). Secondly the soluble, often viscous mucus found in the luminal contents (Allen and Carroll, 1985). Soluble mucus may result from direct secretion, erosion of the adherent gel by proteases or through mechanical shear (Allen, 1989). It is believed that the adherent mucus layer plays a major role in protection of the delicate underlying epithelium against the various endogenous and exogenous insults, such as acidic pH (providing an "unstirred boundary layer"), digestive enzymes (pepsin), pathogens (bacteria) and abrasion, while the soluble mucus may play an important role in acting as a lubricant for ingested food. The requirement for such a protective adherent gel layer becomes obvious since from a physiological point of view the luminal side of the gastrointestinal tract can still be considered as the outer side of the body. These and other aspects regarding the function of mucus have been extensively described by various authors e.g. Allen (1981, 1983, 1989), Silberberg and Meyer (1982) and Bhaskar, Garik, Turner, Bradley, Bansil, Stanley and Lamont (1992).

THE ADHESIVE: ANIONIC, CATIONIC AND NEUTRAL POLYMERS

The general properties a mucoadhesive polymer for oral drug delivery should possess are:

- High molecular weight;
- Good wettability and spreading;
- No undesirable pharmacological or physiological actions or toxicity;
- Capacity for high drug loading;
- Economical.

For bioadhesion to occur, an intimate contact between the adhesive and the substrate (mucus) is a prerequisite, where factors like good wettability and spreading as well as hydration are important (Huntsberger, 1967; Chen and Cyr, 1970; Peppas and Buri, 1985). During the establishment of the adhesive bond the total surface energy between the two materials is diminished, destroying two free surfaces and creating a new interface. Intimate contact is achieved when the spreading coefficient (S) for one of the two phases is positive and wetting of the substrate by the adhesive has occurred (Kaelbe and Maocanin, 1977; Leung and Robinson, 1988; Boddé, 1990). The spreading coefficient (S) represents the difference between the "work of adhesion" (W_a) and the "work of cohesion" (W_c) of one of the two phases

$$S = W_a - W_c \tag{1}$$

and $S \geq 0$ favours wetting of the substrate by the adhesive (Junginger, 1991).This first step is believed to be followed by physical or mechanical bond formation obtained by deposition and inclusion of the adhesive material in the crevices of the mucus and chain entanglement between polymer chains of both phases (also referred to as inter-diffusion) (Boddé, 1990; Jabbari, Wisniewski and Peppas, 1993). Lehr, Bouwstra, Speis, Onderwater, van het Noordeinde, Vermeij-Keers, van Munsteren and Junginger (1992c) have used electron microscopy in an attempt to visualize intermixing between a polyacrylic acid derivative (polycarbophil) and mucus. They were unable to observe intermixing in the micron range but did not exclude this phenomenon for the nanometre range. Sufficient chain flexibility is required to form secondary chemical bonds such as van der Waals forces as well as hydrogen bonding (Leung and Robinson, 1988; Duchêne *et al.*, 1988). The formation of primary (covalent) chemical bonds is important in hard tissue adhesion in orthopaedics and dentistry. However, for mucoadhesion, chemical reactions of this type have not been considered so far, since a long term attachment is not required (Peppas and Buri, 1985).

Polymers with hydroxyl or carboxyl groups on their surface had been recognised as the most desirable candidates for bioadhesion, rather than polymers with other functional groups or cations (Peppas and Buri, 1985). The polyacrylic acid derivatives polycarbophil (Carbopol® EX-55) and carbomer (Carbopol® 934) have been by far the most studied mucoadhesive polymers to date (Table 18.3). Both materials are polyanionic and interaction with mucus has largely been attributed to chain entanglement of the polymer chains with mucin as a result of swelling in water and hydrogen bonding due to the carboxyl groups in their unionised state at low pH (Robinson, Longer and Veillard, 1987; Leung and Robinson, 1988; Ponchel, Touchard, Duchêne and Peppas, 1988; Jabbari *et al.*, 1993). Polycarbophil is described as a water insoluble but swellable polymer of polyacrylic acid crosslinked with divinylglycol used in the treatment of diarrhoea and as a bulk laxative. Carbomer is a water soluble polymer of acrylic acid loosely crosslinked with allylsucrose.

According to Lehr *et al.* (1992b), the need for hydrogen-bonding capabilities and negative charge in bioadhesive materials should not be generalized. They suggested that polycationic polymers might interact with the anionic sites on the mucins more favourably due to their opposite charges providing additional molecular attraction forces. For example, interactions between charged polymeric molecules have been employed in colloid titration (Terayama, 1952; Senju, 1969). The method is based on the principle that positively charged macromolecules will react with negatively charged macromolecules. The neutralisation reaction will proceed stoichiometrically, allowing an estimation of either material if a standard colloid solution is used. Katayama, Takai, Kariyama and Kanemasa (1978) used the method for the titration of heparin using cat-floc (polydiallyldimethyl ammonium chloride) as standard polycation. Van Damme, Blackwell, Murphy and Preston (1992) measured the negative charge content in cartilage using cat-floc as well. Interactions between alginates and pectins with cationic polypeptides such as poly(L-lysine) and poly(Lys-Lys-Ala) have been studied using circular dichroism (Bystricky, Malovíková and Sticzay, 1990). Differences in interaction efficiency between the polymers were attributed to differences in conformational flexibility of the polyanionic chains in solution. Takahashi, Takayama, Machida and Nagai (1990) studied the characteristics of polyion complexes of chitosan with sodium alginate and sodium polyacrylate using viscometry and Fourier transform infra-red spectroscopy (FT-IR). They found that chitosan and alginate reacted with a defined binding ratio which was found to be relatively constant in media of various pH values. In contrast, for polyacrylate–chitosan interactions the unit molecular binding ratio was greatly affected by the pH.

Chitosan appears to be an ideal candidate as a mucoadhesive polycationic polymer — it is being produced on a large scale (Jeuniaux, Voss-Fouchart, Poulicek and Bussers, 1989; Alimuniar and Zainuddin, 1992). Although chitosan has not received regulatory approval by the Food and Drug Administration (FDA) for pharmaceutical and alimentary use, chitosan containing material obtained from the treatment of the waste streams of food processing plants, may be used as livestock feed in the U.S.A. as long as the level of chitosan does not exceed 0.1% (Weiner, 1992).

Chitosan has been approved as a food additive in Japan since 1983 and has been placed on the Japanese Natural Additive List. It is used as a thickener and stabilizer (Weiner, 1992). It is a food ingredient in some dietary cookies and noodles from Hihon Kayaku Inc. and Tanami Foods Inc. as well as in vinegars of Nakano Inc., making use of the hypocholesterolaemic properties (Hirano, 1989). The food industry has also exploited the chelating properties of chitosan for the clarification of beverages such as apple and carrot juices (Imeri and Knorr, 1988; Soto Peralta, Müller and Knoll, 1989).

The lack of acute oral toxicity of chitosan has been supported by experiments in mice (Arai, Kinumaki and Fujita, 1968) who determined an LD50 of > 10g/kg. However the literature lacks adequate scientific studies on long term and widespread human exposure through food and pharmaceutical products (McCurdy, 1992).

Variations in molecular weight and degree of deacetylation together with the ability to form gels and films allow flexibility in formulation design (Acatürk, 1989; Miyayaki, Yamaguchi, Takada, Hou, Takeichi and Yasabuchi, 1990; Errington, Harding, Vårum and Illum, 1993). Chitosan has been shown to possess mucoadhesive properties almost as strong as poly(acrylic acid) (Lehr *et al.*, 1992b). Various types of chitosan have been screened for their mucoadhesive properties together with some other polymers (see Table 18.1). Sedimentation velocity experiments on an analytical ultracentrifuge revealed a complex formation between chitosan (Sea Cure +210) and purified pig gastric mucin (Fiebrig *et al.*, 1994), Table 18.2.

Table 18.1 SURVEY OF MUCOADHESIVE PROPERTIES OF VARIOUS POLYMERS REPRODUCED FROM LEHR *ET AL.* (1992b). MUCOADHESION WAS DETERMINED AS THE FORCE OF DETACHMENT FOR POLYMER-COATED COVER GLASSES FROM PIG INTESTINAL MUCOSA IN ISOTONIC SALINE. THE DRY FILMS WERE SWOLLEN IN THE TEST MEDIUM FOR 5 min AND THEN BROUGHT INTO CONTACT WITH THE TISSUE UNDER VERY SLIGHT PRESSURE (~ 10 mN) AND KEPT IN THIS POSITION FOR AN ADDITIONAL 1 min. A VERTICAL FORCE WAS SLOWLY INCREASED UNTIL THE POLYMER BECAME DETACHED. INDICATED IS THE MEAN (SD) FORCE OF DETACHMENT OF 2 TO 6 MEASUREMENTS.

Polymer	*Force of detachment (mN/cm^2)*
Cationic polymers	
Chitosan (Wella 'low viscosity')	3.9 (1.2)
Chitosan (Wella 'high viscosity')	6.7 (0.7)
Chitosan (Dr. Knapezyk)	5.7 (1.1)
Daichitosan H	8.0 (5.7)
Daichitosan VH	9.5 (2.4)
Sea Cure 240	4.1 (2.9)
Sea Cure 210 +	9.5 (2.5)
Chitosan (Sigma)	6.6 (3.0)
Polycarbophil/Daichitosan VH blend	11.9 (2.5)
DEAE-dextran	0
Aminodextran	0
Non-ionic polymers	
Steroglucan	2.8 (2.8)
HE-starch	0.6 (0.8)
HPC	0
Anionic polymers	
CMC (low viscosity)	1.8 (1.1)
CMC (medium viscosity)	0.3 (0.3)
CMC (high viscosity)	1.3 (1.0)
Pectin	0
Xanthan gum	0
Polycarbophil	17.6 (3.6)

TABLE 18.2 SEDIMENTATION COEFFICIENTS ($s_{20°C,BUFFER}$ ± STANDARD DEVIATION) IN SVEDBERG UNITS, S ARE GIVEN FOR MUCIN AT RESPECTIVE CONCENTRATIONS OF 0.2 mg/ml AND 1 mg/ml IN ACETATE BUFFER pH 4.5, I=0.1 AS WELL AS FOR THEIR COMPLEX RESULTING FROM ADMIXTURE OF THE TWO COMPONENTS. INCREASES IN s-VALUES ARE INDICATIVE OF CONCOMITANT INCREASE IN MOLECULAR WEIGHTS AS A RESULT OF ASSOCIATION.

Mixture ($s_{20,b}$ *for complex)*	*Control* ($s_{20,b}$ *for mucin)*	*Control* ($s_{20,b}$ *for chitosan)*
780.00 (24.00)	52.50 (1.50)	2.10 (0.01)

Is mucus an appropriate target?

There are three physiological aspects which remain critical for the concept of gastrointestinal mucoadhesion: (i) turnover of the adherent mucus layer, (ii) interactions of the formulation with soluble, i.e. non-adherent mucus prior to adhesion and (iii) gastrointestinal motility.

TURNOVER OF THE ADHERENT MUCUS LAYER

Adherent mucus is continuously lost into the gastrointestinal lumen by proteolysis and mechanical sloughing (e.g. Allen, 1981; Allen and Caroll, 1985). The latter, caused by the ingested food and its digestion, is thought to be the major cause of loss of gastric mucin (Waldron-Edward, 1977). A dynamic balance exists at the mucosal surface *in vivo*, between mucus secretion and mucus erosion. Mucus erosion, either by pepsin or by abrasion, must be replenished by the mucosal secretion of new material in order to maintain a protective function (Allen, Flemström, Garner and Kivilaakso, 1993). The difficulties in measuring mucus secretions *in vivo* has been outlined by Allen (1989). Studies on the turnover time of intestinal mucus gel layer in the rat *in situ* loop (Poelma and Tukker, 1987) by Lehr, Poelma, Junginger and Tukker (1991) have attempted to shed some light on the limitations to gastrointestinal mucoadhesion. In view of these authors the maximal residence time of a bioadhesive DDS at the site of adhesion would be limited by the time it takes for the mucus gel layer to be renewed as determined by the steady state of synthesis, secretion and degradation of the mucins (Allen, 1981). Although their estimate for the mucus turnover time is relatively crude (47–270 min), it is interesting to find that this time scale is similar to the mean residence time found for mucoadhesive microspheres (94±18 min) in earlier experiments using the same animal model (rat). Further to this it had been observed that stimulating the mucus output by perfusion with 10 mM sodium taurocholate led to a significant shortening of the mean residence time of microspheres. Of even greater interest is the observation that the microspheres did not become detached from dead mucosal tissue *in vitro* when the system was stirred for more than 18 h. This leads to a further consideration; that of choosing the appropriate model substrate. This is discussed below. Although mucus turnover in an *in situ* isolated gut loop in the rat (which has undergone surgery and has been removed from its normal function) might be quite different from mucus turnover in healthy humans or patients, this physiological factor could substantially limit potential adhesion to the adherent mucus in the gastrointestinal tract.

INHIBITORY INTERACTIONS WITH SOLUBLE MUCINS

Any formulation entering the gastrointestinal tract is likely to interact with soluble mucins of the lumen. This in turn might render the device incapable of adhering to the target surface. That this might indeed be a serious problem has been shown recently by Lehr, Bouwstra, Kok, Noach, de Boer and Junginger (1992d). Tomato lectin, that specifically binds to isolated pig enterocytes and monolayers of human Caco-2 cell cultures, was proposed as a favourable candidate for specific bioadhesion to epithelial cells of the gastrointestinal tract. However, binding was also noticed to occur with crude pig gastric mucus. This would mean that *in vivo* the adhesion of lectins to the mucosal surface may be inhibited by mucus. Fiebrig *et al.* (1994) showed marked interaction between dilute solutions of highly purified pig gastric mucin and the potentially bioadhesive polymer chitosan (SeaCure+210, characterized by Errington *et al.*, 1993), using analytical ultracentrifugation.

In a further study, human gastric mucin, purified from gastric aspirates, showed a similar interaction. The mucin with a weight average molecular weight of M= $(1.8\pm0.1)\times10^6$ g/mol as determined by GPC/MALLS and a sialic acid content of 3.2% (for methods see Fiebrig *et al.*, 1994) at a concentration of 0.3 mg/ml sedimented at a rate of (11.0 ± 0.18) S in an acetate buffer of pH 4.0 and at a temperature of 20°C (Figure 18.1). When mixed with chitosan at a concentration of 1.0 mg/ml a sedimenting species of (222.0 ± 1.8) S was detected (Figure 18.2), corresponding to a complex between mucin and chitosan. Using chitosan alone as control, under the same conditions of concentration, solvent and temperature, a sedimentation coefficient of (2.1 ± 0.03) S was obtained (Table 18.2), a value close to that expected for this macromolecule.

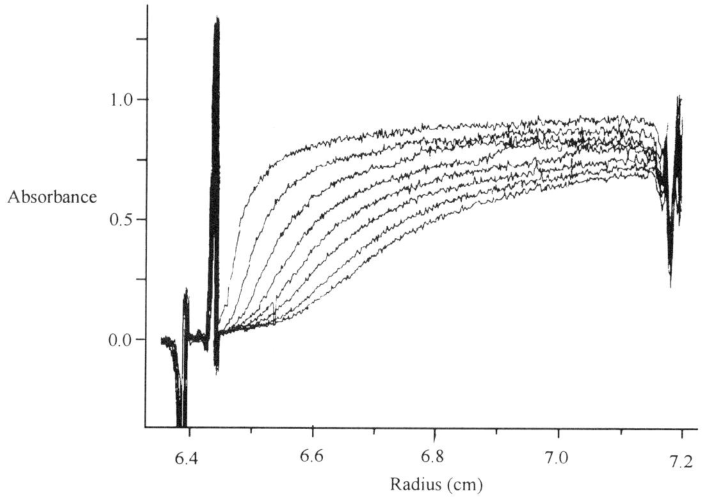

Figure 18.1 Human gastric mucin (control) at a concentration of 0.3 mg/ml in acetate buffer pH 4.5. Scans taken at 20°C, 20,000 rpm and a scanning interval of 11 min with absorption optics at a wavelength of λ = 226 nm using an XL-A Beckman Analytical Ultracentrifuge. Sedimentation coefficient $s_{20^{o}C,\,buffer}$ = (11.0±0.18) S

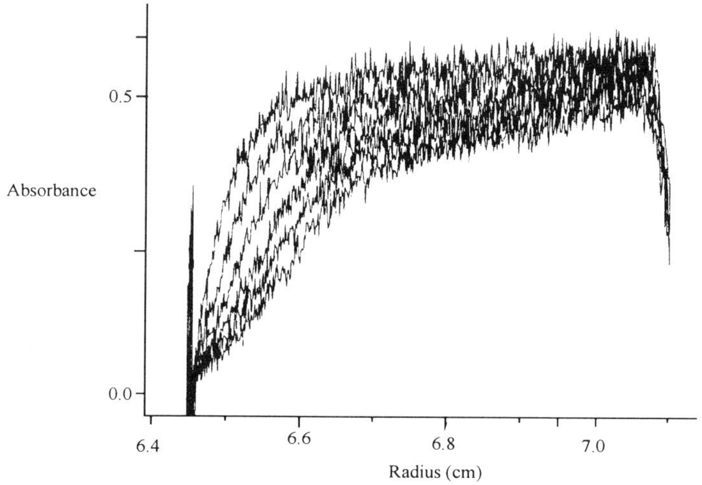

Figure 18.2 Human gastric mucin/chitosan mixture (0.3/1.0 mg/ml) in acetate buffer pH 4.5. Scans taken at 20°C, 7,000 rpm and a scanning interval of 6 min with absorption optics at a wavelength of λ = 256 nm using an XL-A Beckman Analytical Ultracentrifuge. Sedimentation coefficient $s_{20^{o}C,\,buffer}$ = (222.0± 1.8) S

GASTROINTESTINAL MOTILITY

Gastrointestinal motility patterns and in particular the so called "housekeeper wave" which involves strong gastrointestinal contractions, serves as a cleaning mechanism to clear all indigestible materials, including non-disintegrating dosage forms, from the stomach or proximal intestine (Code and Marlett, 1975; Grundy, 1985; Leung and Robinson, 1988). Thus, a good oral mucoadhesive drug delivery system needs to overcome the cleaning action of the "housekeeper wave" and remain in the stomach or proximal small intestine.

Experimental approaches used to study mucoadhesion

MODEL MUCUS' AND MUCINS FOR *IN VITRO* EXPERIMENTATION

In vitro test methods can be divided into (i) those that employ freshly excised tissue from various animals (frog, rat, rabbit, pig, cow, etc.) used either immediately as live or dead tissue or stored frozen and defrosted prior to use and (ii) those methods using mucus or mucin at various degrees of degradation and purity either solubilised or as gel usually from pig stomach or bovine submaxillary glands (Table 18.3). Whatever model material is used, its relevance to the human mucus, whether in health or disease state, has to be considered (MacAdam, 1993). Dead mucosal tissue may not produce any new mucus while degradation of existing mucus will still take place. This will have a marked effect on the rheological characteristics of the substrate, considered to be highly relevant to adhesion phenomena. Mucus thickness may vary from species to species and intersubject as well as intrasubject variability of the mucosal tissue poses problems in terms of reproducibility. Mucin carbohydrate composition varies within the gastrointestinal tract (Allen, 1989). Pig gastric mucin appears to be a good model mucus glycoprotein for the human because its carbohydrate composition is comparable. The content of sialic acid is similar to that of the small intestine in humans and pig mucus is readily available in reasonable quantities from freshly slaughtered pigs. Purification methods allow the removal of other components present in mucus in order to obtain purified mucin which still shows the gel-forming characteristics of native mucus (Bell, Sellers, Allen, Morris, Ross-Murphy, 1985; Allen, 1989).

Commercially available pig gastric mucins or mucus (Table 18.3) are quite different in their composition when compared with freshly prepared and purified material and they may be rather degraded or the freeze drying procedure may have altered the structure in such a way that it becomes difficult to redissolve them completely. Commercially available submaxillary mucins are quite different from the mucins secreted in the gastrointestinal tract. They are secreted in a viscous soluble form rather than as water-insoluble gels (for differences see Gottschalk, Bhargava and Murty, 1972). Nevertheless highly purified mucins can give more accurate information on the actual interaction with the main mucin-forming component. The use of dilute mucin solutions allows the study of mucin–bioadhesive polymer interactions on a fundamental level.

It has been recognized that the degree of hydration of the bioadhesive DDS as well as the amount of water available plays an important role in determining the strength of adhesion or whether adhesion can take place at all (Leung and Robinson, 1988; Chen and Cyr, 1970). The hydration aspect can be controlled in local applications such as mouth or vagina by drying excess water immediately prior to application (Deasy and O'Neill, 1989). The amounts of fluid available are also limited. In the gastrointestinal tract, however, excess water at the site of adhesion as well as excess in the amount of surrounding liquid cannot be controlled. Lehr, Bouwstra, Schacht and Junginger, (1992b) pointed out that numerous so-called mucoadhesive polymers adhere only under conditions where the amount of interstitial liquid is limited. This kind of dry-to-wet adhesion or "blotting adhesion" is due to the capillary forces drawing liquid from the mucus into the delivery system (Huntsberger, 1967; Lehr, Bouwstra, Schacht and Junginger, 1992b; Mortazavi and Smart, 1993). If the polymer involved offers no intrinsic ability to form a bond with the substrate (e.g. some cellulose derivatives), the initial adhesive forces although high at the beginning may become negligible as soon as the material is fully hydrated (Junginger and Lehr, 1990). Therefore adhesion measurements in fully hydrated systems and over a period of time are necessary to avoid attributing a high adhesive force to intrinsic mucoadhesive properties. The adhesion mechanism of capillary attraction between a dry, water-absorbing polymer and a wet, mucosal surface being dehydrated is quite different to the interactions between two hydrogels (polymer and mucus) in equilibrium with a third liquid phase (Mortazavi and Smart, 1993).

IN VITRO METHODS

Tensiometry

The main method used in *in vitro* testing of mucoadhesion is tensiometry (Table 18.3: Entry Numbers 1, 2, 14, 15, 17–32). The method employs putative mucoadhesive polymers that are usually in the form of tablets made by direct compression of the polymer or polymer coated surfaces made by casting of polymer solutions. These are consequently put in contact with a mucus surface usually with a given force applied to the system for a given period of time after which the adhesive joint is destroyed by applying a vertical force in the opposite direction or a shear force in the horizontal direction. The force required to destroy the bond is taken as a qualitative and quantitative parameter for adhesion. If the experiment is done under full hydration of the polymer and in an aqueous environment the likelihood of mimicking *in vivo* conditions is higher, given that a potential formulation, which is usually swallowed with liquid would not arrive at the target site in a totally dry state. However, this method neglects the fact that a swallowed formulation does not make intimate contact with the mucus gel spontaneously. Furthermore, the cohesion of mucus is also related to its thickness and rheological features. The experimental setup does not allow to appreciate if the bonding failure is adhesive or cohesive in nature (Huntsberger, 1967; Lehr *et al.*, 1992c). Although information on the screening of polymers can easily be obtained, the method appears unsuitable for assessing adhesive behaviour of formulations intended for gastrointestinal application. Similar comments apply to the Wilhelmy plate method used by e.g. Smart, Kellaway and Worthington, (1984) (Table 18.3, Entry Numbers 3–5). Tensiometry seems more useful for buccal, vaginal or other applications where liquid is controllable and more limited.

Flow through systems

Flow through systems appears an appropriate alternative in assessing the various aspects of mucoadhesion with a pharmaceutical formulation in the gastrointestinal tract. A flow channel device was first described by Mikos and Peppas (1990). The channel had a length of approximately 30 cm, a width of 4 cm and a height of approximately 0.5 cm, it was thermostatted by a jacket connected to a constant temperature water bath. A cavity inside the channel allowed placement of a mucin gel or the mucosal of a tissue and the placement of a single polymer microparticle on top of it. The channel was connected through a set of valves to a gas cylinder. The volumetric flow rate was gradually increased until the particle, which was observed by an optical microscope, was detached from the mucous surface. The system is suitable as a model for studying nasal mucoadhesion. For gastrointestinal models a fluid should be substituted for the air. Liquid flow methods were developed later (Table 18.3, Entry Numbers 33–35) and a similar technique employed live tissue to monitor intestinal drug absorption at the same time (Junginger, Lehr, Bouwstra, Tukker and Verhoef, 1990). The observation of adhesion of a formulation (which ought to be insoluble so as to avoid dilution and rapid wash-off as well as rapid drug leaching) from a flow of solution directly onto a mucus tissue would be most desirable. A swellable but insoluble formulation is usually achieved by crosslinking of the polymer chains. This would lead to chain rigidity which in turn can limit mucoadhesion since the proposed "interpenetration" or "interdiffusion" mechanism would be restricted.

Colloidal gold staining

A method presented by Park (1989) is particularly interesting because unlike tensiometric methods it does not monitor a destruction process. Instead of measuring the adhesion strength or the duration of adhesion, the "adhesion number" is determined as a direct function of adhesion. Furthermore the technique is claimed to give precise and reproducible values, is simple to perform and the experimental cost is fairly low.

The adherent material (or substrate) consisted of colloidal gold particles of an approximate diameter of 18 nm that carried mucin adsorbed onto their surface. The mucin used was a solution of bovine submaxillary mucin Type I. Colloidal gold (cAu) sols are usually prepared by reducing $HAuCl_4$ with reducing agents like sodium citrate. Particle sizes vary depending on the reducing agent as well as on the preparation procedure.

The bioadhesive material used by Park was a copolymer made from acrylic acid and acrylamide cross-linked with N,N'-methylene-bis-acrylamide [P(AA-co-AM)]. The transparent hydrogels made of this material were cut into rectangular shapes of varying thickness. The polymer strips were incubated with the cAu-mucin conjugates and after a rinsing procedure, the absorbance of the strip was measured at 525 nm with a spectrophotometer using a transparent control polymer strip as a blank. The values obtained were a function of the amount of cAu adsorbed onto the surface. Alternatively the absorbance of the cAu-mucin preparation was measured before and after incubation. In this case, the magnitude of the decrease in the absorbance value from the initial value (ΔAbs) was used as a quantitative parameter indicating the interaction between cAu-mucin and hydrogel. The author also used an image analyser to quantify the intensity of red colour on the polymer surfaces. This alternative is necessary for mucoadhesive polymers which are not transparent.

The cAu-mucin conjugates prepared by Park required the addition of albumin to stabilise the preparation further. Albumin molecules are believed to adsorb onto small bare spots on the cAu particle where mucin molecules do not cover (Horisberger and Rosset, 1977; De Mey, 1984). It can be argued, however, whether any interaction phenomenon observed is due solely to the properties of the mucin. Besides, if the affinity of albumin to the cAu is higher than that of mucin, a displacement of mucin from the cAu is also possible.

As outlined by Park (1989) an alternative approach using the colloidal gold staining technique is that of developing cAu-(bioadhesive polymer) conjugates instead of cAu-mucin conjugates. The polymer coated cAu particles acting as the adhesive this time could be directly applied to the surface of target tissues. In this case the cAu-polymer conjugate would act as a model drug delivery system. The possibility of additional adsorption of drugs, e.g. peptide drugs onto these "nanospheres" would be interesting. Chitosan-stabilised cAu has been successfully prepared by Horisberger and Clerc (1988) to use as a marker for anionic sites on various micro-organisms and by Fiebrig, Harding, Stokke, Vårum, Jordan and Davis (1994) to visualize the sites of interaction of chitosan in a mucin-chitosan complex.

Miscellaneous methods

Viscometry, rheometry, photon correlation spectroscopy and nuclear magnetic resonance as well as analytical ultracentrifugation (Table 18.3, Entry Numbers 7–9, 11–13) are methods that can give information on [bioadhesive polymer]–[glycoprotein] interactions. The use of a standardised material throughout the experiments allows the comparison of results and avoids inter-sample variations. Such methods also allow us to study the influence of factors that may influence interaction (ionic strength, bile salts, temperature, proteins) and hence the elucidation of interaction mechanisms. It is well appreciated that the experimental conditions bear no relation to *in vivo* situation. Concentrations may be very low, as in the case of analytical ultracentrifugation and dynamic light scattering.

IN VIVO

When studying bioadhesion, with the aim of assessing the performance of a bioadhesive DDS, it is usually best to work with an animal model, human volunteers or patients. For buccal, vaginal, cervical or nasal applications (Table 18.3) the residence time of the device can be inspected visually, while the subject can give direct information on aspects of tolerance (discomfort, usefulness, etc.) (Nagai, 1986; Bottenberg, Cleymaet, de Muynck, Remon, Coomans, Michotte and Slop, 1991; Smid-Corbar, Kristl, Cop and Groselj, 1991). Plasma levels of drug or pharmacodynamic effect (for e.g. delivery of insulin) can give direct evidence for bioequivalence or enhanced bioavailability. Aspects of intersubject variability and disease condition may have to be taken into account.

With regards to gastrointestinal bioadhesive DDS we are faced with major experimental difficulties. Once swallowed, the device has to reach the adherent mucus layer. As has been outlined earlier, the process of bioadhesion requires *intimate contact* in its first step. There is little experimental evidence for this prerequisite actually taking place in gastrointestinal bioadhesion. Lodged on the target surface a delivery system has to resist the dislodging forces of gastrointestinal motility. In humans this particular aspect has been examined for the first time in a double blind study by Anderson (1991) using coloured tablets made from DEAE-dextran and ethylcellulose (1:1) with ethylcellulose tablets as controls. Seven patients undergoing routine gastroscopy examination swallowed both tablets with approximately 20 ml of water immediately before endoscopy. Mucoadhesion or the lack of it was assessed using a finger controlled water jet attachment. The clinician sprayed the tablet with water, for a fixed time period and at a constant rate, and assessed adhesion in terms of the number of sprays required to dislodge the tablet. The results did not suggest any significant adhesion of the test formulation to the gastric mucosa. In 50% of the patients examined at longer time intervals (up until 65 min post dose) neither the control nor the test tablet could be found in the stomach. For those patients where tablets could be observed, no significant differences in adhesive behaviour between control and test tablet with the gastric mucosa could be detected when judged using the finger controlled water jet attachment. There was no actual measurement of the water spray properties. However, the results illustrate the probable lack of adhesion for both tablet formulations as well as the intersubject variation in gastric emptying times for tablet formulations in the fasted state as observed by other authors.

The rat as *in vivo* model (Table 18.3) has a mucus layer about half the thickness of that in man (Allen, 1978; Kerss *et al.*, 1982), while little is known about mucus turnover compared to man. It has been suggested that in the rat there is very little soluble gastric mucus when compared with the dog where there are considerable amounts of this material (Robinson *et al.*, 1987). Although the mucin of pig gastrointestine is similar to that of humans with regards to its carbohydrate and protein composition (Allen, 1989), gastric emptying in pigs has been shown to be slower than in man (Aoyagi, Ogata, Kaniwa, Uchiyama, Yasuda and Tanioka, 1992) and consequently this animal may not be the appropriate *in vivo* model.

TABLE 18.1 SUMMARY OF METHODS USED TO STUDY BIOADHESION FOR DRUG DELIVERY SYSTEMS

<table>
<tr><th>Entry Number</th><th>Substrate</th><th>Adhesive</th><th>Drug</th><th>Preparation</th><th>Test</th><th>Application</th><th>Reference</th></tr>
<tr><td colspan="8">In vitro studies using various model substances (mucin, mucus or other) as substrate</td></tr>
<tr><td>1</td><td>Plexiglas®</td><td>PE gel (5% in petrolatum) + Na-CMC or gelatin</td><td>—</td><td>Viscous pastes hydrated with artificial saliva</td><td>Tensiometry</td><td></td><td>Gurny, Meyer and Peppas, 1984</td></tr>
<tr><td colspan="8">Instron 1114 tensile tester, the paste is placed between two Plexiglas® discs, discs are pulled apart and stress—strain curves recorded measuring "adhesive strength".</td></tr>
<tr><td>2</td><td>Filter paper discs + 15% PGM_n solution</td><td>PAA, HA, HAE polymers</td><td>—</td><td>Hydrated circular matrices, 12 mm diameter</td><td>Tensiometry</td><td>Ocular</td><td>Saettone, Chetoni, Torracca, Burgalassi and Giannaccini, 1989</td></tr>
<tr><td colspan="8">Modified tensile tester of Ch'ng et al. is used to determine the force required to separate two mucin coated surfaces; the relevance of mucin-soaked filter papers as mucous substrate can be argued. PGM_n from Biochemo SpA.</td></tr>
<tr><td>3</td><td>Homogenized mucus from guinea-pig intestine</td><td>Numerous polymers</td><td>—</td><td>Polymer-coated glass plate</td><td>Wilhelmy plate method</td><td>—</td><td>Smart, Kellaway and Worthington, 1984</td></tr>
<tr><td colspan="8">The method measures the maximum force required to detach the plate from the mucus gel with a microforce balance.</td></tr>
<tr><td>4</td><td>Crude PGM_s (Sigma)</td><td>PAA and PMAA with various cross-linkers</td><td>—</td><td>Polymer-coated rexin sheet</td><td>Wilhelmey plate method</td><td>—</td><td>Chitnis, Malshe and Lalla, 1991</td></tr>
<tr><td colspan="8">A horizontal polymer-coated rexin sheet is pulled by means of a "small crane" from a 5% mucin solution kept in a beaker on a top-loading balance, the minimum weight is recorded; the polymer was allowed to hydrate for 5 minutes.</td></tr>
</table>

5	Pig small intestinal mucus, native and homogenised	CaPC, NaCMC, HPMCP, HPMC Eudr. RS100, AP	—	Polymer-coated glass plates, either as film or as powder	Wilhelmey plate method	—	Sam, van den Heuij and Tukker, 1992

Based on the method described by Smart *et al.*, polymer powders (non film-forming sample) are adhered to partially dried Eudragit RS100 and force—time diagrams are recorded.

6	Bovine submaxillary mucin TI (Sigma)	P(AA-co-AM)	— —	— —	Colloidal gold staining	— —	Park, 1989

Colloidal gold is stabilised with mucin and hence acts as a marker for light as well as electron microscopy. Determination of the "adhesion number" measures bond formation rather than a destruction process as a function of staining intensity.

7	PGM_s	PAA, low molecular weight	—	Polymer solutions at pH 7.4 and 4.0	Photon correlation spectroscopy (PCS)	—	Kerr, Kellaway, Lewis, Kelly and Parr 1989

Malvern RR144 with a multibit correlator, K7025 and a He—Ne 35 mW laser was used to measure the diffusion coefficients.

8	PGM_s	PAA, low molecular weight	—	Polymer solutions at pH 7.4 and 4.0	Nuclear magnetic resonance (^{13}C-NMR)	—	Kerr *et al.*, 1989

Bruker WM360 FT spectrometer operating at 90 MHz and RT.

9	PGM_n, purified	DEAE—dextran	—	Dilute polymer solutions	Analytical ultracentrifugation	(Gastrointestinal)	Anderson, Harding and Davis, 1989

Based on the assumption that any marked interaction between the two components would lead to co-sedimentation, shown by an increase in the sedimentation coefficient of the sedimenting boundaries.

10	Bovine sub-maxillary mucin gel (Sigma)	PHEMA	—	Microspheres ~156 μm diameter	Flow channel method (air flow method)	—	Mikos and Peppas, 1990

The method observes the adhesive behaviour of a microparticle under the microscope connected to a photo or video camera as air is blown through a channel containing the preparation.

11	PGM_s (Sigma)	Various neutral, anionic and cationic polymers including chitosan	—	Polymer solutions in 0.1 N HCl and acetate buffer pH 5.5	Viscometry	—	Hassan and Gallo, 1990

Brookfield viscometer Model RTV and Ostwald viscometer were used to measure the solutions of individual polymers and mixtures thereof.

12	Crude and homogenised PGM_s; PGM_n	C-934P	—	C-934P gel (5 mg/g)	Rheometry	—	Mortazavi, Carpenter and Smart, 1992, 1993

Determination of mean storage modulus and loss modulus of mucus or mucin—polymer mixtures.

13	PGM_n, purified	Chitosan	—	Dilute polymer solutions	Analytical ultracentrifugation	(Gastrointestinal)	Fiebrig, Harding and Davis, 1994

Same as Anderson *et al.* 1989

Ex vivo techniques (using animal mucosal tissue as substrate in most models)

14	Mouse peritoneal membrane	HPC:C-934, mixing ratios 3:1, 2:1, 1:1 or 1:2	(Insulin)	Tablets made by direct compression of polymer mixtures	Tensiometry	(Buccal)	Ishida, Machida, Nambu and Nagai, 1981

"Stickiness test" measuring "sticking force". Instead of a tensile tester a simple spring balance is use, the tablet is applied dry and the force measured until detachment occurs.

15	Pig oesophagus	Various gelatin capsules and tablets	Not relevant	Various gelatin capsules and tablets	Tensiometry	Oesophagus	Marvola, Vahervuo, Sothmann, Marttila and Rajaniemi, 1982

The is used to study the tendency of formulations to adhere to the oesophagus, measuring the weight required until detachment of formulation.

16	Fresh rabbit stomach tissue	Polymers from AA and MA with various cross-linking agents	—	Prehydrated, insoluble polymer	Tensiometry	—	Ch'ng, Park, Kelly and Robinson, 1985

Measures the force required to detach mucus from bioadhesive polymer i.e. the destruction process of a mucosa—polymer—mucosa "sandwich" in a fully hydrated environment.

17	Fresh rabbit conjunctival membrane (eyelid)	Polymers of AA cross-linked with dvG and dmH	^{14}C-labelled progesterone	Prehydrated, insoluble polymer	Tensiometry	Ocular	Hui and Robinson, 1985

Same as Ch'ng *et al.* (1985).

18	Bovine sublingual mucus tissue	C-934 0%—90% mixed with HPMC	—	12 mm diameter tablets prepared by direct compression	Tensiometry	—	Ponchel, Touchard, Duchêne and Peppas, 1987

Based on the methods of Gurny *et al.* and Ch'ng *et al.* using Instron 1026 tensile apparatus, tablet is prehydrated with a defined amount of water before being placed on mucus tissue, glued to the lower support, relative humidity of 60% in the environment, measurement of detachment vs. elongation.

19	Bovine sublingual mucus tissue	C-934 0%—90% mixed with HPMC	—	12 mm diameter tablets prepared by direct compression	Tensiometry	—	Peppas, Ponchel and Duchêne, 1987

Experimental setup of Peppas *et al.* (1987), repeated straining followed by relaxation measuring stress relaxation.

20	Bovine sublingual mucus tissue	C-934 0%—90% mixed with HPMC	50% metronidazole	12 mm diameter tablets prepared by direct compression	Tensiometry	—	Ponchel, Touchard, Wouessidjewe, Duchêne and Peppas, 1987
Experimental setup of Peppas *et al.* (1987), measuring detachment force vs. elongation.							
21	Fresh rabbit stomach tissue	Copolymers from AA and MM with divinyl glycol as cross-linking agent	—	Prehydrated, insoluble polymer	Tensiometry	—	Leung and Robinson, 1988
Same as Hui and Robinson (1985), with shear stress measurement additional to the tensile stress testing.							
22	Frog oesophagus with and without mucus layer	PNVP, PHEMA, PMMA, NaCMC, C-941, ALA, NaAL	—	Polymer powder stuck onto hydrated PHEMA disc	Tensiometry	—	Robert, Buri and Peppas, 1988
Du Noüy tensiometer measuring detachment force, not conducted in liquid medium, otherwise polymer is removed, reference disc made from PTFE.							
23	Pig oral mucosal tissue	Copolymers of AA and BuAA XL:EDEMA	—	Hydrogel strip	Tensiometry	—	de Vries, Bodde, Busscher and Junginger, 1988
Instron 1122 test unit to determine the peel force pulling at 90° away from the surface.							
24	Pig oral mucosal tissue	Copolymers of AA and BuAA XL:EDEMA	—	Hydrogel strip	Tensiometry and determination of "shear time"	—	de Vries *et al.*, 1988
The epithelial flap was attached vertically with pins to a rigid surface, a hydrogel strip laid on top of the mucosa surface by applying a known force. Upon attaching a 50 g mass to the bottom end of the test strip, the time needed for the test strip to come off the substrate was measured at 34° and 100% humidity.							
25	Bovine vaginal and sublingual mucosa	HPMC and C-934P	Metronidazole	Tablet made by direct compression	Tensiometry	Vaginal and oral	Lejoyeux, Ponchel, Wouessidjewe, Peppas and Duchêne, 1989
Method of Peppas *et al.*, 1987, system dipped (or not) in a liquid medium of variable composition (pH, ionic strength, divalent cations), determination of adhesion work.							

26	Bovine vaginal mucosa	C-934P. NaCMC	$ZnSO_4$	Tablets made by direct compression	Tensiometry	Vaginal	Gürsoy. Sohtorik. Uyanik and Peppas. 1989
Tensilon. UTM11. Japan used for measuring detachment force.							
27	Rat small intestine	C-934. PC. PEG. HPC. HPMC. gelatin	See comments	Discs made from compressed powder	Tensiometry	—	Smart. 1991
Experimental setup of Ch'ng *et al.*, sample surrounded by buffer pH 6.8. commercial buccal preparations were tested as well: Adcortyl in Orabase® (triamcinolone acetonide), Buccastem® (prochlorperazine maleate) and Suscard Buccal® (glyceryl trinitrate), instead of "sandwich" one mucosal surface only.							
28	Pig attached gingiva	Modified starch. C-934. PEG and NaCMC	Sodium fluoride	7 mm diameter flat tablets by direct compression	Tensiometry	Buccal (dental)	Bottenberg. Cleymaet. de Muynck. Remon. Coomans. Michotte and Slop. 1991
Same as Ponchel *et al.*, 1987							
29	Ox sublingual mucosa	C-934	—	Tablet made by direct compression	Tensiometry	—	Jaques and Buri. 1992
Tablet diameter: 10 mm. Schenk—Trebel RM 50 tensile tester. and force and displacement transducer. Adhesion of the connective side to the underlying support: also made of C-934. tablet applied in a dry state. work of elongation of the mucosa is measured.							
30	Healthy human uterine cervix tissue	HPC. HEC. MC C-934P. NaCMC and mixtures thereof	—	Flexible films: casting polymer solutions followed by cold air stream drying	Tensiometry with a LVDT sensor	—	Woolfson. McCafferty. Gorman. McCarron and Price. 1992
Experimental similar to that of Peppas *et al.*, 1987. sensor output: force vs. vertical displacement. both adherents are glued to the respective mounting pedestal. polymers prehydrated as well as dry.							
31	Pig buccal mucosa	HPC—PAA mixtures	Propranolol	Direct compressed discs. 9.5 mm in diameter	Tensiometry	Buccal (systemic)	Chen and Hwang. 1992
Modified tensile test of Peppas *et al.*, 1987 and Leung and Robinson. 1988							

32	Rat jejunum or stomach	PC, NaCMC, HPMC, MC PT	—	Polymer-coated glass spheres or drug crystals	Liquid flow method	—	Ranga-Rao and Buri, 1989
Diameter of glass beads: 0.45 mm — 0.50 mm, crystals of drug > 630 μm, particles are applied in a non-hydrated, dry state; particles adhered to the tissue after washing procedure are counted.							
33	Rat ileal segments	HPMA	Tritiated ostradiol as marker	Polymer nanospheres	Liquid flow method	—	Pimienta, Lenaerts, Cadieux, Raymond, Juhasz, Simard and Jolicoeur, 1990
Ileal segments secured on glass tube (cut lengthwise and inclined at an angle of 12°), then perfused and perfusate fractions analysed (2,4,6,7-^{3}H-estradiol radioactive marker) with â-counter, control: glass tube without mucus tissue.							
34	Rat ileal segments	Poloxamer (407, 238, 403), poloxamine (908)	—	IHCA nanospheres coated with polymers	Liquid flow method	—	Pimienta, Chouinard, Labib and Lenaerts, 1992
Liquid flow method as Pimienta *et al.*, 1992, radioactivity of the collected samples (ZnPC-$^{125}I_4$) label) was measured by γ-scintigraphy.							
35	Rabbit small intestinal tissue	AA or MMAA grafted onto starch	Isosorbide dinitrate	Polymer grafted starch microspheres	Liquid flow method	Buccal (systemic)	Vyas and Jain, 1992
Method of Pimienta *et al.* 1990, ratio of applied to adhered micro-spheres.							
36	Cultured human conjunctival epithelial cells	Numerous anionic, cationic and neutral polymers	—	Polymer solutions	Excimer fluorescence of pyrene	—	Park and Robinson, 1984
Addition of polymer binding to the cell membrane results in compression of the pyrene-containing lipid bilayer of the cell, causing a change in fluorescence measured by a fluorescence spectrophotometer.							
37	Rat intestine	HPMA copolymers with various residues	—	Polymer solutions	Electron microscopy	Gastrointestinal	Bridges, Woodley, Duncan, Kopeckova and Kopecek, 1987
Samples prepared for light and electron microscopy after interaction with colloidal-gold—HPMA copolymer conjugates.							

38	Rat intestine	HPMA copolymers with various residues	--	Polymer solutions	Incubation method	Gastrointestinal	Bridges, Woodley, Duncan and Kopecek, 1988

Intestinal rings were incubated with ^{125}I-labelled copolymers with carbohydrate residues such as galactosamine, mannosamine, glucosamine, fucosylamine or glactose or cationic groups (quaternary ammonium groups) and assayed for radioactivity.

In vivo testing in animal models (rats, rabbits, dogs)

39	Rat	HPC:C-934 ratio 1:2	^{3}H-TAA	Double layered tablet	Microautoradiogram of tongue tissue	Buccal	Kubo, Yamamoto, Yamaguchi, Hashimoto, Ikura and Suzuki, 1982

Ointment used as control.

40	Fasted rat	PC, P(MA-dvB)	--	Polymers labelled with ^{51}Cr in gelatin capsule	Determination of radioactivity in gut segments	--	Ch'ng, Park, Kelly and Robinson, 1985

Surgical insertion of No. 3 gelatin capsules containing the polymer into rat stomach under anaesthesia.

41	Fasted rat	PC	Chlorothiazide	PC-albumin beads in gelatin capsule	Plasma level of the drug via HPLC	Gastrointestinal	Longer, Ch'ng and Robinson, 1985

PC particles and albumin beads mixed at a ratio of 7:3 (w/w) were loaded into gelatin No. 4 capsules, mean particle diameter 505 μm, surgical insertion of gelatin capsules containing the polymer into rat stomach under anaesthesia.

42	Rat	C-934	Hydrochlorothiazide	Polymer solutions containing the dispersed drug	HPLC analysis of urine	Gastrointestinal	Harris, Fell, Taylor, Lynch and Sharma, 1989

Formulations were delivered with a syringe directly into the stomach of the conscious rat.

43	Rat	PC, C-934, HA, PSSA, HEC	--	Polymer solutions with ^{51}Cr-labelled microspheres	Intestinal segments assayed for radio activity	Gastrointestinal	Harris, Fell, Taylor, Lynch and Sharma, 1990

Formulations were delivered directly into the stomach by passing a gavage tube down the oesophagus. Where the formulation was too viscous to be dosed by gavage, gelatin capsules were filled with the suspension immediately prior to administration. Radioactivity determined with γ-counter.

44	Rat	PC, C934P	DGAVP	PHEMA microspheres coated with PC	DGAVP plasma concentration by RIA	Gastrointestinal	Lehr, Bouwstra, Tukker and Junginger, 1990; Lehr, Bouwstra, Kok, de Boer, Tukker, Verhoef, Breimer and Junginger, 1992

Chronically isolated intestinal loop (*in situ*) as described by Poelma and Tukker, 1987, microspheres 315 μm -- 400 μm as well as direct injection of the formulation into the small intestine via catheter.

45	Rabbit	Polymer of AA cross-linked with dmH	Progesterone ^{14}C- labelled	Polymer--drug suspension	Determination of radioactivity in humor samples	Ocular	Hui and Robinson, 1985

Application into the lower cul-de-sac of the rabbit.

46	Rabbit	Na-hyaluronate	Pilocarpine	Aqueous solutions of drug and polymer labelled with ^{99m}Tc	Miotic response and γ-scintigraphy	Ocular	Gurny, Meyer and Peppas, 1987

Saline solutions as well as pH-sensitive latex preparation and temperature setting Pluronic preparation as controls.

47	Rabbit	C-934P PVA (control)	--	Mucoadhesive polymer solution radiolabelled with ^{113m}In	γ-scintigraphy	Ocular	Davies, Farr, Hadgraft and Kellaway, 1988a

Equiviscous control polymer PVA as well as radiolabelled PBS using ^{113m}DTPA were used as controls.

48	Rabbit	Same as Gurny *et al.*, 1987	Pilocarpine	Mucoadhesive polymer solution with 1% pilocarpine	Miotic response	Ocular	Davies, Farr, Hadgraft and Kellaway, 1988b
49	Rabbit	PAA, HPMC, CMC	--	$BaSO_4$ microcapsule coated with polymer, packed into hard gelatin capsule	Radio-opaque material shown on repeated X-ray photographic imaging	Gastrointestinal	Chattaraj, Sudip and Gupta, 1988
Time taken for over 50% microcapsules to empty from stomach and travel to the small intestine. $BaSO_4$ tablets used as control.							
50	Rabbit	PAA, HA, HAE polymers	Pilocarpine and tropicamide	4 mm diameter circular inserts	Miotic and mydriatic activity tests	Ocular	Saettone, Chetoni, Torracca, Burgalassi and Giannaccini, 1989
Inserts were made from films (slow evaporation of polymer solution) or by directly compressing polymer, commercial aqueous eyedrops of pilocarpine and tropicamide were used as reference vehicles.							
51	Rabbit	PAA, HA, HAE polymers	--	50 µl of gels or inserts containing fluorescin	Inspection under long-wave fluorescent lamp	Ocular	Saettone *et al.*, 1989
Measurement of time during which a fluorescent layer was present over the corneal surface.							
52	Rabbit	Solubilised fresh pig gastric mucus	--	^{99m}Tc-labelled PGM_s solution	γ-scintigraphy	Ocular	Greaves and Wilson, 1993
^{99m}Tc-MDP as control solution, preparations instilled into lower fornix.							
53	Beagle dog	HPC:C-934 ratio 1:2	Insulin	Two base tablet -- core: insulin; periphery: bioadhesive polymer	Plasma insulin level and blood sugar level	Buccal	Ishida, Machida, Nambu and Nagai, 1981
The tablet was applied on the oral mucosa of the dog. In order to stick the tablet to the oral mucosa, it was necessary to wipe away the saliva with absorbent cotton previous to application.							

54	Beagle dog	MCC, HPC, C-934	Insulin	Powder dosage form	Plasma glucose level	Nasal	Nagai, Nishimoto, Nambu, Suzuki and Sekine, 1984

For application, a polythene tube was inserted in the dog's nose about 2 cm from the nostril and then was sprayed by a special sprayer, liquid controls were administered by dropping the sample solution in the nostril.

Human Trials

55	Female volunteers	HPC:C-934 ratio 1:2	Bleomycin	Disc-type preparation made by direct compression	Number of cancerous foci found after 2--3 weeks of treatment	Cervical (uterine cancer)	Machida, Masuda, Fujiyama, Ito, Iwata and Nagai, 1979

13 mm in diameter prepared by using the die and punch which is used in preparing KBr tablets samples for infra-red spectroscopy, preparation placed on cervix.

56	Female volunteers	HPC:C-934 ratio 1:2	Bleomycin, carbazilquinone, 5-fluorouracil	Rod-shaped preparation made by direct compression	Number of cancerous foci found after 2--3 weeks of treatment	Cervical (uterine cancer)	Machida, Masuda, Fujiyama, Iwata and Nagai, 1980

40 mm in length prepared with specially designed dies and punches, for application in the cervical canal.

57	Human volunteers	HPC:C-934 ratio 1:2	Triamcinolone acetonide	Double-layered tablet	"Improvement" and "usefulness"	Buccal	Nagai, Machida, Suzuki and Ikura, patent

Upper supporting layer: coloured lactose; lower adhesive layer: contains drug, commercial name: Aphtach®.

58	Human volunteers	HPC:C-934 ratio 1:2	Lidocaine	Two base tablet -- core: lidocaine; periphery: bioadhesive polymer and cap layer	Gas chromatography on lidocaine	Buccal (toothache)	Ishida, Nambu and Nagai, 1982

Application to the human gingiva, the percentage of lidocaine absorbed was calculated from the amount remaining in the dosage form after removing it completely from the gingiva.

59	Human volunteers	PVA	Protirelin or salicylate as	Same as Anders and Merkle, 1989	Amount of residual polymer	Buccal (systemic)	Anders and Merkle, 1983
Polymer and drug remaining on the patch after a given time were assayed, protirelin by RIA or HPLC, sodium salicylate by UV spectrometry, PVP was analysed using a colorimetric method.							
60	Human volunteers	Na-hyaluronate	Pilocarpine	Aqueous solutions of drug and polymer	Miotic response	Ocular	Gurny, Ibrahim, Aebi, Buri, Wilson, Washington, Edman and Camber, 1987
Saline solutions with drug as controls.							
61	Human volunteers	PC	--	Pellets in hard gelatin capsules	γ-scintigraphy	Gastrointestinal	Khosla and Davis, 1987
Pellets of Amberlite IRA410 anionic resin (0.5 mm -- 1.0 mm), ^{99m}Tc-labelled, mixed in size 0 hard gelatin capsules with 100 mg polycarbophil (0.5 mm -- 1.0 mm), three fasted subjects swallowed one capsule with 100 ml of water, control formulation without PC.							
62	Human volunteers	HEC, HPC, PVP, PVA	--	Two-ply laminates (adhesive patches)	Time until patch lost its adhesive contact with the mucosa	Buccal	Anders and Merkle, 1989
Laminates consisting of an impermeable backing layer and hydrocolloid polymer layer containing the drug made by casting and drying.							
63	Mongrel dogs and human volunteers	HPC:C-934, 2:1	Timolol base	Two base tablet -- core: drug; periphery: bioadhesive polymer and cap layer	Residual drug content assayed by spectrophotometer after periodical removal	Buccal (systemic)	Deasy and O'Neill, 1989
Sedated dogs: the oral mucosa adjacent to the upper first molar was dried with absorbent tissue prior to application of a compact; human volunteers: after drying the gingival mucosa below the first molar, a compact was located there by applying finger pressure to the cap for approximately 20 s. Sodium lauryl sulphate used as a penetration enhancer in some cases.							

64	Male volunteers	PC, C-934P	--	Amberlite resin beads and polymer mixture in hard gelatin capsule	γ-scintigraphy with ^{99m}Tc	Gastrointestinal	Harris, Fell, Sharma and Taylor, 1990
Formulations similar to those of Longer *et al.*, 1985, size 0 hard gelatin capsule filled with radiolabelled resin beads (Amberlite, particle size 20 μm) in admixture with bioadhesive powder or lactose (control).							
65	Healthy volunteers	Modified starch, C-934, PEG and NaCMC	--	7 mm diameter flat tablets made by direct compression	Self-assessment by subject	Buccal (dental)	Bottenberg, Cleymaet, de Muynck, Remon, Coomans, Michotte and Slop, 1991
Assessment of irritation, hindrance, taste alteration, etc. and end of adhesion after application of one tablet on the attached gingiva in the region of the upper canine pressed directly onto the mucosa for 30 s.							
66	Human volunteers	Eudispert®	Metronidazole	Films prepared by casting and drying	Duration of adhesion until complete wash-off from the buccal mucosa	Buccal (local)	Smid-Corbar, Kristl, Cop and Groselj, 1991
Plasticizers incorporated in films, such as glycerol, PEG or sorbitol; in addition to the duration of adhesion assessment, metronidazole in saliva was determined via HPLC.							
67	Female patients with bacterial vaginosis	C-934	Metronidazole	Tablet containing 5% bioadhesive polymer; remainder: maize starch and 10% drug	Bacterial vaginosis evaluated clinically, by microscopy and bacteriologically	Vaginal	Bouckaert, Temmerman, Dhont and Remon, 1994
Control: placebo and Flagyl® conventional oral tablets, it does not show the effect of the bioadhesive itself.							

Abbreviations

AA	acrylic acid
ALA	alginic acid
AO	absorptive ointment
AP	cross-linked amylopectin
BuAA	butyl acrylate
C934	Carbopol 934
C934P	Carbopol 934 pharmaceutical grade
C941	Carbopol 941
CaPC	calcium polycarbophil
CMC	carboxymethyl cellulose
DGAVP	9-desglycinamide, 8-arginine vasopressin
dmH	2,5-dimethyl-1,5-hexadiene
DTPA	diethylenetriaminepentaacetic acid
dvG	divinyl glycol
EDEMA	ethylene glycol dimethacrylate
Eudispert	copolymer of poly(methylmethacrylic) acid and its methylester
HA	hyaluronic acid
HAE	HA ethyl ester
HEC	hydroxyethyl cellulose
HP	hydrophilic petrolatum
HPC	hydroxypropyl cellulose
HPLC	high performance liquid chromatography
HPMA	polyhydroxypropylmethacrylate
HPMC	hydroxypropylmethyl cellulose
HPMCP	hydroxypropylmethylcellulose phthalate
^{3}H-TTA	tritium-labelled triamcinolone acetonide
IHCA	isohexylcyanoacrylate
LVDT	linear variable differential transformer
MA	methacrylic acid
MC	Methocel
MCC	microcrystalline cellulose
MDP	methylene diphosphate
MM	methyl methacrylate
MMAA	methyl methacrylic acid
NaAL	sodium alginate
NaCMC	sodium carboxymethylcellulose
PAA	polyacrylic acid
P(AA-co-AM)	copolymer of acrylic acid and acrylamide cross-linked with N,N'methylene-bis-acrylamide
PBS	phosphate-buffered saline
PC	polycarbophil (Carbopol EX55)
PE	polyethylene
PEG	polyethylene glycol
PGM_n	pig gastric mucin
PGM_s	pig gastric mucus
PHEMA	poly(2-hydroxyethylmethacrylate)
PMAA	poly(methacrylic acid)
P(MA-dvB)	poly(methacrylic acid-divinylbenzene)
PNVP	poly(N-vinyl-2-pyrrolidone)
PSSA	poly(styrenesulphonic) acid
PT	pectin
PTFE	poly(tetrafluoroethylene)
PVA	polyvinyl alcohol
PVP	polyvinyl pyrrolidone
RIA	radioimmunoassay
RT	room temperature
WP	white petrolatum
XL	cross-linker
ZnPC-$^{125}I_4$	^{125}I-phthalocyanine-Zn complex

References

Acatürk, F. (1989) Preparation of a prolonged-release tablet formulation of diclofenac sodium. Part 1: Using chitosan. *Pharmazie*, **44**, 547–549

Al-Dujaili, H., Salole, E.G. and Florence, A.T. (1983) Drug formulation and oesophageal injury. *Adverse Drug Reactions and Acute Poisoning Reviews*, **2**, 235–256

Alimuniar, A. and Zainuddin, R. (1992) An economical technique for producing chitosan. In *Advances in Chitin and Chitosan*, edited by Brine C.J., Sandford, P.A. and Zikakis, J.P., London: Elsevier Applied Science, pp 627–632

Allen, A. (1978) Structure of gastrointestinal mucus glycoproteins and the viscous and gel-forming properties of mucus. *British Medical Bulletin*, **34**, 28–33

Allen, A. (1981) Structure and function of gastrointestinal mucus. In *Physiology of the Gastrointestinal Tract*, edited by Johnson, L.R., New York: Raven Press, Vol. 1, pp 617–639

Allen, A. (1983) Mucus – a protective secretion of complexity. *Trends in Biochemical Sciences*, **8**, 169–173

Allen, A. and Caroll, N.J.H. (1985) Adherent and soluble mucus in the stomach and duodenum. *Digestive Diseases and Sciences*, **30**, 55S–62S

Allen, A. (1989) Gastrointestinal mucus. In *Handbook of Physiology – The Gastrointestinal Physiology. Salivary, Gastric and Hepatobiliary Secretions*, edited by Bethesda, M.D., American Physiological Society, Section 6, Vol. III, pp 359–382

Allen, A., Flemström, G., Garner, A. and Kivilaakso, E. (1993) Gastroduodenal mucosal protection. *Physiological Reviews*, **73**, 823–857

Anders, R. and Merkle, H.P. (1983) Buccal absorption of protirelin: an effective way to stimulate thyrotropin and prolactin. *Journal of Pharmaceutical Sciences*, **72**, 1481–1483

Anders, R. and Merkle, H.P. (1989) Evaluation of laminated muco-adhesive patches for buccal drug delivery. *International Journal of Pharmaceutics*, **49**, 231–240

Anderson, M.T., Harding, S.E. and Davis, S.S. (1989) On the interaction in solution of a candidate mucoadhesive polymer, diethylaminoethyl-dextran, with pig gastric mucus glycoprotein. *Biochemical Society Transactions*, 631st meeting Guildford, **17**, pp 1101–1102

Anderson, M.T. (1991). *The interaction of mucous glycoproteins with polymeric materials*, Ph.D. thesis, University of Nottingham, U.K.

Aoyagi, N., Ogata, H., Kaniwa, N., Uchiyama, M., Yasuda, Y. and Tanioka, Y. (1992) Gastric emptying of tablets and granules in humans, dogs, pigs, and stomach-emptying-controlled rabbits. *Journal of Pharmaceutical Sciences*, **81**, 1170–1174

Arai, K., Kinumaki, T. and Fujita, T. (1968) Toxicity of chitosan. *Bulletin of the Tokai Regional Fisheries Research Laboratory*, **56**, 89–94

Beachey, E.H. (1980). *Bacterial Adherence*, Series B, Volume 6, London and New York: Chapman and Hall

Bechgaard, H. and Ladefoged, K. (1978) Distribution of pellets in the gastrointestinal tract. The influence on transit time exerted by the density and diameter of pellets. *Journal of Pharmacy and Pharmacology*, **30**, 690–692

Beermann, B., Groschinsky-Grind, M. and Rosen, A. (1976) Absorption, metabolism and excretion of hydrochlorothiazide. *Clinical Pharmacology and Therapeutics*, **19**, 718–721

Beerman, B. and Groschinsky-Grind, M. (1978) Enhancement of the gastrointestinal absorption of hydrochlorothiazide by propantheline. *European Journal of Clinical Pharmacology*, **13**, 385–387

Bell, A.E., Sellers, L.A., Allen, A., Morris, E., Ross-Murphy, S. (1985) Properties of gastric and duodenal mucus: effects of proteolysis, disulphide reduction, bile, acid, ethanol and hypertonicity on mucus gel structure. *Gastroenterology*, **88**, 269–280

Bhaskar, K.R., Garik, P., Turner, B.S., Bradley, J.D., Bansil, R., Stanley, H.E. and Lamont, J.T. (1992) 'Viscous fingering of HCl through gastric mucin. *Nature*, **360**, 458–461

Boddé, H.E. (1990) Principles of Bioadhesion. In *Bioadhesion – Possibilities and Future Trends*, edited by Gurny, R. and Junginger, H.E., Wissenschaftliche Verlagsgesellschaft mbH Stuttgart, APV, Vol. 25, pp 44–64

Boedecker, E.C. (1984) *Attachment of Organisms to the Gut Mucosa*. CRC Press, Florida, Volumes I and II

Booth, C.C. (1967) Sites of absorption in the small intestine. *Federation Proceedings*, **28**, 1583–1588

Bottenberg, P., Cleymaet, R., de Muynck, C., Remon, J.P., Coomans, D., Michotte, Y. and Slop, D. (1991) Development and testing of bioadhesive, fluoride-containing slow-release tablets for oral use. *Journal of Pharmacy and Pharmacology*, **43**, 457–464

Bouckaert, S., Temmerman, M., Dhont, M. and Remon, J.P. (1994) The treatment of bacterial vaginosis with a bioadhesive vaginal slow-release tablet with metronidazole. *Proceedings of the International Symposium on Controlled Release of Bioactive Materials*, **21**, 585–586

Bowman, W.C. and Rand, M.J. (1980). *Textbook of Pharmacology*, Oxford: Blackwell Scientific Publications, Chapter 25

Brahams, D. (1984) Death of a patient participating in trial of oral morphine for relief of postoperative pain. *Lancet*, **1**, 1083–1084

Bremecker, K.-D. (1990) Semi-solid aqueous gels as nasal and buccal bioadhesives. In *Bioadhesion – Possibilities and Future Trends*, edited by Gurny, R. and

Junginger, H.E., Wissenschaftliche Verlagsgesellschaft mbH Stuttgart, APV, Vol. 25, pp 191–201

Bridges, J.F., Woodley, J.F., Duncan, R., Kopeckova, P. and Kopecek, J. (1987) *In vitro* and *in vivo* evaluation of N-(2-hydroxypropyl)-methacrylamide copolymers as a potential oral drug delivery system. *Proceedings of the International Symposium on Controlled Release of Bioactive Materials*, **14**, 14–15

Bridges, J.F., Woodley, J.F., Duncan, R. and Kopecek, J. (1988) Soluble N-(2-hydroxypropyl)-methacrylamide copolymers as a potential oral, controlled-release drug delivery system. I. Bioadhesion to the rat intestine *in vitro*. *International Journal of Pharmaceutics*, **44**, 213–223

Bystricky, S., Malovíková, A. and Sticzay, T. (1990) Interaction of alginates and pectins with cationic polypeptides. *Carbohydrate Polymers*, **13**, 283–294

Cargill, R., Caldwell, I.J., Engle, K., Fix, J.A., Porter, P.A. and Gardner, C.R. (1988) Controlled gastric emptying. I. Effects of physical properties on gastric residence times of non-disintegrating geometric shapes in beagle dogs. *Pharmaceutical Research*, **5**, 533–536

Cargill, R., Engle, K., Gardner, C.R., Porter, P., Sparer, R.V. and Fix, J.A. (1989) Controlled gastric emptying. II. *In vitro* erosion and gastric residence times of an erodible device in beagle dogs. *Pharmaceutical Research*, **6**, 506–509

Carlstedt, I. and Sheehan, J.K. (1984) Is the macromolecular architecture of cervical, respiratory and gastric mucins the same? *Biochemical Society Transactions*, **12**, 615–617

Carlstedt, I. and Sheehan, J.K. (1988) Structure and macromolecular properties of mucus glycoproteins. *Monographs in Allergy*, Basel, **24**, 16–24

Chattaraj, S.C., Sudip, K.D. and Gupta, B.K. (1988) The development of bioadhesive microcapsules and assessment for delaying the gastro-intestinal transit in rabbits. *Proceedings of the International Symposium on Controlled Release of Bioactive Materials*, **15**, 444–445

Chen, J.L. and Cyr, G.N. (1970) Compositions producing adhesion through hydration. In *Adhesion in Biological Systems*, edited by Manly, R.S., New York: Academic Press, pp 163–181

Chen, W.-G. and Hwang, G.C.-C. (1992) Adhesive and *in vitro* release characteristics of propranolol bioadhesive disc system. *International Journal of Pharmaceutics*, **82**, 61–66

Chitnis, V.S., Malshe, V.S. and Lalla, J.K. (1991) Bioadhesive polymers – synthesis, evaluation and application in controlled release tablets. *Drug Development and Industrial Pharmacy*, **17**, 879–892

Ch'ng, H.S., Park, H., Kelly, P. and Robinson, J.R. (1985) Bioadhesive polymers as platforms for oral controlled drug delivery II: Synthesis and evaluation of some swelling, water-insoluble bioadhesive polymers. *Journal of Pharmaceutical Sciences*, **74**, 399–405

Code, C.F. and Marlett, J.A. (1975) The interdigestive myo-electric complex of the stomach and small bowel of dogs. *Journal of Physiology*, **246**, 289–309

Creeth, J.M. (1978) Constituents of mucus and their separation. *British Medical Bulletin*, **34**, 17–24

Davies, N.M., Farr, S.J., Hadgraft, J. and Kellaway, I.W. (1988a) Gamma scinitigraphic evaluation of a mucoadhesive polymer as a vehicle for ophthalmic agents in the rabbit. *Journal of Pharmacy and Pharmacology*, **40**, 15P

Davies, N.M., Farr, S.J., Hadgraft, J. and Kellaway, I.W. (1988b) The effect of a mucoadhesive polymer on the bioavailability of pilocarpine. *Proceedings of the International Symposium on Controlled Release of Bioactive Materials*, **15**, 388

Davis, S.S. (1985) The design and evaluation of controlled release systems for the gastrointestinal tract. *Journal of Controlled Release*, **2**, 27–38

Davis, S.S., Stockwell, A.F., Taylor, M.J., Hardy, J.G., Whalley, D.R., Wilson, C.G., Bechgaard, H. and Christensen, F.N. (1986) The effect of density on gastric emptying of single- and multiple-unit dosage forms. *Pharmaceutical Research*, **3**, 208–213

Davis, S.S. (1989) Small intestine transit. In *Drug Delivery to the Gastrointestinal Tract*, edited by Hardy, J.G., Davis, S.S., Wilson, C.G., Chichester: Ellis Horwood Ltd., pp 49–61

Deasy, P.B. and O'Neill, C.T. (1989) Bioadhesive dosage form for peroral administration of timolol base. *Pharmaceutica Acta Helvetiae*, **64**, 231–235

De Mey, J. (1984) Colloidal gold as marker and tracer in light and electron microscopy. *Electron Microscopy Society of America*, **14**, 54–66

Duchêne, D., Touchard, F. and Peppas, N.A. (1988) Pharmaceutical and medical aspects of bioadhesive systems for drug administration. *Drug Development and Industrial Pharmacy*, **14**, 283–381

Errington, N., Harding, S.E., Vårum, K.M., Illum, L. (1993) Hydrodynamic characterization of chitosans varying in degree of acetylation. *International Journal of Biological Macromolecules*, **15**, 113–117

Fiebrig, I., Harding, S.E. and Davis, S.S. (1994) Sedimentation analysis of potential interactions between mucins and a putative bioadhesive polymer. *Progress in Colloid and Polymer Science*, **93**, 66–73

Fiebrig, I., Harding, S.E., Stokke, B.T., Vårum, K.M., Jordan, D. and Davis, S.S. (1994) The potential of chitosan as mucoadhesive drug carrier: studies on its interaction with pig gastric mucin on a molecular level. *European Journal of Pharmaceutical Sciences*, **2**, 185

Förster, H. and Lippold, B.C. (1982) Conception of peroral sustained-release dosage forms: calculation of initial and maintenance dose with consideration of accumulation, *Pharmaceutica Acta Helvetiae*, **57**, 345–349

Gibbons, R.A. (1972) Physico-chemical methods for the determination of the purity, molecular size and shape of glycoproteins. In *Glycoproteins: Their Composition, Structure and Function*. edited by Gottschalk, A., Amsterdam: Elsevier Publishing Company, pp 31-141

Gottschalk, A., Bhargava, A.S. and Murty, V.L.N. (1972) Submaxillary gland glycoproteins. In *Glycoproteins, Their Composition, Structure and Function*, edited by Gottschalk, A., Amsterdam: Elsevier Publishing Company, pp 810–829

Greaves, J.L. and Wilson, C.G. (1993) Treatment of diseases of the eye with mucoadhesive delivery systems. *Advanced Drug Delivery Reviews*, **11**, 349–383

Gröning, R. and Heun, G. (1984) Oral dosage forms with controlled gastrointestinal transit. *Drug Development and Industrial Pharmacy*, **10**, 527–539

Gröning, R. and Heun, G. (1989) Dosage forms with controlled gastrointestinal passage – studies on the absorption of nitrofurantoin. *International Journal of Pharmaceutics*, **56**, 111–116

Gruber, P., Longer, M.A. and Robinson, J.R. (1987) Some biological issues in oral controlled drug delivery. *Advanced Drug Delivery Reviews*, **1**, 1–18

Grundy, D. (1985). *Gastrointestinal Motility*, Lancaster: MTP Press Ltd

Gu, J.-M., Robinson, J.R. and Leung, S.-H.S. (1988) Binding of acrylic polymers to mucin/epithelial surfaces: structure-property relationships. *Critical Reviews in Therapeutic Drug Carrier Systems*, **5**, 21–67

Gurny, R., Meyer, J.-M. and Peppas, N.A. (1984) Bioadhesive intraoral release systems: design, testing and analysis. *Biomaterials*, **5**, 336–340

Gurny, R., Ibrahim, H., Aebi, A., Buri, P., Wilson, C.G., Washington, N., Edman, P. and Camber, O. (1987) Design and evaluation of controlled release systems for the eye. *Journal of Controlled Release*, **6**, 367–373

Gürsoy, A., Sohtorik, I., Uyanik, N. and Peppas, N.A. (1989) Bioadhesive controlled release systems for vaginal delivery. *S.T.P. Pharma*, **5**, 886–892

Harding, S.E. (1989) The macrostructure of mucus glycoproteins in solution. *Advances in Carbohydrate Chemistry and Biochemistry*, **47**, 345–381

Harper, C.M. and Ralston, M. (1983) Isobutyl 2-cyanoacrylate as an osseous adhesive in the repair of osteochondral fractures. *Journal of Biomedical Materials Research*, **17**, 167–177

Harris, D., Fell, J.T., Taylor, D.C., Lynch, J. and Sharma, H.L. (1989) Oral availability of a poorly absorbed drug, hydrochlorothiazide, from a bioadhesive formulation in the rat. *International Journal of Pharmaceutics*, **56**, 97–102

Harris, D., Fell, J.T., Sharma, H.L. and Taylor, D.C. (1990) GI transit of potential bioadhesive formulations in man: a scintigraphic study. *Journal of Controlled Release*, **12**, 45–53

Harris, D., Fell, J.T., Taylor, D.C., Lynch, J. and Sharma, H.L. (1990) GI transit of potential bioadhesive systems in the rat. *Journal of Controlled Release*, **12**, 55–65

Hassan, E.E. and Gallo, J.M. (1990) A simple rheological method for the *in vitro* assessment of mucin-polymer bioadhesive bond strength. *Pharmaceutical Research*, **7**, 491–495

Hayton, W.L. (1980) Rate-limiting barriers to intestinal drug absorption: A review. *Journal of Pharmacology and Biopharmaceutics*, **8**, 321–334

Helliwell, (1993) The use of bioadhesives in targeted delivery within the gastrointestinal tract. *Advanced Drug Delivery Reviews*, **11**, 221–251

Hirano, S. (1989) Production and application of chitin and chitosan in Japan. In *Chitin and Chitosan*, edited by Skjåk-Bræk, G., Anthonsen, T. and Sandford, P., London: Elsevier Applied Science, 37–43

Hollingsbee, D.A. and Timmins, P. (1990) Topical adhesive systems. In *Bioadhesion - Possibilities and Future Trends*, edited by Gurny, R. and Junginger, H.E., Wissenschaftliche Verlagsgesellschaft mbH Stuttgart, APV, Vol. 25, pp 140–164

Horisberger, H. and Rosset, J. (1977) Colloidal gold, a useful marker for transmission and scanning electron microscopy. *Journal of Histochemistry and Cytochemistry*, **25**, 295–305

Horisberger, M. and Clerc, M.-F. (1988) Chitosan-colloidal gold complexes as polycationic probes for the detection of anionic sites by transmission and scanning electron microscopy. *Histochemistry*, **90**, 165–175

Hörstedt, P., Danielsson, A., Nyhlin, H., Stenling, R. and Suhr, O. (1989) Adhesion of bacteria to the human small-intestinal mucosa. *Scandinavian Journal of Gastroenterology*, **24**, 877–885

Hui, H.-W. and Robinson, J.R. (1985) Ocular delivery of progesterone using a bioadhesive polymer. *International Journal of Pharmaceutics*, **26**, 203–213

Huntsberger, J.R. (1967) Mechanisms of adhesion. *Journal of Paint Technology*, **36**, 199–211

Imeri, A.G. and Knorr, D. (1988) Effects of chitosan on yield and compositional data of carrot and apple juice. *Journal of Food Science*, **53**, 1707–1709

Ishida, M., Machida, Y., Nambu, N. and Nagai, T. (1981) New mucosal dosage form of insulin. *Chemical and Pharmaceutical Bulletin*, **29**, 810–816

Ishida, M., Nambu, N. and Nagai, T. (1982) Mucosal dosage form of lidocaine for toothache using hydroxypropyl cellulose and Carbopol. *Chemical and Pharmaceutical Bulletin*, **30**, 980–984

Ito, S. (1981) Functional gastric morphology. In *Physiology of the Gastrointestinal Tract*, edited by Johnson, L.R., New York: Raven, Vol. 1, pp 517–550

Jabbari, E., Wisniewski, N. and Peppas, N.A. (1993) Evidence of mucoadhesion by chain interpenetration at a poly(acrylic acid)/mucin interface using ATR-FTIR spectroscopy. *Journal of Controlled Release*, **26**, 99–108

Jacques, Y. and Buri, P. (1992) Optimization of an *ex vivo* method for bioadhesion quantification. *European Journal of Pharmacy and Biopharmaceutics*, **38**, 195–198

Jeuniaux, C., Voss-Fouchart, M.-F., Poulicek, M. and Bussers, J.-C. (1989) Sources of chitin, estimated from new data on chitin biomass and production. In *Chitin and Chitosan*, edited by Skjåk-Bræk, G., Anthonsen, T. and Sandford, P., London: Elsevier Applied Science, pp 3–11

Junginger, H.E., Lehr, C.-M., Bouwstra, J.A., Tukker, J.J. and Verhoef, J. (1990) Site specific intestinal absorption using bioadhesives: Improved oral delivery of peptide drugs by means of bioadhesive polymers. In *Bioadhesion – Possibilities and Future Trends*, edited by Gurny, R. and Junginger, H.E., Wissenschaftliche Verlagsgesellschaft mbH Stuttgart, APV, Vol. 25, pp 177–190

Junginger, H.E. and Lehr, C.-M. (1990) Bioadhäsive Arzneistoffabgabesysteme und Arzneiformen für perorale und rektale Anwendung. *Deutsche Apotheker Zeitung*, **130**, 791–801

Junginger, H.E. (1991) Mucoadhesive hydrogels. *Pharmazeutische Industrie*, **53**, 1056–1065

Junginger, H.E. and Verhoef, J.C. (1992) Perorale Applikation von Peptiden und Proteinen. *Deutsche Apotheker Zeitung*, **132**, 1279–1290

Kaelbe, D.H. and Moacanin, J. (1977) A surface energy analysis of bioadhesion. *Polymer*, **18**, 475–481

Katayama, T., Takai, K.-I., Kariyama, R. and Kanemasa, Y. (1978) Colloid titration of heparin using cat-floc (polydiallyldimethyl ammonium chloride) as standard polycation. *Analytical Biochemistry*, **88**, 382–387

Kaus, L.C. (1987) The effect of density on the gastric emptying and intestinal transit of solid dosage forms: comments on the article by Davis et al., *Pharmaceutical Research*, **4**, 78

Kerr, L.J., Kellaway, I.W., Lewis, J.D., Kelly, D.R. and Parr, G.D. (1989) The interaction of polyacrylic acid with gastric glycoprotein. *Proceedings of the International Symposium on Controlled Release of Bioactive Materials*, **16**, 400–401

Kerss, S., Allen, A., Garner, A. (1982) A simple method for measuring thickness of the mucus gel layer adherent to rat, frog and human gastric mucosa: influence of feeding, prostaglandin, N-actylcysteine and other agents. *Clinical Science*, **63**, 187–195

Khosla, R. and Davis, S.S. (1987) The effect of polycarbophil on the gastric emptying of pellets. *Journal of Pharmacy and Pharmacology*, **39**, 47–49

Kikendall, J.W., Friedman, A.C., Oyewole, M.A., Fleischer, D. and Johnson, L.F. (1983) Pill-induced esophageal injury. *Digestive Diseases and Sciences*, **28**, 174–181

Kubo, J., Yamamoto, M., Yamaguchi, H., Hashimoto, Y., Ikura, H. and Suzuki, Y. (1982) *Kiso-to-Rinsho*, **16**, 4599

Lehr, C.-M., Bouwstra, J.A., Tukker, J.J. and Junginger, H.E. (1990) Intestinal transit of bioadhesive microspheres in an *in situ* loop in the rat. A comparative study with copolymers and blends based on poly(acrylic acid). *Journal of Controlled Release*, **13**, 51–62

Lehr, C.-M., Poelma, F.G.J., Junginger, H.E. and Tukker, J. (1991) An estimate of turnover time of intestinal mucus gel layer in the rat *in situ* loop. *International Journal of Pharmaceutics*, **70**, 235–240

Lehr, C.-M., Bouwstra, J.A., Kok, W., de Boer, A.G., Tukker, J.J., Verhoef, J.C., Breimer, D.D. and Junginger, H.E. (1992a) Effects of the mucoadhesive polymer polycarbophil on the intestinal absorption of a peptide drug in the rat. *Journal of Pharmacy and Pharmacology*, **44**, 402–407

Lehr, C.-M., Bouwstra, J.A., Schacht, E.H. and Junginger, H.E. (1992b) *In vitro* evaluation of mucoadhesive properties of chitosan and some other natural polymers. *International Journal of Pharmaceutics*, **78**, 43–48

Lehr, C.-M., Bouwstra, J.A., Speis, F., Onderwater, J., van het Noordeinde, J., Vermeij-Keers, C., van Munsteren, C.J. and Junginger, H.E. (1992c) Visualization studies of the mucoadhesive interface. *Journal of Controlled Release*, **18**, 249–260

Lehr, C.-M., Bouwstra, J.A., Kok, W., Noach, A.B.J., de Boer, A.G. and Junginger, H.E. (1992d) Bioadhesion by means of specific binding of tomato lectin. *Pharmaceutical Research*, **9**, 547–553

Lejoyeux, F., Ponchel, G., Wouessidjewe, D., Peppas, N.A. and Duchêne, D. (1989) Bioadhesive tablets: Influence of the testing medium composition on bioadhesion. *Drug Development and Industrial Pharmacy*, **15**, 2037–2048

Leung, S.S.-H. and Robinson, J.R. (1988) The contribution of anionic polymer structural features to mucoadhesion. *Journal of Controlled Release*, **5**, 223–231

Lippold, B.C. and Günther, J. (1991) *In vivo* Prüfung einer multipartikulären Retard-Schwimmarzneiform, *European Journal of Pharmacy and Biopharmaceutics*, **37**, 254–261

Longer, M.A., Ch'ng, H.S. and Robinson, J.R. (1985) Bioadhesive polymers as platforms for oral-controlled drug delivery III: Oral delivery of chlorothiazide using bioadhesive polymer. *Journal of Pharmaceutical Sciences*, **74**, 406–411

Lynch, J., Pownall, R.E. and Taylor, D.C. (1987) Site-specific absorption of hydrochlorothiazide in the rat intestine. *Journal of Pharmacy and Pharmacology*, **39**, 55P

MacAdam, A. (1993) The effect of gastro-intestinal mucus on drug absorption. *Advanced Drug Delivery Reviews*, **11**, 201–220

Machida, Y., Masuda, H., Fujiyama, N., Ito, S., Iwata, M. and Nagai, T. (1979) Preparation and phase II clinical examination of topical dosage forms for the treatment of carcinoma colli containing bleomycin with hydroxypropyl cellulose. *Chemical and Pharmaceutical Bulletin*, **27**, 93–100

Machida, Y., Masuda, H., Fujiyama, N., Iwata, M. and Nagai, T. (1980) Preparation and phase II clinical examination of topical dosage forms for the treatment of carcinoma colli containing bleomycin, carboquone or 5-fluorouracil with hydroxypropyl cellulose. *Chemical and Pharmaceutical Bulletin*, **28**, 1125–1130

Manninen, V., Apajalahti, A., Melin, J. and Karesoja, M. (1973) Altered absorption of digoxin in patients given propantheline and metoclopramide. *Lancet*, **1**, 398–400

Marvola, M., Vahervuo, K., Sothmann, A., Marttila, E. and Rajaniemi, M. (1982) Development of a method for study of the tendency of drug products to adhere to the esophagus. *Journal of Pharmaceutical Sciences*, **71**, 975–977

McCurdy, J.D. (1992) FDA and the use of chitin and chitosan derivatives. In *Advances in Chitin and Chitosan*, edited by Brine, C.J., Sandford, P.A. and Zikakis, J.P., London: Elsevier Applied Science, pp 659–662

Mergenhagen, S.E. and Rosan, B. (1985). *Molecular Basis of Oral Microbial Adhesion*. American Society for Microbiology, Washington, D.C.

Mikos, A.G. and Peppas, N.A. (1990) Bioadhesive analysis of controlled-release systems. IV. An experimental method for testing the adhesion of microparticles with mucus. *Journal of Controlled Release*, **12**, 31–37

Minami, H. and McCallum, R.W. (1984) The physiology and pathophysiology of gastric emptying in humans. *Gastroentrology*, **86**, 1592–1610

Miyayaki, S., Yamaguchi, H., Takada, M., Hou, W.-M., Takeichi, Y. and Yasabuchi, H. (1990) Pharmaceutical application of biomedical polymers. *Acta Pharmaceutica Nordica*, **2**, 401–406

Moës, A.J. (1993) Gastroretentive dosage forms. *Critical Reviews in Therapeutic Drug Carrier Systems*, **10**, 143–195

Mortazavi, S.A., Carpenter, B.G. and Smart, J.D. (1992) An investigation of the rheological behaviour of the mucoadhesive/mucosal interface. *International Journal of Pharmaceutics*, **83**, 221–225

Mortazavi, S.A., Carpenter, B.G. and Smart, J.D. (1993) A comparative study on the role played by mucus glycoproteins in the rheological behaviour of the mucoadhesive/mucosal interface. *International Journal of Pharmaceutics*, **94**, 195–201

Mortazavi, S.A. and Smart, J.D. (1993) An investigation into the role of water movement and mucus gel dehydration in mucoadhesion. *Journal of Controlled Release*, **25**, 197–203

Müller-Lissner, S.A. and Blum, A.L. (1981) The effect of specific gravity and eating on gastric emptying of slow-release capsules. *New England Journal of Medicine*, **304**, 1365–1366

Nagai, T., Nishimoto, Y., Nambu, N., Suzuki, Y., Sekine, K. (1984) Powder dosage form of insulin for nasal administration. *Journal of Controlled Release*, **1**, 15–22

Nagai, T. (1986) Topical mucosal adhesive dosage forms. *Medicinal Research Reviews*, **6**, 227–242

Nagai, T., Machida, Y., Suzuki, Y. and Ikura, H., Japan Pat. 1,177,734, May 13, 1983

Neutra, M.R. and Forstner, J.F. (1987) Gastrointestinal mucus: Synthesis, secretion and function. In *Physiology of the Gastrointestinal Tract* edited by Johnson, L.R., New York: Raven Press, pp 975–1009

Park, K. and Robinson, J.R. (1984) Bioadhesive polymers as platforms for oral-controlled drug delivery: method to study bioadhesion. *International Journal of Pharmaceutics*, **19**, 107–127

Park, K. and Park, H. (1987) Enzyme-digestible balloon hydrogels for long-term oral drug delivery: synthesis and characterization. *Proceedings of the International Symposium on Controlled Release of Bioactive Materials*, **14**, 41–42

Park, K. (1989) A new approach to study mucoadhesion: colloidal gold staining. *International Journal of Pharmaceutics*, **53**, 209–217

Peppas, N.A. and Buri, P.A. (1985) Surface, interfacial and molecular aspects of polymer bioadhesion on soft tissues. *Journal of Controlled Release*, **2**, 257–275

Peppas, N.A., Ponchel, G. and Duchêne, D. (1987) Bioadhesive analysis of controlled-release systems. II. Time-dependent bioadhesive stress in poly(acrylic acid)-containing systems. *Journal of Controlled Release*, **5**, 143–149

Pimienta, C., Lenaerts, V., Cadieux, C., Raymond, P., Juhasz, J., Simard, M.A. and Jolicoeur, C. (1990) Mucoadhesion of hydroxypropylmethacrylate nanoparticles to rat intestinal ileal segments *in vitro*. *Pharmaceutical Research*, **7**, 49–53

Pimienta, C., Chouinard, F., Labib, A. and Lenaerts, V. (1992) Effect of various poloxamer coatings on *in vitro* adhesion of isohexylacyanoacrylate nanospheres to rat ileal segments under liquid flow. *International Journal of Pharmaceutics*, **80**, 1–8

Poelma, F.G.J. and Tukker, J.J. (1987) Evaluation of a chronically isolated internal loop in the rat for the study of drug absorption kinetics, *Journal of Pharmaceutical Sciences*, 76, 433–436

Ponchel, G.F., Touchard, F., Duchêne, D. and Peppas, N.A. (1987) Bioadhesive analysis of controlled-release systems. I. Fracture and interpenetration analysis in poly(acrylic acid)-containing systems. *Journal of Controlled Release*, **5**, 129–141

Ponchel, G.F., Touchard, F., Wouessidjewe, D., Duchêne, D. and Peppas, N.A. (1987) Bioadhesive analysis of controlled-release systems. III. Bioadhesive and release behaviour of metronidazole-containing poly(acrylic acid)-hydroxypropyl methylcellulose systems. *International Journal of Pharmaceutics*, **38**, 65–70

Ranga Rao, K.V. and Buri, P. (1989) A novel *in situ* method to test polymers and coated microparticles for bioadhesion. *International Journal of Pharmaceutics*, **53**, 265–270

Robert, C., Buri, P. and Peppas, N.A. (1988) Experimental method for bioadhesive testing of various polymers. *Acta Pharmaceutica Technologica*, **34**, 95–98

Robinson, R., Longer, M.A., Veillard, M. (1987) Bioadhesive polymers for controlled drug delivery. In *Annals New York Academy of Science*, edited by Juliano, R.L., New York, Vol. 507, pp 307–314

Robinson, J.R. (1990) Rationale of bioadhesion/mucoadhesion. In *Bioadhesion – Possibilities and Future Trends* edited by Gurny, R. and Junginger, H.E., Wissenschaftliche Verlagsgesellschaft mbH Stuttgart, APV, Vol. 25, pp 13–28

Russell, J. and Bass, P. (1985) Canine gastric emptying of polycarbophil: an indigestible, particulate substance. *Gastroenterology*, **89**, 307–312

Saettone, M.F., Chetoni, P., Torracca, M.T., Burgalassi, S. and Giannaccini, B. (1989) Evaluation of muco-adhesive properties and *in vivo* activity of ophthalmic vehicles based on hyaluronic acid. *International Journal of Pharmaceutics*, **51**, 203–212

Sam, A.P., van den Heuij, J.T.M. and Tukker, J.J. (1992) Mucoadhesion of both film-forming and non-film-forming polymeric materials as evaluated with the Wilhelmy plate method. *International Journal of Pharmaceutics*, **79**, 97–105

Sangekar, S., Vadino, W.A., Chaudry, I., Parr, A., Beihn, G. and Digenis, G. (1987) Evaluation of the effect of food and specific gravity of tablets on gastric retention time. *International Journal of Pharmaceutics*, **35**, 187–191

Schauer, R. (1992) Sialinsäurereiche Schleime als bioaktive Schmierstoffe. *Nachrichten aus Chemie Technik und Lababoratorium*, **40**, 1227–1232

Schor, J.M., Davis, S.S., Nigalaye, A. and Bolton, S. (1983) Susadrin transmucosal tablets (nitroglycerin in synchron controlled release base). *Drug Development and Industrial Pharmacy*, **9**, 1359–1377

Senju, R. (1969). *Koroido Tekireiho [Colloid Titration]*, Tokyo: Nankodo

Sheehan, J.K. and Carlstedt, I. (1989) Models for the macromolecular structure of mucus glycoproteins. In *Dynamic Properties of Biomolecular Assemblies* edited by Harding, S.E. and Rowe, A.J., Cambridge: Royal Society of Chemistry, pp 256–275

Sheth, P.R. and Tossounian, J. (1984) The hydrodynamically balanced system (HBS®): a novel drug delivery system for oral use. *Drug Development and Industrial Pharmacy*, **10**, 313–339

Silberberg, A. and Meyer, F.A. (1982) Structure and function of mucus. In *Mucus in Health and Disease-II* edited by Chantler, E.N., Elder, J.B. and Elstein, M., New York: Plenum Press, pp 53–74

Slomiany, B.L. and Meyer, K. (1972) Isolation and structural studies of sulfated glycoproteins of hog gastric mucosa. *Journal of Biological Chemistry*, **247**, 5062–5070

Smart, J.D., Kellaway, I.W. and Worthington, H.E.C. (1984) An *in-vitro* investigation of mucosa-adhesive materials for use in controlled drug delivery. *Journal of Pharmacy and Pharmacology*, **36**, 295–299

Smart, J.D. (1991) An *in vitro* assessment of some mucosa-adhesive dosage forms. *International Journal of Pharmaceutics*, **73**, 69–74

Smart, J.D. (1993) Drug delivery using buccal-adhesive systems. *Advanced Drug Delivery Reviews*, **11**, 253–270

Smid-Korbar, J., Kristl, J., Cop, L. and Groselj, D. (1991) Formulation and evaluation of oral mucoadhesive films containing metronidazole. *Acta Pharmaceutica Jugoslavica*, **41**, 251–258

Soto Peralta, N.V., Müller, H. and Knorr, D. (1989) Effects of chitosan treatments on the clarity and colour of apple juice. *Journal of Food Science*, **54**, 495–496

Takahashi, T., Takayama, K., Machida, Y. and Nagai, T. (1990) Characteristics of polyion complexes of chitosan with sodium alginate and sodium polyacrylate. *International Journal of Pharmaceutics*, **61**, 35–41

Terayama, H. (1952) Method of colloid titration (a new titration between polymer ions). *Journal of Polymer Science*, **8**, 243–253

Timmermans, J. and Moës, A.J. (1990) How well do floating dosage forms float? *International Journal of Pharmaceutics*, **62**, 207–216

Timmermans, J. (1991) Floating Hydrophilic Matrix Dosage Forms for Oral Use: Factors Controlling Their Buoyancy and Gastric Residence Capabilities, Ph.D. thesis, Université Libre de Bruxelles, Brussels, Belgium

Van Damme, M.-P.I., Blackwell, S.T., Murphy, W.H. and Preston, B.N. (1992) The measurement of negative charge in cartilage using a colloid titration technique. *Analytical Biochemistry*, **204**, 250–257

Vere, D. (1984) Death from sustained release morphine sulphate. *Lancet*, **1**, 1477 de Vries, M.E., Bodde, H.E., Busscher, H.J. and Junginger, H.E. (1988) Hydrogels for buccal drug delivery: properties relevant for muco-adhesion. *Journal of Biomedical Materials Research*, **22**, 1023–1032

Vyas, S.P. and Jain, C.P. (1992) Bioadhesive polymer-grafted starch microspheres bearing isosorbide dinitrate for buccal administration. *Journal of Microencapsulation*, **9**, 457–464

Waldron-Edward, D. (1977) The turnover of mucin glycoprotein in the stomach. In *Mucus in Health and Disease*, edited by Elstein, M. and Parke, D.V., New York: Plenum Press, pp 301–307

Wang, P.Y. (1974) Surgical adhesives and coatings. In *Medical Engineering* edited by Ray, C.D., Chicago, U.S.A.: Year Book Medical Publishers, pp 1123–1129

Wearley, L.L. (1991) Recent progress in protein and peptide delivery by noninvasive routes. *Critical Reviews in Therapeutic Drug Carrier Systems*, **8**, 331–394

Webster's Encyclopedic Unabridged Dictionary of the English Language (1989). Avenel (New Jersey, U.S.A.): Gramercy Books

Weiner, M.L. (1992) An overview of the regulatory status and of the safety of chitin and chitosan as food and pharmaceutical ingredients. In *Advances in Chitin and Chitosan* edited by Brine, C.J., Sandford, P.A. and Zikakis, J.P., London: Elsevier Applied Science, pp 663–672

Wilson, (1990) *In vivo* testing of bioadhesion. In *Bioadhesion – Possibilities and Future Trends*, edited by Gurny, R. and Junginger, H.E., Wissenschaftliche Verlagsgesellschaft mbH Stuttgart, APV, Vol. 25, pp 93–108

Winkler, R. (1986). *Stoma Therapy: An Atlas and Guide for Intestinal Stomas*, Stuttgart (Germany): Georg Thieme Verlag

Woolfson, A.D., McCafferty, D.F., Gorman, S.P., McCarron, P.A. and Price, J.H. (1992) Design of an apparatus incorporating a linear variable differential transformer for the measurement of type III bioadhesion to cervical tissue. *International Journal of Pharmaceutics*, **84**, 69–76

Wright, P.S. (1981) Composition and properties of soft lining materials for acrylic dentures. *Journal of Dentistry*, **9**, 210–223

19

COVALENT ATTACHMENT OF PEG TO PROTEIN AND LIPOSOMAL SURFACES

D. FISHER, C. DELGADO, B. SELISKO, F. MALIK, A.K. AGRAWAL and G.E. FRANCIS
Molecular Cell Pathology, Royal Free Hospital School of Medicine, Rowland Hill Street, London. NW3 2PF U.K.

Introduction

Polyethylene glycol (PEG) is a linear, sometimes branched, neutral polymer available in a range of molecular weights, and hence in a range of chain lengths. It is made up of repeating ethylene oxide units with terminal hydroxyl groups and is soluble in water (Bailey and Koleske, 1991). Many of its applications in biotechnology and medicine arise from its extreme water solubility and participation in phase separation phenomena. In particular, in recent years there has been considerable interest in the covalent attachment of PEG to proteins and liposomes: PEGylation increases circulation lifetimes with benefits to the therapeutic and diagnostic uses of these constructs. In this chapter we will briefly review this work, and make some comparisons between the different methods of PEGylation available, and the organisation of PEG chains on PEG-proteins and PEG-lipids.

Polyethylene glycol and phase separation

In aqueous solutions of PEG the polymer is extensively hydrated with approximately 2—3 water molecules being bound by each ethylene oxide unit (Tilcock and Fisher, 1979; Antonsen and Hoffman, 1992). Additional water molecules are also structured by the PEG (up to 16 water molecules per ethylene oxide unit (Baran, Solomentseeva, Mank and Kurilenko, 1972)) and PEG solutions show decreased polarity and dielectric constant compared with water (Arnold, Herrmann, Gawrisch and Pratsch, 1988). High concentrations of PEG cause membrane fusion (Ahkong, Fisher, Tampion and Lucy, 1975) and this has been related to dehydration effects of the PEG solutions (Blow, Botham, Fisher, Goodall, Tilcock and Lucy, 1978; Tilcock and Fisher, 1979, 1982).

PEG is particularly effective at excluding other polymers or particles from its presence in an aqueous environment (Harris, 1992). Thus PEG will precipitate proteins and viruses (Juckes, 1971) and liposomes (Tilcock and Fisher, 1982). Furthermore, when mixed with other water-soluble polymers above critical concentrations, phase separation generally occurs. Albertsson's early observation of phase separation with dextran and the distribution of macromolecules and bioparticles between the two aqueous phases so formed (Albertsson, 1956; Albertsson, 1958) has led to the development of aqueous polymer two-phase partitioning as an analytical and preparative technique (Albertsson, 1960; Albertsson, 1971; Albertsson, 1986; Walter, Brooks and Fisher, 1985; Fisher and Sutherland, 1989; Walter and Johansson, 1994; Rogers and Eitman, 1995).

PEG-modified surfaces: PEG-proteins and PEG-liposomes

The pioneering work of Davis and Abuchowski (Abuchowski, McCoy, Palczuck, Van Es and Davis, 1977) on the covalent attachment of PEG 5000 (~100 ethylene oxide units) to enzymes to render them non-antigenic and non-immunogenic was based on the concept that the covalent attachment of a linear, flexible, uncharged and hydrophilic polymer would provide a "shell around the enzyme that would cover antigenic determinants and, by presenting a flexible unbranched, hydrophilic surface for inspection by the immune processes, would prevent recognition of the interior enzyme as a foreign substance against which an immune response would be provoked". PEG-enzymes showed enhanced circulation lifetimes in the blood (Davis, Abuchowski, Van Es, Palczuk, Savoca, Chen and Pyatak, 1980).

In recent years there has been increased activity in the PEG modification (PEGylation) of other proteins. Modification of recombinant and native proteins with polyethylene glycol (PEG) almost invariably provides the generic benefits of increased plasma half-life, increased resistance to proteolysis and substantial reduction in antigenicity/immunogenicity (Francis, Delgado and Fisher, 1992; Delgado, Francis and Fisher, 1992). PEG-modification of antibodies raised against tumour antigens has been reported to enhance their tumour localisation (Kitamura, Takahashi, Takashina, Yamaguchi, Noguchi, Tsurumi, Toyokuni and Hakomori, 1990; Kitamura, Takahashi, Yamaguchi, Noguchi, Takashina, Tsurumi, Inagake, Toyokuni and Hakamori, 1991; Delgado, Pedley, Herraez, Boden, Boden, Keep, Chester, Fisher, Begent and Francis, 1995).

The excellent screen effect of PEG chains from the immune process led Nagaoka and colleagues (Nagaoka, Mori, Takiuchi, Yokota, Tanzawa and Nishiumi, 1984) to consider the idea that polymers with PEG chains "may be promising as a blood-contacting material which reduces the adhesion of the blood components"; they observed decreased platelet and plasma protein adsorption on hydrogels containing

PEG chains, the effect increasing with PEG chain length. They accounted for this behaviour by volume exclusion effects and the high mobility of the PEG chains (demonstrated by decreased line width in ^{13}C-NMR studies) of which the latter effect would shorten the time available for adsorbing material to form irreversible adhesion. PEG-modification of surfaces is now well established as reducing protein adsorption and is frequently described as providing protein rejecting surfaces (Merrill, 1992; Gölander, Herron, Lim, Claesson, Stenius and Andrade, 1992; Gombotz, Guanghui, Horbett and Hoffman, 1992; Park and Kim, 1992).

PEG-modification of liposomes also increases plasma half-life (Blume and Cevc, 1990; Klibanov, Maruyama, Torchilin and Huang, 1990; Allen, Hansen, Martin, Redemann and Yau-Young, 1991; Fisher, 1990; Fisher, Delgado, Morrison, Yeung and Tilcock, 1991; Senior, Delgado, Fisher, Tilcock and Gregoriadis, 1991; Woodle and Lasic, 1992; Tilcock, Ahkong and Fisher, 1993) and this aspect has been extensively reviewed recently (Woodle and Lasic, 1992; Allen, 1992; Lasic, 1994). PEG-liposomes also show a passive tumour localising effect (Papahadjopoulos, Allen, Gabizon, Mayhew, Matthay, Huang, Lee, Woodle, Lasic, Redemann and Martin, 1991).

How these beneficial effects of PEGylation arise is the subject of much current interest. While in some cases the reduction of interactions with serum components and cell receptors that normally lead to destabilisation, destruction, and/or removal might be the critical factor, in other examples the increase in size (in the case of small proteins) is more important (e.g. reduced renal clearance of small proteins). How the PEG chain are organised on these surfaces is clearly of importance.

The organisation of PEG chains on PEG-proteins and on PEG-liposomes

INCREMENTS IN HYDRODYNAMIC RADIUS OF PROTEINS BY PEGYLATION

Since proteins contain a number of lysines in addition to the N-terminal amino-acid, PEGylation can be expected to produce species with one, two, three and higher numbers of PEG chains attached. Figure 19.1 shows FPLC profiles of granulocyte-macrophage colony stimulating factor (GM-CSF) treated with increasing ratios of TMPEG 5000 (Malik, Delgado, Knusli, Irvine, Fisher and Francis, 1992). Such preparations have plasma half-lives 7—15-fold greater than unmodified GM-CSF (Malik *et al.*, 1992). The single peak of unmodified GM-CSF is progressively replaced by peaks of lower retention volume. Partition coefficients of the fractions through the elution profile were measured in PEG-dextran two phase systems (Delgado, Malik, Selisko, Fisher and Francis, 1994) and the unresolved FPLC peaks were deconvoluted into Gaussian curves representing single PEG-conjugate species, by mathematical modelling using the partition coefficient as an additional parameter to characterise the material in the component fractions (Delgado and Francis, unpublished). The positions of the PEG_1-GM-CSF, PEG_2-GM-CSF and PEG_3-GM-CSF are shown in Figure 19.1. The apparent hydrodynamic radius of these three species was calculated from the elution volume, the relationship between hydrodynamic radius and elution volume having been established using globular proteins and polyethylene glycol standards of known molecular weight. The hydrodynamic radii for the standards were calculated using the relationships given by Hagel (1988):

For proteins: $R_h = 0.794 \times (\text{molecular weight})^{0.333}$
For PEGs: $R_h = 0.255 \times (\text{molecular weight})^{0.517}$

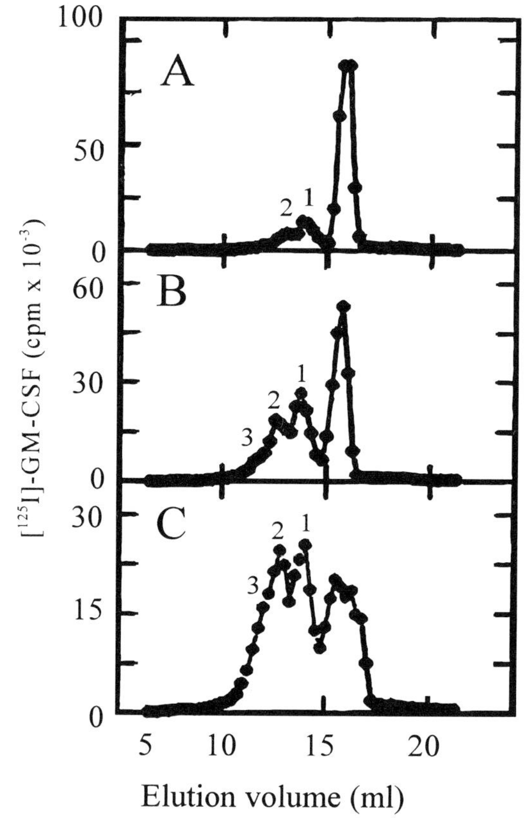

Figure 19.1 FPLC profile of ^{125}I-GM-CSF incubation with TMPEG (A: 5 mg/ml; B: 20 mg/ml; C: 40 mg/ml) for 2 hours at room temperature. Peaks 1, 2 and 3 indicate the positions of PEG_1-GM-CSF, PEG_2-GM-CSF and PEG_3-GM-CSF obtained as described in text. After Delgado *et al.* (1994)

Figure 19.2 shows a progressive increase in the hydrodynamic radius (R_h) of PEG-GM-CSF with molecular weight (calculated as the sum of the protein and the PEG chains attached) (Malik, 1993; Selisko, unpublished observations). The dependence of R_h on molecular weight for the PEG-GM-CSF constructs is quite different from that shown by globular proteins or by free (unattached) PEG molecules. The increases in the R_h of GM-CSF that occur on PEGylation are of the order of 1 nm for each PEG chain added (Figure 19.2). As the R_h of both unattached PEG-5000 and GM-CSF is about 2 nm, it appears that in this case the radius of the conjugated is smaller than the sum of the components. The protein surface available to PEG chains in PEG_1-GM-CSF, PEG_2-GM-CSF and PEG_3-GM-CSF can readily be calculated to be ~50, ~25 and ~12.5 nm^2 per PEG chain.

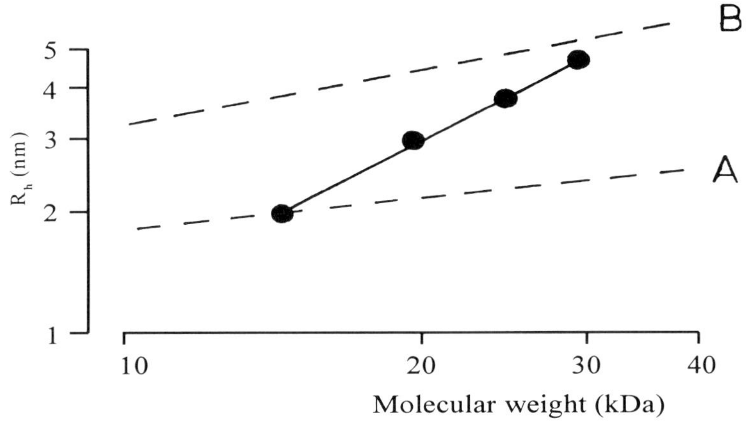

Figure 19.2 Dependence of hydrodynamic radius (R_h) on molecular weight of GM-CSF, and PEG_1-GM-CSF, PEG_2-GM-CSF and PEG_3-GM-CSF. Dotted lines show the dependence of globular proteins (A) and polyethylene glycols (B) as reported by Hagel (1988)

POLYMER LAYER THICKNESS IN PEG-LIPOSOMES

The organisation of PEG chains in PEG-lipids and on the surface of PEG-liposomes has received considerable attention since the first reports on PEG-liposomes appeared in 1990 (Blume and Cevc, 1993; Lasic, 1994; Lasic, Martin, Gabizon, Huang and Papahadjopoulos, 1991; Needham, Hristova, McIntosh, Dewhirst, Wu and Lasic, 1992; Torchilin, Omelyanenk, Papsivov, Bogdanov, Trubetskoy, Herron and Gentry, 1994). X-ray diffraction measurements have shown that when PEG 1900-distearoylphosphatidylethanolamine (PEG 1900-DSPE) is incorporated with bilayer-forming lipids, the chain extends about 5 nm from the lipid surface (Needham, McIntosh and Lasic, 1992).

Changes in the polymer thickness as the concentration of PEG 1900-DSPE is increased in the bilayer surface have been examined by surface force measurements between bilayers containing the PEG-lipid: at 1.3% surface content the PEG chains do not overlap and adopt a "mushroom" configuration with a PEG layer thickness of 3.5 nm. At 4.5% there is weak overlap and the thickness is 3.5–4 nm; by 9% content there is strong overlap with the chains adopting a "brush" organisation with the chains more extended, the thickness increasing to 6.5–7.5 nm (Kuhl, Leckband, Lasic and Israelachvili, 1994). As a 100 nm diameter liposome has about 4250 lipid molecules in its outer layer (Torchilin and Papisov, 1994), liposomes containing 1.3, 4.5 and 9 mol% PEG 1900-DSPE can be calculated to have surface-availabilities of the PEG chains of about 57, 16 and 8 nm^2 per PEG chain respectively, which are similar to those present at the surface of PEGylated GM-CSF that we have obtained.

Similar results on thickness of the PEG chains have been obtained with dynamic light scattering measurements of liposomes PEGylated at the outer surface with TMPEG-5000 which indicated a polymer thickness of about 15 nm (about twice the thickness obtained for PEG 1900 species) (Tilcock, Ahkong, Koenig, Brown, Kabalka and Fisher, 1992).

As the R_h of PEG-5000 is ~2 nm and of PEG 1900 is ~1.3 nm (as obtained by Hagel, 1988), it is clear that the PEG at liposomal surfaces is considerably extended.

COMPARISON OF ORGANISATION OF PEG CHAINS ON PROTEINS AND LIPOSOMES

Although PEG-modification is used to extend the plasma half-lives of both proteins and liposomes and a number of different methods have been used to give insight into the organisation of PEG chains present, there has been no detailed and comparative analysis of the impact of surface organisation (as opposed to PEGylation *per se*) both between liposomes and proteins and between different proteins.

However, Figure 19.2 shows that the increase in R_h as GM-CSF is PEGylated is of the order of 1 nm per PEG chain. Thus it is possible that in PEG-GM-CSF the PEG chains are not as extended as they are on liposomal surfaces. The surface availability of the PEG chains at the PEG-protein and PEG-liposome junctions are similar, and this suggests that interactions other than just exclusion between adjacent PEG molecules must operate to give rise to the more extended chains in the PEG-liposomes. The surface for the protein is more curved than that for the larger liposome and this might have an impact on the organisation of the PEG chains. In addition there might be some association of the PEG chains with the protein surface. The ability of PEG to aid the refolding of proteins has indicated that PEG can associate with hydrophobic sites on proteins (Cleland and Wang, 1990; Cleland and Randolph,

1992; Cleland, Builder, Swartz, Winkler, Chang and Wang, 1992; Cleland, Hedgepeth and Wang, 1992). If PEG-protein associations are important, we might expect to see benefits of PEGylation to vary with the protein modified. Interestingly, other recombinant proteins appear to show differences in the effect of PEG on their hydrodynamic radii when non-glycosylated and glycosylated equivalents of the same protein were examined (Malik, Delgado, Fisher and Francis, unpublished).

Current approaches to the production of PEG-proteins and PEG-liposomes

PEGylation of proteins is achieved by reaction of a suitably activated PEG (Davis *et al.*, 1980; Harris, 1985; Zalipsky and Lee 1992; Delgado *et al.*, 1992) with a group on the protein, generally the amino groups of N-terminal amino acids and lysyl side-chains. The monomethoxy PEG is used so that only one free OH group is available for activation. Molecular weights of 1900 and 5000 have most generally been used, but it must be appreciated that PEGs are polydisperse. In addition, PEGylation will generally produce a mixture of PEG-proteins with variation in the number and sites of PEG chains attached. This and the polydispersity of the PEG produce rather heterogeneous products.

PEG-liposomes can be prepared by treating pre-formed liposomes containing amino-containing lipids such as phosphatidylethanolamine (PE) or phosphatidylserine in aqueous solution using tresylmonomethoxy-PEG (TMPEG) (Fisher *et al.*, 1991). This method locates PEG-chains only at the outer surface. PEG-liposomes can also be prepared by incorporation of a PEG-lipid (generally a PEG-PE derivative), synthesised in organic solvents, using TMPEG (Tilcock *et al.*, 1993) or other activated PEG derivatives (Blume and Cevc, 1990; Klibanov *et al.*, 1990; Allen *et al.*, 1991). However, this has the disadvantage that the PEG is also present on the interior of the liposomes and depending on the type of contents present, may decrease the volume available for entrapment (e.g. approximately 30% of the interior volume of a 100 nm diameter unilamellar vesicles containing PEG 1900-PE could be unavailable because of the PEG chains).

Nature of the PEG-protein and PEG-lipid linkage

Figure 19.3 shows the most common linkages in PEG-proteins made by modification of primary amino-groups (N-terminal and lysyl side-chains). Similar linkages can be made to the amino-group of phosphatidylethanolamine to make PEG-lipids. Importantly only the secondary amine linkage (obtained with tresyl-activated PEG) provides a direct linkage of the polymer to the amino group; with all the other

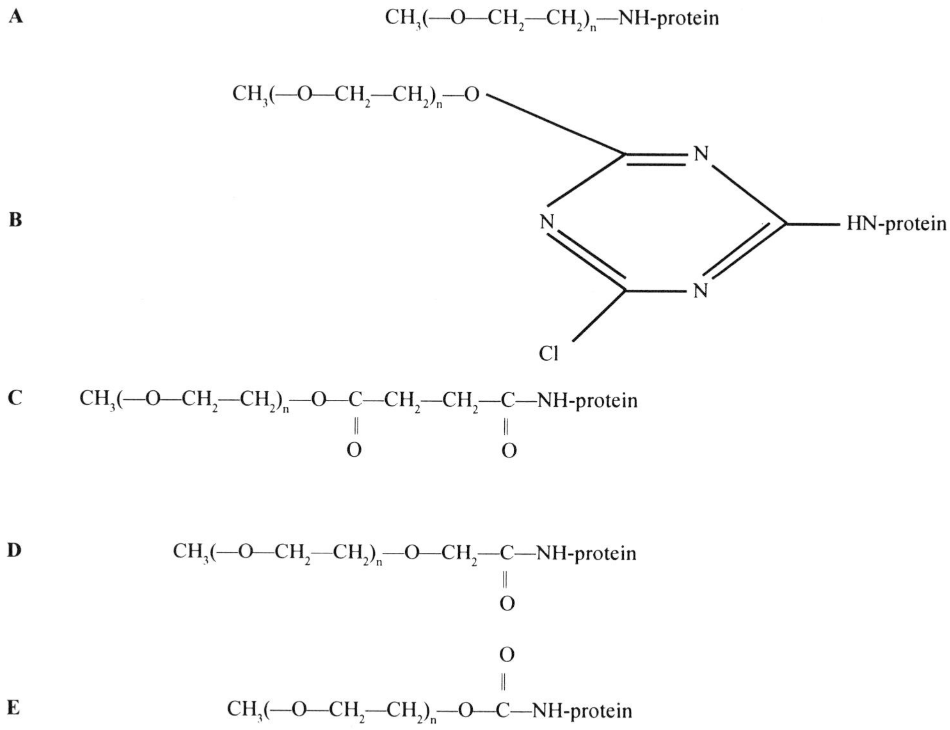

Figure 19.3 Typical linkages of PEG to amino groups on proteins. A, secondary amine linkage as obtained by tresyl PEG; B, secondary amine linkage as obtained with cyanuric acid activated PEG; C, ester and amide linkages, produced by the active ester method; D, amide linkage; E, carbamate (urethane) linkage. Similar linkages can be made to the amino group of phosphatidylethanolamine (PE) to form PEG-lipids

methods some part of the activated PEG is included in the linkage area. We have termed this part the coupling moiety. Furthermore, there is variation in terms of charge and stability: only the secondary amine bond does not alter the charge status of the nitrogen atom; both the amide and the carbamate bonds remove the ability of the nitrogen to be protonated.

Several consequence are to be expected from these difference in linkage:

(1) The presence of coupling moieties may prove antigenic/immunogenic, particularly in repeatedly administered preparations.
(2) Loss of the amine nature of the modified nitrogen alters the charge status of the protein or lipid. In the case of PEG-proteins, this may alter important charge interactions. In the case of PEG-lipids, if the positive nature of the amino nitrogen in PE is lost the molecule has a net negative charge arising from the phosphate group. This charge status has been measured for carbamate-linked PEG 1900-DSPE (Woodle, Collins, Sponsler, Kossovsky,

Papahadjopolous and Martin, 1992). It has yet to be demonstrated if the negative charge found in some PEG-PEs but not in others has any impact on the behaviour of PEG-liposomes, but it may be important not to assume all PEG-PE constructs as being necessarily similar in behaviour.

(3) A particular problem of the triazine linkage is that the coupling moiety can react further. In practice we have found formation of extremely large conjugates which suggests the formation of cross-linked products when using this method (Fisher, Delgado, Tejedor, Malik and Francis, 1995).

(4) The susceptibility of the polymer linkage to chemical or enzymatic hydrolysis will vary. It has recently been reported that PEG-PE linked by a succinate linkage is very labile in serum, at 37°C rapidly releasing a product that corresponded to free polymer (Parr, Ansell, Choi and Cullis, 1994). In studies we have made on erythropoietin (EPO) modified with PEG linked by coupling by the active ester method (Product C in Figure 19.3) (Fisher *et al.*, 1995), there appeared to be loss of bioactivity of over one order of magnitude. Although this PEGylation method has been claimed to have good preservation of bioactivity (Katre, Knauf and Laird, 1987), such claims have been made on the basis of short term assays. In the assay we used for PEG-EPO, the EPO needed to be available throughout the 14–16 days of the bioassay. Hydrolysis of the ester bond would remove the PEG, but leave the EPO with lysine groups modified with carboxy terminated moieties. The stability of other linkages to physiological fluids is, however, more controversial. PEG-PE constructed with a carbamate linkage showed release of free polymer after 24 h incubation in serum at 37°C (Parr *et al.*, 1994). This contrasts with observations by Zalipsky and Lee (1992) who report the carbamate linkage in PEG-proteins to be "completely stable under a variety of physiological conditions". PEG-PE containing the amide linkage showed no release of free polymer on incubation with serum for 24 h (Parr *et al.*, 1994). However, the stability of the amide bonds is likely to depend on the availability of serum amidases and Sears (Sears, 1983; Sears, 1984) has predicted that the linkage would be biodegradable. The secondary amine linkage, as produced by the tresylated PEG method currently used in our laboratory, is expected to be very stable to physiological breakdown.

(5) The different activated PEGs used to PEGylate proteins by these different methods may have differential effects on the retention of biological properties. For example, using erythropoietin as the target protein, we have compared the tresylation method with three other common methods of PEG activation: cyanuric acid, phenylchloroformate and succinimidyl succinate (Fisher *et al.*, 1995). We found that either conservation of biological activity or lack of toxic contaminants (or both) was inferior to the other methods. We presume that the reduced activity with cyanuric acid is due to the presence of highly cross-linked material. The succinyl succinate method gave loss of bioactivity

(described in (4) above). With the phenylchloroformate method (which forms a carbamate linkage) there appeared to be inhibitory material and/or reduced activities of the constructs.

CONCLUDING REMARKS

Polyethylene glycol is a very versatile polymer with numerous applications in biomedical and biotechnical fields. PEGylation of proteins and liposomes is now of proven benefit and the development of many polymer-modified biopharmaceuticals is underway. As we seek to apply PEGylation practically it is important to understand the molecular basis of its effects. As we attempt to optimise PEGylation for defined applications it is essential to appreciate that there are critical differences in the methods by which this can be achieved.

Acknowledgements

C.D. is supported by the Cancer Research Campaign.

References

Abuchowski, A., McCoy, J.R., Palczuk, N.C., Van Es, T. and Davis, F.F. (1977) Effect of covalent attachment of polyethylene glycol on immunogenicity and circulating life of bovine catalase. *Journal of Biological Chemistry*, **252**, 3582–3586

Ahkong, Q.F., Fisher, D., Tampion, W. and Lucy, J.A. (1975) Mechanisms of membrane fusion. *Nature*, **253**, 194–195

Albertsson, P.-Å. (1956) Chromatography and partition of cells and cell fragments. *Nature*, **177**, 711–774

Albertsson, P.-Å. (1958) Partition of proteins in liquid polymer-polymer two-phase systems. *Nature*, **182**, 709–711

Albertsson, P.-Å. (1960). *Partition of Cell Particles and Macromolecules*, 1st edition, Stockholm: Almqvist and Wiksells

Albertsson, P.-Å. (1971). *Partition of Cell Particles and Macromolecules*, 2nd edition, New York: J. Wiley and Sons, Inc.

Albertsson, P.-Å. (1986). *Partition of Cell Particles and Macromolecules*, 3rd edition, New York, John Wiley and Sons

Allen, T.M. (1992) Stealth liposomes: five years on. *Journal of Liposome Research*, **2**, 289–305

Allen, T.M., Hansen, C., Martin, F., Redemann, C. and Yau-Young, A. (1991) Liposomes containing synthetic lipid derivatives of poly(ethylene glycol) show prolonged circulation half-lives in vivo. *Biochimica et Biophysica Acta*, **1066**, 29–36

Antonsen, K.P. and Hoffman, A.S. (1992) Water structure of PEG solutions by differential scanning calorimetry measurements. In *Poly(Ethylene Glycol) Chemistry – Biotechnical and Biomedical Applications*, edited by Harris, J.M., New York and London: Plenum Press, pp 15–28

Arnold, K., Herrmann, A., Gawrisch, K. and Pratsch, L. (1988) Water-mediated effects of PEG on membrane properties and fusion. In *Molecular Mechanisms of Membrane Fusion*, edited by Ohki, S., Doyle, D., Flanagan, T.D., Hui, S.W. and Mayhew, E., New York and London: Plenum Press, pp 255–272

Bailey, F.E. and Koleske, J.V (1995). *Alkylene Oxides and their Polymers*, New York: Marcel Dekker, Inc.

Baran, A.A., Solomentseeva, I.M., Mank, V.V. and Kurilenko, O.D. (1972) Role of solvation factor in stabilizing disperse systems containing water soluble polymers. *Doklady Akademii Nauk SSSR*, **207**, 363–366

Blow, A.M.J., Botham, G.M., Fisher, D., Goodall, A.H., Tilcock, C.P.S. and Lucy, J.A. (1979) Calcium ions and cell fusion. *FEBS Letters*, **94**, 305–310

Blume, G. and Cevc, G. (1990) Liposomes for the sustained drug release in vivo. *Biochimica et Biophysica Acta*, **1029**, 91–97

Blume, G. and Cevc, G. (1993) Molecular mechanism of the lipid vesicle longevity *in vivo*. *Biochimica Biophysica Acta*, **1146**, 157–168

Cleland, J.L., Builder, S.E., Swartz, J.R., Winkler, M., Chang, J.Y. and Wang, D.I.C. (1992) Polyethylene glycol enhanced protein refolding. *Biotechnology*, **10**, 1013–1019

Cleland, J.L., Hedgepeth, C. and Wang, D.I. (1992) Polyethylene glycol enhanced refolding of bovine carbonic anhydrase B. Reaction stoichiometry and refolding model. *Journal of Biological Chemistry*, **267**, 13327–13334

Cleland, J.L. and Randolph, T.W. (1992) Mechanism of polyethylene glycol interaction with the molten globule folding intermediate of bovine carbonic anhydrase-B. *Journal of Biological Chemistry*, **267**, 3147–3153

Cleland, J.L. and Wang, D.I. (1990) Refolding and aggregation of bovine carbonic anhydrase B: quasi-elastic light scattering analysis. *Biochemistry*, **29**, 11072–11078

Davis, F.F., Abuchowski, A., Van Es, T., Palczuk, N.C., Savoca, K., Chen, R.H.-L. and Pyatak, P. (1980). Soluble, nonantigenic polyethylene glycol-bound enzymes. In *Biomedical Polymer, Polymeric Materials and Pharmaceuticals for Biomedical Use*, edited by Goldberg, E.P. and Nakajima, A., New York: Academic Press, pp 441–452

Delgado, C., Francis, G.E. and Fisher, D. (1992) The uses and properties of PEG-linked proteins. *Critical Reviews in Therapeutic Drug Carrier Systems*, **9**, 249–304

Delgado, C., Patel, J.N., Francis, G.E. and Fisher, D. (1990) Coupling of poly(ethylene glycol) to albumin under very mild conditions by activation with tresyl chloride: characterization of the conjugate by partitioning in aqueous two-phase systems. *Biotechnology and Applied Biochemistry*, **12**, 119–128

Delgado, C., Malik, F., Selisko, B., Fisher, D. and Francis, G.E. (1994) Quantitative analysis of polyethylene glycol (PEG) in PEG-modified proteins/cytokines by aqueous two-phase systems. *Journal of Biochemical and Biophysical Methods*, **29**, 237–250

Delgado, C., Pedley, R.B., Herraez, A., Boden, R., Boden, J.A., Keep, P.A., Chester, K.A., Fisher, D., Begent, R.H.J. and Francis, G.E. (1995) Enhanced tumour specificity of an anti-carcinoembrionic antigen Fab' fragment by poly(ethylene glycol)(PEG)-modification. British Journal of Cancer (in press)

Fisher, D. (1990) Liposomes. International Patent Application WO 90/04384

Fisher, D. and Sutherland, I.A. (1989). *Separations Using Aqueous Phase Systems – Applications in Cell Biology and Biotechnology*, New York and London: Plenum Press

Fisher, D., Delgado, C., Morrison, J., Yeung, G. and Tilcock, C. (1991) PEGylation of membrane surfaces. In *Cell and Model Membrane Interactions*, edited by Ohki, S., New York: Plenum Press, pp 47–62

Fisher, D., Delgado, C., Tejedor, M.C., Malik, F. and Francis, G.E. (1995) PEG–protein constructs for clinical use. In *Perspectives on Protein Engineering and Complementary Technologies*, edited by Geisow, M.J. and Epton, R., Mayflower Worldwide Ltd, pp 223–226

Francis, G.E., Delgado, C. and Fisher, D. (1992) PEG-modified proteins. In *Stability of Protein Pharmaceuticals, Part B: In Vivo Pathways of Degradation and Strategies for Protein Stabilization*, edited by Ahern, T.J. and Manning, M.C., New York: Plenum Press, pp 235–263

Gölander, C.-G., Herron, J.M., Lim, K., Claesson, P., Stenius, P. and Andrade, J.D. (1992) Properties of immobilized PEG films and the interaction with proteins: experiments and modeling. In *Poly(Ethylene Glycol) – Biotechnical and Biomedical Applications*, edited by Harris, J.M., New York: Plenum Press, pp 221–245

Gombotz, W.R., Guanghui, W., Horbett, T.A. and Hoffman, A.S. (1992) Protein adsorption to and elution from polyether surfaces. In *Poly(Ethylene Glycol) Chemistry – Biotechnical and Biomedical Applications*, edited by Harris, M.J., New York: Plenum Press, pp 247–261

Hagel, L. (1988) Pore size distribution. In *Aqueous Size Exclusion Chromatography*, edited by Dubin, P.L., Amsterdam: Elsevier, pp 119–155

Harris, J.M. (1985) Laboratory synthesis of polyethylene glycol derivatives. *Journal of Macromolecular Science, Reviews in Macromolecular Chemistry and Physics*, **C25**, 325–373

Harris, J.M. (Editor) (1992) *Poly(Ethylene Glycol) Chemistry – Biotechnical and Biomedical Applications*, New York: Plenum Press

Juckes, I.R.M. (1971) Fractionation of proteins and viruses with poly(ethylene glycol). *Biochimica et Biophysica Acta*, **229**, 535–546

Katre, N.V., Knauf, M.J. and Laird, W.J. (1987) Chemical modification of recombinant interleukin 2 by polyethylene glycol increases its potency in the murine Meth A sarcoma model. *Proceedings of the National Academy of Sciences of the United States of America*, **84**, 1487–1491

Kitamura, K., Takahashi, T., Takashina, K., Yamaguchi, T., Noguchi, A., Tsurumi, H., Toyokuni, T. and Hakomori, S. (1990) Polyethylene glycol modification of the monoclonal antibody A7 enhances tumor localization. *Biochemical and Biophysical Research Communications*, **171**, 1387–1394

Kitamura, K., Takahashi, T., Yamaguchi, T., Noguchi, A., Takashina, K., Tsurumi, H., Inagake, M., Toyokuni, T. and Hakamori, S. (1991) Chemical engineering of the monoclonal antibody A7 by polyethylene glycol for targeting cancer chemotherapy. *Cancer Research*, **51**, 4310–4315

Klibanov, A.L., Maruyama, K., Torchilin, V.P. and Huang, L. (1990) Amphipathic polyethyleneglycols effectively prolong the circulation time of liposomes. *FEBS Letters*, **268**, 235–237

Kuhl, T.L., Leckband, D.E., Lasic, D.D. and Israelachvili, J.N. (1994) Modulation of interaction forces between bilayers exposing short-chained ethylene oxide headgroups. *Biophysical Journal*, **66**, 1479–1488

Lasic, D.D. (1994) Sterically stabilized vesicles. *Angewandte Chemie – International Edition in English*, **33**, 1685–1698

Lasic, D.D., Martin, F.J., Gabizon, A., Huang, S.K. and Papahadjopoulos, D. (1991) Sterically stabilized liposomes: a hypothesis on the molecular origin of the extended circulation times. *Biochimica et Biophysica Acta*, **1070**, 187–192

Malik, F. (1993). *Construction and characterisation of polyethylene glycol modified cytokines*, Ph.D. thesis, Open University, Milton Keynes, U.K.

Malik, F., Delgado, C., Knusli, C., Irvine, A.E., Fisher, D. and Francis, G.E. (1992) Polyethylene glycol (PEG)-modified granulocyte-macrophage colony-stimulating factor (GM-CSF) with conserved biological activity. *Experimental Hematology*, **20**, 1028–1035

Malik, F., Delgado, C., Fisher, D. and Francis, G.E. (1995) Polyethylene glycol (PEG) modification of granulocyte-macrophage colony stimulating factor (GM-CSF): influence on pharmacokinetics and tissue distribution. (Submitted)

Merrill, E.W. (1992) Poly(ethylene oxide) and blood contact: a chronicle of one laboratory. In *Poly(Ethylene Glycol) – Biotechnical and Biomedical Applications*, edited by Harris, J.M., New York: Plenum Press, pp 199–220

Nagaoka, S., Mori, Y., Takiuchi, H., Yokota, K., Tanzawa, H. and Nishiumi, S. (1984) Interaction between blood components and hydrogels with poly(oxyethylene) chains. In *Polymers as Biomaterials*, edited by Shalaby, S.W., Hoffman, A.S., Ratner, B.D. and Horbett, T.A., New York: Plenum Press, pp 361–374

Needham, D., Hristova, K., McIntosh, T.J., Dewhirst, M., Wu, N. and Lasic, D.D. (1992) Polymer-grafted liposomes: physical basis for the "stealth" properties. *Journal of Liposome Research*, **2**, 411–430

Needham, D., McIntosh, T.J. and Lasic, D.D. (1992) Repulsive interactions and mechanical stability of polymer-grafted lipid membranes. *Biochimica et Biophysica Acta*, **1108**, 40–48

Papahadjopoulos, D., Allen, T.M., Gabizon, A., Mayhew, E., Matthay, K., Huang, S.K., Lee, K.D., Woodle, M.C., Lasic, D.D., Redemann, C. and Martin, F.J. (1991) Sterically stabilized liposomes: improvements in pharmacokinetics and antitumor therapeutic efficacy. *Proceedings of the National Academy of Sciences of the United States of America*, **88**, 11460–11464

Park, K.D. and Kim, S.W. (1992) PEO-modified surfaces – *in vitro*, *ex vivo*, and *in vivo* blood compatibility. In *Poly(Etheylene Glycol) Chemistry – Biotechnical and Biomedical Applications*, edited by Harris, J.M., New York: Plenum Press, pp 283–301

Parr, M.J., Ansell, S.M., Choi, L.S. and Cullis, P.R. (1994) Factors influencing the retention and chemical stability of poly(ethylene glycol)-lipid conjugates incorporated into large unilamellar vesicles. *Biochimica et Biophysica Acta*, **1195**, 21–30

Rogers, R.D. and Eitman, M.A. (1995). *Aqueous Biphasic Separations: from Biomolecules to Metal Ions*, New York: Plenum Press

Sears, B.D. (1983) Synthetic phospholipid compounds and their preparation. European Patent Application A1 0 072 111

Sears, B.D. (1984) Synthetic phospholipid compounds. United States Patent 4,426,330

Senior, J., Delgado, C., Fisher, D., Tilcock, C. and Gregoriadis, G. (1991) Influence of surface hydrophilicity of liposomes on their interaction with plasma protein and clearance from the circulation: studies with poly(ethylene glycol)-coated vesicles. *Biochimica et Biophysica Acta*, **1062**, 77–82

Tilcock, C.P.S. and Fisher, D. (1979) Interaction of phospholipid membranes with PEG. *Biochimica et Biophysica Acta*, **577**, 53–61

Tilcock, C.P.S. and Fisher, D. (1982) The interaction of phospholipid membranes with poly(ethylene glycol). Vesicle aggregation and lipid exchange. *Biochimica et Biophysica Acta*, **688**, 645–652

Tilcock, C., Ahkong, Q.F., Koenig, S.H., Brown, R.D., Kabalka, G. and Fisher, D. (1992) Nuclear magnetic relaxation dispersion and ^{31}P-NMR studies of the effect of covalent modification of membrane surfaces with poly(ethylene glycol). *Biochimica et Biophysica Acta*, **1110**, 193–198

Tilcock, C., Ahkong, Q.F. and Fisher, D. (1993) Polymer-derivatized technetium Tc(99m)-labeled liposomal blood pool agents for nuclear medicine applications. *Biochimica et Biophysica Acta*, **1148**, 77–84

Torchilin, V.P. and Papisov, M.I. (1994) Why do polyethylene glycol-coated liposomes circulate so long? (Molecular mechanism of liposome steric protection with polyethylene glycol: role of polymer chain flexibility). *Journal of Liposome Research*, **4**, 725–739

Torchilin, V.P., Omelyanenk, V.G., Papsivov, M.I., Bogdanov, Jr., A.A., Trubetskoy, V.S., Herron, J.N. and Gentry, C.A. (1994) Poly(ethylene glycol) on the liposome surface: on the mechanism of polymer-coated liposome longevity. *Biochimica et Biophysica Acta*, **1195**, 11–20

Walter, H. and Johansson, G. (1994). *Aqueous Two-Phase Systems*, San Diego: Academic Press

Walter, H., Brooks, D.E. and Fisher, D. (1985). *Partitioning in Aqueous Two-Phase Systems – Theory, Methods, Uses, and Applications to Biotechnology*, Orlando: Academic Press

Woodle, M.C. and Lasic, D.D. (1992) Sterically stabilized liposomes. *Biochimica et Biophysica Acta*, **1113**, 171–199

Woodle, M.C., Collins, L.R., Sponsler, E., Kossovsky, N., Papahadjopoulos, D. and Martin, F.J. (1992) Sterically stabilized liposomes. Reduction in electrophoretic mobility but not electrostatic surface potential. *Biophysical Journal*, **61**, 902–910

Zalipsky, S. and Lee, C. (1992) Use of functionalized poly(ethylene glycol)s for modification of polypeptides. In *Poly(Ethylene Glycol) Chemistry – Biotechnical and Biomedical Applications*, edited by Harris, J.M., New York: Plenum Press, pp 347–370

20

CHEMICAL INTERACTIONS IN MIXED BIOPOLYMER SYSTEMS

S.E. HILL and J.R. MITCHELL
Department of Applied Biochemistry and Food Science, University of Nottingham, Sutton Bonington Campus, Loughborough, Leicestershire. LE12 5RD U.K.

Introduction

The behaviour of biopolymer mixtures is generally interpreted in terms of non-covalent interactions between the polymers. Within this chapter it is the intention to show that the covalent interactions that can take place between large molecules or between macromolecules and small compounds can also have an important influence on the gelation, thickening, emulsifying and preservative properties of biopolymers. The examples used in this discussion will be based on two types of reactions of interest to the authors. First the use of the Maillard reaction between proteins and reducing sugars, or proteins and polysaccharides, and secondly the control of oxidative degradation of polysaccharides.

Biopolymer functionality and the Maillard reaction

The covalent interactions involving the carbonyl groups of reducing sugars and the amino groups of proteins (Maillard or non-enzymatic browning) are often viewed in terms of the production of colour, the loss of essential amino acids, the development of potential carcinogens and the loss of enzyme activity due to glycosylation of the protein. Of particular recent interest is protein crosslinking promoted by the Maillard reaction.

PROTEIN CROSSLINKING

The formation of covalent crosslinks associated with the Maillard reaction is dependent on time, temperature and substrates, but it is now clear that they can play an important role in the quality and function of biologically based materials. Crosslinking reactions are thought to be the reason for the undesirable changes in long-lived proteins in the body such as collagen and lens crystallin proteins (Cerami, 1985; Brownlee, Vlassara, Kooney, Ulrich and Cerami, 1986). In these cases sugar in solution reacts with immobilised protein. For food materials there are examples of changes in functional behaviour due to interactions of soluble protein with soluble sugars or, of particular interest to the food industry, the changes that occur when storing mixtures of reducing sugar and proteins at low water levels.

Some of the complexes that can form on storing the sugars with protein show enhanced solubility (Kato, Matsuda, Kato, Watanabe and Nakamura, 1986a). However, long storage times or higher temperatures produce products which are more difficult to solubilise (Lea and Hannan, 1950; Reynolds, 1965; Watanabe, Sato and Kato, 1980; Eble, Thorpe and Batnes, 1983; Boissel, Kasper and Bunn, 1986; Kato, Matsuda, Kato, Watanabe and Nakamura, 1986b; Kato, Matsuda, Kato and Nakamura, 1988). This loss in solubility would seem to be associated with an increase in molecular weight due to crosslinks being formed between protein molecules. A well known example of this is the deterioration in the quality of dried egg white on storage. Kato *et al.* (1988) studied freeze dried samples of ovalbumin with glucose and lactose, stored at 50°C and 65% relative humidity. Polymerisation of the protein was monitored by SDS–polyacrylamide gel electrophoresis. With time (> 4 days), the amount of the band associated with the ovalbumin decreased and bands of lower mobility, i.e. greater molecular weight, increased. Colour development and increase in molecular weight was greater with glucose compared with lactose. In some studies free sugar was removed from the mixture after a few days (2–4 days) and the degree of polymerization followed during storage of the resulting samples. Molecular weight increases were still observed, but there was no colour development. This demonstrated that polymerisation occurred due to the subsequent actions of the products formed in the early stages of the Maillard reaction and free sugars do not form part of the crosslink.

Although a number of studies have been carried out, the structure of Maillard induced protein crosslinks is still not clear. It has been shown that complete sugar molecules can be adducts to the protein, for example this has been clearly demonstrated using matrix assisted laser desorption–ionization mass spectrometry techniques (Lapolla, Fedele, Gerhardinger, Baldo, Crepaldi, Seraglia, Catinella and Traldi, 1994). Many other workers have shown that the crosslinking agent is likely to be an intermediate in the Maillard reaction. The compound 3-deoxyglucosone (3DG) is often implicated in the crosslinking mechanism (Shin, Hayase and Kato, 1988). It can be generated by the action of protein with a hexose and will form directly from fructose. Hayase, Kim and Kato (1991) demonstrated that 3DG was generated in a lysozyme–glucose system and a lysozyme–fructose system and that the degree of polymerisation was higher in the fructose system. Further evidence that 3DG was important in the crosslinking mechanism came from the observation that materials (aminoguanidine and semicarbazide) that react readily with 3DG, limit the polymerisation of the protein.

In systems as complex as food, other reactions may also be responsible for protein crosslinking in addition to chemical reactions between proteins and reducing sugars. The byproducts of the Maillard reaction are numerous and many resemble the products of lipid oxidation. Oxidative degradation of polyunsaturated fatty acids also produces many compounds which are capable of interacting with the proteins. The major compounds include aldehydes, and ketohydroxy and epoxy compounds, but perhaps the most important is malonaldehyde. For example, Buttkus (1967) demonstrated that malonaldehyde acted as a crosslinking agent with myosin. Maillard reactions are favoured by heat and an alkaline environment and these conditions also encourage protein crosslinking by ß-elimination. Proteins can be crosslinked either intra- or intermolecularly by the conjugated amino acids lysinoalanine, lanthionine or histinoalanine. These compounds can often be detected in food materials and other crosslinking proteins might also occur but to a lesser extent (Walter, Henle, Haeßner and Klostermeyer, 1994).

PROTEIN GELATION

In addition to the generally undesirable effects of protein crosslinking outlined above, protein polymerisation, as a result of the Maillard reaction, might be expected to improve the properties of thermally set protein gels, since these additional covalent linkages would reinforce the non-covalent bonds and the disulphide interactions normally considered to be responsible for protein gelation. Enhancement of gelation would be expected to be particularly significant when heating was carried out at high temperatures in the presence of reducing sugars that show a high level of Maillard reactivity.

Our initial work suggested that this was indeed the case. Using a range of food proteins, the minimum levels of protein that could be used to form a self-supporting gel on autoclaving unbuffered protein solutions was determined in the presence of sucrose or a reducing sugar (Hill, Mitchell and Armstrong, 1992). The results are shown in Table 20.1 and indicate that the inclusion of the reducing sugar lowers the minimum concentration for gelation.

The reactivity series for the sugars in promoting the Maillard reaction, as judged by colour formation, is also followed by their ability to promote protein gelation as measured by the gel break-strength (Table 20.2). Gel formation can be induced at temperatures below 100°C by heating in the presence of the more reactive sugars such as xylose. For example, when mixtures of 3% ribose and 1% bovine serum albumen (BSA) are heated at 90°C, gelation is observed after 80 minutes (Figure 20.1).

Table 20.1 MINIMUM PROTEIN CONCENTRATION FOR GEL FORMATION

Protein	*Protein concentration*	*Protein concentration in the presence of 3% xylose*
Soya isolate Purina 500E (Ralson Purina)	15% – 20%	6% – 7.5%
Soya isolate Purina 530E (Ralson Purina)	15% – 20%	6% – 7.5%
Pea protein (Grinsted Products)	15% – 20%	3% – 4%
Egg albumin (Sigma)	3% – 4%	0.25% – 0.5%
Bovine serum albumin (BSA) (Sigma)	9% – 10%	2% – 2.5%

(Samples were mixed with distilled water in glass bottles and heated to 120°C for 30 minutes. The samples were said to have gelled if the bottle could be inverted without movement of material and the free water did not exceed 10% of the total volume. Hill *et al.*, 1992.)

Table 20.2 STRENGTH OF COLOUR OF GELS FORMED FROM 2% BOVINE SERUM ALBUMIN (BSA) AND 2% SUGARS

Sugar	*Gel strength (N)*	*Colour (L)*
No sugar	No gel	ND
Sucrose	No gel	ND
Lactose	1.27	77.60
Fructose	1.50	69.13
Mannose	1.96	64.53
Sorbose	2.20	62.60
Xylose	2.91	43.16

(Gels were retorted at 120°C for 60 minutes and penetration strength assessed. Colour measurements were carried using a Hunter Colourquest spectrophotometer and are expressed as L values, where white = 100 and black = 0. Colour of liquid samples were not determined (ND). Hill *et al.*, 1992.)

It was soon appreciated that the apparent improvement in gelation behaviour cannot be ascribed to covalent crosslinking alone. This is because heating a protein in the presence of a reducing sugar will cause a fall in pH as illustrated in Figure 20.2. One factor causing this pH change is the loss of positive charges on the protein, but organic acid formation due to sugar fragmentation would seem to be responsible for the major increase in hydrogen ions. As a consequence the reducing sugar gels have a lower pH than the gels formed in the absence of sugars or with the non-

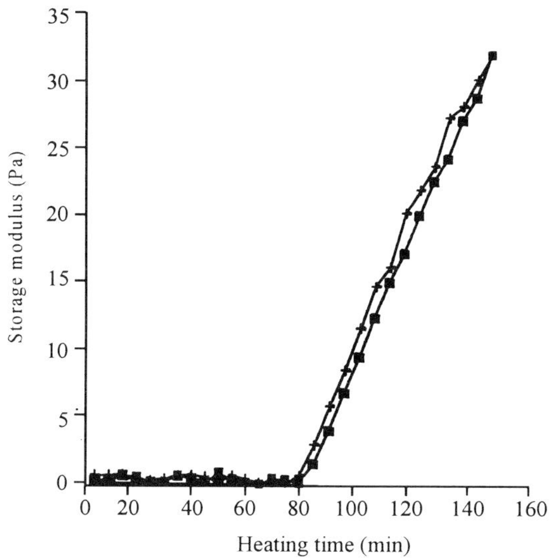

Figure 20.1 Storage modulus as a function of heating time at 90°C for 1% BSA plus 3% ribose measured at 1 Hz. The two sets of data points refer to measurements carried out on different days with different solutions. (After Mat Easa, Hill, Mitchell and Taylor, 1995.)

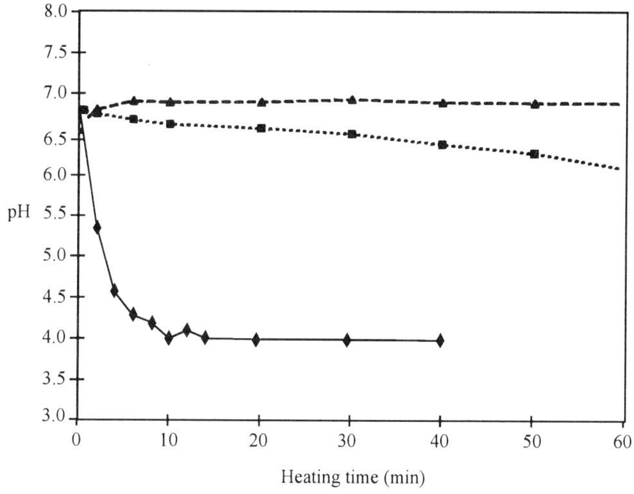

Figure 20.2 Changes in pH as a function of heating time for 3% bovine serum albumin at 95°C (▲) and 3% bovine serum albumin and 2% xylose at 95°C (■) and 120°C (♦). (After Hill *et al.*, 1992; Mat Easa *et al.*, 1995.)

reducing sugar, sucrose. It is difficult to prevent this pH fall even by the addition of strong buffers. When compared with gels prepared by heating a mixture of protein and glucono-δ-lactone (GDL), so that as far as possible the pH fall on heating is similar to that of the protein–reducing sugar mixture, the Maillard gels still have some significant advantages. These are (a) a much higher rupture strength; (b) greater clarity; and (c) substantially less syneresis during formation and subsequent storage (Armstrong, Hill, Schrooyen and Mitchell, 1994; Cabodevila, Hill, Armstrong, Sousa and Mitchell, 1994).

Studies in dissociating solvents clearly indicated that Maillard induced covalent crosslinking does occur. The solubility of the gel in solvents that break hydrophobic, electrostatic interactions and disulphide bridges, e.g. a mixture of sodium dodecyl sulphate (SDS) and ß-mercaptoethanol (ME), are much lower for the Maillard gels than gels prepared in the absence of the reducing sugars (Figure 20.3) confirming that additional covalent crosslinks induced by the Maillard reaction contribute to the structure of the gel. For soya isolate gels prepared in the presence of reducing sugars approximately 35% of the gel remains soluble in SDS plus ME (Cabodevila *et al.*, 1994). Since this is equivalent to the total solubility found for the GDL gel in SDS alone, it was suggested that the Maillard induced crosslinks occur in the regions of the protein where interchain disulphide bonds are present. In the case of the soya isolate gel the soluble component would be the 7 S globulin which contains very little cysteine and therefore would not participate in Maillard crosslinking if this is specific to regions close to interchain disulphide bonds. If this interpretation is correct then it would appear that a disulphide bond is required for the formation of a Maillard crosslink either because this bond participates in some way in the linkage

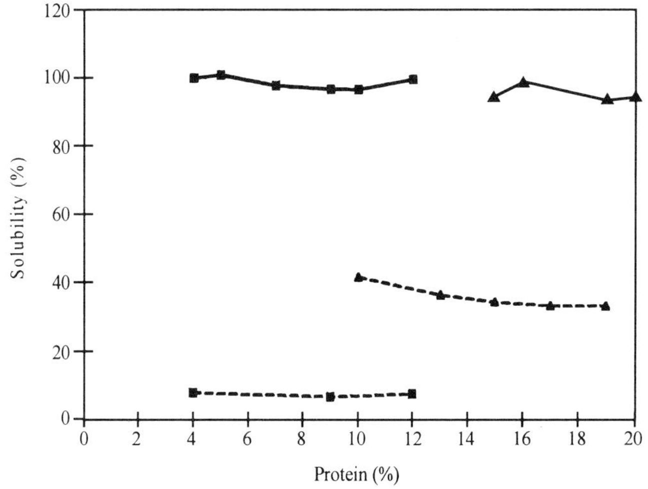

Figure 20.3 The amount of protein solubilized from gels (formed from solutions containing a range of either soya or bovine serum albumin (BSA) concentrations, by heating at 120°C for 60 minutes) by a mixed solvent containing 1% ß-mercaptoethanol and 1% sodium dodecyl sulphate. Soya gels (▲) formed with 3% xylose (- - -) and without sugar (—). BSA gels (■) formed with 3% xylose (- -) and without a reducing sugar (—). (After Armstrong *et al.*, 1994; Cabodevila *et al.*, 1994.)

or because the disulphide bond holds regions of the chains together allowing the Maillard linkages to form.

In addition to the pH fall and the formation of additional crosslinks a third factor that may be important in Maillard gels is the change in protein charge. A distinction is frequently made between fine stranded and aggregated globular protein gels (Clark and Lee Tuffnell, 1986; Hermansson, 1988). The latter form under conditions where electrostatic repulsion between protein molecules is low, that is to say at pHs close to the protein isoelectric point and/or at high salt concentrations. Such gels have poor water holding capacity and are opaque. The relative clarity of the Maillard gels and their superior water holding capacity at pHs close to the original isoelectric point would suggest that they are more akin to fine stranded gels despite their low pH. The reason for this is almost certainly the relatively high net negative charge possessed by the protein at pHs close to the original isoelectric point as a consequence of the loss of positively charged sites due to the participation of amino groups in the Maillard reaction. The isoelectric point of gels formed by autoclaving BSA–xylose mixtures was about 1 pH unit lower than conventional heat-set gels (Mat Easa, unpublished results). The isoelectric point of the latter was very close to native BSA.

Studies on dilute solutions of BSA heated for less than 80 minutes at 90°C in the presence of xylose have indicated that the molecular size of the protein aggregates increases throughout the duration of heating. Dynamic light scattering and ultracentrifugation studies would suggest that the protein molecules associate to form linear structures (Mat Easa, Armstrong, Mitchell, Hill, Harding and Taylor, 1995). This would be compatible with the Maillard gels at higher concentrations possessing a fine stranded character. The brown colour of Maillard gels limits their utility in many systems, though undoubtedly Maillard crosslinking plays a part in promoting texture in many proteinaceous food products which are processed at high temperatures.

PROTEIN–POLYSACCHARIDE INTERACTIONS

Production of improved emulsifiers

Not only will single sugars form adducts to the protein, but many polysaccharides can be added by linking the terminal carboxyl group of the sugar to a nucleophilic protein site. These adducts could form naturally or be encouraged by storage of proteins with polysaccharides under suitable conditions. Such materials would seem to have industrial potential since they are formed from natural ingredients. It would be expected that they would be effective emulsifiers since the protein will provide a strong attachment to the oil–water interface and the highly solvated polysaccharide component would be expected to provide effective steric stabilisation. Kato and co-workers (Kato, Murata and Kobayashi, 1988; Kato, Sato and Kobayashi, 1989; Kato, Sasaki, Furuta and Kobayashi, 1990) reported that they could produce ovalbumin–dextran conjugates by incubation of lyophilized mixtures of dextran and protein at a relative humidity of over 75% at 60°C for up to three weeks. The compounds formed had excellent emulsifying properties. Even better emulsifying properties were reported by conjugating lysozyme with dextran (Nakamura, Kato and Kobayashi, 1991) and a product of lysozyme with a mannase hydrolysate of guar gum also produced an effective emulsifier suitable for food use (Nakamura, Kato and Kobayashi, 1992). These conjugates acted to stabilise emulsions at acidic pH values and high salt concentrations where commercial emulsifiers were not so effective.

The emulsifying properties of protein–polysaccharide complexes formed as a result of the Maillard reaction have also been extensively studied by Dickinson and co-workers (Dickinson and Galazka, 1991, 1992; Dickinson and Semenova, 1992; Dickinson, 1993). Some of these studies are briefly alluded to in Chapter 17. Proteins investigated by this group include bovine serum albumen, ß-lactoglobulin, 11 S globulin and ß-casein. With the exception of the latter it was possible to form compounds that had useful emulsifier ability by heating with dextran or, in some cases, amylopectin, propylene glycol alginate or dextran sulphate. Figure 20.4 shows an example of the greatly improved stability against coalescence for emulsion stabilized by the conjugate (complex) compared with unheated mixtures of 11 S globulin and dextran. The inability to form useful complexes with casein (heating in the presence of dextran produced complexes with reduced emulsifying ability

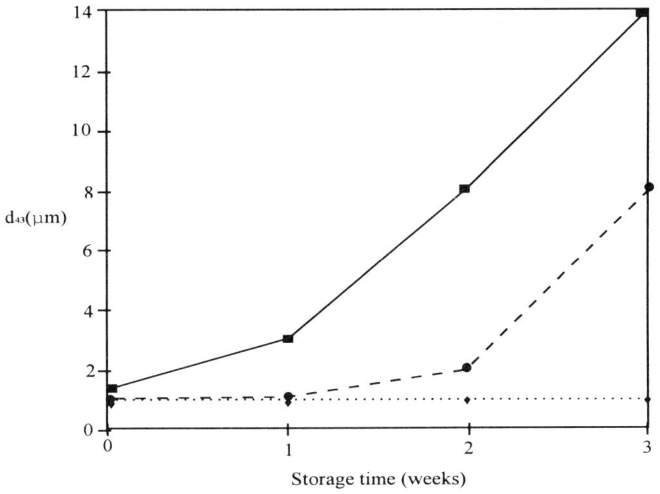

Figure 20.4 Average droplet size d_{43} of a stored oil-in-water emulsions (10% vol% n-tetradecane, 0.5% 11 S globulin, pH 8.0, ionic strength 0.1 M): (■) 11 S globulin; (●) simple mixture of 11 S globulin and dextran T40; and (♦) conjugate of the globulin and dextran made by dry heating at 60°C for three weeks. (After Dickinson, 1993.)

compared with the protein alone) is surprising. A possible interpretation was that this disordered protein entangled with the polysaccharide thus shielding the protein hydrophobic groups from the interface (Dickinson, 1993). Low levels of polysaccharide are known to destabilise oil-in-water emulsions by depletion flocculation (see Chapter 9) although in some cases bridging flocculation may play a role. It would therefore seem to be important that all the polysaccharide reacts with the protein. The results from studies with different ratios of the protein and the polysaccharide would suggest however that systems containing high polysaccharide levels are quite effective emulsifiers provided the mixture has been first heated (Dickinson, 1993). Jumel, Harding, Mitchell, and Dickinson (1993) attempted to determine the molecular weight of dry heated BSA–dextran mixtures using gpc–MALLS and ultracentrifugation and found a maximum at a molar ratio of polysaccharide to protein of 2:3 whereas higher ratios are more effective emulsifiers (Dickinson, 1993). Such investigations are complicated by the fact that the protein also associates under these heating conditions (Jumel, personal communication). It has however been clearly demonstrated that heating the protein alone does not result in improved emulsification behaviour. Heating these low mobility systems where the efficiency of mixing may be difficult to control might be expected to make it hard to obtain reproducible complexes. If solutions of non-interacting biopolymers are freeze-dried prior to heat treatment then phase separation may be expected to occur as the water content reduces. Interactions will then only occur at the border of the domains within the dry powder. The size and shape of the domains would be expected to be dependent on the kinetics of the drying process. Clearly there is still substantial work to be carried out in this area. Extrusion at relatively low temperatures might be an interesting way of making these complexes on a continuous basis.

PRESERVATIVE EFFECTS

It has been reported that the protein–polysaccharide complexes described above exhibited an antimicrobial effect against gram-negative bacteria (Nakamura, Kato and Kobayashi, 1991) (Figure 20.5). Another interesting example of a preservative and antimicrobial effect associated with polysaccharide–protein interactions is the role of sphagnan in preserving animal tissues in peat bogs (Painter, 1991, 1995). Sphagnan is structurally related to the complex pectins of vascular plants, but in addition to residues of α-D-galacturonic acid, it also contains residues of 5-keto-D-mannuronic acid (D-lyxo-5-hexosulouronic acid). It is therefore a naturally occurring glycuronoglycan which contains numerous highly reactive carbonyl groups distributed along its chains in addition to the usual reducing end-group. Its chemical properties resemble those of oxidised (dialdehyde) starch, whose tanning properties are well established.

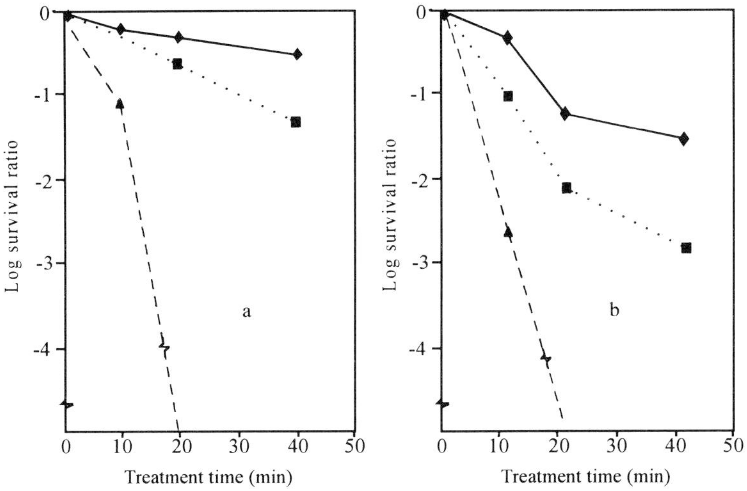

Figure 20.5 Antimicrobial activity of conjugates of lysozyme and dextran for (a) *Escherichia coli* IFO 12713 and (b) *Proteus mirabilis* IFO 12668: (♦) control, no addition of lysozyme or conjugate; (■) native lysozyme; (▲) conjugates of lysozyme and dextran. (After Nakamura *et al.*, 1991.)

In its native state sphagnan occurs in the hyaline cell walls of *Sphagnum* mosses, where it is covalently linked to cellulose and an amyloid-like xyloglucomannan. Upon the death and decay of the moss however it is slowly liberated in its soluble form and is then free to react with the collagen and other tissue proteins in the animal remains. This tannin reaction shows all the characteristics of a typical Maillard reaction.

In its insoluble, cellulose bound form, sphagnan acts in more subtle ways to suppress microbial activity. Exo-enzymes secreted by putrefactive bacteria are immobilised and deactivated on the surfaces of insoluble fragments of moss and prevented from attacking their substrate. Any ammonia or amino acids that do not get liberated are sequestered by the Maillard reaction and thereby become inaccessible to bacteria. Essential metal cations are also sequestered by sphagnan. The antimicrobial action therefore consists of extreme nutrient limitation rather than toxicity.

Polysaccharide degradation by oxidative reductive depolymerisation

Subjecting the polysaccharides to heat can result in substantial loss in viscosity and gel strength, the decrease being strongly dependent on pH (Pilnik and MacDonald, 1968; Bradley and Mitchell, 1988). The loss on heating is generally far lower at neutral pHs then acid pHs which is indicative of the importance of acid hydrolysis. Pectin is an exception as depolymerisation is dominated by ß-elimination. At neutral pHs most polysaccharides will degrade as a result of what is known as “oxidative reductive depolymerisation" (ORD) reactions. The reaction is oxygen, transition metal and free radical mediated and thus should be able to be controlled by the presence of other materials in the matrix. Antioxidants are incorporated in polysaccharide systems used in oil field applications, e.g. Wellington (1983), but little attempt has been made to extend this approach to food systems using non-toxic antioxidants or to try and determine if these types of reactions are important in retorted food products. The next section discusses some of the work carried out by the authors that addresses this area.

LINEAR POLYSACCHARIDES

The loss of functionality occurring on heating a range of polymers was investigated when antioxidants were incorporated into the system. The most promising food allowed "antioxidants" that prevented loss of viscosity were a mixture of sulphite, normally considered to scavenge oxygen from the system, and propyl gallate, a polar phenolic material that can form stable free radicals (Mitchell, Reed, Hill and Rogers, 1991). It is well recognized that sulphite alone does have some protective effect on galactomannans (Rodriguez, 1985). If the two additives are used in combination a synergistic effect results in much greater stabilization than is achieved with comparable levels of sulphite alone. For example when guar gum solutions (0.8%) are retorted at a pH of 7.0 in the presence of a total additive concentration of 0.02%, optimum stability is obtained at a 3:1 ratio of sulphite to propyl gallate (Figure 20.6) (Mitchell, Hill, Jumel, Harding and Aidoo, 1992). This synergism between an oxygen scavenger and a free radical terminator has previously been demonstrated for the protection of xanthan gum by Wellington (1983).

Having established that the viscosity and hence the molecular weight of the galactomannans could be protected by the use of additives, it was interesting to predict what would occur to gels formed from mixtures of locust bean gum and other polysaccharides. It is known that a reduction in molecular weight of a polysaccharide results in a decrease in the break-strength before the small deformation elastic modulus is changed (Mitchell, 1980; Rochas, Rinaudo and Landry, 1990). Thus if the incorporation of the antioxidants prevents molecular weight loss, the gel should not become brittle as a result of the autoclaving process.

Retorted gels containing carrageenan alone are not stabilised by the inclusion of the antioxidants, but autoclaved samples of locust bean gum plus carrageenan are. The break-strengths for the mixed polysaccharide gels containing different levels of antioxidants are shown in Figure 20.7. It is clear that the gel strength can be enhanced by a factor of about 4 compared with the system containing no additives although the strength of the non-autoclaved system is not achieved. It would appear that the sulphite:gallate ratio required for optimum gel strength was higher (17:3 sulphite to gallate) than that required for control of guar viscosity. It is well

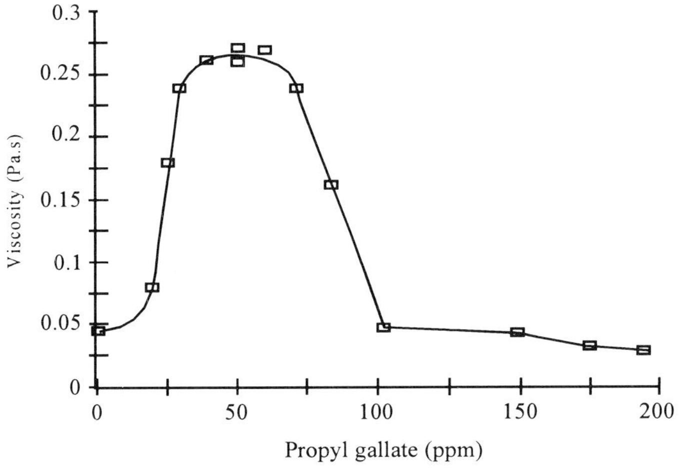

Figure 20.6 Effect of the ratio of propyl gallate to sodium sulphite on the viscosity (measured at 20°C and 50 sec^{-1}) of a retorted (120°C for 60 minutes) 0.8% guar gum solution at pH 7.0. The total concentration of the antioxidants was 200 ppm and the numbers on the abscissa are the gallate concentration. The unretorted guar (containing no additives) had a viscosity of 0.55 Pa.s. (After Mitchell *et al.*, 1992.)

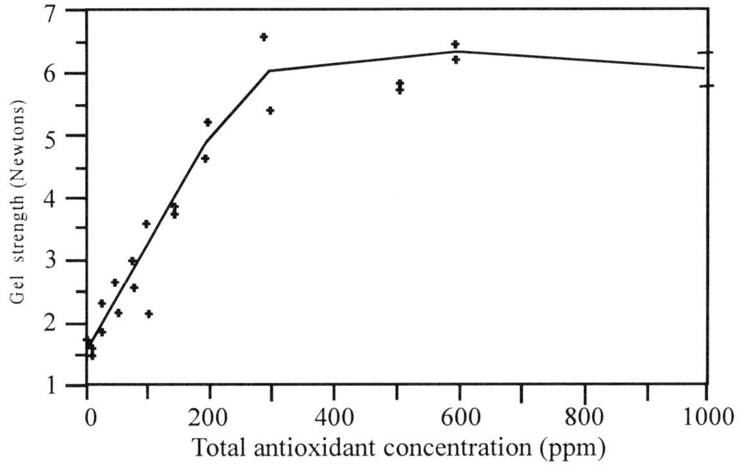

Figure 20.7 The effect of the addition of propyl gallate and sulphite (at a ratio of 17:3) on the break-strength of retorted (120°C for 60 minutes) gels (0.4% locust bean gum plus 0.4% carrageenan). (After Mitchell *et al.*, 1992.)

recognised that the strength of these systems is strongly influenced by pH and K^+ levels. The gel strength is enhanced by the presence of potassium chloride; however even in the presence of 0.1 M KCl the antioxidant systems still have a substantial beneficial affect. Maximum gel strength is obtained near neutral pH, and below a pH of 5 a solid gel is not obtained after retorting whether or not the antioxidants are present. This is presumably because at the lower pHs acid hydrolysis becomes far more important as a degradation mechanism compared with ORD reactions (Mitchell *et al.*, 1992).

A number of models have been suggested to explain the synergistic interaction between carrageenan and locust bean gum (see Chapters 13 and 14). A structure can be envisaged where locust bean gum associates with the carrageenan junction zones at only a small number of points on the galactomannan chain. The long flexible galactomannan chains would then make only a small entropic contribution to the modulus but a major contribution to the rupture strength by holding the network together while the carrageenan junctions reform. This structure would seem easier to reconcile with the strong dependence of brittleness on locust bean gum molecular weight implied by this work than would be the case if the two polysaccharides occupied separate domains in the gel.

Neither xanthan gum nor locust bean gum gels by itself. However, a mixture of the two gums form "elastic" gels of low modulus. This implies that the crosslink density is low and the distance between the junction zones along the locust bean gum chain will be large. A reduction in locust bean gum molecular weight will rapidly reduce the functionality (number of regions of the chain participating in crosslinks) below the critical value of two required for gel formation. It would therefore be expected that autoclaving will have a drastic effect on the mixed locust bean gum–xanthan gel. After autoclaving solutions of 0.3% xanthan and 0.3% locust bean gum, at pH 7.0, for one hour at 120°C there was no gel formation when the sample was cooled to room temperature. However, if the antioxidant system was included, at similar levels to those used for the carrageenan–locust bean gum gels, a self supporting gel was formed (Mitchell and Hill, 1995).

STARCH

Although many linear polysaccharides are important functional food additives, these materials are of relatively minor importance compared with starch. In addition to its use as a thickener and gelling agent, starch plays the major role in determining the structure of baked products. On heating in excess water the starch granule gelatinizes. During this process water enters the granule which swells and polysaccharide is released from the granule matrix. The predominantly linear amylose is more likely to be lost from the granule than the high molecular weight and highly branched amylopectin. If ordered regions of the granule (arrays of double helical associates between amylopectin A chains) are lost, then non-specific molecular entanglements would be expected to play the major role in preserving granule integrity. For a branched, very high molecular weight material such as amylopectin, the extent and lifetime of these entanglements would be expected to be high. The entanglement density will determine the degree of polysaccharide loss and the extent to which the granule swells. The granule swelling can be regarded as analogous to the swelling of a gel which is strongly influenced by crosslink and/or entanglement density. An additional factor for the starch granules would be the presence of amylose–lipid complexes which will not melt until temperatures of 100°C are approached. There is a correlation between the extent of bound lipid and the degree of granule swelling (Morrison, Tester, Snape, Law and Gidley, 1993). If the polysaccharides are depolymerised then the number of entanglements per molecule will be reduced. This will reduce granule integrity and enhance the amount of polysaccharide found outside the granule phase.

The hypothesis that at high temperatures and high shear rates some polysaccharide degradation occurs and this could be controlled by the addition of antioxidant systems with a resultant effect on granule swelling and the amount of material released from the granule was tested. As sulphite and propyl gallate stabilise some neutral polysaccharides it was assumed that they would protect starch polysaccharides against degradation. The amount of carbohydrate coming out from the granule and the volume occupied by the swollen granules were assessed after the starches were suspended in aqueous solutions and heated for 1 hour at 95°C. The incorporation of gallate and sulphite, in combination, had little affect on the parameters being assessed. Potato starch was an exception as the antioxidant combination caused a dramatic decrease in the swelling volume (Paterson, Mat Hashim, Hill, Mitchell and Blanshard, 1994). The different behaviour of this starch can be explained in terms of general salt effects on a swollen polyelectrolyte system as, in contrast to the other starches, potato amylopectin contains charged phosphate groups (Muhrbeck and Eliasson, 1987).

Sulphite addition protects guar gum from degradation (Rodriguez, 1985). If it had the same effect on starch polysaccharides it would be expected that less polysaccharide would be lost from the granule following gelatinization and swelling. However, this was not the case – sulphite addition seemed to promote polysaccharide release and reduce swelling volume. The level of sulphite was important (Mat Hashim, Moorthy, Mitchell, Hill, Linfoot and Blanshard, 1992). Figure 20.8 shows that for 1% cassava starch there is a minimum in swelling volume after pasting and a maximum in polysaccharide release at an active sulphite concentration of 0.007%. The action of low levels of sulphite in promoting polysaccharide release was observed for all the starch varieties tested (Table 20.3).

The increased polysaccharide release would suggest that the sulphite is acting to promote ORD reactions and thus degrade the starch polysaccharides. To confirm this, samples were solubilized by mixing the remnants of the gelatinized granule structure and the solubilized material with 0.5 M KOH. This rendered the whole of the polysaccharide soluble so that the intrinsic viscosity could be determined as a predictor of the average molecule weight. Table 20.4 shows that the starch polysaccharides treated in this way had similar intrinsic viscosities to the control if sulphite plus propyl gallate were added. High levels of sulphite incorporation (0.07%) gave intrinsic viscosity just statistically different from the controls. The inclusion of the low levels of sulphite (0.007%) markedly reduced the intrinsic viscosity. No protection of the starch polysaccharides was observed in any example.

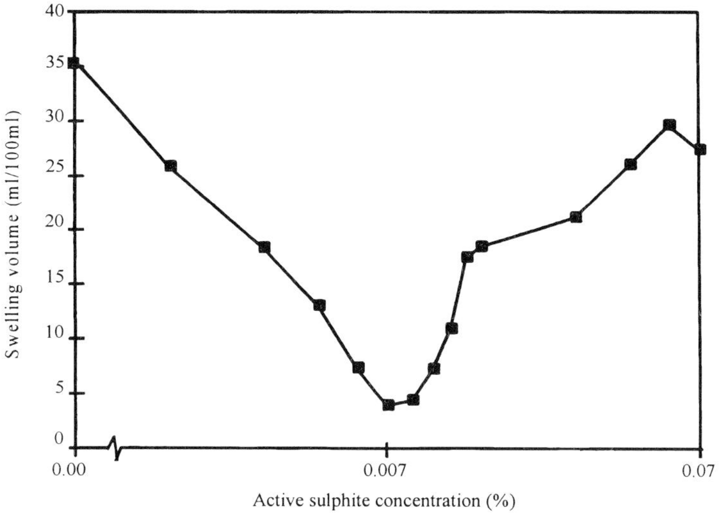

Figure 20.8 The swelling volume (the volume in ml/100 ml) occupied by granules after heating a 1 g/100 ml (w/v) cassava starch suspension at 95°C for 60 minutes in the presence of sulphite. "Active sulphite" are the levels as assessed by the 5,5'-dithiobis-(nitrobenzoic acid) method and was approximately 70% of the weight of compound added. (After Mat Hashim *et al.*, 1992.)

Table 20.3 THE EFFECT OF LOW LEVELS OF ADDITIVES ON THE SOLUBILITY OF A RANGE OF STARCHES

Starch	*No sulphite* *% solubility*	*0.007% sulphite* *% solubility*	*0.007% sulphite plus 0.005% gallate* *% solubility*
Potato	13.9±3.2	56.0±6.4	27.4±2.3
Sago	42.6±11.0	77.0±3.6	39.7±1.5
Cassava	42.6±4.2	95.4±2.4	51.2±2.4
Rice	19.3±6.6	53.3±3.5	24.1±2.6
Wheat	31.9±3.1	58.2±4.3	45.9±0.5
Maize	21.4±4.7	37.0±7.3	24.9±1.3

(15ml of 1% (w/v) starch suspensions in distilled water, with and without 0.007% "active" sulphite (see Figure 20.8 for details) or with sulphite and 0.005% propyl gallate, were pasted at 95°C for 60 minutes. The suspensions were centrifuged and the % solubility was expressed as (total solids in the supernatant / total solids in whole system). (After Paterson, 1995.))

Table 20.4 INTRINSIC VISCOSITY (ml/g) OF CASSAVA, WHEAT AND POTATO STARCH AFTER PASTING IN THE PRESENCE OF ADDITIVES

Starch	*Intrinsic viscosity (ml/g)*			
	No sulphite	*0.007% sulphite*	*0.07% sulphite*	*0.007% sulphite plus 0.05% gallate*
Potato	224±8.9	182±8.5	193±7.1	220±1.0
Cassava	180±2.9	115±2.8	158±9.1	174±4.1
Wheat	137±4.2	94±4.9	113±0.0	132±1.0

(15ml of 1% (w/v) starch suspensions in distilled water, with and without 0.007% or 0.07% "active" sulphite or with sulphite and 0.005% propyl gallate, were pasted at 95°C for 60 minutes. The resulting material was solubilized in 0.5 M KOH and the intrinsic viscosity determined. (After Paterson *et al.*, 1994.))

It therefore appears that the control of oxidative degradation may be important in the behaviour of starch granules on pasting. Low levels of sulphite in the presence of some oxygen promote polysaccharide degradation. This may be relevant to some food systems as the inclusion of low levels of sulphite is not uncommon. Evidence is available that other materials, e.g. ascorbic acid, also have a pro-oxidant effect on the starch when assessed by the methods used for the sulphite. An explanation for the changes in the baking properties of cake flours that are chlorinated is often based on free radical attack and ORD of the starch polysaccharides (Varriano-Marston, 1985).

What is yet to be understood is why ORD reactions are retarded in most linear polysaccharide system by the inclusion of an oxygen scavenger but are promoted in starch. It is perhaps worth adding that we have confined our discussion of degradation to polysaccharides. There is little doubt that protein degradation can also be important in some systems. An obvious example is the limitations in the use of gelatin as a gelling agent in canned meat products because of degradation during autoclaving. For globular proteins it is less clear what happens to the main chain during high temperature processing as any measurements are complicated by protein–protein associations.

Conclusion

In the studies discussed, the functionality of the biopolymer is strongly influenced by its chemical interaction with other compounds. Although model systems have been used to demonstrate this chemical dimension, these reactions will also occur in real food manufacturing processes. It is apparent that in more complex food systems other ingredients appear to have an antioxidant effect in preventing galactomannan degradation even when antioxidants are omitted (Fronimos, 1990). It may be that this is the explanation for some of the synergy reported, for example, that found between galactomannans and milk proteins. Thus the protein will scavenge free radicals and protect the galactomannan against degradation. Another example where these ideas could have some broader significance is in processes involving heat treatment of proteins where reducing sugars are naturally present. For example the heat treatment of milk products where the reducing sugar lactose is available in quantity.

References

Armstrong, H.J., Hill, S.E., Schrooyen, P. and Mitchell, J.R. (1994) A comparison of the viscoelastic properties of conventional and Maillard protein gels. *Journal of Texture Studies*, **25**, 285–298

Boissel, J.P., Kasper, T. and Bunn, H.F. (1986) Protein crosslinking during *in vivo* non-enzymatic glycation. In *Amino-Carbonyl Reactions in Food and Biological Systems*, edited by Fujimaki, M., Namiki, M. and Kato, H., Amsterdam: Elsevier Applied Science, pp 433–438

Bradley, T.D. and Mitchell, J.R. (1988) The determination of the kinetics of polysaccharide thermal degradation using high temperature viscosity measurements. *Carbohydrate Polymers*, **9**, 257–267

Brownlee, M., Vlassara, A., Kooney, A., Ulrich, P. and Cerami, A. (1986) Aminoguanidine prevents diabetes-induced arterial wall protein cross-linking. *Science*, **232**, 1629–1632

Buttkus, H. (1967) The reaction of myosin with malonaldehyde. *Journal of Food Science*, **32**, 432–434

Cabodevila, O., Hill, S.E., Armstrong, H.J., Sousa, I. and Mitchell, J.R. (1994) Gelation enhancement of soya isolates using the Maillard reaction and high temperature. *Journal of Food Science*, **59**, 872–875

Cerami, A. (1985) Hypothesis: Glucose as a mediator of aging. *Journal of the American Geriatrics Society*, **33**, 626–634

Clark, A.H. and Lee Tuffnell, C.D. (1986) Gelation of globular protein. In *Function Properties of Food Macromolecules*, edited by Mitchell, J.R. and Ledward, D.A., Barking: Elsevier Applied Science Publishers Ltd., pp 203–272

Dickinson, E. (1993) Protein–polysaccharide interactions in food colloids. In *Food Colloids and Polymers: Stability and Mechanical Properties*, edited by Dickinson, E. and Walstra, P., London: Royal Society of Chemistry, pp 77–93

Dickinson, E. and Galazka, V.B. (1991) Emulsion stabilization by ionic and covalent complexes of ß-lactoglobulin with polysaccharides. *Food Hydrocolloids*, **5**, 281–296

Dickinson, E. and Galazka, V.B. (1992) Emulsion stabilization by protein/polysaccharide complexes. In *Gums and Stabilisers for the Food Industry 6*, edited by Phillips, G.O., Williams, P.A. and Wedlock, D.J., Oxford: IRL Press, pp 471–478

Dickinson, E. and Semenova, M.G. (1992) Emulsifying behaviour of protein in the presence of polysaccharide under conditions of thermodynamic incompatibility. *Journal of the Chemical Society, Faraday Transactions*, **88**, 849–854

Eble, A.S., Thorpe, S.R. and Batnes, J.W. (1983) Nonenzymatic glucosylation and glucose-dependent crosslinking of protein. *Journal of Biological Chemistry*, **258**, 9406–9412

Fronimos, P. (1990). *Physico-chemical Properties of Two Important Food Polysaccharides*, M.Phil. thesis, University of Leicester, U.K.

Hayase, F., Kim, H.Y. and Kato, H. (1991) Inhibition of the protein crosslinking in Maillard reaction by aminoguanidine and semicarbazide. *Agricultural and Biological Chemistry*, **55**, 1435–1436

Hermansson, A.-M. (1988) Gel structure of food biopolymers. In *Food Structure – Its Creation and Evaluation*, edited by Blanshard, J.M.V. and Mitchell, J.R., London: Butterworth, pp 25–40

Hill, S.E., Mitchell, J.R. and Armstrong, H.J. (1992) The production of heat stable gels at low protein concentrations by the use of the Maillard reaction. In *Gums and Stabilisers for the Food Industry 6*, edited by Phillips, G.O., Williams, P.A. and Wedlock, D.J., Oxford: IRL Press, pp 471–478

Jumel, K., Harding, S.E., Mitchell, J.R. and Dickinson, E. (1993) Evidence for protein–polysaccharide complex formation as a result of dry-heating of mixtures. In *Food Colloids and Polymers: Stability and Mechanical Properties*, edited by Dickinson, E. and Walstra, P., London: Royal Society of Chemistry, pp 157–160

Kato, Y., Matsuda, T., Kato, N., Watanabe, K. and Nakamura, R. (1986a) Maillard reaction of some aldohexoses with ovalbumin. In *Amino-Carbonyl Reactions in Food and Biological Systems*, edited by Fujimaki, M., Namiki, M. and Kato, H., Amsterdam: Elsevier Science Publishers, pp 115–124

Kato, Y., Matsuda, T., Kato, N., Watanabe, K. and Nakamura, R. (1986b) Browning and insolubilization of ovalbumin by the Maillard reaction with some aldohexoses. *Journal of Agricultural and Food Chemistry*, **34**, 351–355

Kato, Y., Matsuda, T., Kato, N. and Nakamura, R. (1988) Browning and protein polymerization induced by amino–carbonyl reaction of ovalbumin with glucose and lactose. *Journal of Agricultural and Food Chemistry*, **36**, 806–809

Kato, A., Murata, K. and Kobayashi, K. (1988) Preparation and characterization of ovalbumin–dextran conjugate having excellent emulsifying properties. *Jour-*

nal of Agricultural and Food Chemistry, **36**, 421–425

Kato, A., Sato, T. and Kobayashi, K. (1989) Emulsifying properties of protein polysaccharide complexes and hybrids. *Agricultural and Biological Chemistry*, **53**, 2147–2152

Kato, A., Sasaki, Y., Furuta, R. and Kobayashi, K. (1990) Functional protein polysaccharide conjugate prepared by controlled dry-heating of ovalbumin dextran mixtures. *Agricultural and Biological Chemistry*, **54**, 107–112

Lapolla, A., Fedele, D., Gerhardinger, C., Baldo, L., Crepaldi, G., Seraglia, R., Catinella, S. and Traldi, P. (1994) An investigation on *in vivo* and *in vitro* glycated proteins by matrix assisted laser desorption/ionization mass spectroscopy. In *Maillard Reactions in Chemistry, Food and Health*, edited by Labuza, T.P., Reineccius, G.A., Monnier, V., O'Brien, J. and Baynes, J., Cambridge: Royal Society of Chemistry, pp 101–105

Lea, C.H. and Hannan, R.S. (1950) Studies of the reaction between proteins and reducing sugars in the "dry" state. *Biochemica et Biophysica Acta*, **5**, 433–440

Mat Easa, A., Hill, S.E., Mitchell, J.R. and Taylor, A.J. (1995) Bovine serum albumin gelation as a result of the Maillard reaction. *Food Hydrocolloids* (in press)

Mat Easa, A., Armstrong, H.J., Mitchell, J.R., Hill, S.E., Harding, S.E. and Taylor, A.J. (1995) Characterisation of BSA/xylose complex in dilute solutions. (Submitted to *International Journal of Biological Macromolecules*)

Mat Hashim, D.B., Moorthy, S.N., Mitchell, J.R., Hill S.E., Linfoot, K.J. and Blanshard, J.M.V. (1992) The effect of low levels of antioxidants on the swelling and solubility of cassava starch. *Starch/Stärke*, **44**, 471–475

Mitchell, J.R. (1980) The rheology of gels. *Journal of Texture Studies*, **11**, 315–337

Mitchell, J.R. and Hill, S.E. (1995) The use and control of chemical reactions to enhance the functionality of macromolecules in heat processed foods. *Trends in Food Science and Technology*, **6**, 219–224

Mitchell, J.R., Hill, S.E., Jumel, K., Harding, S.E. and Aidoo, M. (1992) The use of anti-oxidants to control viscosity and gel strength loss on heating of galactomannan systems. In *Gums and Stabilisers for the Food Industry 6*, edited by Phillips, G.O., Williams, P.A. and Wedlock, D.J. Oxford: IRL Press, pp 303–310

Mitchell, J.R., Reed, J., Hill, S.E. and Rogers, E. (1991) Systems to prevent loss of functionality on heat treatment of galactomannans. *Food Hydrocolloids*, **5**, 141–144

Morrison, W.R., Tester, R.F., Snape, C.E., Law, R. and Gidley, M.J. (1993) Swelling and gelatinization of cereal starches. IV. Some effects of lipid-complexed amylose and free amylose in waxy and normal barley starches. *Cereal Chemistry*, **70**, 385–391

Muhrbeck, P. and Eliasson, A.C. (1987) Influence of pH and ionic strength on the viscoelastic properties of starch gels – A comparison of potato and cassava starches. *Carbohydrate Polymers*, **7**, 291–300

Nakamura, S., Kato, A. and Kobayashi, K. (1991) New antimicrobial characteristics of lysozyme–dextran conjugate. *Journal of Agricultural and Food Chemistry*, **39**, 647–650

Nakamura, S., Kato, A. and Kobayashi, K. (1992) Bifunctional lysozyme galactomannan conjugate having excellent emulsifying properties and bactericidal effect. *Journal of Agricultural and Food Chemistry*, **40**, 735–739

Pilnik, W. and McDonald, R.A. (1968) The stability of some hydrocolloids. *Gordian*, **12**, 531–535

Painter, T.J. (1991) Lindow man, Tollund man, and other peat-bog bodies: The preservative and antimicrobial action of sphagnan, a reactive glycuronoglycan with tanning and sequestering properties. *Carbohydrate Polymers*, **15**, 123–142

Painter, T.J. (1995). In *Bog Bodies – New Discussions and Perspectives*, edited by Turner, R.C. and Scaife, R.G., London: British Museum Press

Paterson, L.A., Mat Hashim, D.B., Hill, S.E., Mitchell, J.R. and Blanshard, J.M.V. (1994) The effect of low levels of sulphite on the swelling and solubility of starches. *Starch/Stärke*, **46**, 288–291

Paterson, L.A. (1995). *The Effects of Low Levels of Sulphite on Starch*, thesis submitted for the degree of Ph.D., University of Nottingham

Reynolds, T.M. (1965) Chemistry of nonenzymic browning II. *Advances in Food Research*, **14**, 167–283

Rochas, C., Rinaudo, M. and Landry, S. (1990) Role of the molecular weight on the mechanical properties of kappa carrageenan gels. *Carbohydrate Polymers*, **12**, 255–266

Rodriguez, P. (1985). Stabilising polysaccharide – Thickened aqueous medium and stable aqueous gel, U.S. Patent 4514318

Shin, D.B., Hayase, F. and Kato, H. (1988) Polymerization of proteins caused by reaction with sugars and the formation of 3-deoxyglucosone under physiological conditions. *Agricultural and Biological Chemistry*, **52**, 1451–1458

Varriano-Marston, E. (1985) Flour chlorination: New thoughts on an old topic. *Cereal Food World*, **30**, 339–343

Walter, A.W., Henle, T., Haeßner, R. and Klostermeyer, H. (1994) Studies on the formation of lysinomethylalanine and histidinomethylalanine in milk products. *Zeitschrift für Lebensmittel-Untersuchung und -Forschung*, **199**, 243–247

Watanabe, K., Sato, Y and Kato, Y. (1980) Chemical and conformational changes of ovalbumin due to the Maillard reaction. *Journal of Food Processing and*

Preservation, **3**, 263–274
Wellington, S.L. (1983) Biopolymer solution viscosity stabilization – Polymer degradation and antioxidant use. *Society of Petroleum Engineers Journal*, **9296**, 901–912

21

BIOPOLYMER MIXTURES AT LOW WATER CONTENT

J.M.V. BLANSHARD
Department of Applied Biochemistry and Food Science, University of Nottingham, Sutton Bonington Campus, Loughborough, Leicestershire. LE12 5RD U.K.

Introduction

The study of biopolymer mixtures at low water content is a relatively unexplored area though of enormous practical significance and unfortunately fraught with a number of experimental difficulties. It is, for example, uncertain in many systems as to whether or not the components are truly miscible, and whether or not they are in thermodynamic or merely metastable equilibrium awaiting the precise conditions or time when phase separation can occur. Since the biopolymer molecules themselves are large molecules and the system is almost, by definition, of high viscosity, diffusion (which is usually the mechanism of phase separation) will be very slow. A further complication is that, because we are dealing with biopolymers, in many situations the ubiquitous water molecule will have a substantial effect in plasticizing the molecules and may in fact be promoting the solubility of smaller molecules which may be admixed with them.

The intention of this short chapter is to point out some areas where a variety of techniques have given insight into the molecular interactions and behaviour that are taking place at low water content. It is hoped the outcome will be to point the way to the reader as well as emphasizing the need for furthering research in an area which is crucial to the understanding of the behaviour of low water food materials. This article will partly review and partly reflect some interesting recent developments in our laboratory.

Techniques

A number of techniques have proved valuable in providing information about the state of the molecules and changes that occur in low water biopolymer systems.

DIFFERENTIAL SCANNING CALORIMETRY

The differential scanning calorimeter has a distinguished record (Levine and Slade, 1988) as a means for recording exothermic and endothermic changes associated with phase changes as well as monitoring the glass transition temperature and a variety of dynamic effects where these proceed rapidly. It is particularly useful for systems of smaller molecules and for polymers which have a highly repetitive structure along the macromolecular chain e.g. polysaccharides. However, it is not surprising that in proteins the T_g is more difficult to see, presumably because the variety of structures along the polypeptide chain permits sections of the polymer chain to display individual "micro-T_gs", resulting in a diffuse, smeared change rather than a pronounced steep change in the heat capacity, C_p.

MECHANICAL METHODS

The "three-point bend test" (Ablett, Attenburrow and Lillford, 1986) clearly demonstrates the transition from brittle fracture to rubbery decay and is useful in relating mechanical properties to human perception of such events. However, it is too clumsy for systematic experimental investigations of the glass phase transition and associated phenomena. A much more useful technique is dynamic mechanical thermal analysis which records the elastic modulus (E'), the loss modulus (E'') and tan δ ($=E''/E'$) (Kalichevsky, Jaroszkiewicz and Blanshard, 1992a). The initial fall in E', the maximum in E'' and the peak of tan δ do not coincide but they do follow a similar pattern. The fall in E' appears to coincide with the rigid lattice limit (RLL) recorded by pulse NMR spectroscopy (Kalichevsky, Jaroszkiewicz, Ablett, Blanshard and Lillford, 1993) while the maximum in E'' coincides with the T_g determined by differential scanning calorimetry.

NMR SPECTROSCOPY

As already indicated pulse NMR spectroscopy can be used to detect the mobility of the protons of the biopolymer chain and a considerable change in their mobility is recorded at regions close to the T_g; in practice this is termed the rigid lattice limit. The situation needs to be considered critically as the above relation does not hold for smaller molecules such as oligosaccharides (Ablett, Darke, Izzard and Lillford, 1993). It is interesting that high frequency NMR can be used to record the spectral changes as molecules assume greater mobility as well as to detect the emergence of crystalline phases from amorphous glasses (Kalichevsky *et al.*, 1992a). Magic angle spinning techniques, of course, resolve the crystalline solids but can do little to handle the broad lines of amorphous, glassy materials.

INFRA-RED SPECTROSCOPY

Infra-red spectroscopy can be used, particularly in the FT-IR ("Fourier transform infra-red") mode, as an effective means not only of monitoring the spectral changes when materials move from the glassy to rubbery states but also of studying interactions between plasticizing agents and biopolymers as well as between macromolecules and other molecules (Orset, 1994; Farhat, Moreau, Mitchell, Derbyshire and Blanshard, 1996).

X-RAY DIFFRACTION

X-ray diffraction has proved useful for recording the ordering of domains within systems, and particularly for the detection of whether or not specific products are responsible for crystallinity and phase separation.

Phase separation using dynamic mechanical thermal analysis

Two methods of preparation have been employed in this area and are as follows.

1. The most commonly used method has been to sprinkle either the requisite amount of water or concentrated sugar solution into liquid nitrogen, to grind the frozen product to a fine powder under liquid nitrogen and to add the requisite amount of biopolymer to give the desired water content. The various components are mixed under liquid nitrogen and the combined material stored in a freezer at -20°C until required. The product is pressed for five minutes at 60°C into a rectangular slab approximately 8 cm × 1 cm × 0.2 cm under a pressure of 20 tons per square inch (≡3.1 × 10^8 Pa).
2. An alternative method, which is especially useful for gluten, is to mix the biopolymer in a pin mixer with water or sugar solution into a range of 24% to 48% relative to the gluten. The sample is pressed as in "1." above and dried in an oven at 30°C.

Both of the above techniques are of particular value where the samples are to be studied by dynamic mechanical thermal analysis but clearly they can also be used for DSC, NMR or other measurements.

In the next section we shall concentrate on investigating the impact of water upon the glass transition of biopolymers and in exploring the effect of different sugars at increasing concentration on amylopectin, gluten and casein biopolymers which differ in their hydrophobicity. The effect of water on the glass transition temperature of amylopectin, gluten and casein is illustrated in Figure 21.1. It will be evident that whereas gluten and amylopectin parallel each other in their response to water, casein appears to be somewhat less sensitive to increasing water content.

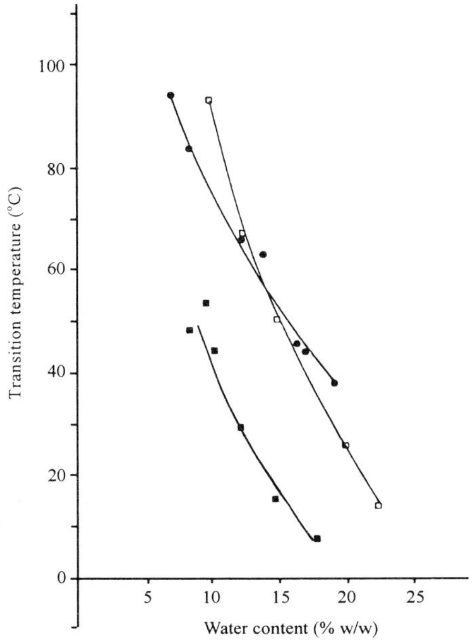

Figure 21.1 Variation of T_g with water content: T_g measured by temperature of change in slope of log E'. Casein (●); gluten (■); and amylopectin (□)

It is therefore not surprising that with such results there results the need to describe the behaviour of these two-phase systems by some form of physical expression. The initial success of the "free volume theory" with synthetic polymers (Bueche, 1962) was not paralleled with the same success in biopolymer—water systems. The Couchman—Karasz equation (Couchman and Karasz, 1978) which assumes continuity of configurational entropy at T_g, full miscibility and the absence of any crystallinity or cross-linking, has proved much more successful and for a two-component system takes the general form

$$\ln T_g = \frac{W_1 \Delta C_{p1} \ln T_{g1} + W_2 \Delta C_{p2} \ln T_{g2}}{W_1 \Delta C_{p1} + W_2 \Delta C_{p2}} \quad (1)$$

The subscripts 1 and 2 refer to the two components where W_1 and W_2 are the weight fractions of 1 and 2 while ΔC_{p1} and ΔC_{p2} are the differences in capacity heat between the rubbery and glassy states at T_g. This particular equation is theoretically rigorous and assumes that ΔC_p is independent of temperature. If, however, we assume that ΔC_p is proportional to $1/T_{abs}$ then a simpler form is possible, namely

$$T_g = \frac{W_1 \Delta C_{p1} T_{g1} + W_2 \Delta C_{p2} T_{g2}}{W_1 \Delta C_{p1} + W_2 \Delta C_{p2}} \tag{2}$$

Furthermore this equation can be extended to three components such that

$$T_g = \frac{W_1 \Delta C_{p1} T_{g1} + W_2 \Delta C_{p2} T_{g2} + W_3 \Delta C_{p3} T_{g3}}{W_1 \Delta C_{p1} + W_2 \Delta C_{p2} + W_3 \Delta C_{p3}} \tag{3}$$

The real experimental problem with this, as will be seen later, is in finding the T_gs and ΔC_ps of all the components. In two-component systems the Gordon-Taylor equation (Gordon and Taylor, 1952), which has an empirical basis, has proved valuable. This assumes the form

$$T_g = \frac{W_1 T_{g1} + \kappa W_2 T_{g2}}{W_1 + \kappa W_2} \tag{4}$$

It can immediately be seen that this is equivalent to the Couchman-Karasz equation if

$$\kappa = \frac{\Delta C_{p2}}{\Delta C_{p1}} \tag{5}$$

It will be evident, therefore, that if T_{g1} and T_{g2} are known as well as W_1 and W_2 then the value of κ can be established by fitting procedures and if one C_p is known then the other less accessible ΔC_p can also be determined. Clearly, however, this equation cannot be extended beyond two components.

At levels of sugar below 10% there is a pronounced plasticization effect by both water and sugar which broadly fits in with the Couchman-Karasz equation and the dynamic mechanical thermal analysis (DMTA) traces are typical of amylopectin-water. However, as the sugar content is increased a more complex thermomechanical spectrum appears, see Figure 21.2. It will be evident that instead of one tan δ peak there are now two and there are similar processes which match the pronounced tan δ peaks on the log E' and log E'' traces. It would seem a reasonable assumption that total miscibility has not been attained and phase separation is present within

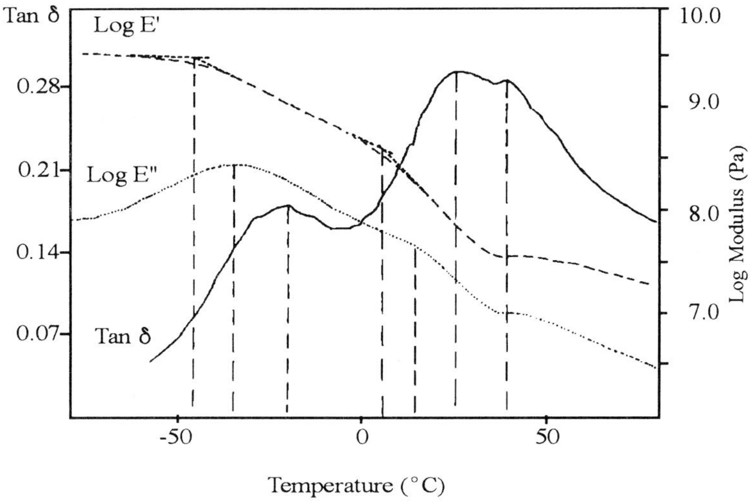

Figure 21.2 DMTA tan δ, log E′ and log E″ vs temperature for amylopectin + 25% fructose (85% RH; 18.6% aq)

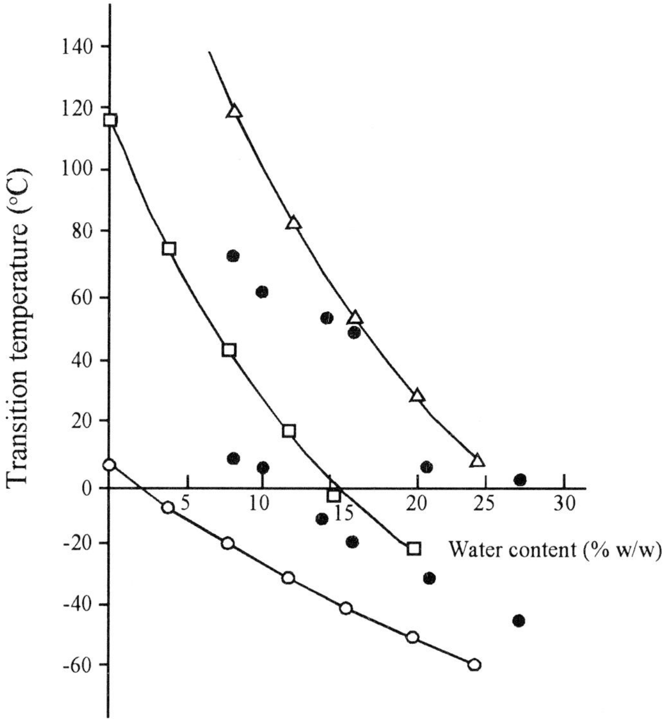

Figure 21.3 DMTA tan δ peaks (●) for amylopectin—fructose (2:1) as a function of water content. Also shown are the theoretical DSC curves for amylopectin alone (△), for fully miscible amylopectin—fructose (2:1) (□) and for fructose alone (○)

the system (Kalichevsky, Jaroszkiewicz and Blanshard, 1993). Figure 21.3 illustrates a more extended series of results in which the actual peaks are plotted relative to what would be anticipated for pure amylopectin, pure fructose and a fully miscible fructose–amylopectin mixture of the given composition. It appears that the peaks, in fact, fall between the two extremes and we would conclude that the high temperature one reflects an amylopectin rich phase which has been plasticized by fructose while the lower peaks represent fructose rich systems where the amylopectin has acted as an antiplasticizer. A somewhat more detailed study of this type of system (for example of a mixture of amylopectin plus 25% fructose) has revealed the interesting feature that the phase with the higher temperature T_g is defined by strong tan δ peaks but the reduction in E′ and the height of the log E″ associated with this T_g are fairly small. In contrast the phase with lower T_g has a very pronounced log E″ peak but only modest reduction in E′ and tan δ peaks. Whether the relative intensities of these parameters are significant is uncertain at present.

In a further series of studies where the DMTA transition in an amylopectin–fructose system was investigated at a fixed water concentration of 10%, but with increasing fructose contents, a number of notable features have emerged (Figure 21.4). First, it appeared that both the log E″ and the tan δ peaks were less plasticized by the fructose at higher concentrations than was anticipated and secondly, that the low temperature fructose rich phase was more antiplasticized than was originally anticipated. If we examine the tan δ peaks of this type of system where the water content is being maintained between 7.3% and 9.9% but the fructose concentration is being increased, it is evident that the modest tan δ peak at +100°C for 0% fructose is dramatically changed both in position and the area encompassed as the fructose

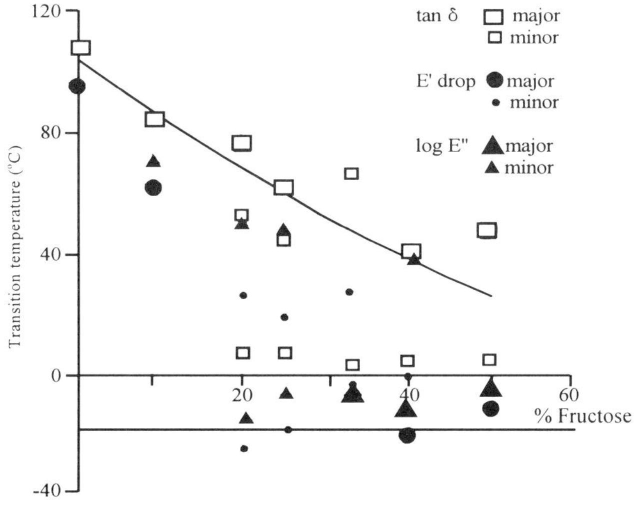

Figure 21.4 DMTA peak tan δ, log E′ drop, log E″ maximum values for amylopectin containing various fructose contents at 10% aq.

concentration increases to 50% (Figure 21.5). The inevitable conclusion from this is that a much greater proportion of the amylopectin is being mobilized by the presence of the fructose.

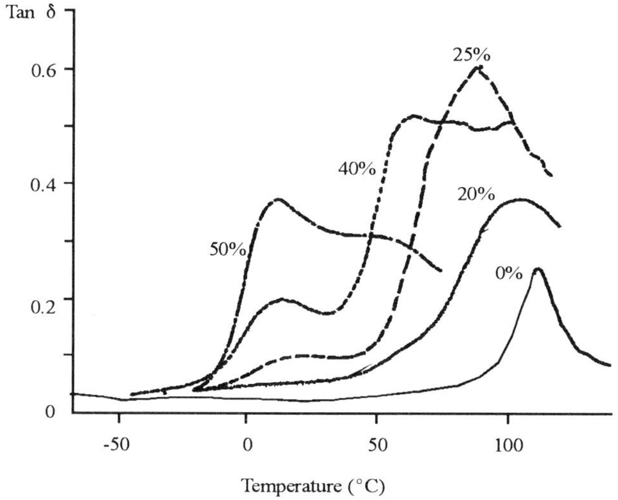

Figure 21.5 DMTA tan δ plots for samples containing 0%–50% fructose, stored at 12° RH (containing between 7.3–9.9% aq.)

If we now examine the effect of sugar on the three-point bend test modulus for amylopectin, as we might expect the addition of fructose at modest concentrations, e.g. 33%, leads to a reduction in the elastic modulus over the whole moisture range from 8 to 25 but also enhanced ductility in comparison to the brittle character of the pure amylopectin (Kalichevsky and Blanshard, 1993). When the concentration of fructose is 50%, we can assume a different form presumably because the fructose is the dominating matrix. Again the elastic modulus is substantially reduced.

GLUTEN

Overall the DMTA trace for gluten by itself is somewhat similar in terms of its tan δ, E' and E" (Figure 21.6). As with amylopectin it is also possible to explore the interaction of gluten with various sugars and to determine whether the behaviour follows the Couchman-Karasz pattern. It is interesting that even in a 10:1 gluten–fructose system there is evidence that both the DSC and the log E″ deviate from anticipated values at water contents greater than 10%. With a 2:1 gluten–fructose system the deviation is substantial and the only conclusion that can be drawn is that pronounced phase separation has occurred and that the fructose is exercising an

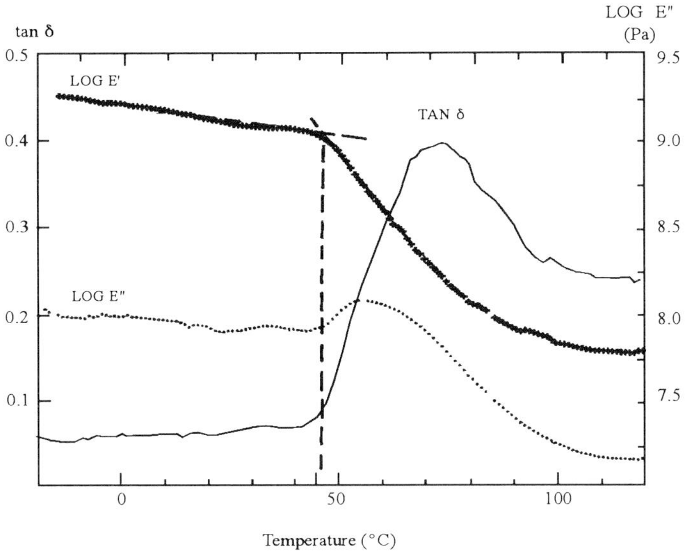

Figure 21.6 DMTA plot for gluten (RH = 65%) vs temperature

antiplasticizing effect upon the gluten at high water contents (Figure 21.7). An examination of Table 21.1 will show how pronounced this effect is when the water content is increased from 10% to 20%. We conclude that the gluten is exhibiting pronounced hydrophobic behaviour (Kalichevsky, Jaroszkiewicz and Blanshard, 1992b).

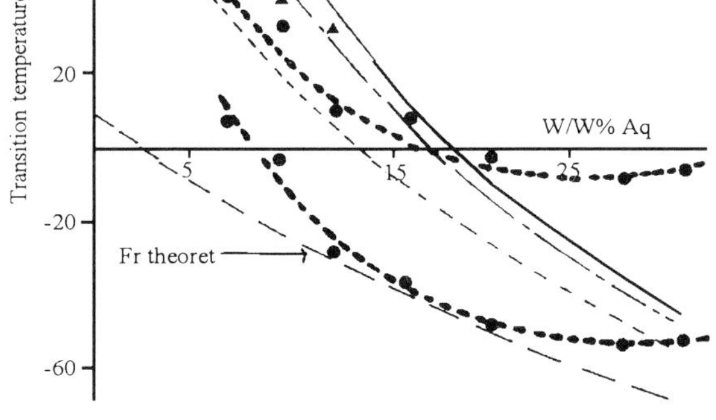

Figure 21.7 DSC — 2:1 gluten—fructose: midpoint (▴); peaks in log E″ (●)

Table 21.1 SUMMARY OF CHANGES IN T_G MEASURED BY DSC AND DMTA FOR GLUTEN—SUGAR AND GLUTEN—GLYCOL MIXTURES

Sample	*Values used for theoretical curve (A or B)*	*Change (°C) in T_g due sugar at w/w% aq.* 10%	20%
Gluten	(A and B are equivalent)	—	—
Gluten—sucrose (10:1)	B	-5	+9
Gluten—glucose (10:1)	—	0	+8
Gluten—glucose (5:1)	B	-4	+5
Gluten—fructose (10:1)	A	-7	+16
Gluten—fructose (2:1)	A	-22	+2
Gluten—glycerol (10:1)	—	-10	—
Gluten—glycerol (3:1)	—	-45	-12

It is interesting, and not surprising, that if amylopectin and gluten are mixed together, they have no plasticizing effect upon each other; indeed the DMTA picture of a mixed system contains peaks which coincide with those of the pure separate components. Studies of Sigma (Poole, UK) casein have shown that glucose, sucrose and fructose at 2:1 concentration have no effect upon the T_g of the casein water contents between 7 and 30. In sharp contrast, fructose does have a pronounced plasticizing effect upon sodium caseinate (Kalichevsky, Blanshard and Tokarczuk, 1993).

Water Redistribution Between Biopolymers

The method of preparation adopted was to vigorously shake binary mixtures of biopolymers in a large glass bottle. Samples were dried at 70°C *in vacuo* for 16 hours and subsequently hydrated isopiestically over saturated salt solutions for seven days at 25°C.

The critical issue here is to determine the manner of water redistribution between two biopolymers and how can this be determined. The samples that were investigated in this instance were an egg white–starch mix. The procedure adopted was to exploit water sorption isotherms. The power of this approach can be more fully exploited by taking the “Guggenheim-Anderson-de Boer (GAB) sorption” isotherm which is a more general sorption model than that of Brunauer-Emmett-Teller (BET). It has

It has the form

$$\frac{W}{W_m} = \frac{cka_w}{(1 - ka_w)\ (1 - ka_w + cka_w)} \quad (6)$$

where W is the water content on a dry weight basis, W_m is the "monolayer" moisture content, a_w the water activity and c and k are constants.

It will be evident that if k = 1 then this model assumes the BET equation as a limiting case. The virtue of the GAB sorption model is that it claims to describe isotherms very accurately up to water activities of 0.9 (Van den Berg, 1984) and this was shown to be the case for both egg white and starch (Hartley, Chevance, Hill, Mitchell and Blanshard, 1996). When these are mixed it is necessary to determine the moisture content of the mixture (M_{MIX}). This may be defined by

$$M_{MIX} = \frac{\Sigma W_i\ m_i}{\Sigma W_i} \quad (7)$$

where W_i are the individual component masses of components i expressed on a dry mass basis and M_i are the equilibrium water contents on a dry mass basis of each of the components i at the specified a_w. If variations occur between different components we can introduce an interaction parameter which is equivalent to an activity coefficient such that the expression now takes the form

$$M_{MIX} = \frac{\Sigma W_i\ \xi_i\ M_i}{\Sigma W_i} \quad (8)$$

(Kaminski and Al Bezweni, 1994). If we now combine this equation and the GAB equation for each component in the system ($M_i = f_i \times (A_w)$) this will allow us to formulate a predictive equation of the water absorption isotherm for the mixture. Figure 21.8 in fact shows the absorption isotherms for egg white, starch and a 50:50 mixture with the GAB fitted curve for which it is necessary to give ξ a value of 0.79 to get an appropriate fit. It is now, of course, possible to calculate the water content of each of the materials in the mix. The interesting feature of this system, however, is that it is possible to check these results since the X-ray crystallinity of starch is a function of the water content (Bear and French, 1941). This is evident from Figure 21.9. Using this technique we can observe the dynamics of equilibration. For example, if we mix perfectly dry egg white with starch containing 26% H_w on a 50:50 basis we would anticipate that if no change takes place then the starch should retain a crystallinity characteristic of a water content of ~13%. If we monitor the change over time, however, we can observe the equilibration process and after 25 hours the actual value has substantially reached the anticipated equilibrium value

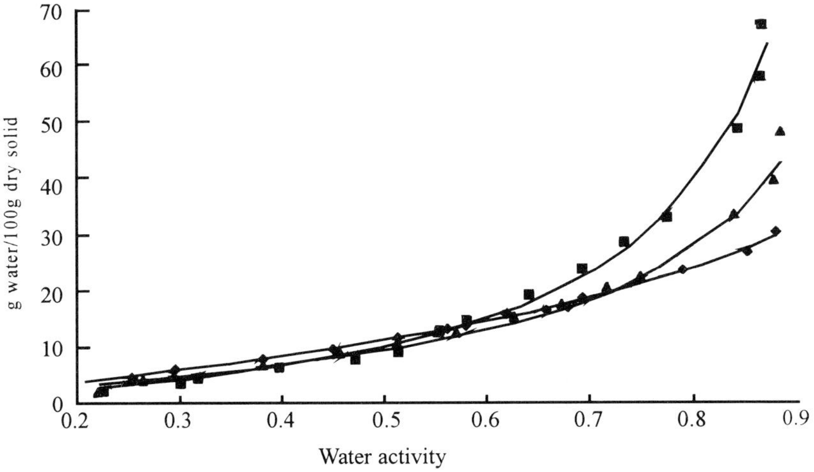

Figure 21.8 Adsorption isotherms (experimental results and GAB fitted isotherms) for egg white (■), starch (♦) and a 50/50 mixture (▲)

based on the ξ value of 0.79. Similarly, if we monitor the X-ray crystallinity of a 50:50 egg white–starch mixture at different mixture moisture contents and compare the predicted real water content obtained using a value of $\xi = 0.79$ and associated crystallinity and compare this with a predicted ideal value we find that the predicted real water content and the observed water content determined by the X-ray crystallinity coincide (Figure 21.10). This technique can also be used to calculate the water distribution in a 30:70 egg white–starch mixture and again experiment confirms predictions when we employ $\xi = 0.79$. The exact nature of the interaction which gives rise to the non-ideal behaviour is not yet clear (Hartley *et al.*, 1996).

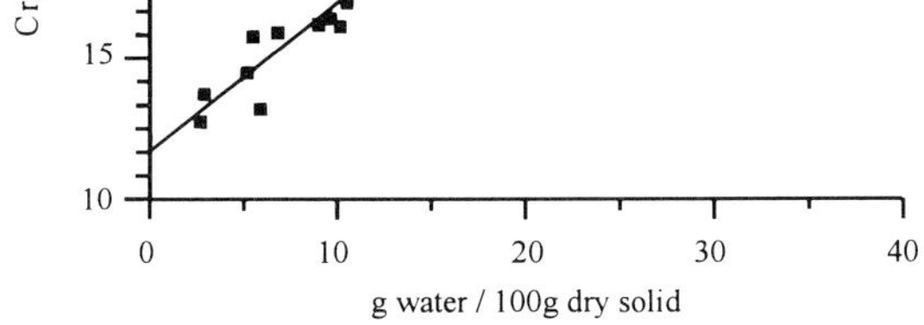

Figure 21.9 Variation of X-ray crystallinity of wheat starch with water content

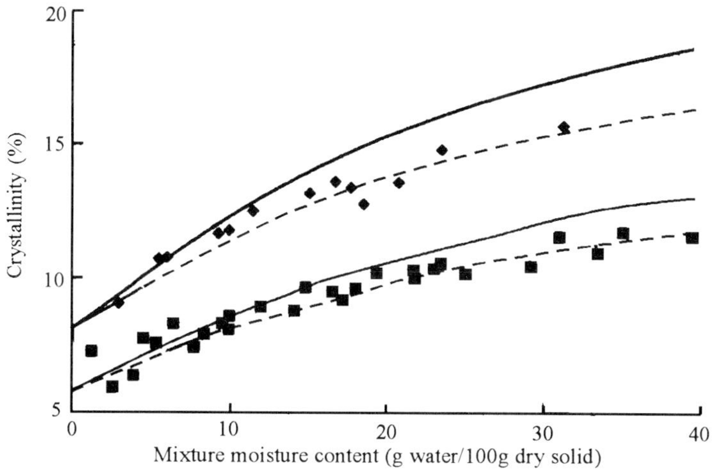

Figure 21.10 Plot of predicted and experimental values of the X-ray crystallinity of 30:70 (▴) and 50:50 (■) egg-white–starch mixtures at varying water contents. Predicted values are drawn as solid lines (— for ξ = 1.0; - - - - for ξ = 0.79)

Protein—Sugar—Water Interactions

Previous studies in our department have suggested that different sugars behave differently in terms of their interactions with a protein, e.g. ovalbumin. FT-IR had been used to investigate the protein–sugar interactions and the competition for water between the two components during the process of hydration. More recent studies (Orset, 1994) have resulted in the construction of a sorption cell which permitted the sequential hydration of the protein–sugar film *in situ* in the chamber of the spectrometer. The hydration was effected by a flow of nitrogen bubbling through various saturated salts solutions which provided a wide range of relative humidities in the cell where the film was placed.

The analytical method employed was a direct difference method (Poole and Finney, 1982, 1984) in which the spectrum of the dry system was subtracted from that of the hydrated material, the process of hydration being performed *in situ* so that the same area of protein–sugar film was being monitored by the spectrometer beam throughout the experiment.

The bands that are potentially of interest are the amide I and amide II bands at 1640 cm^{-1} and 1540 cm^{-1} respectively. It is believed that the carbonyl group contributes some 80% to the amide I while in the amide II that is reduced to 60% with a contribution of 50% from the imino group. The sorption isotherms for ovalbumin and the three sugars under consideration (sucrose, fructose and glucose) are all known and these are plotted in Figure 21.11. The spectra of, for example, the ovalbumin–glucose film on a 1:1 ratio and the difference spectrum between the wet and dry films are both shown in Figure 21.12. The peaks 1157 cm^{-1} and 1034 cm^{-1} are a consequence of hydration of the sugar molecules.

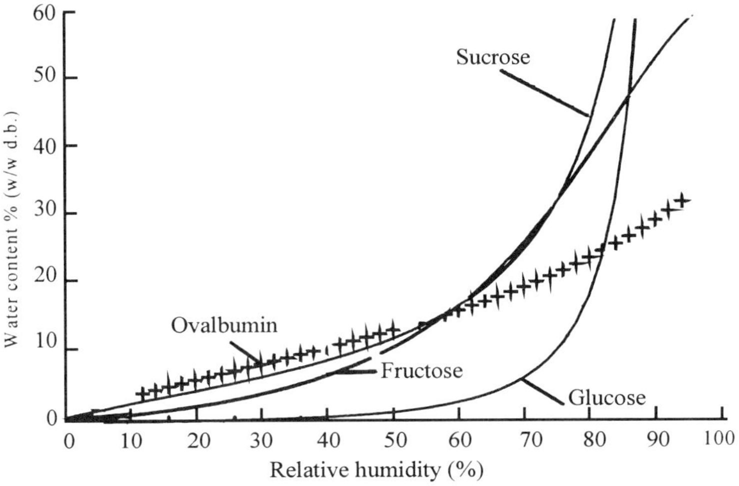

Figure 21.11 Sorption isotherms for amorphous sugars (Iglesias and Chirife, 1982) and freeze-dried ovalbumin

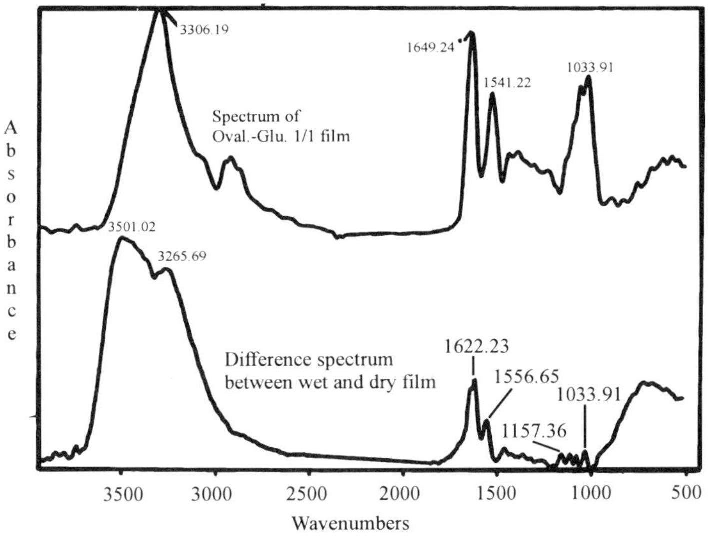

Figure 21.12 Comparison of dry film of ovalbumin—glucose with the direct difference spectrum of the same film at RH 100% over the range 4000 cm^{-1}—500 cm^{-1}

It is possible to analyze the potential effects of different interactions by plotting the percentage increase of either the amide I or II band versus the water content of these films (g water(W)/g solid(S)). The slope of the curve (fractional intensity = f(g water(W)/g dry solid(S)) is related to the rate of hydration of the site under investigation, and the W/S value where a levelling-off is observed is equivalent to the saturation of the site by hydrogen bonding. Figure 21.13a illustrates the typical pattern for hydration of a protein–sugar system where there is no interaction between the protein and sugar and equal partitioning of water. In comparison Figure 21.13b shows the situation where, although there is no interaction between the sugar and the protein, unequal partitioning of the water between the protein and sugar may be observed. In consequence at a given water content within the system the amide bands are less hydrated and therefore show a reduced intensity. In contrast, in Figure 21.13c we have the situation where there is interaction between protein and sugar and no preferential partitioning of water. In that case some amide bonds are blocked and the amide bonds for hydration are rapidly hydrated.

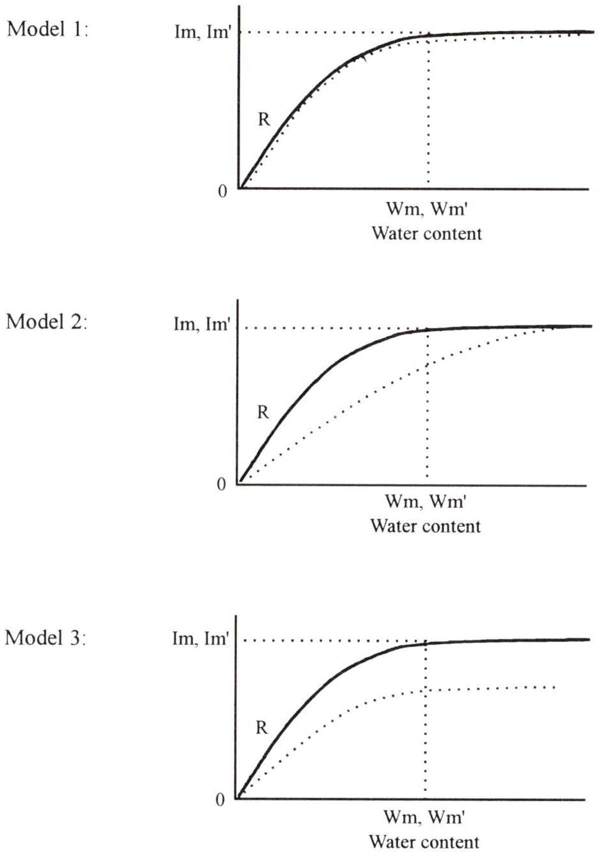

Figure 21.13 Models representing presence or absence of preferential competition for water between ovalbumin as a function of increasing hydration represented by g water / g solid. The effect of interactions between the protein and sugar is also illustrated.

In the light of these theoretical models it is interesting to observe how different sugars behave. In Figure 21.14 we observe the response shown by the amide I band. It will be noted that the ovalbumin—sucrose shows two significant features. First, there is a cross-over point at about 0.11% water content which is what would be envisaged from the sorption isotherm. It is also clear, however, that there is no interaction and equal distribution of water between the two components at higher water contents. In contrast with both fructose and glucose there is some degree of interaction between the protein and sugar and presumably also preferential partitioning of water to the sugar. The depression is particularly noticeable for glucose.

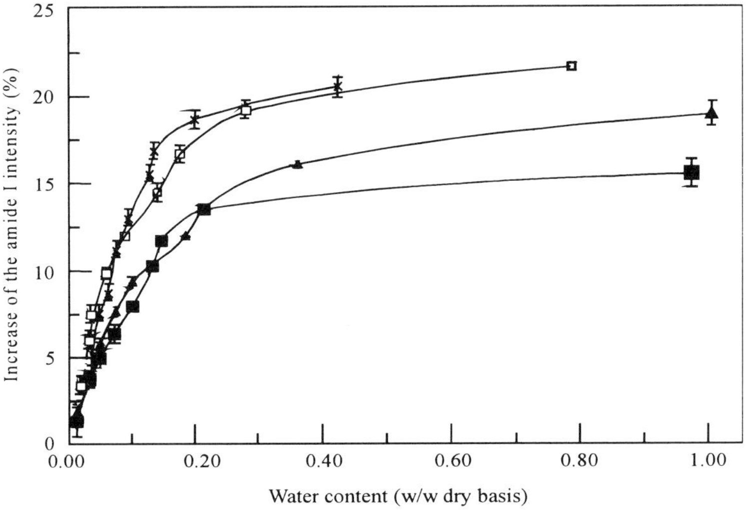

Figure 21.14 Fractional increase of amide I band of ovalbumin x (1), ovalbumin—glucose ■ (2), ovalbumin—fructose (▲) and ovalbumin—sucrose (□) with increasing hydration

In contrast when the amide II band is examined no interactions with the sucrose, fructose and glucose are observed but there is preferential partitioning of water to, particularly, the fructose and glucose.

In conclusion it is evident that low moisture systems offer an important and fruitful area of research which has particular relevance to the food industry.

Acknowledgements

The author warmly thanks his colleagues, especially Prof. J.R. Mitchell, Prof. W. Derbyshire, Dr. S.E. Hill, Lee Hartley, Imad Farhat, Fabienne Chevance and Sandra Orset for the use of their results, prior to publication and for the financial support from the HYDRA and ACTIF I and II programmes which made much of this work possible.

References

Ablett, S., Attenburrow, G.E. and Lillford, P.J. (1986) Significance of water in the baking process. In *Chemistry and Physics of Baking*, edited by Blanshard, J.M.V., Frazier, P.J. and Galliard, T., London: Royal Society of Chemistry, pp 30–41

Ablett, S., Darke, A., Izzard, M. and Lillford, P.J. (1993) Studies of the glass transition in malto-oligomers. In *The Glassy State in Foods*, edited by Blanshard, J.M.V. and Lillford, P.J., Nottingham (U.K.): Nottingham University Press, pp 189–206

Bear, R.S. and French, D. (1941) The significance of X-ray diffraction patterns obtained from starch granules. *Journal of the American Chemical Society*, **63**, 2198–2235

Van den Berg, C. (1991) Food–water relations, progress and integration, comments and thoughts. In *Water Relationships in Food*, edited by Levine, H. and Slade, L., New York: Plenum Press, pp 21–29

Bueche, F. (1962). *Physical Properties of Polymers*, New York: Wiley and Sons Inc.

Couchman, P.R. and Karasz, F.E. (1978) A classical thermodynamic discussion of the effect of composition on glass-transition temperatures. *Molecules*, **11**, 117–119

Farhat, I.A., Moreau, P., Mitchell, J.R., Derbyshire, W. and Blanshard, J.M.V. (1996) Spectroscopic studies on hydrocolloid water interactions. In *Gums and Stabilisers for the Food Industry 8* (in press)

Gordon, M. and Taylor, J.S. (1952) Ideal copolymers and the second order transitions of synthetic rubbers. 1. Non-crystalline copolymers. *Journal of Applied Chemistry*, **2**, 493–500

Hartley, L., Chevance, F., Hill, S.E., Mitchell, J.R. and Blanshard, J.M.V. (1996) Partitioning of water in binary mixtures at low water content. *Carbohydrate Polymers* (in press)

Iglesias, H.A. and Chirife, J. (1982). *Handbook of Isotherms*, Academic Press, pp 152–225

Kalichevsky, M.T., Jaroszkiewicz, E.M., Ablett, S., Blanshard, J.M.V. and Lillford, P.J. (1992a) The glass transition of amylopectin measured by DSC, DMTA and NMR. *Carbohydrate Polymers*, **18**, 77–88

Kalichevsky, M.T., Jaroszkiewicz, E.M. and Blanshard, J.M.V. (1992b) Glass transition of gluten. I: Gluten and gluten–sugar mixtures. *International Journal of Biological Macromolecules*, **14**, 257–266

Kalichevsky, M.T. and Blanshard, J.M.V. (1993) The effect of fructose and water on the glass transition of amylopectin. *Carbohydrate Polymers*, **20**, 107–113

Kalichevsky, M.T., Blanshard, J.M.V. and Tokarczuk, P. (1993) Effect of water content and sugars on the glass transition of casein and sodium caseinate. *International Journal of Food Science and Technology*, **28**, 139–151

Kalichevsky, M.T., Jaroszkiewicz, E.M. and Blanshard, J.M.V. (1993) A study of the glass transition of amylopectin–sugar mixtures. *Polymer*, **34**, 346–358

Kaminski, E., Al-Bezweni, M. (1994) Calculation of water sorption isotherms for multi-component protein-containing mixtures. *International Journal of Food Science and Technology*, **29**, 129–136

Levine, H. and Slade, L. (1988) Water as a plasticizer: physicochemical aspects of low-moisture polymeric systems. In *Water Science Reviews*, edited by Franks, F., Volume 3, Cambridge: Cambridge University Press, pp 79–185

Orset, S. (1994). *Hydration study of protein—sugar films using Fourier transform infra red spectroscopy: optimization of method of analysis*, DEA – Mention Européenne, University of Nottingham

Poole, P.L. and Finney, J.L. (1982) A direct difference infrared cell for studying the hydration of protein glasses. *Journal of Physics E*, **15**, 1073–1076

Poole, P.L. and Finney, J.L. (1984) Sequential hydration of dry proteins – A direct difference IR investigation of sequence homologies lysozyme and alpha-lactalbumin. *Biopolymers*, **23**, 1647–1666

LIST OF PARTICIPANTS

Abeysekera, Dr. R.	University of York, Heslinton, York YO1 5DD, U.K.
Albertsson, Prof. P-Å.	Dept. of Biochemistry, Chemical Centre, University of Lund, P.O.Box 124, S-22100 Lund, Sweden
Annable, Mrs. P.	Centre for Water Soluble Polymers, North East Wales Institute, Connah's Quay, Deeside, Clwyd CH5 4BR
Antipova, Ms. A.	Institute of Food Substances of RAS, 28, Vavilov Str, Moscow V 334 GSP 1, 117813 Russia
Ball, Dr. A.	University of Cambridge, Cavendish Laboratory, Madingley Road, Cambridge CB3 0HE
Barraclough, Mr. J.	Unilever Research, Colworth House, Sharnbrook, Bedford MK44 1LQ
Baumann, I.	Meyhall Rhone-Poulenc Group, Sonnenwiesenstrasse 18, 8280 Kreuzlingen, Switzerland
Biffen, Miss. A.	Centre for Water Soluble Polymers, North East Wales Institute, Connah's Quay, Deeside, Clwyd CH5 4BR
Blanshard, Prof. J. M. V.	Department of Applied Biochemistry & Food Science, University of Nottingham, Sutton Bonington Campus, Sutton Bonington, Loughborough, Leics LE12 5RD
Bogracheva, Miss. T.	Institute of Food Substances of RAS, 28, Vavilov Str, Moscow V 334 GSP 1, 117813 Russia

Borgström, Mr. J.	Physical Chemistry 1, University of Lund, 22100 Lund, Sweden
Bowler, Dr. P.	RHM Research & Engineering, Lincoln Road, High Wycombe, Bucks HP12 3QR
Budolfsen, G.	Novo Nordisk, Novo Allé, DK-2880 Bagsvaerd, Denmark
Chen, Mr. J.	Procter Department of Food Science, University of Leeds, Leeds, LS2 9JT
Chronakis, Mr. I.	Department of Food Research & Technology, Cranfield University, Silsoe College, Silsoe, Bedford MK45 4DT
Clark. Dr. A.	Unilever Research, Colworth House, Sharnbrook, Bedford MK44 1LQ
Clewlow, Mr. A.	National Centre for Macromolecular Hydrodynamics, Department of Biochemistry, Leicester University, Adrian Building, University Road, Leicester LE1 7RH
Colfen, Dr. H.	Department of Applied Biochemistry & Food Science, University of Nottingham, Sutton Bonington Campus, Sutton Bonington, Loughborough, Leics LE12 5RD
Comfort, Miss. S.	University of Surrey, School of Biological Sciences, Guildford, Surrey GU2 5XH
Craig, Ms.,N.	Pedigree Petfoods, Mill Street, Melton Mowbray, Leics LE13 1BB
Cruijsen, Mr. H.	Nutricia Research, P.O.Box 1, 2700 MA Zoetermeer, Netherlands
Cutts, Miss. L.	Dept of Pharmaceutical Sciences, University of Nottingham, University Park, Nottingham
Davis, Prof. S. S	Dept. of Pharmaceutical Sciences, University of Nottingham, University Park, Nottingham
Dickinson, Prof. E.	Procter Department of Food Science, University of Leeds, Leeds LS2 9JT

Dickinson, Dr. M.	Department of Physiology & Environmental Science, University of Nottingham, Sutton Bonington Campus, Sutton Bonington, Loughborough, Leics LE12 5RD
Donald, Dr. A.	Polymers & Colloids Group, University of Cambridge, Cavendish Laboratory, Madingley Road, Cambridge CB3 0HE
Doublier, Dr. J.-L.	INRA, Laboratoire de Physico-Chimi des Macromolécules, Rue de la Géraudière, B.P.527, 44026 Nantes Cédex 03, France
Draget, Dr. K.	Norwegian Biopolymer Laboratory (Nobipol), Dept. of Biotechnology, Norwegian Institute of Technology, University of Trondheim, N-7034 Trondheim, Norway
Durrani, Dr. M.	Cavendish Laboratory, Univesity of Cambridge, Madingley Road, Cambridge CB3 0HE
Duynslager, Ms. B.	Amycor, Coordination Centre of the Amylum Group, Burchtstraat 10, B-9300 Aalst, Belgium
Elie, Mrs. U.	United Biscuits (UK) Ltd., Group Research & Development, Lane End Road, Sands, High Wycombe HP12 4JX
Ellis, P.	Division of Life Sciences, Kings College London, Campden Hill Road, London W8 7AH
Errington, Mr. N.	Department of Applied Biochemistry & Food Science, University of Nottingham, Sutton Bonington Campus, Sutton Bonington, Loughborough, Leics LE12 5RD
Falshaw, Dr. R.	Industrial Research Ltd., P. O. Box 31310, Lower Hutt, New Zealand
Farhat, Mr. I.	Department of Applied Biochemistry & Food Science, University of Nottingham, Sutton Bonington Campus, Sutton Bonington, Loughborough, Leics LE12 5RD
Fernandes, Dr. P.	Nestlé Research Centre, P.O.Box 44, 100 Lausanne 26, Switzerland

Fiebrig, Mr. I.	Department of Applied Biochemistry & Food Science, University of Nottingham, Sutton Bonington Campus, Sutton Bonington, Loughborough, Leics LE12 5RD
Fisher, Dr. D.	Royal Free Hospital Medical School, Dept. of Biochemistry, Molecular Cell Pathology Laboratory, Rowland Hill Street, London NW3 2PF
Fitzpatrick, Mr. P.	Centre for Water Soluble Polymers, North East Wales Institute, Connah's Quay, Deeside, Clwyd CH5 4BR
Foster, Dr. T.	Unilever Research Laboratories, Colworth House, Sharnbrook, Bedfordshire
Friedli, Mr. G.-L.	University of Surrey, School of Biological Sciences, Guildford, Surrey GU2 8XH
Fritsche, S.	Meyhall Rhone-Poulenc Group, Sonnenwiesenstrasse 18, 8280 Kreuzlingen, Switzerland
Garnier, Dr. C.	Institut National la Recherche Agronomique, Laboratoire de Physico-Chimie des Macromolécules, Rue de la Géraudière, B.P. 527, 44026 Nantes Cedex 03, France
Georget, Mr. D.	Institute of Food Research, Norwich Research Park, Colney Lane, Norwich NR4 7UA
Gilchrist, Miss. J.	Stable Microsystems, Vienna Court, Lammas Road, Godalming, Surrey GU7 1JG
Goncalves, Dr. P.	Faculdade de Engenmaria DA, Universidade do Porto, Rua dos Bragas, 4099 Porto Codex, Portugal
Good, Miss. L.	Meyhall, Poleacre Lane, Woodley, Stockport, Cheshire SK6 1PQ
Goycoolea, Dr. F.	Department of Food Research & Technology, Silsoe College, Silsoe, Bedford MK45 4DT
Halstead, Dr. T.	Department of Chemistry, University of York, Heslington, York YO1 5DD
Harding, Dr. S.E.	Department of Applied Biochemistry & Food Science, University of Nottingham, Sutton Bonington Campus, Sutton Bonington, Loughborough, Leics LE12 5RD

Hartley, Mr. L.	Department of Applied Biochemistry & Food Science, University of Nottingham, Sutton Bonington Campus, Sutton Bonington, Loughborough, Leics LE12 5RD
Heeney, Mr. L.	Unilever Research, Colworth House, Sharnbrook, Bedford MK44 1LQ
Hermansson, Prof. A-M.	The Swedish Institute for Food Research, P.O.Box 5401, S-402, 29, Göteborg, Sweden
Hey, Dr. M.	Chemistry Dept, University of Nottingham, University Park, Nottingham
Hill, Dr. S.	Department of Applied Biochemistry & Food Science, University of Nottingham, Sutton Bonington Campus, Sutton Bonington, Loughborough, Leics LE12 5RD
Hockaus, Ms. M.	Novo Nordisk, Novȯ Allé, DK-2880 Bagsvaerd, Denmark
Horton, Dr. J.	University of Nottingham, Sutton Bonington Campus, Sutton Bonington, Loughborough, Leics LE12 5RD
Hoskins, Mr. R.	Centre for Water Soluable Polymers, North East Wales Institute, Connah's Quay, Deeside, Clwyd, CH5 4BR
Howell, Dr. N.	University of Surrey, School of Biological Sciences, Guildford, Surrey GU2 5XH
Ipsen, Dr. R.	DEN KGL., Veterinaer-og Landbohøjskole, Department of Dairy and Food Science, Thorvaldsensvej 40, DK 1871 Frederiksberg C, Denmark
Jauregi, Ms. P.	Department of Food Science & Technology, Biotechnology and Biochemistry Engineering Group, University of Reading, Whiteknights, P.O.Box 226, Reading RG6 2AP
Johansson, Mr. H-O.	Dept of Biochemistry, University of Lund, P. O. Box 124, Sweden S-221 00
Jones, Dr. R.	Cavendish Laboratory, Univesity of Cambridge, Madingley Road, Cambridge CB3 0HE

Jumel, Dr. K.	Department of Applied Biochemistry & Food Science, University of Nottingham, Sutton Bonington Campus, Sutton Bonington, Loughborough, Leics LE12 5RD
Kasapis, Dr. S.	Cranfield Institute of Technology, Dept of Food Research & Technology, Silsoe College, Silsoe, Bedford MK45 4DT
Keenan, Mr. D.	Department of Chemistry, University of York, Heslington, York YO1 5DD
Kelly, Miss. R.	Department of Applied Biochemistry & Food Science, University of Nottingham, Sutton Bonington Campus, Sutton Bonington, Loughborough, Leics LE12 5RD
Kennedy, Dr. D.	Ministry of Agriculture, Food and Fisheries, Room 203, Nobel House, 17, Smith Square, London SW1P 3JR
Langley, Mr. K.	Institute of Food Research, Earley Gate, Whiteknight Road, Reading, RG6 2EF
Lindfors, Dr. L.	Drug Delivery Research, Astra Hässle AB, S-43183 Mölndal, Sweden
Lundin, Mr. L.	SIK, The Swedish Institute for Food Research, Box 5401, S-402 2, Gothenburg, Sweden
Magnani, Ms. K.	Nestec SA, Vers-chez-les-Blanc, P.O.Box 44, 1000 Lausanne 26, CH-1000, Switzerland
Manoj, Miss. P.	Department of Food Research & Technology, Cranfield University, Silsoe College, Silsoe, Bedford MK45 4DT
Marsh, Miss. R.	Tastemaker Ltd., Chippenham Drive, Kingston, Milton Keynes MK10 0AE
Matser, Ir. A.	Netherland Institute for Carbohydrate Research, Rouaanstraat 27,9723 CC Groningen, Netherlands
Melia, Dr. C.	Dept of Pharmaceutical Sciences, University of Nottingham, University Park, Nottingham

Michon, Ms. C.	Laboratoire de Biophysique, Food Science Dept, ENSIA 1, Avenue des Olympiades, 91305 Massey, France
Mitchell, Prof. J.	Department of Applied Biochemistry & Food Science, University of Nottingham, Sutton Bonington Campus, Sutton Bonington, Loughborough, Leics LE12 5RD
Moerenhout, Ms. H.	Amycor, Coordination Centre of the Amylum Group, Burchtstraat 10, B-9300 Aalst, Belgium
Mohammed, Ms. Z.	Department of Applied Biochemistry & Food Science, University of Nottingham, Sutton Bonington Campus, Sutton Bonington, Loughborough, Leics LE12 5RD
Morris, Prof. E.	Cranfield Institute of Technology, Dept of Food Research & Technology, Silsoe College, Silsoe, Bedford MK45 4DT
Morris, Dr. V.	AFRC, Norwich Laboratory, Norwich Research Park, Colney, Norwich NR4 7UA
Navarini, Dr. L.	Research Centre, Poly-tech, Area Science Park, Padriciano 99, 34012 Trieste, Italy
Norton, Dr. I.	Unilever Research, Colworth House, Sharnbrook, Bedford MK44 1LQ
Ong, Miss. M.	Department of Applied Biochemistry & Food Science, University of Nottingham, Sutton Bonington Campus, Sutton Bonington, Loughborough, Leics LE12 5RD
Parker, Mr. A.	Research & Development, Food Additives Division, Sanofi Bio-Industries, Baupte, 50500 Carentan, France
Parovuori, Mr. P.	VTT Biotechnology and Food Research, P. O. Box 1500, Finland
Pelan, Dr. E.	Unilever Research, Colworth House, Sharnbrook, Bedford MK44 1LQ
Penders, Dr. M.	Unilever Research, Colworth House, Sharnbrook, Bedford MK44 1LQ

Piculell, Dr. L.	Physical Chemistry 1, Chemical Centre, University of Lund, P.O.Box 124, Lund, Sweden
Ptitchkina, Dr. N. M	Chemistry Institute of Saratov State University, Astrakhanskaya Str 83, Saratov 410026, Russia
Rayment, Miss. P.	Biopolymers Group, Division of Life Sciences, Kings College London, Campden Hill Road, London W8 7AH
Roefs, S.	NIZO, Netherlands Institute for Dairy Research, P. O.Box 20, 6710 BA Ede, Netherlands
Ross-Murphy, Prof. S.	Kings College, University of London, Kensington Campus, Campden Hill Road, Kensington, London
Rowe, A.	Department of Biochemistry, University of Leicester, University Road, Leicester LE1 7RH
Schorsch,Ms. C.	Institut National la Recherche Agronomique, Laboratoire de Physico-Chimie des Macro-molécules, Rue de la Géraudière, B.P. 527, 44026 Nantes Cedex 03, France
Semenova, Dr. M.	Institute of Food Substances of RAS, 28, Vavilov Str, 117813 Moscow, Russia
Simpkin, Mr. N.	Department of Applied Biochemistry & Food Science, University of Nottingham, Sutton Bonington Campus, Sutton Bonington, Loughborough, Leics LE12 5RD
Skantze, Mr. U.	Drug Delivery Research, Astra Hässle AB, S-43183 Mölndal, Sweden
Somers, P.	University of Surrey, School of Biological Sciences, Guildford, Surrey GU2 5XH
Speirs, Dr. C.	Pedigree Petfoods, Mill Street, Melton Mowbray, Leics LE13 1BB
Stewart, Prof. G.	Department of Applied Biochemistry & Food Science, University of Nottingham, Sutton Bonington Campus, Sutton Bonington, Loughborough, Leics LE12 5RD

Stone, Dr. A.	United Biscuits (UK) Ltd., Group Research & Development Centre, Lane End Road, Sands, High Wycombe HP12 4JEX
Syrbe, Mr. A.	Nestec SA, Vers-chez-les-Blanc, P.O.Box 44, 1000 Lausanne 26, CH-1000, Switzerland
Taylor, Dr. A.	Department of Applied Biochemistry & Food Science, University of Nottingham, Sutton Bonington Campus, Sutton Bonington, Loughborough, Leics LE12 5RD
Therkelsen, Mr. G.	Copenhagen Pectin, Ved Banen 16, DK-4623 Lllle Skensved, Denmark 4623
Tjerneld, Dr. F.	Dept of Biochemistry, University of Lund, P. O. Box 124, S-221 00 Lund, Sweden
Tobitani, Mr. A.	Division of Life Sciences, Kings College London, Campden Hill Road, London W8 7AH
Tombs, Prof. M.	Department of Applied Biochemistry & Food Science, University of Nottingham, Sutton Bonington Campus, Sutton Bonington, Loughborough, Leics LE12 5RD
Towers, Dr. P.	Weetabix Ltd., Station Road, Burton Latimer, Kettering, Northants NN15 5JR
Tromp, Dr. R.	Cavendish Laboratory, Univesity of Cambridge, Madingley Road, Cambridge, CB3 0HE
van Eendenburg, Mr. J.	Unilever Research Laboratory, Oliver van Noortlaam 120, 3133 At Vlaardingen, Netherlands
Viebke, Mr. C.	Physical Chemistry 1, Chemical Centre, University of Lund, P.O.Box 124, Lund, Sweden
Walkenström, Ms. P.	SIK, The Swedish Institute for Food Research, Box5401, S-402 29 Gothenburg, Sweden
Walter, H.	Lab of Chem. Biol-151, VA Medical Centre, Long Beach 90822, California , USA
Ward-Smith, Mr. S.	Lab A30, Chemistry Dept, University of Nottingham, University Park, Nottingham
Wasserman, Ms. L.	Institute of Food Substances of RAS, 28, Vavilov Str, Moscow V 334 GSP 1, 117813 Russia

Williams, Dr. M.	Department of Chemistry, University of York, Heslington, York YO1 5DD
Williams, Dr. P.	Faculty of Science, The North East Wales Institute, Connah's Quay, Deeside, Clwyd CH5 4BR

INDEX

A

B

C

D

H

I

J

K

L

M

Y

Z

MIXTURES INDEX BY CHAPTER NUMBER

Chapter 11

Agar: Gelatin (g)
Gelatin : Gellan (g)
Gelatin : Gellan : Small sugars (g)
Gelatin : Small sugars (g)
Gellan : Small sugars (g)
Gelatin : Maltodextrin (g)
Maltodextrin : Milk proteins (g)
Milk proteins: Soya bean proteins
Polybutadiene : Styrene-acrylonitrile (g)

Chapter 12

Amylose : amylopectin
Amylose : amylopectin (g)

Chapter 13

Acetan (Xylinan) : Konjac glucomannan
Agar : Konjac glucomannan
Alginate : Gellan
Alginate : Pectin (g)
Alginate : Poly-L-lysine
Anionic polysaccharides : κ-casein
Carrageenans : κ-casein
Carrageenans : Konjac glucommanan
Carrageenans : Locust bean gum
De-acetylated xanthan: Konjac glucomannan
Furcelleran: Konjac glucomannan
Furcellaran: Locust bean gum
Furcelleran : Tara gum
Guar gum : Gum arabic
κ-carrageenan : Tara gum
Konjac glucomannan : Rhamsan
Konjac glucomannan : Whelan
Konjac glucomannan: Xanthan
Konjac glucomannan : Xanthan polytetramer
Locust bean gum : Rhamsan
Locust bean gum : Welan
Locust bean gum : Xanthan (g)
Locust bean gum : Xanthan polytetramer
Poly-D-galacturonate : Poly-L-guluronate

Chapter 14

Acetan (xylinan) : Carob gum
Acetan (xylinan) : Konjac glucomannan
Agarose : Konjac glucomannan
Alginate : Pectin
Bovine serum albumin : Pectin
Carob gum : Dextran
Carob gum : Furcelleran
Carob gum : κ-carrageenan
Carob gum : Xanthan
Dextran : Konjac glucomannan
Konjac glucomannan : Xanthan
Tara gum : Xanthan

Chapter 15

Bovine serum albumin : Hydroxyethyl cellulose
Bovine serum albumin : Polyethylene oxide

Chapter 16

α-lactalbumin : Lysozyme
β-lactoglobulin : Lysozyme
Blood plasma proteins : Egg albumen proteins (g)
Gluten proteins : Lysozyme
Soya isolate : Whey proteins

Chapter 17

α-casein : Gelatin
β-casein : b-lactoglobulin
β-casein : Gelatin (g)
β-lactoglobulin : Dextran sulphate
Bovine serum albumin : Alginate
Bovine serum albumin : Dextran sulphate
Blood plasma proteins
Lipid : Protein

Chapter 18

Chitosan : Alginate
Chitosan : Mucin
Chitosan : Polyarylate
Mucin : Alginate
Mucin : Amylopectin
Mucin : Amino-dextran
Mucin : Carbopol
Mucin : Carbomer
Mucin : Carboxymethyl cellulose
Mucin : DEAE-dextran
Mucin : Hyaluronate
Mucin : Hydroxyethyl starch
Mucin : Hydroxypropyl cellulose
Mucin : Hydroxypropylmethyl cellulose
Mucin : Pectins
Mucin : Polyacrylic acid
Mucin : Polycarbophil
Mucin : Polyethylene glycol
Mucin : Poly-L-lysine
Mucin : Poly-L-(Lys-Lys-Ala)
Mucin : Polyvinyl propylene
Mucin : Sleroglucan
Mucin : Xanthan

Chapter 19

Chapter 20

Chapter 21